# Agricultural Electrification

**Truman C. Surbrook**
Associate Professor
Michigan State University

**Ray C. Mullin**
Regional Manager
Bussman Division
McGraw-Edison

*Published by*

IE55

**SOUTH-WESTERN PUBLISHING CO.**

CINCINNATI    WEST CHICAGO, IL    DALLAS    PELHAM MANOR, NY    PALO ALTO, CA

Copyright © 1985
by South-Western Publishing Co.,
Cincinnati, Ohio

ALL RIGHTS RESERVED

The text of this publication, or any part thereof, may not be reproduced or transmitted in any form or by any means, electronic, or mechanical, including photocopying, recording, storage in an information retrieval system, or otherwise, without prior written permission of the publisher.

ISBN: 0-538-33550-5

Library of Congress Catalog Card Number: 84-50214

1 2 3 4 5 D 9 8 7 6 5

Printed in the United States of America

# CONTENTS

Preface .................................................... v

UNIT 1 INTRODUCTION TO FARM WIRING................ 1

UNIT 2 ELECTRICAL FUNDAMENTALS .................... 13

UNIT 3 FARM ELECTRICAL SYSTEMS ..................... 41

UNIT 4 FARM BUILDING SERVICES ...................... 54

UNIT 5 FARM MAIN SERVICE ENTRANCE ................ 92

UNIT 6 OVERCURRENT PROTECTION: FUSES AND CIRCUIT BREAKERS ...... 104

UNIT 7 VOLTAGE DROP ................................ 123

UNIT 8 FEEDERS ..................................... 138

UNIT 9 FARM WIRING METHODS AND MATERIALS ....... 153

UNIT 10 EQUIPMENT GROUNDING AND BONDING ....... 212

UNIT 11 FARM LIGHTING ............................... 224

UNIT 12 STANDBY ELECTRICAL POWER SYSTEM ......... 252

UNIT 13 PHASE CONVERTERS ........................... 262

UNIT 14 TRANSFORMERS ............................... 273

| UNIT 15 | HEATING AND COOLING | 296 |
|---|---|---|
| UNIT 16 | ELECTRIC MOTORS | 317 |
| UNIT 17 | MOTOR CONTROL | 341 |
| UNIT 18 | CORD- AND PLUG-CONNECTED EQUIPMENT | 373 |
| UNIT 19 | READING WIRING DIAGRAMS AND BLUEPRINTS | 381 |
| UNIT 20 | GENERAL FARM WIRING | 388 |
| UNIT 21 | WIRING FOR LIVESTOCK HOUSING | 394 |
| UNIT 22 | DAIRY FARM WIRING | 398 |
| UNIT 23 | FEEDING SYSTEM WIRING | 406 |
| UNIT 24 | SWINE HOUSING WIRING | 412 |
| UNIT 25 | POULTRY FARM WIRING | 417 |
| UNIT 26 | GRAIN-DRYING AND STORAGE WIRING | 421 |
| UNIT 27 | IRRIGATION SYSTEM WIRING | 428 |
| INDEX | | 434 |

# *PREFACE*

Agricultural wiring is unique in that it often requires protection from physical damage, moisture, dust, and corrosion. A knowledge of the use of special wiring materials is required for an acceptable installation in some locations. Knowledge of agricultural operations is necessary for a safe, reliable, and adequate farm wiring system.

The AGRICULTURAL ELECTRIFICATION text discusses specific types of farms, with recommendations for wiring installations. The many illustrations and examples, along with problems and solutions help the reader to easily understand the application of the *National Electrical Code*®[1] (*NEC*) to agricultural wiring. Numerous references to specific sections of the Code make this text an excellent tool for learning to use and understand the Code.

Several sections in the *National Electrical Code* are devoted exclusively to agricultural wiring conditions, and other sections will most likely be added in the future. This text explains agricultural references now in the Code, and refers to some installations that are now lacking in the Code, such as phase converter installation and central pole electrical distribution.

The 1984 edition of the *National Electrical Code* uses metric (SI) measurements in addition to the traditional English measurements. Accordingly, metric measurements are included in this text, where applicable.

*Truman C. Surbrook*
*Associate Professor*
*Michigan State University*
*East Lansing, Michigan*

*Ray C. Mullin*
*Regional Manager*
*Bussman Division*
*McGraw-Edison Company*
*Northbrook, Illinois*

---

[1] National Electrical Code® and NEC® are Registered Trademarks of the National Fire Protection Association, Inc., Quincy, MA.

# ABOUT THE AUTHORS

The authors of AGRICULTURAL ELECTRIFICATION, Truman C. Surbrook and Ray C. Mullin, have extensive practical experience in the areas of agricultural, residential, and commercial wiring, as well as many years of experience as instructors in the electrical trade.

Truman C. Surbrook, Ph.D., is an Associate Professor of Agricultural Engineering at Michigan State University, a registered Professional Engineer, and Master Electrician. Dr. Surbrook developed an Agricultural Electrical Apprenticeship Program at Michigan State University, and later served as chair for curriculum development for a statewide electrical inspector training short course. Dr. Surbrook has spent his entire professional career teaching; working with youths, farmers, and contractors; and conducting research in the area of agricultural wiring and electronics. He is an active member of several professional organizations, and has served on numerous technical committees. Organizations in which he is most active are the American Society of Agricultural Engineers, International Association of Electrical Inspectors, and the Institute of Electrical and Electronics Engineers, Inc. For 15 years, Dr. Surbrook was the producer for a weekly radio program dealing with rural electrification. He has written many bulletins, articles, and technical publications on various subjects related to agricultural wiring.

Ray C. Mullin was an electrical circuit instructor for the Electrical Trades, Wisconsin Schools of Vocational, Technical, and Adult Education. Mr. Mullin, a former member of the International Brotherhood of Electrical Workers, is now a member of the International Association of Electrical Inspectors, the Institute of Electrical and Electronics Engineers, Inc., and the National Fire Protection Association, Electrical Section.

Mr. Mullin completed his apprenticeship training and has worked as a journeyman and supervisor. He has taught both day and night electrical apprentice and journeyman trade extension courses, and has conducted engineering seminars. Mr. Mullin is now Regional Manager for a large electrical components manufacturer.

# UNIT 1
# INTRODUCTION TO FARM WIRING

## OBJECTIVES

After studying this unit, the student will be able to

- describe an adequate farm electrical system.
- explain the electrical shock hazard that exists for animals.
- discuss the need for farm electrical power reliability.
- describe the problems resulting from farm long wiring distances.
- explain the effects of damp, wet, and corrosive farm conditions on wiring and equipment.
- state how animals and machines can damage wiring.
- list sources of help in solving farm wiring problems.
- name some of the metric units used in the *National Electrical Code*.

Farms today require well-designed electrical systems which may range in size from 200 amperes (A) to 800 amperes. The electrical installer serving agricultural communities must understand the problems and needs of a farm enterprise to ensure that appropriate wiring materials are used and that the wiring is adequate for the load. The electrical contractor is often given farm building plans which show only the locations and sizes of electrical equipment. It is left to the contractor to specify and plan for the size and type of service equipment, wires, panelboards, and controls. Compare this situation to the case of new commercial or industrial construction, where an architect or an engineer first develops basic plans and equipment specifications for the contractor to follow.

## ADEQUATE FARM ELECTRICAL WIRING SYSTEM

*Section 90-1(b)* of the *National Electrical Code (NEC)* refers to provisions for an initial adequate wiring system which, when properly installed and maintained, is basically free from hazard. An adequate electrical wiring system is essential for an efficient, safe, and productive farm enterprise. The system must provide adequate electrical energy for present and future farm needs. *Adequate* wiring is considered to be

- able to serve present loads.
- safe for humans and animals.
- reliable.

- convenient.
- economical.
- able to supply some future loads.
- easily and economically expanded.

Each of these requirements is described in more detail in the following sections.

## Serving Present Farm Loads

A farm wiring system installed to comply with the *National Electrical Code* is capable of serving the loads at the time of the installation. However, some unique situations in agricultural wiring are not covered adequately by the *NEC*. Examples of such situations include the following.

1. The farm installation of central electrical distribution, Figure 1-1
2. Multiple taps from the main service to feed out buildings, Figure 1-2
3. Sizing the wires used for long feeders and circuits to minimize the voltage drop
4. Installing phase converters to supply 3-phase motors from a single-phase service
5. Services and wiring systems exposed to the weather and physical abuse.

In time, the *NEC* will provide additional information concerning these subjects.

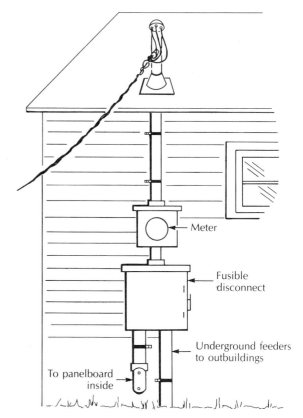

Figure 1-2  A fusible disconnect may be installed on the outside of a building to provide a convenient means of tapping the main service to provide power for outbuildings.

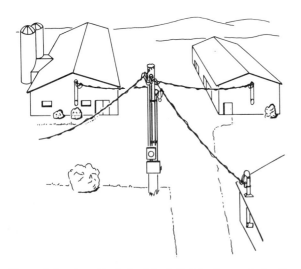

Figure 1-1  A central electrical distribution pole is often the main service for a farm.

## Safety for Humans and Animals

Normally, wet or damp conditions exist on most farms in areas where electrical energy is used. These conditions increase the electrical shock hazard, or the likelihood that a person or an animal may complete a path to ground from faulty equipment. It is essential that proper grounding be provided for all electrically operated equipment, piping systems, and metal structures likely to become energized. Of special concern is electrically powered portable machinery. Such machinery must be supplied with adequate grounding cords from a properly grounded electrical system.

Animals are naturally grounded, and thus are much more likely to experience electrical shocks than are humans. People often are not aware of a low-voltage potential between metal equipment and the ground. Humans, with their dry skin and such foot coverings as shoes and boots, frequently have a resistance to ground of tens or hundreds of thousands of ohms ($\Omega$). The current flowing through a human body having this high resistance from metal equipment to ground is generally not sensed. However, a well-grounded animal, drinking water and standing on a damp concrete slab or on damp ground, has a resistance to ground of only a few hundred ohms.

The shock hazard that exists for animals is shown in Figure 1-3. Animals can detect electrical potentials of less than 10 volts (V) on a watering system. Because of the shock, the animal may be reluctant to drink, resulting in reduced water intake. In the case of dairy cows, reduced water intake results in less milk production and a financial loss to the farmer. As shown in Figure 1-3, 350 Ω is a typical resistance value from the cow's muzzle in a watering cup to the earth under the hooves. This resistance allows 14 milliamperes (mA), or 0.014 ampere (A), to flow when there is only a 5-V potential between the watering cup and the ground. A current of 14 mA causes a painful shock in humans. However, a human wearing dry shoes when touching the same watering cup is not affected, and may not be aware of the electrical shock hazard that exists for the cow.

## *Reliability*

The electrical supply system of a farm must be reliable, especially in the case of livestock operations. Poultry, beef, and hogs housed in environmentally controlled buildings are dependent upon electricity for ventilation, water, waste removal, and lighting. The loss of ventilation can cause animal stress, which may result in sickness or even death. A reliable milking system is required on a dairy farm, where the cows must be milked regularly and the milk cooled immediately.

A reliable ventilation system is extremely important in a poultry operation, Figure 1-4. In the modern poultry egg-laying house, several hens are kept in each cage, with cages two to four layers high. If the ventilation system fails, the oxygen in the building can be used up in only a few minutes, resulting in the suffocation of large numbers of fowl.

Standby electrical power is essential for many farm enterprises. In addition, it is good insurance against a power outage. Provisions for standby power should be included for most farms at the time service equipment is installed initially or when it is upgraded. Generally, power outages are the result of damage to the electrical power supply system from storms or accidents. Power outages to circuits or an entire building on a farm may be due to an inadequate or improperly installed farm electrical system. To ensure reliability, the farm wiring system must be planned and installed properly.

## *Convenience*

A well-planned farm wiring system provides safe, convenient service and control for the operator and is

**Figure 1-4** A reliable ventilation system is extremely important in a poultry egg-laying house.

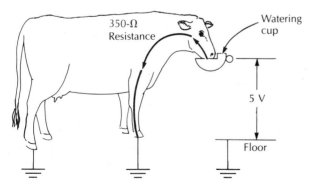

**Figure 1-3** Shock hazard to animals

Unit 1  Introduction to Farm Wiring  3

energy efficient. As farm buildings and machines are added, the wiring requirements change. Soon the service equipment may become overloaded and the voltage drop excessive. Control equipment may not be located to take the best advantage of the farm operator's working patterns. The farm operator and the installer must cooperate in planning the wiring system to ensure that it provides convenience of control to the operator.

There are two primary advantages to a convenient wiring system:

1. It is safer, because the operator is more likely to disconnect the electrical supply before working on equipment.
2. It tends to be more energy efficient, because equipment and lights are turned off when not in use.

## Economy Factors

A farm wiring system must be economical and still provide an adequate supply of electricity. Achieving these goals requires good planning and an adequate understanding of conditions that cause the rapid deterioration of equipment. The service equipment must be located close to the major loads to reduce the cost of circuit wiring.

The electrical installer should prepare diagrams of proposed circuits so as to minimize the amount of wire used to supply the outlets. An example of a wiring diagram for a proposed circuit is shown in Figure 1-5. A simple diagram can be used to plan a circuit and determine the most efficient route. The physical structure of the building and obstacles must be considered in planning the circuit.

Electrical equipment repair and replacement costs are also factors in planning farm wiring. Equipment and materials suitable for one location may not be suitable for another location. High humidity, condensation, corrosive materials, and physical abuse can lead to the rapid deterioration of electrical equipment and materials. In many cases, the proper placement of electrical equipment and the routing of wires can prevent exposure to the conditions that cause deterioration. However, when these conditions cannot be avoided, suitable materials are required. The equipment must be identified for use in this type of adverse environment, *NEC Section 110-11*.

## Future Loads and Expansion

An adequate electrical system is one which provides for the addition of a limited number of loads. Service equipment is sized exactly to the load to be served only for the case of a very specific load. The general practice is to leave several extra spaces in the panelboard for circuit breakers for future loads. When the service equipment demand load is calculated, the feeders and equipment are sized to allow for additional loads. Generally, it is not economical to install a small panelboard in a farm building with provisions for only a few circuits.

By allowing for future expansion when the initial installation is made, the cost of labor and materials required is less than the cost of replacing and rewiring the electrical system in the future. Space should be allowed around the service equipment for the installation of future electrical equipment. If conduit is run in concrete or under a portion of a building, it should be sized large enough to accept additional wires. Auxiliary gutters and pull boxes should be sized large enough so that wires and splices can be added for future electrical loads. *Section 90-7(a)* of the *NEC* recommends that provisions be made for future loads.

The method by which all farm buildings are supplied from a main service may itself limit future expansion. There are situations where central point distribution is preferred to a main service at one building, Figure 1-6. This is true if several farm buildings are to be served, or if additions are to be made to the operation. A future change from a main farm service at the residence to central point distribution may be expensive.

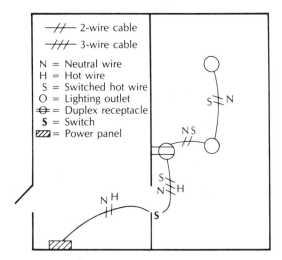

**Figure 1-5  Wiring diagram of a proposed circuit**

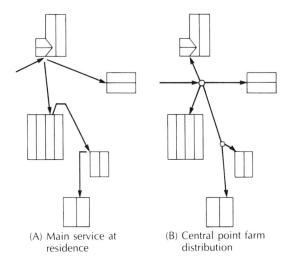

(A) Main service at residence

(B) Central point farm distribution

**Figure 1-6** As a farm enterprise grows, a main service (A) at one building, such as a residence, limits farm growth. Central point electrical distribution (B) is required for most farm enterprises.

## LONG WIRING DISTANCES

The buildings of a farmstead generally are spread over a large area to allow for expansion, to permit the movement of machinery around the buildings, and to reduce the risk of fire spreading between the buildings. These long distances, however, make it difficult to provide ample electrical power without excessive voltage drop.

When planning the layout of the farm, buildings with a high electrical demand should be located as close as possible to the electrical load center. Generally, the electrical installer has little choice in the arrangement of buildings and must work with the existing layout. Under these conditions, the installer will try to place the main service equipment for the farm in the best location to serve both the residence and the buildings where the major loads are expected.

Voltage drop must be considered when sizing overhead or underground feeders from the main service equipment to each farm building. Since future buildings may receive their electrical supply from the service in an existing building, these new loads may create excessive voltage drop.

As an example of the long distances involved, consider animal confinement buildings where the branch circuits for lighting and ventilation fans may be several hundred feet long. In many cases, the wire size must be increased to reduce the voltage drop. Motor branch circuits for electrically operated irrigation equipment are often more than half a mile (0.8 kilometer) in length.

## DAMP AND WET LOCATIONS

Water is a problem in and around buildings housing livestock and in areas where many farm products are processed. Water may condense on and within electrical equipment in areas of high humidity. Watertight electrical equipment must be installed where water is sprayed for cleaning. Booster pumps which raise the pressure of water to more than 100 pounds per square inch (690 kilopascals) are common. Electrical equipment exposed to the weather must be protected from rain and snow. In many cases, cords and conduits containing wires are in direct contact with puddles of water on floors and on the ground.

Rust and corrosion are major problems in areas where humidity is high or water comes into contact with electrical equipment. Materials deteriorate and fail after prolonged exposure to moisture. If moisture condenses inside contactors or other moving metal parts, the resulting rust restricts movement. Condensed moisture in equipment can freeze in cold weather and foul moving parts.

Moisture-laden air from a building containing livestock can migrate through conduits, and condense in areas where the conduit is cold, Figure 1-7. For this reason, cable wiring is often more acceptable than raceway wiring for wiring farm buildings. When an interior raceway is exposed to great temperature variations, sealing prevents air circulation from a warmer section to a cooler section, *NEC Section 300-7(a)*.

When raceways which contain service-entrance conductors are exposed to the weather and must be arranged to drain, *NEC Section 230-53,* then the service riser shown in Figure 1-7 cannot be sealed at the bottom. Sealing prevents water which accumulates in the conduit from draining. Therefore, it is recommended that a service panelboard or other electrical equipment *not* be placed inside rooms or buildings with high humidity, unless there is no other acceptable alternative. The service-entrance arrangements shown in Figure 1-8 are more acceptable than the one shown in Figure 1-7 for areas where humidity may be a problem.

When a panelboard must be placed in a warm area with high humidity, the two methods shown in Figure 1-8 are acceptable ways of bringing the service conductors into the panelboard to prevent condensation in the

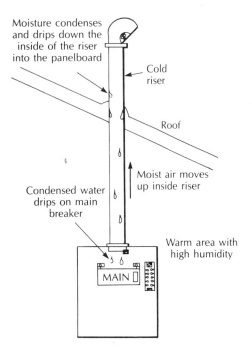

**Figure 1-7** Moist warm air from an area of high humidity rises in a service-entrance raceway. Moisture condenses on the inside surface of the cold raceway, and water then drips down into the panelboard.

service raceway. A third method is the use of service-entrance cable.

Raceway systems must be sealed on the warm end when they extend from an area containing warm, moist air to an area where the temperature is much colder *NEC Section 300-7(a)*. In addition, the raceway must be installed so that condensed water can drain without coming into contact with live or moving parts.

## CORROSIVE CONDITIONS

Corrosive materials are present on most farms. *NEC Section 300-6(c)* states that stable areas (buildings housing animals) may present corrosive conditions. Electrical equipment and wiring materials must be protected from these conditions. Animal waste produces ammonia vapors which combine with condensed water in enclosed areas of high humidity to form a corrosive substance. Another corrosive mixture is formed when some animal feeds combine with moisture. In general, chemicals and fertilizer are corrosive to most metal electrical equipment and wiring materials. Animal waste combined with water on floors will cause deterioration of most metal wiring materials.

The proper planning of wiring can usually overcome the problem of running wiring materials near or through the floor where animal waste is present. When corrosive conditions cannot be avoided, electrical equipment and metal wiring materials must be resistant to corrosion and

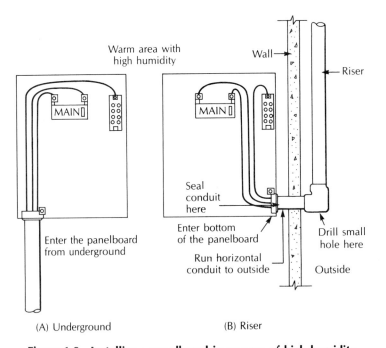

**Figure 1-8** Installing a panelboard in an area of high humidity

6   Unit 1   Introduction to Farm Wiring

suitable for the environment, *NEC Sections 300-6* and *300-6(a)*.

## SETTLING PARTICLES

*Article 547*, "Agricultural Buildings," was added to the *National Electrical Code*, partly as a fire precaution. An increased fire hazard results when flammable materials enter electrical equipment installed in certain farm buildings, especially in high-humidity areas where animals are housed.

Dust and settling particles are often present in the air in farm buildings because of cleaning, feed processing, and materials handling. Such particles present a fire hazard when they enter electrical equipment which is not dustproof or does not have a well-fitted cover. Most metal wiring boxes have openings through which dust, particles, and insects can enter. When utility openings such as knockouts are not filled, rodents can enter the box and build nests of flammable materials.

## DAMAGE

Electrical equipment and wiring materials used in farm systems are exposed to damage from animals, rodents, and machines. The electrical installer must be aware of the problem and take steps to protect the wiring system at the time of installation. Wires within reach can be rubbed, kicked, or chewed by animals or hit by machines. Such wires must be protected by rigid raceway or a suitable guard. Rigid steel conduit can be used for protection, but rapid corrosion may be a problem. Impact strength is provided by rigid nonmetallic conduit, but chewing animals may damage the conduit. *NEC Section 300-5(d)* allows Schedule 80 rigid nonmetallic conduit to be used in these areas.

Electrical controls and equipment should be installed out of reach of animals, or such devices should be fitted with protective covers. Many animals are more clever than most humans are willing to admit. Animals have been known to operate handles and switches.

Some farm wiring locations will present major problems to electrical contractors. The solution to such problems requires careful planning and a thorough knowledge of the types of wiring materials available.

## CODES AND STANDARDS

The electrician responsible for agricultural wiring must be aware of the various codes which apply to the installation. In addition, equipment and installation specifications are developed by various organizations to help make the wiring system safe and effective. The *National Electrical Code (NEC)* is the minimum standard for wiring installations in areas where the Code is accepted by the governmental unit having jurisdiction. *Section 90-1(a)* of the *National Electrical Code* states the purpose of the Code, which was developed in 1897 and is now revised every three years.

When a farm falls within the jurisdiction of the Occupational Safety and Health Act (OSHA), the wiring system of the farm may be required to comply with the *NEC*. In general, wiring installed to meet the most recent edition of the *National Electrical Code* will comply with OSHA guidelines.

### National Fire Protection Association (NFPA)

The National Fire Protection Association (NFPA) was organized in 1896 and assumed sponsorship of the *National Electrical Code* in 1911. The NFPA is a nonprofit organization dedicated to 1) promoting the science of fire protection, and 2) improving fire protection methods. A multiple-volume series covering the national fire codes, including the *National Electrical Code*, is published by the NFPA. A copy of the *National Electrical Code* can be obtained by writing to:

National Fire Protection Association
Batterymarch Park
Quincy, MA 02269

### American National Standards Institute, Inc. (ANSI)

After the *National Electrical Code* is adopted by the NFPA, it is submitted to the American National Standards Institute, Inc. (ANSI) for adoption. Once adopted by ANSI, the Code is considered a national standard. Other organizations developing standards also submit them to ANSI. The adoption of national standards leads to uniformity of equipment manufacture and minimum safety standards.

### Canadian Standards Association (CSA)

The *Canadian Electrical Code*, which is similar to the *National Electrical Code*, is adopted as a standard by the Canadian Standards Association (CSA). A copy of the *Canadian Electrical Code* can be obtained by writing to:

> Canadian Standards Association
> 178 Rexdale Boulevard
> Rexdale, Ontario, Canada M9W 1R3

## Underwriters' Laboratories, Inc. (UL)

*NEC Section 110-2* states that conductors and equipment required or permitted by the *National Electrical Code* shall be acceptable only when approved. The Code defines *approved* as being acceptable to the authority having jurisdiction. Approval may be for a specific purpose, environment, or application. Wiring materials which are labeled and/or listed are considered approved.

Wiring equipment and materials are labeled or listed only

- after evaluation according to a nationally recognized standard; or
- when found suitable for use in a specified manner after testing by a nationally recognized testing laboratory, inspection agency, or similar organization.

The Underwriters' Laboratories, Inc. (UL) is a recognized organization that provides a product testing and listing program. The UL was founded in 1894 as a nonprofit organization. A *Catalog of Standards for Safety* and other publications can be obtained by writing to:

> Underwriters' Laboratories, Inc.
> Public Information Office
> 333 Pfingsten Road
> Northbrook, IL 60062

Underwriters' Laboratories scientists and engineers develop tests for electrical materials, devices, and equipment that determine the capability of a product to operate safely in order to prevent the hazards of electrical shock and/or fire. The manufacturer submits the product to Underwriters' Laboratories for testing. Upon passing tests, the product is listed in a UL listing publication. The manufacturer is permitted to display on the product an appropriate UL label such as the one shown in Figure 1-9. (Notice UL labels on materials and equipment pictured in this text.)

UL provides manufacturers with minimum design and construction specifications for materials, devices, and equipment. These specifications have been developed after extensive testing. A typical example of a man-

**Figure 1-9** A typical Underwriters' Laboratories label

ufacturer specification is the minimum allowable distance between an electrically live terminal and the metal enclosure. UL specifications must be met if the manufacturer is to use the UL label.

UL also conducts periodic inspections after a product has been listed. Inspections of manufacturing are conducted, and products are selected at random for follow-up testing. These follow-up inspections and testing ensure that electrical products continue to meet minimum standards for safe operation, and do not create a shock or fire hazard when used properly.

## National Electrical Manufacturers Association (NEMA)

The National Electrical Manufacturers Association (NEMA) is a nonprofit organization. NEMA develops and maintains specifications for uniform equipment manufacture. These specifications are in the form of NEMA "Standards" or "Suggested Standards for Future Design." NEMA is supported in its efforts by the manufacturers and suppliers of electrical equipment.

Electricians are most concerned with NEMA standards for motor control device size ratings, enclosure types, and motor frame specifications and enclosures. A thorough understanding of this information is particularly important when selecting motors and control equipment for agricultural applications. A list of NEMA standards may be obtained by writing to:

> National Electrical Manufacturers Association
> 2101 L Street, N.W.
> Washington, DC 20037

## American Society of Agricultural Engineers (ASAE)

The American Society of Agricultural Engineers (ASAE) is a professional nonprofit organization whose nearly ten thousand members work in industry, government agencies, educational institutions, and private engineering practice. Several technical committees within ASAE encourage meetings and research on farm wiring problems. In addition, the organization sponsors the development of farm wiring recommendations and standards which are published annually in the *Yearbook of Agricultural Engineering*. This publication is available from:

> American Society of Agricultural Engineers
> 2950 Niles Road, P.O. Box 410
> St. Joseph, MI 49085

Agricultural engineering departments in many state universities have at least one faculty member working on rural electrification problems. Electrical contractors are encouraged to contact these individuals for help in solving farm wiring problems. In Canada, the Canadian Society of Agricultural Engineers and agricultural engineering departments in several Canadian universities can provide assistance to contractors and electricians.

## Institute of Electrical and Electronics Engineers, Inc. (IEEE)

The Institute of Electrical and Electronics Engineers (IEEE) is another professional organization which develops recommendations and standards for electrical wiring and equipment and for electrical system and equipment safety. In addition, IEEE develops testing procedures for electrical equipment and materials. Many IEEE standards have been adopted as national standards by ANSI. A rural electrification group within the IEEE is responsible for the development of standards for delivering electrical power to rural areas and for farm wiring. A list of IEEE standards may be obtained by writing to:

> IEEE Standards Office
> 345 East 47th Street
> New York, NY 10017

## Illuminating Engineering Society (IES)

The Illuminating Engineering Society (IES) makes illumination level and luminaire recommendations for residential, school, commercial, and industrial lighting applications. The IES and the American Society of Agricultural Engineers work together to develop and maintain farm lighting standards and recommendations.

## International Association of Electrical Inspectors (IAEI)

The International Association of Electrical Inspectors (IAEI) strives to bring about improved understanding and application of the *National Electrical Code*. The membership of this organization consists of electrical inspectors, electricians, contractors, and manufacturers. Representatives of *National Electrical Code* committees attend all regional and many state meetings to answer questions about the interpretation of the Code.

> International Association of Electrical Inspectors
> 930 Busse Highway
> Park Ridge, IL 60068

## National Food and Energy Council (NFEC)

The membership of the National Food and Energy Council (NFEC) consists of representatives from electrical power suppliers and electrical equipment manufacturers with an interest in the application of electrical energy for the production of food, feed, and fiber. A planning guide for farm wiring, the *Agricultural Wiring Handbook,* is prepared by the NFEC in cooperation with the Edison Electric Institute, the National Rural Electric Cooperative Association, the National Association of Electrical Distributors, the National Electrical Contractors Association, the ASAE, the IAEI, and the IES. The handbook and list of available publications and materials may be obtained by writing to:

> National Food and Energy Council
> Vandiver West Business Park
> Columbia, MO 65202

## Local Electrical Associations

Local electrical associations can provide technical information for wiring system installation and interpretation of local, state, and national codes. Membership in these organizations includes electricians, contractors, apprenticeship instructors, inspectors, electrical power supplier representatives, electrical distributors, and manufacturers' representatives. Contractors and electricians are encouraged to participate in their local electrical association. Membership information can be obtained by contacting a local electrical inspector.

# NEC USE OF METRIC (SI) MEASUREMENTS

Metric measurements were included with English measurements for the first time in the 1981 *National Electrical Code*. The metric system is known as the *International System of Units* (SI). Metric units in this text are referred to as *SI units, metric SI,* or simply *SI*. Electrical values such as volts (V), amperes (A), watts (W), and ohms (Ω), are SI units with which we are already familiar. The SI system uses Celsius degrees (°C) rather than Fahrenheit degrees (°F). However, the Code has been giving values in both degrees Celsius and Fahrenheit for many years.

A conversion from °F to °C is shown in Figure 15-1. English-to-SI conversions are contained at appropriate locations in the Code.

Metric measurements appear in the Code as follows:

- In the Code paragraphs, the approximate metric measurement appears in parentheses following the English measurement.
- In the Code tables, a footnote shows the SI conversion factors. The tables themselves do not show the metric measurement.

A metric measurement is not shown for conduit size, box size, wire size, horsepower designation for motors, and other "trade sizes" that are not actual measurements.

# HOW TO USE METRICS

All of the units of measurements in the metric system are based on multiples of 10. Thus, they increase by 10, 100, 1000, etc. They decrease by 1/10, 1/100, 1/1000, etc. For instance, a kilowatt is 1 000 times greater than a watt.

Just as the English system has names for measurements, such as inches, feet, yards, so does the metric system have names. But to make the metric system much simpler, a prefix may be added to the name of the unit. If we add *kilo* to *watt*, then we have *kilowatt*, which is the same as 1 000 watts. If we add *kilo* to *meter*, then we have *kilometer*, which is the same as 1 000 meters. Common prefixes for the SI system of measurements are shown in Figure 1-10.

| nano  | 0.000 000 001 | (one-billionth) (1/1 000 000 000) |
| micro | 0.000 001     | (one-millionth) (1/1 000 000)     |
| milli | 0.001         | (one-thousandth) (1/1 000)        |
| centi | 0.01          | (one-hundredth) (1/100)           |
| deci  | 0.1           | (one-tenth) (1/10)                |
| the unit | 1          | (one)                             |
| deka  | 10            | (ten)                             |
| hecto | 100           | (one hundred)                     |
| kilo  | 1 000         | (one thousand)                    |
| mega  | 1 000 000     | (one million)                     |

**Figure 1-10** SI measurement system prefixes and their values

The dimensions most frequently used in the Code are units of length and area. Other units used in several locations are volume, weight, temperature, and density. The electrical installer will encounter other English-to-SI unit conversions in the future. Motor power ratings given in the 1984 *NEC* are in horsepower (hp) but in the future, the SI unit, kilowatt (kW), will be encountered.

The gram (g) and the kilogram (kg) are actually units of mass but, on the surface of the earth, they are usually used as units of weight. The unit of force in the English system is the pound (lb); in SI the unit of force is the newton (N). On the surface of the earth, a 1-kg mass weighs 9.8 N. It would take a force of 9.8 N to lift a 1-kg mass.

The term *newton* may be encountered when making electrical connections. A torque wrench is sometimes used by an electrical installer to tighten bolts and other pressure connectors at wire terminations. Usually, the amount of torque is stated in pound-feet (lb-ft) but, in the future, torque may be stated in newton-meters (N·m). The amount of torque developed by motors is also sometimes stated in newton-meters rather than in pound-feet.

Table 1-1 gives conversion factors for the English and SI units that are most likely to be important to an electrical installer now and in the future.

**Table 1-1** Converson factors for common English and SI units.

| Length | |
|---|---|
| centimeters (cm) × 0.3937 | = inches (in) |
| inches (in) × 0.0254 | = meters (m) |
| inches (in) × 0.254 | = decimeters (dm) |
| inches (in) × 2.54 | = centimeters (cm) |
| inches (in) × 25.4 | = millimeters (mm) |

**Table 1-1 Continued**

| | |
|---|---|
| feet (ft) × 0.3048 | = meters (m) |
| kilometers (km) × 1000 | = meters (m) |
| kilometers (km) × 0.621 | = miles (mi) |
| meters (m) × 3.2808 | = feet (ft) |
| miles (mi) × 1.609 | = kilometers (km) |
| millimeters (mm) × 0.03937 | = inches (in) |

### Area

| | |
|---|---|
| square centimeters (cm$^2$) × 0.155 | = square inches (in$^2$) |
| square inches (in$^2$) × 6.452 | = square centimeters (cm$^2$) |
| square feet (ft$^2$) × 0.093 | = square meters (m$^2$) |
| square meters (m$^2$) × 10.764 | = square feet (ft$^2$) |
| square meters (m$^2$) × 1.196 | = square yards (yd$^2$) |
| square yards (yd$^2$) × 0.8361 | = square meters (m$^2$) |

### Volume

| | |
|---|---|
| gallons (gal) × 3.785 | = liters (L) |
| liters (L) × 0.264 | = gallons (gal) |
| cubic inches (in$^3$) × 16.39 | = cubic centimeters (cm$^3$) |
| cubic feet (ft$^3$) × 0.0283 | = cubic meters (m$^3$) |
| cubic yards (yd$^3$) × 0.765 | = cubic meters (m$^3$) |
| cubic centimeters (cm$^3$) × 0.061 | = cubic inches (in$^3$) |
| cubic meters (m$^3$) × 35.3 | = cubic feet (ft$^3$) |
| cubic meters (m$^3$) × 1.31 | = cubic yards (yd$^3$) |
| cubic feet (ft$^3$) × 28.32 | = liters (L) |

### Weight

| | |
|---|---|
| pounds (lb) × 0.454 | = kilograms (kg) |
| kilograms (kg) × 2.2 | = pounds (lb) |
| ounces (oz) × 28.35 | = grams (g) |
| grams (g) × 0.035 | = ounces (oz) |

### Force

| | |
|---|---|
| pounds (lb) × 4.45 | = newtons (N) |
| newtons (N) × 0.225 | = pounds (lb) |
| ounces (oz) × 0.278 | = newtons (N) |

### Torque

| | |
|---|---|
| foot-pounds (ft-lb) × 1.356 | = newton meters (N·m) |
| newton meters (N·m) × 0.7375 | = foot-pounds (ft-lb) |

### Density

| | |
|---|---|
| pounds per cubic foot (lb/ft$^3$) × 16.02 | = kilograms per cubic meter (kg/m$^3$) |

### Power

| | |
|---|---|
| horsepower (hp) × 0.746 | = kilowatts (kW) |
| kilowatts (kW) × 1.34 | = horsepower (hp) |

### Pressure

| | |
|---|---|
| pounds per square inch (lb/in$^2$) × 6.895 | = kilopascals (kPa) |

(One kilopascal is equal to a pressure of one newton per square meter: 1 kPa = 1 N/m$^2$)

### Heat

| | |
|---|---|
| British thermal unit (Btu) × 1.055 | = kilojoules (kJ) |

# REVIEW

Refer to the *National Electrical Code* when necessary to complete the following review material. Write your answers on a separate sheet of paper.

1. List the characteristics of a wiring system installed according to the *National Electrical Code*.

2. Is an animal more subject to electrical shock than a human? Explain.

3. Give an example demonstrating how the loss of electrical power on a farm may cause serious problems.

4. When farm buildings are spaced far apart, long wires are required to deliver electrical energy to the buildings. This situation often results in an excessive ___?___ in the voltage along the wires.

5. According the *NEC Section 300-___?___*, where portions of an interior raceway system are exposed to widely different temperatures, the circulation of air from a warmer to a colder section shall be prevented by ___?___.

6. List the common types of damage animals may cause to electrical equipment and wiring materials.

7. What association sponsors the *National Electrical Code*, and how often is the Code revised?

8. Where can electrical contractors and installers obtain help in solving farm wiring problems?

9. A certain machine has all installation dimensions given in SI measurements. The machine must be mounted at least 60 cm from the wall. Convert 60 cm to inches, rounded to the nearer ¼ inch.

10. You are asked how many watts of fluorescent lighting are required per square meter for a particular farm building. The recommended lighting level is ¾ watt (W) per square foot. Determine the lighting level in watts per square meter.

# UNIT 2
# ELECTRICAL FUNDAMENTALS

## OBJECTIVES

After studying this unit, the student will be able to

- describe direct current and alternating current.
- define voltage, current, and resistance.
- measure voltage, current, and resistance.
- describe the relationships among voltage, current, and resistance in a circuit.
- state the effects in a circuit of loads arranged in series and in parallel.
- determine electrical power
- describe power factor and its effect in electrical circuits.
- define electrical work and efficiency.

An electrical installer must understand the behavior of electricity in a circuit to be able to solve electrical problems. When electrical equipment or an electrical circuit fails to operate properly, an electrical installer must be able to pinpoint the problem by applying the basic laws of electricity.

## WHAT IS ELECTRICITY?

An understanding of electricity begins with an understanding of the atom and the behavior of electrical charges. Everything which occupies space and has weight is called *matter*. Matter includes gases, liquids, and solid materials. Matter is composed of *atoms*, as represented in Figure 2-1. Ninety-two basic types of atoms are known to occur naturally on the earth. These basic types of atoms are called *elements*.

Each atom has a core, or *nucleus*, which contains one or more positively (+) charged particles called *protons*. The nucleus of most atoms also contains one or more uncharged particles. These particles are called *neutrons*. Negatively (−) charged particles called *electrons* move in orbits around the nucleus.

There is an electron for every proton in the nucleus. Some electrons orbit close to the nucleus, and others orbit at a greater distance. The atoms of elements such as silver, copper, aluminum, and other metals have one to three electrons moving in an orbit far from the nucleus. These electrons are not held tightly to the atom. Because the electrons can move freely from one atom to another, as shown in Figure 2-2, they are called *free electrons*.

The negatively charged free electrons are attracted to a positive charge, such as the positive (+) terminal of a battery, Figure 2-3. Negative charges repel one another.

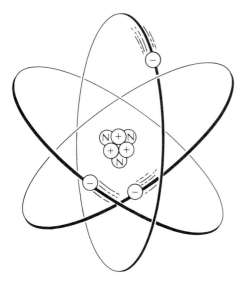

**Figure 2-1** Matter is composed of atoms having a positively charged nucleus. Negatively charged electrons move in orbits around the nucleus.

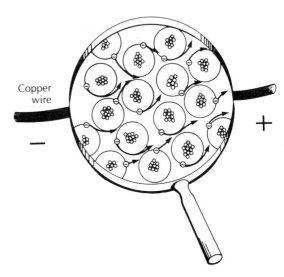

**Figure 2-3** The free electrons of the copper wire are repelled by the negative (−) charge and attracted to the positive (+) charge. This flow of electrons in a wire is the flow of electrical current.

Therefore, the free electrons are repelled by the negative (−) terminal of a battery. If a battery is connected to the ends of a piece of wire, the free electrons will flow away from the negative terminal and toward the positive terminal. As the free electrons flow from the negative terminal of a battery to the positive terminal, they must flow through any load, such as a lamp, causing it to light. See Figure 2-4.

The flow of free electrons through a conductor is defined as the flow of electrical current.

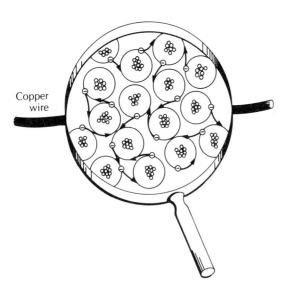

**Figure 2-2** The atoms of copper and other metals have loosely attached free electrons which move freely from one atom to another.

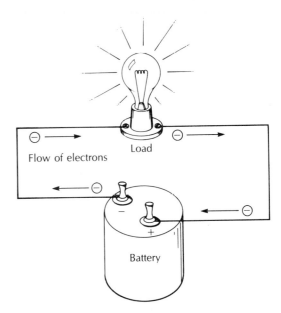

**Figure 2-4** Electrons flow from the negative (−) terminal to the positive (+) terminal of the battery. They must flow through any load, such as the light bulb, which is placed in the electrical path between the negative and positive battery terminals.

## Current

*Direct current* (dc) travels in one direction only. The electrons flow from the negative terminal of a battery to the positive terminal, as shown in Figure 2-4. Common sources of direct current include batteries, solar photovoltaic cells, dc generators, rectifiers, and dc power supplies used in electronic equipment. The electrician generally is not concerned with direct current in agricultural wiring, except in electronic solid-state control systems and variable-speed motors such as the ones used to meter concentrates into animal feed.

*Alternating current* (ac) travels first in one direction, stops, then reverses and flows in the opposite direction, Figure 2-5. The electricity is then said to have completed one cycle. The electricity delivered to homes, farms, businesses, and industry in the United States completes 60 cycles in one second. The term *hertz* (Hz) represents one cycle per second. Therefore, in the United States, normal electrical energy is delivered at 60 Hz. Outside of the United States, the electrical energy is often generated at 50 Hz.

Alternating current changes its direction of flow 120 times every second because of the way in which it is generated. A magnetic field spins inside a coil of wire to produce electricity, Figure 2-6. This device is known as an electric generator. When the north pole of the magnet nears the top coil, the electricity generated flows in one direction in the circuit. When the south pole of the mag-

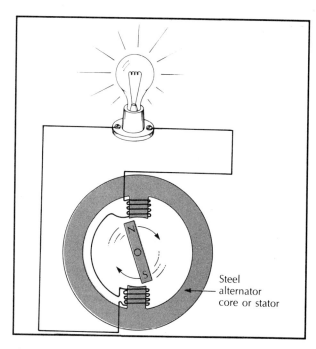

Figure 2-6 Electricity used for homes, farms, and businesses in the United States is in the form of alternating current which is produced by an alternator consisting of a magnet spinning inside a coil of wire.

net nears the top coil, the electricity reverses its direction and flows in the opposite direction in the circuit.

## VOLTAGE

Voltage is the push that forces the electrons to flow in a circuit. The higher the voltage, the greater is the push on the electrons. More electrons or electricity will flow through a circuit if the push or voltage is increased.

A battery or direct-current source of electricity has a fixed voltage. Flashlight batteries produce about 1.5 (V). Automobile batteries are rated at 12 V. If a battery is completely charged, it will produce a constant voltage.

Alternating electricity does not have a constant voltage. As the electricity begins to move in one direction, the voltage starts at zero and builds up to a maximum value. Then the voltage decreases back to zero. The electricity reverses direction and voltage builds up to a maximum, but this time in the negative direction, Figure 2-7. The voltage pattern shown is characteristic of alternating electricity, and is called a *sine wave*. An *oscilloscope* is a voltage measuring instrument which displays the voltage pattern, or traces, on a screen similar to that of a television set.

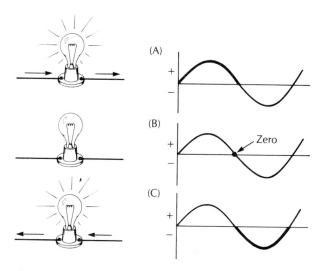

Figure 2-5 Alternating current travels (A) first in one direction, (B) stops, and (C) reverses, and flows in the opposite direction.

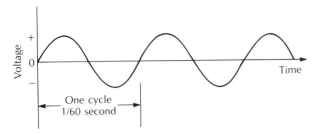

**Figure 2-7** The voltage producing alternating current travels first in one direction (above the horizontal line), and then reverses to travel in the opposite direction (below the line). The resulting graph of the voltage is a sine wave.

## Measuring Voltage

Voltage is always measured between two wires, between a wire and the ground, or across an electrical device using electricity. The instrument used to measure voltage is the *voltmeter*. The voltmeter shown in Figure 2-8 is measuring the voltage across the lamp.

Direct-current voltage measurements are made using a *dc voltmeter*. An *ac voltmeter* is required for alternating-voltage measurements. Meters designed to measure both ac and dc voltages are in common use. A selector switch on such a meter is set for the type of electrical voltage (ac or dc) to be measured, Figure 2-9.

*Ac voltage measurements.* An ac voltmeter measures the *root mean square* (rms) value of the alternating-voltage sine wave, Figure 2-10. This value is also known as the *effective voltage*. The rms voltage is 0.707 times the

**Figure 2-9** Electricians use an ac-dc multimeter in the field to measure voltage or to determine when a circuit is hot: *(Courtesy of Amprobe Instruments)*

value of the peak voltage of the wave. The standard values used in ac electrical systems, such as 120, 208, 240, 277, and 480 volts, are all rms voltages.

The determination of the polarity and the connection of the voltmeter to maintain the proper polarity are not a concern when making ac voltage measurements. A proper reading is always obtained.

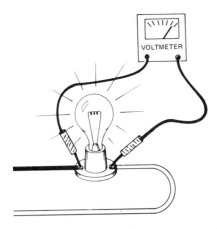

**Figure 2-8** To measure the voltage across an electrical load, such as a light bulb, the voltmeter is connected in parallel with the load.

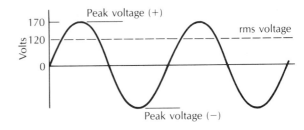

**Figure 2-10** An ac voltmeter measures the rms voltage, or 0.707 times the peak voltage of the alternating wave.

***Dc voltage measurements.*** When dc voltage measurements are to be made, the voltmeter probes must be connected to the circuit so as to maintain the proper polarity. The positive probe (generally red) is connected to the positive (+) side of the circuit. The negative probe (generally black) is connected to the negative (−) side of the circuit. If the probes are reversed, the needle of the voltmeter deflects in the wrong direction and tries to indicate a reading less than zero. This can damage the meter.

## CURRENT

Electrical *current* is the flow of electrons through a conductor. The unit of measurement of electrical current is the ampere. The *ampere* (A) is the measure of the rate of flow of electrons. For a dc circuit, the current flows in one direction at a constant rate. The current in an ac circuit changes direction 120 times each second. Alternating current looks like the voltage wave shown in Figure 2-10.

## Measuring Current

An *ammeter* is used to measure the current flowing in a circuit. Direct current is measured using a *dc ammeter*. An *ac ammeter* is used to measure the rms (root mean square) value of alternating current. In other words, the reading indicated by an ac ammeter is 0.707 times the peak current value of the alternating wave. Meters designed to measure both alternating and direct current are in common use.

To measure electrical current flowing in a circuit, the current must pass through the meter. Therefore, the circuit must be opened and the ammeter wired in series with the circuit, Figure 2-11. It does not matter if the ammeter is installed in the hot wire or in the grounded wire of the circuit, since the same current flows in each wire.

> **CAUTION:** *Never* connect an ammeter across an electrical device, or between a hot wire and a grounded wire. An ammeter has a very low internal resistance, resulting in a "dead short" when the device is connected between a hot wire and the neutral. This type of faulty connection will ruin the meter.

As current flows through a wire, a magnetic field is built up in the air around the wire. For ac, a *clamp-on ammeter* is used to obtain a current measurement. This type of meter is clamped around one of the circuit wires, Figure 2-12. The meter measures the strength of the magnetic field produced by the current flowing in the

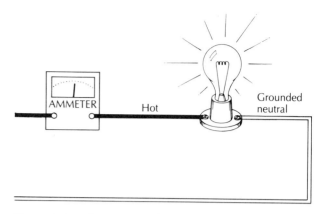

**Figure 2-11** An ammeter is installed in series with the electrical load to ensure that the circuit current flows through the ammeter.

wire, and converts this value to a current reading. To measure a small current, the wire can be looped through the meter a second time, Figure 2-13. In this case, the ammeter reading is two times the current flowing in the circuit.

Most meters will provide the most accurate readings near the center of the scale. Meter readings close to zero are hard to read, and are often inaccurate.

$$\text{Amperes flowing} = \frac{\text{Ampere reading on meter}}{\text{Number of wire loops through meter}} \qquad \text{Eq. 2.1}$$

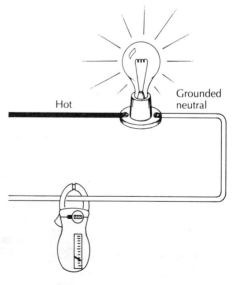

**Figure 2-12** A clamp-on ammeter is clamped around one circuit wire to measure the current flowing in the wire.

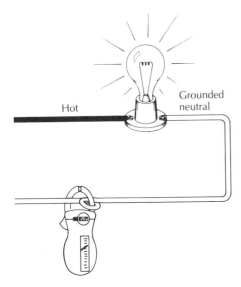

**Figure 2-13** Looping the circuit wire through the clamp-on ammeter two times doubles the current reading on the ammeter.

## Problem 2-1

A clamp-on ammeter is used to measure the current in an ac circuit. The electrician finds that the current flowing in the circuit is less than one ampere (1 A), and desires to obtain a more accurate reading. To increase the meter reading, the electrician loops one circuit wire through the meter two times, Figure 2-13. The meter reads 1.4 A. How much current is flowing in the circuit? (Current is measured in amperes.)

## Solution

Equation 2.1 is used to determine the current in amperes in the wire.

$$\text{Amperes flowing} = \frac{1.4 \text{ A}}{2 \text{ loops}} = 0.7 \text{ A}$$

## WIRE DIMENSIONS

Wire sizes are expressed in circular mils. A *mil* is equal to 0.001 inch (in). A *circular mil* (CM) is defined as the area of a circle having a diameter of 1 mil. Equation 2.2 is used to find the area of a circle in square inches (in²). (The *square* of a number is the number multiplied by itself.)

$$\text{Area of a circle} = \frac{\pi d^2}{4} \qquad \text{Eq. 2.2}$$

where: Area = area, in square inches
$\pi$ = 3.14
d = diameter of circle, in inches

To avoid using Equation 2.2, and thus simplify problems, the circular mil (CM) area is used to express wire sizes. The area of a wire in circular mils is simply the square of the diameter, as shown in Equation 2.3. (The *square* of a number is the number multiplied by itself.)

$$\text{Circular mil area} = (\text{Diameter in mils})^2 \qquad \text{Eq. 2.3}$$

For No. 14 wire, the diameter is 0.064 1 inch, or 64.1 mils. The area of this wire, the circular mils, is 4 110:

$$\begin{aligned}\text{Circular mil area of No. 14 wire} \\ = (64.1 \text{ mils})^2 \\ = 64.1 \times 64.1 = 4\ 110 \text{ CM}\end{aligned}$$

**Note:** Answers in this text are rounded.

The sizes of electrical wire commonly used are expressed by American Wire Gauge (AWG) numbers. The numbers range from 36 (wire diameter of 5 mils) to 0000 (wire diameter of 460 mils). As the AWG numbers decrease, the circular mil area increases. The area of a wire nearly doubles as the AWG number decreases by three. Another way of expressing this relationship is to say that the area of a wire is nearly halved as the AWG number is increased by three. The following example illustrates this relationship.

No. 4 AWG  41 740 CM
No. 3 AWG  —
No. 2 AWG  —
No. 1 AWG  83 690 CM

No. 1 AWG wire has an area about twice that of No. 4 AWG wire. The area of a wire increases ten times as the AWG number is decreased ten sizes. For example, compare No. 4 AWG wire (having an area of 41 740 CM) to No. 14 AWG wire (having an area of 4 110 CM). It can be seen that as the wire number decreases ten sizes from No. 14 to No. 4, the area of the wire increases ten times.

Wire sizes larger than No. 0000 AWG are expressed according to their cross-sectional area, in thousands of circular mils (MCM). (The first M in this abbreviation is the Roman numeral for 1 000.) AWG and MCM wire sizes are listed in *Table 8, Chapter 9* of the *National Electrical Code*. The information given includes diameter, circular mil area, and resistance values for copper and aluminum wire. *NEC Table 8* is reproduced in this text as Table 2-1. The diameters listed in the table are expressed in inches. To convert from inches to mils, simply multiply the value in inches by one thousand (1 000).

## Problem 2-2

Determine the circular mil area of No. 8 AWG solid wire having a diameter of 0.128 5 inch (in).

## Solution

$$\text{Diameter, in mils} = 0.128\ 5\ \text{in} \times 1\ 000$$
$$= 128.5\ \text{mils}$$
$$\text{Circular mil area} = 128.5 \times 128.5$$
$$= 16\ 510\ \text{CM}$$

Large wires are made up of many individual strands of wire. To determine the cross-sectional area of the wire, it is necessary to find the area of each individual strand. Then, assuming that the strands are of the same size, the area of one strand is multiplied by the number of strands. Table 2-1 gives the number of strands and the diameter of each strand for the common sizes of electrical wire.

## RESISTANCE

As electricity flows through a wire, it encounters *resistance*, an electrical friction that tries to prevent the flow of current. Electrical resistance is a property resulting from the atomic structure of the material. Such metals as copper and aluminum have a low resistance to the flow of electricity. Materials with a very high resistance include porcelain and glass. These materials have such a high resistance that they are used as insulators. The unit of measure of resistance is the *ohm*. The Greek capital letter omega ($\Omega$) is the symbol used to mean ohms of resistance; for example, 5 $\Omega$ (ohms) of resistance. Copper and aluminum wires are common electrical conductor materials encountered by electricians.

Steel or aluminum in equipment and raceways serves as the grounding conductor for an electrical circuit. Brass and bronze are often used to conduct electricity in electrical equipment and devices. The liquid metal mercury, inside a glass tube, is used to make a quiet, sensitive switch for many applications, such as low-voltage thermostats. Other metals, such as zinc and cadmium, are used as coatings which must conduct electricity. The electrical resistance of some common metals compared to copper is shown in Table 2-2.

The resistance of a conductor depends upon the cross-sectional area, the length of the conductor, and the

Table 2-1  (*NEC Table 8, Chapter 9*) Conductor Properties

| Size AWG/ MCM | Area Cir. Mils | Stranding | | Overall | | DC Resistance at 75°C, 167°F | | |
|---|---|---|---|---|---|---|---|---|
| | | Quantity | Diam. In. | Diam. In. | Area. In.² | Copper | | Aluminum |
| | | | | | | Uncoated ohm/MFT | Coated ohm/MFT | ohm/MFT |
| 18 | 1620 | 1 | — | 0.040 | 0.001 | 7.77 | 8.08 | 12.8 |
| 18 | 1620 | 7 | 0.015 | 0.046 | 0.002 | 7.95 | 8.45 | 13.1 |
| 16 | 2580 | 1 | — | 0.051 | 0.002 | 4.89 | 5.08 | 8.05 |
| 16 | 2580 | 7 | 0.019 | 0.058 | 0.003 | 4.99 | 5.29 | 8.21 |
| 14 | 4110 | 1 | — | 0.064 | 0.003 | 3.07 | 3.19 | 5.06 |
| 14 | 4110 | 7 | 0.024 | 0.073 | 0.004 | 3.14 | 3.26 | 5.17 |
| 12 | 6530 | 1 | — | 0.081 | 0.005 | 1.93 | 2.01 | 3.18 |
| 12 | 6530 | 7 | 0.030 | 0.092 | 0.006 | 1.98 | 2.05 | 3.25 |
| 10 | 10380 | 1 | — | 0.102 | 0.008 | 1.21 | 1.26 | 2.00 |
| 10 | 10380 | 7 | 0.038 | 0.116 | 0.011 | 1.24 | 1.29 | 2.04 |

**Table 2-1 Continued**

| | | | | | | | | |
|---|---|---|---|---|---|---|---|---|
| 8 | 16510 | 1 | — | 0.128 | 0.013 | 0.764 | 0.786 | 1.26 |
| 8 | 16510 | 7 | 0.049 | 0.146 | 0.017 | 0.778 | 0.809 | 1.28 |
| 6 | 26240 | 7 | 0.061 | 0.184 | 0.027 | 0.491 | 0.510 | 0.808 |
| 4 | 41740 | 7 | 0.077 | 0.232 | 0.042 | 0.308 | 0.321 | 0.508 |
| 3 | 52620 | 7 | 0.087 | 0.260 | 0.053 | 0.245 | 0.254 | 0.403 |
| 2 | 66360 | 7 | 0.097 | 0.292 | 0.067 | 0.194 | 0.201 | 0.319 |
| 1 | 83690 | 19 | 0.066 | 0.332 | 0.087 | 0.154 | 0.160 | 0.253 |
| 1/0 | 105600 | 19 | 0.074 | 0.373 | 0.109 | 0.122 | 0.127 | 0.201 |
| 2/0 | 133100 | 19 | 0.084 | 0.419 | 0.138 | 0.0967 | 0.101 | 0.159 |
| 3/0 | 167800 | 19 | 0.094 | 0.470 | 0.173 | 0.0766 | 0.0797 | 0.126 |
| 4/0 | 211600 | 19 | 0.106 | 0.528 | 0.219 | 0.0608 | 0.0626 | 0.100 |
| 250 | — | 37 | 0.082 | 0.575 | 0.260 | 0.0515 | 0.0535 | 0.0847 |
| 300 | — | 37 | 0.090 | 0.630 | 0.312 | 0.0429 | 0.0446 | 0.0707 |
| 350 | — | 37 | 0.097 | 0.681 | 0.364 | 0.0367 | 0.0382 | 0.0605 |
| 400 | — | 37 | 0.104 | 0.728 | 0.416 | 0.0321 | 0.0331 | 0.0529 |
| 500 | — | 37 | 0.116 | 0.813 | 0.519 | 0.0258 | 0.0265 | 0.0424 |
| 600 | — | 61 | 0.992 | 0.893 | 0.626 | 0.0214 | 0.0223 | 0.0353 |
| 700 | — | 61 | 0.107 | 0.964 | 0.730 | 0.0184 | 0.0189 | 0.0303 |
| 750 | — | 61 | 0.111 | 0.998 | 0.782 | 0.0171 | 0.0176 | 0.0282 |
| 800 | — | 61 | 0.114 | 1.03 | 0.834 | 0.0161 | 0.0166 | 0.0265 |
| 900 | — | 61 | 0.122 | 1.09 | 0.940 | 0.0143 | 0.0147 | 0.0235 |
| 1000 | — | 61 | 0.128 | 1.15 | 1.04 | 0.0129 | 0.0132 | 0.0212 |
| 1250 | — | 91 | 0.177 | 1.29 | 1.30 | 0.0103 | 0.0106 | 0.0169 |
| 1500 | — | 91 | 0.128 | 1.41 | 1.57 | 0.00050 | 0.00883 | 0.0141 |
| 1750 | — | 127 | 0.117 | 1.52 | 1.83 | 0.00735 | 0.00756 | 0.0121 |
| 2000 | — | 127 | 0.126 | 1.63 | 2.09 | 0.00643 | 0.00662 | 0.0106 |

These resistance values are valid ONLY for the parameters as given. Using conductors having coated strands, different stranding type, and especially, other temperatures, change the resistance.

Formula for temperature change: $R_2 = R_1 [1 + \alpha(T_2 - 20)]$ where: $\alpha_{cu} = 0.00393$, $\alpha_{AL} = 0.00403$.

Class B stranding is listed as well as solid for some sizes. Its overall diameter and area is that of its circumscribing circle. The construction information is per NEMA WC8-1976 (Rev 5-1980). The resistance is calculated per National Bureau of Standards Handbook 100, dated 1966, and Handbook 109, dated 1972.

Conductors with compact and compressed stranding have about 9 percent and 3 percent, respectively, smaller bare conductor diameters than those shown.

The IACS conductivities used: bare copper = 100%, aluminum = 61%.

Reprinted with permission from NFPA 70-1984, *National Electrical Code®*, Copyright © 1983, National Fire Protection Association, Quincy, Massachusetts 02269. This reprinted material is not the complete and official position of the NFPA on the referenced subject, which is represented only by the standard in its entirety.

temperature. Increasing the cross-sectional area of a conductor reduces the resistance to the flow of electricity. Think of a wire as a highway. Two lanes offer less resistance to the flow of traffic than one lane does. When the cross-sectional area of a conductor is doubled, the resistance is reduced to one-half, and the conductor carries twice as much current. When the resistance of a wire is known, the resistance of a second wire with a different cross-sectional area can be determined by using Equation 2.4.

$$\text{Resistance of second wire} = \frac{\text{Area of first wire}}{\text{Area of second wire}} \times \text{Resistance of first wire} \qquad \text{Eq. 2.4}$$

The area of a circular cross-section wire is determined by the square of the diameter of the wire. Equation 2.4 can be written in terms of the square of the diameter of two wires, Equation 2.5. Actually, if the diameter of one

Table 2-2  Approximate electrical resistance of some common metals compared to copper. The actual resistance varies according to the purity of the metal and the composition of alloy metals.

| Metal | Resistance Compared to Copper |
|---|---|
| Silver | 0.92 |
| Copper | 1.00 |
| Gold | 1.38 |
| Aluminum | 1.59 |
| Bronze | 2.41 |
| Tungsten | 3.20 |
| Zinc | 3.62 |
| Brass | 4.40 |
| Cadmium | 4.51 |
| Platinum | 5.80 |
| Iron | 6.67 |
| Nickel | 7.73 |
| Tin | 8.20 |
| Steel | 8.62 |
| Lead | 12.76 |
| Mercury | 54.6 |
| Nichrome | 60.0 |

wire is twice the diameter of another wire, the resistance of the larger wire will be only one-quarter the resistance of the smaller wire. A wire with double the diameter of another wire will carry four times as much current.

$$\text{Resistance of 2d wire} = \frac{(\text{Diameter of 1st wire})^2}{(\text{Diameter of 2d wire})^2} \times \text{Resistance of 1st wire} \quad \text{Eq. 2.5}$$

The resistance of a conductor is increased as the length is increased. If the length of a conductor is doubled, the resistance will be doubled. The resistance of a wire is generally given for 1 000 feet (ft) of wire. Resistance values from Table 2-1 are used in Equation 2.6.

$$\text{Resistance of new length} = \frac{\text{New length}}{1\ 000\ \text{ft}} \times \text{resistance of 1 000 ft} \quad \text{Eq. 2.6}$$

## Problem 2-3

A copper wire has a resistance of 1.21 Ω per 1 000 ft. Determine the resistance of 400 ft of the wire.

## Solution

$$\text{Resistance of 400 ft} = \frac{400\ \text{ft}}{1\ 000\ \text{ft}} \times 1.21\ \Omega = 0.48\ \Omega$$

## Problem 2-4

The circular mil area of No. 3 AWG aluminum wire (52 620 CM) is approximately twice the circular mil area of No. 6 AWG wire (26 240 CM). If 500 ft of No. 6 AWG aluminum wire has a resistance of 0.404 Ω, determine the resistance of 500 ft of No. 3 AWG wire.

## Solution

$$\text{Resistance of No. 3} = \frac{26\ 240\ \text{CM}}{52\ 620\ \text{CM}} \times 0.404\ \Omega = 0.201\ \Omega$$

The resistance of an electrical conductor increases as the temperature of the wire increases. The tungsten filament of an incandescent light bulb has a higher resistance when the lamp is burning than when it is turned off. An electrical wire will heat up a few degrees when an electrical current flows through it. A wire exposed to sunlight will heat up. When a building gets very warm, the resistance of the wire increases.

The resistance values for electrical wire given by the *National Electrical Code* in *Table 8, Chapter 9* (Table 2-1 of this text) are based upon a temperature of 75°C (167°F). In a practical situation on a farm, this is a maximum resistance that is likely to be encountered. Usually, the temperature of the wire and the wire resistance will be lower than the values given in Table 2-1.

## Measuring Resistance

An instrument called an *ohmmeter* is used to measure the resistance of a conductor. The resistance of a wire or electrical device must be measured with the power turned off. The ohmmeter contains a battery which sends a small amount of current through the conductor. The amount of deflection of the ohmmeter's needle depends upon the resistance of the conductor. An ohmmeter with a moving needle must be adjusted to read zero resistance before making a resistance measurement.

The zero adjustment is accomplished by touching the leads together and turning the adjust knob until the meter reads zero on the ohm scale. A digital ohmmeter generally does not require a zero adjustment, but it is a good idea to follow this procedure to make sure that the ohmmeter is working properly.

Resistance determined with an ohmmeter is dc resistance. Alternating current in a wire behaves differently than direct current. One difference is that alternating current tries to flow close to the outside surface of a wire. This action is known as the *skin effect*. The result is that a wire has a slightly higher resistance to ac than it does to dc. The skin effect is greater when the wires are in metal conduit or in metal-sheathed cable.

For small wires, the difference in resistance is not important, but the difference must be taken into account for sizes larger than 4/0.

*Table 9* in *Chapter 9* of the *National Electrical Code* provides factors that can be used to determine ac skin effect as well as other current-opposing effects caused by the inductance in the circuit wiring. Inductance is explained later in this unit. These factors are included in easy-to-use voltage-drop tables provided in Unit 7.

## ELECTRICAL CIRCUITS

A complete electrical circuit is required for current to flow. It is easier to troubleshoot electrical problems if the electrician keeps in mind the principle of an electrical circuit. A proper electrical circuit consists of:

1. A source of electrical pressure to force the electrical current to flow
2. A path for the electrical current to follow
3. A load which converts electrical energy into heat, light, and power, or performs an electronic function
4. A switch to control the flow of current

A typical electrical circuit is shown in Figure 2-14. If the switch is opened, the circuit is broken and the electrical current stops flowing. The circuit will also be broken if the lamp burns out or if a wire comes loose from a terminal.

*Note:* In Figure 2-14, electron flow is shown to represent current flow. Conventional current flow is assumed to be from positive (+) to negative (−) by many individuals.

The circuit in Figure 2-14 is a simple one-path circuit. Every component of the circuit is connected in *series*, or in a row. If a second lamp is added to the circuit, as in Figure 2-15, the two lamps are said to be connected

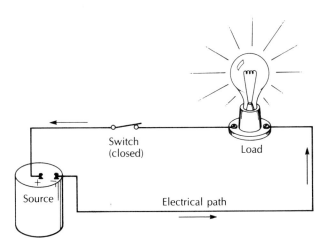

**Figure 2-14** An electrical current will flow in a complete circuit consisting of a voltage source, closed switch, electrical load, and a path to connect these components.

in series. (*Note:* Electron flow is shown to represent current flow.) When one lamp burns out, the circuit is broken, and the other lamp also goes out. An understanding of a series circuit is important, particularly when working with motor control systems. An example of a series control circuit is when two stop buttons are used to stop a machine from two locations, as shown in Figure 2-16.

A parallel circuit is the most common type of electrical circuit encountered in the field. A parallel path occurs in a circuit when it is possible for the electrical current to follow more than one path as it travels through the circuit. Figure 2-17 shows a typical circuit of electric lights, all connected in parallel. The total electrical current flowing in the circuit is the sum of the currents flowing through each light bulb.

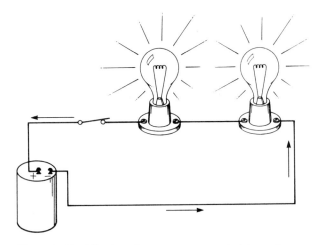

**Figure 2-15** The two lamps are connected in series.

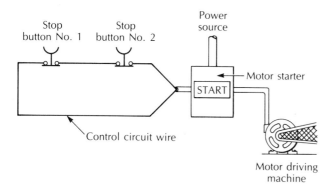

**Figure 2-16** Components in a control system are often connected in series.

## OHM'S LAW

*Ohm's Law* is a simple formula used to show the relationship of voltage (E), resistance (R), and current (I) in an electrical circuit, Figure 2-18. It is important that the electrician understand this relationship. The person learning electrical fundamentals sometimes finds it hard to understand the logic involved when E is used to represent voltage, and I is used to represent current. These symbols date back to the time when investigators were first learning about electricity. The E for voltage derives from *emf*, which is the abbreviation for *electromotive force*. The current flow, I, measured in amperes, is the rate of flow of electrons. You will soon become accustomed to using these standard symbols.

The amount of electrical current flowing in a circuit with a constant resistance depends upon the voltage. When the voltage (E) is increased, the current (I) increases. Similarly, when the voltage is decreased, the flow of current decreases. It is easy to remember that the more push (voltage) behind the electrons, the more current that flows. The flow of current is said to be directly proportional to the voltage of the circuit.

Setting the voltage at a constant level and changing the resistance results in a change in current flow. When the resistance is increased, less current flows. Similarly, decreasing the resistance allows more current to flow. The flow of current is considered to be inversely proportional to the resistance. When resistance goes up, current goes down, and vice versa.

When any two of the quantities, current (I), voltage (E), or resistance (R) are known, the remaining quantity can be determined using Ohm's Law. Equation 2.7 is Ohm's Law.

$$\text{Current} = \frac{\text{Voltage}}{\text{Resistance}}, \text{ or } I = \frac{E}{R} \qquad \text{Eq. 2.7}$$

The equation can be rearranged to solve for the voltage when the circuit resistance and current are known. Equation 2.8 solves for voltage, and Equation 2.9 solves for resistance.

$$\text{Voltage} = \text{Current} \times \text{Resistance}$$
$$\text{or } E = I \times R \qquad \text{Eq. 2.8}$$

$$\text{Resistance} = \frac{\text{Voltage}}{\text{Current}}, \text{ or } R = \frac{E}{I} \qquad \text{Eq. 2.9}$$

## Problem 2-5

Determine the resistance of an electrical heater which draws 8.7 A when operating at 120 V.

## Solution

$$\text{Resistance (R)} = \frac{120 \text{ V}}{8.7 \text{ A}} = 13.8 \text{ }\Omega$$

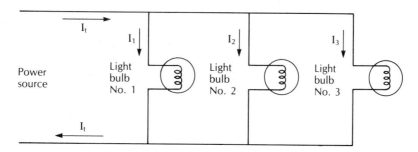

**Figure 2-17** The light bulbs in this circuit are connected in parallel.

Unit 2 Electrical Fundamentals 23

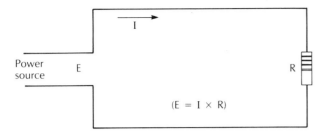

**Figure 2-18** The relationship of voltage (E), current (I), and resistance (R) in an electrical circuit, according to Ohm's Law

## Problem 2-6

How much current will flow through a 100-W light bulb operating at 120 V if the bulb filament resistance is 144 Ω?

## Solution

$$\text{Current (I)} = \frac{120 \text{ V}}{144 \text{ Ω}} = 0.83 \text{ A}$$

## SERIES CIRCUIT

Ohm's Law is useful in understanding series and parallel circuits. In the case of the series circuit in Figure 2-15, the current has only one path to follow. Therefore, the current can be determined easily if the voltage of the source and the resistance of the circuit are known. <u>The current in a series circuit is the same anywhere it is measured</u>, as shown in Figure 2-19. Equation 2.10 shows this relationship for the series circuit in Figure 2-19.

$$I_t = I_1 = I_2 = I_3 \qquad \text{Eq. 2.10}$$

<u>The resistance of a series circuit is the sum of the resistances of each component or part of the circuit.</u> Figure 2-19 has three components. Therefore, the total resistance ($R_t$) is determined by adding the individual resistances, as shown in Equation 2.11.

$$R_t = R_1 + R_2 + R_3 \qquad \text{Eq. 2.11}$$

## Problem 2-7

Determine the current flowing in the circuit of Figure 2-15, which operates at 120 V if one light bulb has a

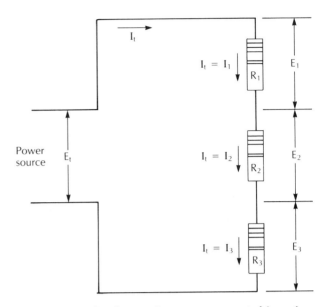

**Figure 2-19** The three resistors are connected in series.

resistance of 92 Ω and the other has a resistance of 138 Ω.

## Solution

First determine the total resistance of the circuit, using Equation 2.11.

$$R_t = 92 \text{ Ω} + 138 \text{ Ω} = 230 \text{ Ω}$$

The current flowing in the circuit can be determined by using Equation 2.7.

$$I = \frac{E}{R} = \frac{120 \text{ V}}{230 \text{ Ω}} = 0.52 \text{ A}$$

The total voltage ($E_t$) of a series circuit is equal to the voltage across each resistance in the circuit. For the circuit of Figure 2-19, with three resistors in series, the total voltage is the sum of the voltages across each resistor, as stated by Equation 2.12.

$$E_t = E_1 + E_2 + E_3 \qquad \text{Eq. 2.12}$$

## Problem 2-8

Determine the voltage across each light bulb in the circuit of Figure 2-15, using the values of current and resistance provided in Problem 2-7 and its solution. Confirm that the total circuit voltage is equal to the sum of the voltage across each light bulb.

## Solution

The voltage across each light bulb can be determined using Equation 2.8, because the resistance of the bulb and the current through the bulb are known.

$$E = I \times R$$

Light bulb No. 1:

$$E_1 = 0.52 \text{ A} \times 92 \text{ } \Omega = 48 \text{ V}$$

Light bulb No. 2:

$$E_2 = 0.52 \text{ A} \times 138 \text{ } \Omega = 72 \text{ V}$$

The total circuit voltage is determined by using Equation 2.12.

$$E_t = 48 \text{ V} + 72 \text{ V} = 120 \text{ V}$$

## PARALLEL CIRCUIT

The total current ($I_t$) flowing in a circuit with parallel components is the sum of the currents flowing in each component. In Figure 2-20, three resistors are arranged in parallel. The total current flowing in the circuit is determined by using Equation 2.13.

$$I_t = I_1 + I_2 + I_3 \qquad \text{Eq. 2.13}$$

Notice that the voltage across each resistor in Figure 2-20 is the same. The voltages across the resistors in the parallel circuit are the same as the supply voltage. Equation 2.14 represents the voltage of a circuit with all components connected in parallel. This is the case with most circuits wired by an electrician.

$$E_{supply} = E_1 = E_2 = E_3 \qquad \text{Eq. 2.14}$$

The total circuit resistance is a little harder to calculate when the resistances are connected in parallel rather than in series. The electronic calculator, however, makes the solution easy. The total circuit resistance can be determined by using either Equation 2.15 or 2.16.

$$\frac{1}{R_t} = \frac{1}{R_1} + \frac{1}{R_2} + \frac{1}{R_3} \qquad \text{Eq. 2.15}$$

$$R_t = \frac{1}{\frac{1}{R_1} + \frac{1}{R_2} + \frac{1}{R_3}} \qquad \text{Eq. 2.16}$$

The total resistance of a parallel circuit is always less than the resistance of any of the circuit's individual resistance values. These equations can be best understood by use of an example.

## Problem 2-9

The resistors in Figure 2-20 have the following values: $R_1 = 4 \text{ }\Omega$, $R_2 = 6 \text{ }\Omega$, and $R_3 = 12 \text{ }\Omega$. Determine the total resistance of the circuit.

## Solution

The total resistance of the parallel circuit is determined by using Equation 2.16.

$$R_t = \frac{1}{\frac{1}{R_1} + \frac{1}{R_2} + \frac{1}{R_3}} = \frac{1}{\frac{1}{4} + \frac{1}{6} + \frac{1}{12}}$$

$$R_t = \frac{1}{0.250 + 0.167 + 0.083} = \frac{1}{0.50}$$

then, $R_t = \dfrac{1}{0.50} = 2 \text{ }\Omega$

or,

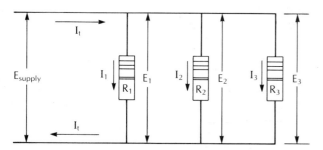

Figure 2-20  The three resistors are connected in parallel.

$$R_t = \cfrac{1}{\cfrac{3}{12} + \cfrac{2}{12} + \cfrac{1}{12}}$$

$$R_t = \cfrac{1}{\cfrac{6}{12}}$$

$$R_t = \frac{12}{6} = 2\ \Omega$$

## Problem 2-10

Determine the current flowing through each resistor of the circuit of Figure 2-20 if the circuit operates with a 120-V supply.

## Solution

Equation 2.7 is used to determine the current flowing through each resistor. According to Equation 2.14, the voltage across each resistor is the same as the supply voltage.

Resistor 1:

$$I_1 = \frac{120\ V}{4\ \Omega} = 30\ A$$

Resistor 2:

$$I_2 = \frac{120\ V}{6\ \Omega} = 20\ A$$

Resistor 3:

$$I_3 = \frac{120\ V}{12\ \Omega} = 10\ A$$

## Problem 2-11

Determine the total current flowing in the parallel circuit of Figure 2-20 using Ohm's Law, Equation 2.7, and Equation 2.13.

## Solution

$$\text{Ohm's Law}\quad I_t = \frac{E}{R_t} = \frac{120\ V}{2\ \Omega} = 60\ A$$

also,

$$I_t = 30\ A + 20\ A + 10\ A = 60\ A$$

## POWER AND WORK

*Power* is the rate of doing work, as shown in Equation 2.17.

$$\text{Power} = \frac{\text{Work}}{\text{Time}} \qquad \text{Eq. 2.17}$$

Assume that 240 units of work are required to lift an electric motor from the floor to its mounting position 6 ft (1.83 m) above the floor. If this lifting is accomplished in 1 minute (min), the power expended lifting the motor would be 240 units of work per minute. But, what if the motor is lifted into position in 2 minutes rather than in 1 minute? The power expended would be 240 units of work per 2 minutes, or only 120 units of work per minute. It takes the same amount of work to get the job done in both cases, but, in the second case, twice as much time is used. Therefore, the per-minute power requirement is only half.

The unit of mechanical work is the *foot-pound* (ft-lb), or the *joule* (J). Assume that the motor in the previous example weighs 40 pounds (lb), or 18.1 kilograms (kg). See Figure 2-21. Lifting the 40-lb (18.1-kg) motor a distance of 6 ft (1.83 m) requires 240 ft-lb (40 lb × 6 ft), or 325 J, of work. If the lifting is done in 1 minute, the power required would be 240 ft-lb, or 163 J, per minute (ft-lb/min, or J/min).

The *horsepower* (hp) is most commonly used as a unit of measure of mechanical power.

$$1\ hp = 33\ 000\ \text{ft-lb/min}\ (44\ 742\ \text{J/min})$$

Using the conversion factor, the power requirement for lifting the motor a distance of 6 ft (1.83 m) in 1 minute is 0.007 3 hp (0.005 kW).

$$\text{horsepower} = \frac{240}{33\ 000} = 0.007\ 3\ hp$$

The unit of measure of electrical energy or work is the *watthour* (Wh). The electric meter installed on a farm service entrance measures electrical energy or work in *kilowatt-hours* (kWh). Power is work divided by time;

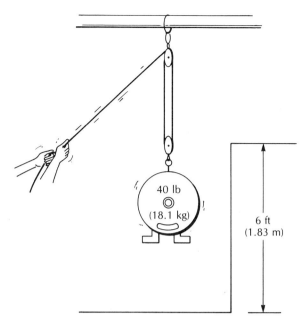

**Figure 2-21** The work required to lift the load depends upon the weight of the load and the height the motor is to be lifted. The power required depends upon the speed at which the motor is lifted.

therefore, the *watt* (W) and *kilowatt* (kW) are the units of measure of electrical power.

$$\text{Electrical power} = \frac{\text{Electrical energy}}{\text{Time}} \quad \textbf{Eq. 2.18}$$
$$= \frac{\text{Watthours}}{\text{Hours}} = \text{Watts}$$

Most electric lights and electrical equipment (with the exception of motors) are rated in units of watts or kilowatts. The amount of electrical energy consumed is determined by multiplying the power rating of the equipment by the length of time in hours the equipment is used, as shown in Equation 2.19.

$$\text{Electrical energy (kWh)} = \quad \textbf{Eq. 2.19}$$
$$\text{Kilowatts} \times \text{hours of use}$$

A typical kilowatt-hour meter used by the utility to measure farm electrical energy is shown in Figure 2-22.

## Problem 2-12

Determine the electrical energy in kilowatt-hours (kWh) used by a 3 200-W radiant heater in a farm shop if the heater is operated for 3 hours (hr).

**Figure 2-22** The power supplier installs a kilowatt-hour meter on each consumer's service to record in kilowatt-hours the electrical energy used.

## Solution

3 200 W = 3.2 kW
Electrical energy = 3.2 kW × 3 h = 9.6 kWh

While electrical power is usually measured in watts and kilowatts, the electrical power supplier often rates the output of a generating plant in *megawatts* (MW), with one megawatt being equal to one million watts. Mechanical power is rated in horsepower. One horsepower (hp) is equal to 746 watts (W).

Heating and cooling equipment is often rated in heating units called *British thermal units per hour* (Btu/hour). One watt is equal to 3.413 Btu/hour. Refrigeration units are sometimes rated in *tons*, with one ton equalling a cooling capacity of 12 000 Btu/hour. These common values of power encountered by the electrician are summarized in Table 2-3.

Electrical power is measured with an instrument called a *wattmeter*. A wattmeter measures the current flowing in a circuit and the voltage of the circuit. The current passes through a stationary coil within the meter, and the voltage is applied to a moving coil. The meter dial is calibrated to read the number of watts drawn by the electrical load connected to the meter. Figure 2-23 shows the connection of a wattmeter to measure the power drawn by a piece of electrical equipment.

Measuring electrical power requires making several connections to the wattmeter. Wattmeters are available

Unit 2 Electrical Fundamentals 27

**Table 2-3** Factors for converting between electrical and mechanical units, and heating and cooling power units

| Multiply | By | To Obtain |
|---|---|---|
| Kilowatts (kW) | 1 000. | Watts (W) |
| Watts (W) | 0.001 | Kilowatts (kW) |
| Horsepower (hp) | 746. | Watts (W) |
| Kilowatts (kW) | 1.34 | Horsepower (hp) |
| Watts (W) | 3.413 | Btu/hour |
| Btu/hour | 0.293 | Watts (W) |
| Kilowatts (kW) | 3 413. | Btu/hour |
| One ton of cooling | 12 000. | Btu/hour |

with all of the connections made by the meter manufacturer. A cord is generally provided to plug into the power source. The load is generally plugged into a receptacle provided on the cord or on the meter.

The power in watts drawn by an electrical load is sometimes equal to the voltage of the circuit times the current in amperes drawn by the load. Equation 2.20 is used to determine the *volt-amperes* (VA) of a load when the voltage of the circuit and the current flowing are known. Watts are equal to volt-amperes for direct-current and certain single-phase, alternating-current loads, such as incandescent lights, electric heaters, and electrical heating appliances.

$$VA = E \times I \quad \text{Eq. 2.20}$$
where: VA = volt-amperes
E = volts
I = current, in amperes

The complete equation for determining the power drawn by a single-phase, alternating-current electrical load is discussed later in this unit. The term *volt-amperes* is used in the *National Electrical Code*. The number of watts drawn by a load such as an incandescent lamp are equal to the number of volt-amperes. An ac electric motor draws fewer watts than volt-amperes. The term volt-amperes is sometimes used instead of watts to avoid the confusion of having to be concerned about the type of load served. For example, *NEC Section 220-2(c)* states that for other than dwellings, a receptacle outlet shall be considered a 180-VA load for the purpose of sizing the building service-entrance conductors. Transformers *(NEC Article 450)* are rated in *kilovolt-amperes* (kVA), not in kilowatts (kW).

## Problem 2-13

Determine the number of watts drawn by a toaster connected to a 120-V circuit if the current flowing to the toaster is 8.33 A. This is a heating appliance, and volt-amperes are equal to watts.

## Solution

$$E \times I = 120 \text{ V} \times 8.33 \text{ A} = 1\,000 \text{ W}$$

## Problem 2-14

A 2-kW electric heater is rated for a 240-V circuit. Determine the current which will be drawn by this electric heater. Volt-amperes are equal to watts for an electric heater, unless there is an electric motor powering a fan or blower.

$$2 \text{ kW} = 2\,000 \text{ W}$$

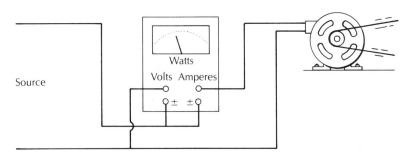

**Figure 2-23** A wattmeter measures both the voltage and the current of the circuit.

## Solution

Rearranging Equation 2.20, the current can be determined when the watts and voltage are known.

$$VA = E \times I = W$$
$$I = \frac{W}{E} = \frac{2\,000\text{ W}}{240\text{ V}}$$
$$I = 8.3\text{ A}$$

A 100-A household service-entrance load center provides electrical power at both 120 and 240 V. The 100-A load center can deliver up to 24 000 W.

$$240\text{ V} \times 100\text{A} = 24\,000\text{ W}$$

or,

$$\begin{aligned}120\text{ V} \times 100\text{ A} &= 12\,000\text{ W}\\ +120\text{ V} \times 100\text{ A} &= 12\,000\text{ W}\\ &\quad 24\,000\text{ W}\end{aligned}$$

A 200-A service can deliver up to 48 000 W.

The *National Electrical Code* in *Article 220* uses electrical power in watts and kilowatts as the basis for determining the size of a building service-entrance requirement. In the case of farm buildings, and some commercial and industrial buildings, amperes instead of watts are counted to size the service. (This procedure is covered in a later unit.)

If current is used for service calculations, the voltage must be known. Ten amperes at 240 V are equal in power to 20 A at 120 V. However, both situations represent 2 400 W.

If an electric heater is rated at 2 kW, but operates at 120 V, the power rating is the same as it is for a 2-kW electric heater operating at 240 V. Together, the two heaters will draw 4 kW. The difference between the two electric heaters is that the 120-V electric heater draws 16.7 A, while the 240-V electric heater draws 8.3 A. For this reason, it is desirable to connect a load, such as an electric motor, for operation at 240 V rather than 120 V whenever possible. The current drawn is only half as much at the higher voltage for exactly the same power output. This means that the wires feeding the electrical equipment can be smaller for the higher voltage.

## Problem 2-15

A residential service entrance supplies power to the following loads:
  1 200 W to the lights at 120 V
  7 500 W to the electric range at 240 V
  5 000 W to the electric water heater at 240 V
  1 100 W to the toaster at 120 V
  900 W to the electric iron at 120 V
Determine the total load in watts supplied by the service.

## Solution

| | |
|---|---|
| Lights: | 1 200 W |
| Range: | 7 500 W |
| Water heater: | 5 000 W |
| Toaster: | 1 100 W |
| Iron: | 900 W |
| Total: | 15 700 W |

## HEAT

Electricity is converted into heat for thousands of useful applications. The *kilowatt-hour* (kWh) can be a measure of the amount of heat produced by electricity. The kilowatt-hour can then be used as a measure of electrical work, electrical energy, or electrical heat. The electrical energy used in kilowatt-hours is equal to power in kilowatts multiplied by time, as shown in Equation 2-21.

Eq. 2.21
$$\text{Heat} = \text{Electrical energy} = \text{Power} \times \text{Time}$$
where: Heat = W × t
  W = power, in watts
  t = time, in hours

Heat is actually measured in *British thermal units* (Btu) in the English system of measurement. In the International System (SI) of metric measurement, the *joule* (J) is the unit of measure of heat. One kWh is equal to 3 413 Btu, or 3.6 million J (3.6 megajoules [MJ]). For the purpose of this discussion, it is easier for the electrician to understand the concept in terms of kWh as a unit of measure of electric heat.

The equation for determining electrical heating can be made more useful by substituting Ohm's Law (E = I × R, Equation 2.8) into the equation, as shown in Equations 2.22 and 2.23.

Unit 2  Electrical Fundamentals

$$\text{Heat} = E \times I \times t$$
$$\text{Heat} = (I \times R) \times I \times t$$
$$\text{Heat} = I^2 \times R \times t$$

**Eq. 2.22**

where: Heat = watthours
I = current, in amperes
R = resistance, in ohms
t = time, in hours

also,

$$\text{Heat} = \frac{E^2}{R} \times t$$

**Eq. 2.23**

where: E = voltage

The relationship in Equation 2.22 is important for understanding the heating of electrical conductors, such as the heating element of a water heater, or the electrical wire leading to the water heater. The resistance of the wire is considered constant. No. 10 AWG copper wire has a resistance of about 1 Ω per 1 000 ft (from Table 2-1 of this text). The resistance of a water heater element is established at the time of manufacture. A 5 000-W, 240-V electrical element has a resistance of about 11.5 Ω.

The heat produced in a circuit depends upon the length of time current is flowing, provided the resistance of the circuit remains constant. The circuit resistance is made up of the resistance of the load and the resistance of the wire. The heat produced in a wire and the load depends upon the square of the current. Doubling the current increases the rate of heat production four times.

## Example

A 5 000-W, 240-V electric water heater draws 20.8 A. Figure 2-24 shows the water heater with two 2 500-W heating elements. Assume that the wire supplying the water heater has a resistance of 0.25 Ω. The heat produced in the wire in 1 hour is determined with Equation 2.22

$$\text{Heat} = I^2 \times R \times t$$
$$\text{Heat} = (20.8 \text{ A})^2 \times 0.25 \text{ Ω} \times 1 \text{ hr}$$
$$\text{Heat} = 108 \text{ Wh, or } 0.108 \text{ kWh}$$

How much heat is produced if two 5 000-W water heaters are supplied by the same No. 10 AWG wire? The current flowing through the wire is 41.6 A.

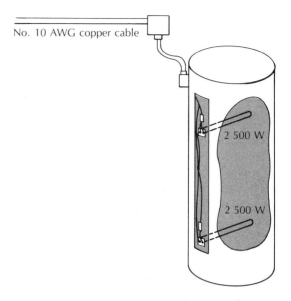

**Figure 2-24** A 5-kW electric water heater with an upper and a lower heater element, each rated at 2.5 kW.

$$\text{Heat} = (41.6 \text{ A})^2 \times 0.25 \text{ Ω} \times 1 \text{ hr}$$
$$\text{Heat} = 433 \text{ Wh, or } 0.433 \text{ kWh}$$

The rate of heating is increased four times by doubling the current, as shown by the example. This is why it is important not to overload an electrical wire. Overloading a wire only 40% (21 A flowing on a 15-A, No. 14 AWG wire) will double the rate of heating in the wire. If the wire is operated long enough at this overload, the maximum safe operating temperature of the wire will most likely be exceeded. Fuses for protecting electrical circuits are designed using the principle of Equation 2.22.

## RESISTANCE IN AN AC CIRCUIT

The power in watts produced when electrical current flows in a circuit is equal to the volts times the amperes when the circuit contains only resistance. The volt-amperes of Equation 2.20 are equal to watts for such resistive loads electric heaters, incandescent lights, or various types of heating appliances not containing motors.

If a picture of the voltage of a circuit containing a resistive load is made with an oscilloscope, the voltage will look like a sine wave (Figure 2-7). The current flowing in the circuit also looks like a sine wave. The voltage and the current are shown together in Figure 2-25. Notice that the voltage and the current are zero at the same time. They also reach their maximum values at the same

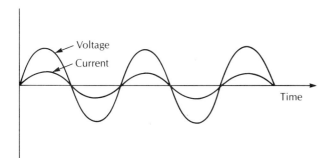

**Figure 2-25** In a circuit containing only resistance as the load, the voltage and the current are in phase with each other.

time. The voltage and the current are considered to be in phase with each other in a circuit with only resistance for the load.

Multiplying the volts and amperes of Figure 2-25 produces a picture of the volt-amperes of the circuit. All of the volt-amperes are positive (above the horizontal line), as shown in Figure 2-26. The net positive volt-amperes is the real power in watts produced by the current flowing in the circuit. In this case, with only resistance as the load, the volt-amperes are equal to the power in watts.

## INDUCTANCE IN AN AC CIRCUIT

Electrical current passing through a coil of wire encounters the resistance of the wire to oppose the flow of current. Alternating current also encounters an opposition produced by a magnetic field around the wire. This opposition is called *inductive reactance*. Electrical equipment containing coils of wire produces magnetic fields and thus inductive reactance. Examples of this type of equipment are transformers, motors, and electric-discharge lamp ballasts (fluorescent, mercury-vapor, and similar lamps).

An effect of the inductive reactance is that the current and voltage sine waves are no longer in phase with each other, as seen in Figure 2-27. The voltage starts at zero and begins to increase, but there is a delay before the current begins to increase from its zero point. The amount of this delay is represented in Figure 2-27 by the distance between the zero point for the voltage wave and the current wave. Notice also that the voltage reaches its maximum peak before the current reaches its maximum peak. The current is said to *lag* the voltage.

Inductive reactance works something like a traffic light in heavy traffic. When the light turns green, there is a delay before traffic begins flowing. The flow of traffic lags the green go-ahead signal. When the light turns red, cars continue to approach the light for a short time. The flow of traffic does not stop abruptly when the light turns red.

A circuit serving a load such as an electric motor contains resistance and inductive reactance. These two types of resistance combined are called the *impedance* of the circuit. The method of calculating the impedance is not discussed in this text. Impedance is mentioned in the *National Electrical Code* and an electrician must understand the concept. Impedance is used in *NEC Section 430-32(c)(4)* where, for example, a clock motor with sufficiently high impedance is subject to less stringent circuit protection requirements.

The volt-amperes produced in a circuit containing resistance and inductive reactance are represented by the wave of Figure 2-28. Both positive and negative volt-amperes are produced. The amount of negative volt-amperes produced depends upon the amount of inductive reactance of the circuit. The greater the inductive reactance, the greater is the amount of negative volt-amperes. The useful power in watts produced in the circuit is the

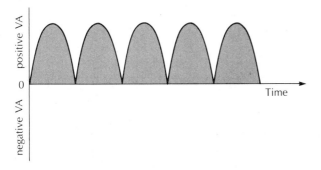

**Figure 2-26** All of the volt-amperes are positive when the circuit load is resistive. The net positive volt-amperes is the real power in watts.

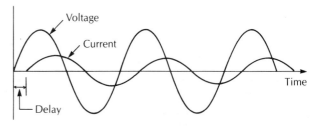

**Figure 2-27** In a circuit containing resistance and inductive reactance, the current lags the voltage. The voltage and current are out of phase.

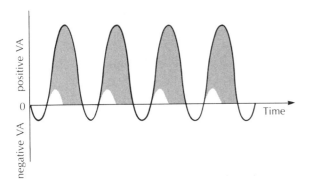

**Figure 2-28** When there is inductive reactance in the circuit, some of the volt-amperes are negative. Subtracting the amount of negative volt-amperes from the amount of positive voltamperes gives the amount of net positive volt-amperes, which is the real power.

sum of the positive and negative portions of the volt-ampere wave, or the net positive volt-amperes. The nonuseful portion of the volt-amperes is referred to as the *reactive volt-amperes* or *VAR* (volt-amperes reactive) or, as it is referred to sometimes, the *reactive power*. One thousand VAR equal one *kilovar* (kVAR).

Increasing the inductive reactance of the circuit causes the current to lag farther and farther behind the voltage. This lag results in negative volt-amperes which reduce the net real power produced. In order to produce a certain amount of useful power in the circuit, more and more current flow is needed with this high inductive reactance present. With more current flowing to produce the needed power, the wire size must be increased.

## POWER AND POWER FACTOR

The term *power factor* (pf) is introduced in Equation 2.20 to account for the reduction in power output due to inductive reactance. The power factor ranges from 0 to 1.0. When the circuit is all resistance, such as in an electric water heater, the power factor is 1.0. When the circuit contains a coil which produces inductive reactance, such as in an electric motor, the power factor is less than 1.0. Power in watts produced in a single-phase, alternating-current circuit can be determined with Equation 2.24.

$$W = E \times I \times pf \qquad \text{Eq. 2.24}$$
where: W = power, in watts
E = volts
I = current, in amperes
pf = power factor

The equation for determining 3-phase power is discussed in unit 3.

The amount of current that lags the voltage is measured in *electrical degrees*. It takes 360 electrical degrees for alternating voltage or current to make one complete cycle. The term *phase angle* represents the degrees of shift between the voltage and the current, and is designated by the Greek letter theta ($\theta$). In the mathematics of trigonometry, the phase angle can be converted to power factor with the cosine function (cos $\theta$). Table 2-4 provides conversion factors from phase angle to power factor. Many electronic calculators make this conversion with the press of a single button. Equation 2.25 shows the conversion from phase angle to power factor. Electricians seldom have to make power factor conversions and calculations, but all electricians must understand the concept of phase angle and power factor.

$$\text{Power factor} = \cos \theta \qquad \text{Eq. 2.25}$$
where: $\theta$ = phase angle, in degrees

## Problem 2-16

The current drawn by an electric motor lags the voltage by 52 electrical degrees, as shown in Figure 2-29. Determine the power factor of the motor.

## Solution

According to Equation 2.25, the power factor is the cosine of the phase angle. Refer to Table 2-4 and find the phase angle 52°.

$$\cos 52° = 0.616$$

therefore,

$$pf = 0.616$$

## Problem 2-17

Determine the power drawn by the single-phase electric motor of Problem 2-16 if the motor operates at 230 V and draws 8 A.

## Solution

Equation 2.24 is used to determine the power drawn by the motor.

Table 2-4  These values for the cosine and tangent of an angle are used to determine and correct the power factor.

| Phase Angle Degrees | Power Factor cos θ | tan θ | Phase Angle Degrees | Power Factor cos θ | tan θ |
|---|---|---|---|---|---|
| 0  | 1.000 | 0.000 | 32 | 0.848 | 0.625 |
| 2  | 0.999 | 0.035 | 34 | 0.829 | 0.675 |
| 4  | 0.997 | 0.070 | 36 | 0.809 | 0.727 |
| 6  | 0.995 | 0.105 | 38 | 0.788 | 0.781 |
| 8  | 0.990 | 0.141 | 40 | 0.766 | 0.839 |
| 10 | 0.985 | 0.176 | 42 | 0.743 | 0.900 |
| 12 | 0.978 | 0.213 | 44 | 0.719 | 0.966 |
| 14 | 0.970 | 0.249 | 46 | 0.695 | 1.036 |
| 16 | 0.961 | 0.287 | 48 | 0.669 | 1.111 |
| 18 | 0.951 | 0.325 | 50 | 0.643 | 1.192 |
| 20 | 0.940 | 0.364 | 52 | 0.616 | 1.280 |
| 22 | 0.927 | 0.404 | 54 | 0.588 | 1.376 |
| 24 | 0.914 | 0.445 | 56 | 0.559 | 1.483 |
| 26 | 0.899 | 0.488 | 58 | 0.530 | 1.600 |
| 28 | 0.883 | 0.532 | 60 | 0.500 | 1.732 |
| 30 | 0.866 | 0.577 |    |       |       |

$$W = E \times I \times pf$$
$$W = 230 \text{ V} \times 8 \text{ A} \times 0.616$$
$$W = 1\ 133 \text{ W}$$

## Measuring Power Factor

Meters are available to measure power factor directly. These meters are often installed on industrial service equipment. Portable power factor meters are available to determine the power factor of a motor circuit on a farm. Occasionally, it may be necessary to determine the power factor of a circuit serving an irrigation system or drainage pump motor. The electric power supplier can provide assistance in making these measurements.

The power factor of a single-phase circuit can be determined using the voltmeter, ammeter, and wattmeter discussed earlier in this unit. First, rearrange Equation 2.24 to solve for the power factor.

$$pf = \frac{W}{E \times I} \qquad \text{Eq. 2.26}$$

The power factor can be determined by measuring the power in watts, the voltage of the circuit, and the load current.

## Problem 2-18

An electrician makes the following measurements for a single-phase electric motor circuit.

Power drawn = 1 180 W
Voltage = 115 V
Current = 13.8 A

Solve for the power factor of this circuit.

## Solution

The power factor is determined with Equation 2.26.

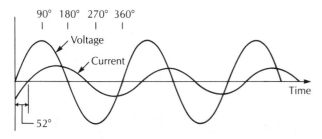

Figure 2-29  The current lags the voltage by 52 electrical degrees for the electric motor in Problem 2-16. This is typical for single-phase, fractional-horsepower motors.

Unit 2   Electrical Fundamentals   33

$$\text{pf} = \frac{W}{E \times I} = \frac{1\,180\text{ W}}{115\text{ V} \times 13.8\text{ A}}$$
$$\text{pf} = 0.744$$

An electrician generally has the equipment available to measure the voltage and current. Making a measurement of the load power may seem difficult at first, because electricians generally do not have a wattmeter available. Even when a wattmeter is available, it is difficult to connect to the circuit and can often be too small to measure the power of the load. However, there is a simple way to measure the power of a load. The power reading can be determined from the power supplier's kilowatthour meter.

It takes a certain number of watthours to move the meter disc one complete revolution. This number of watthours is printed on the face of the meter as the factor $K_h$.

$K_h$ = Number of watthours per disc revolution

The kilowatthour meter of Figure 2-30 has a Kh factor of 7.2. This means that 1 revolution of the disc represents 7.2 watthours (Wh) of electrical energy. If it takes 7.2 Wh to make 1 disc revolution, then it takes 138.89 disc revolutions to make 1 kWh.

$$\frac{1\,000\text{ Wh/kWh}}{7.2\text{ Wh/revolution}} = 138.89\text{ revolutions/kWh}$$

The power of the load can be determined by counting the number of revolutions of the disc in 1 minute. Equation 2.27 is used to determine the power in watts drawn by the load.

**Eq. 2.27**
$$W = 60 \times K_h \times (\text{disc revolutions per min})$$

The procedure for measuring the power factor of a particular circuit is as follows.

1. Shut off all circuits except the one for which power will be measured.
2. Turn on the load and count the number of meter disc revolutions in 1 minute. The meter disc is marked in divisions of 0.01 revolution, so a fraction of a revolution can be determined if necessary.
3. Measure the circuit voltage and the load current. A clamp-on ammeter is most convenient for making the load current measurement.

**Figure 2-30** The $K_h$ factor on the front of the kilowatt-hour meter gives the number of watthours required to move the disc one complete revolution.

4. Calculate the power using equation 2.27. The kilowatthour meter measures the true power. Therefore, power factor is not part of the equation.
5. Calculate the power factor using Equation 2.26.

## Problem 2-19

Determine the power in watts drawn by a 230-V circuit if the $K_h$ of the meter is 7.2, and the meter disc makes 5.79 revolutions per minute.

## Solution

The power is solved using equation 2.27.

$$W = 60 \times K_h \times \text{(disc revolutions per minute)}$$
$$W = 60 \times 7.2 \times 5.79$$
$$W = 2\ 500\ W$$

## Problem 2-20

Determine the power factor for the circuit in Problem 2-19 if the current flowing in the circuit is 12 A.

## Solution

Equation 2.26 is used to solve the problem.

$$\text{pf} = \frac{W}{E \times I} = \frac{2\ 500\ W}{230\ V \times 12\ A} = 0.91$$

## CAPACITANCE IN AN AC CIRCUIT

A low power factor can be corrected back to 1.0 or close to 1.0 by installing a device called a *capacitor*. A capacitor consists of two electrical plates separated by an insulating material called a *dielectric*. Current does not flow through a capacitor, although in an ac circuit it appears as though current is flowing. Current flows in both sides of the circuit, but the transfer at the capacitor plates is by means of an electric field.

Consider the case of a capacitor connected to the terminals of a battery. Electrons flow from the negative terminal of the battery and build up on the left-hand capacitor plate, Figure 2-31. This causes a positive charge to build up on the right-hand plate. Electrons in the right-hand plate flow to the positive terminal of the battery. Electrons flow until the voltage across the plates of the capacitor is identical to the voltage of the battery. The capacitor is considered charged, and no more current flows. The battery can be removed and a charge remains on the plates of the capacitor, Figure 2-32. If the two wires are touched together, the charge is released with a sudden flow of current, Figure 2-33. An electrician can suffer a shock by accidentally touching the terminals of a charged capacitor.

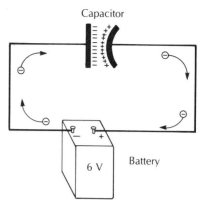

**Figure 2-31** A capacitor becomes charged when it is connected to a battery. Current flows until the capacitor is fully charged.

> **CAUTION:** Before working on the wiring of a machine containing a capacitor, care must be taken to be sure that the capacitor is discharged.

A capacitor has the opposite effect of a coil in an alternating-current circuit. The voltage actually lags the flow of current, due to the capacitive reactance of the circuit. The voltage and current waves of a circuit containing resistance and capacitive reactance are shown in Figure 2-34. Instead of considering the voltage lagging the current, the current is generally considered to be leading the voltage. Capacitive reactance is opposite to inductive reactance. One substracts from the other. A

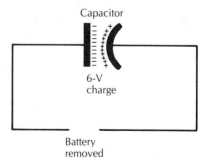

**Figure 2-32** Once the capacitor is charged, the battery can be removed and the charge remains on the capacitor.

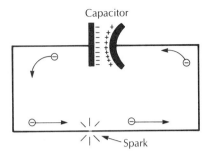

**Figure 2-33** The capacitor is discharged when the ends of the wires are touched together.

capacitor can then be installed on a motor circuit to correct the power factor.

The coil of a motor causes the current to lag the voltage. A capacitor installed on the circuit causes the current to lead the voltage. If the size of the capacitor is chosen correctly, the current and voltage can once again be brought in phase with each other, and the power factor will become 1.0. The ballast of a fluorescent light is often constructed with a built-in capacitor to correct to a high power factor. Capacitors are most commonly rated in microfarads ($\mu$F) or 0.000 001 farad (F). Capacitors for power factor correction are generally rated in kilovar (kVAR).

## IMPROVING POWER FACTOR

The electric power supplier may encourage the farm operator to improve the power factor of a large motor or group of motors, such as an irrigation pumping system. Less current flows on the wires to produce the required power output of the motor with a high power factor as compared to a low power factor. Less current means a smaller voltage drop on the circuit wires. In addition, it may mean smaller electric power supplier distribution wires.

A decision must be made as to how much correction in power factor is desired. Generally, it is expensive to buy enough capacitors to correct the power factor to 1.0. The power factor is often corrected to 0.90 or 0.95. The power supplier will provide assistance in deciding how much correction is adequate.

The original uncorrected power factor can be determined directly with a power factor meter, or it can be calculated with Equation 2.26. The desired power factor is selected by the electrician. Remember that the power factor is cos $\theta$. A value called the *tangent of the phase angle* (tan $\theta$) is needed to calculate the kVAR required to correct the power factor. Equation 2.28 is used to calculate the kVAR.

> kVAR = kW × (tan $\theta_1$ − tan $\theta_2$)   **Eq. 2.28**
> where: kW = power of the load, in kilowatts
> $\theta_1$ = the uncorrected phase angle
> $\theta_2$ = the corrected phase angle

If the power factor is corrected to 1.0, the corrected phase angle is zero degrees and tan 0° is zero.

The tangent of the phase angle can be found from Table 2-4. As an example, assume the power factor is 0.695. This means that cos $\theta_1$ is 0.695. The tangent of the phase angle appears next to the cosine in the table. For the example, the tangent is 1.036.

The size of the capacitor, in kVAR, required to correct the power factor of a motor can be determined by making several measurements.

1. Determine the voltage of the circuit, the current drawn by the load, and the power drawn by the load.
2. Using Equation 2.26, calculate the power factor. If a power factor meter is available, this can be measured directly.
3. Decide upon the desired power factor. This is often between 0.90 and 0.95.
4. Using Equation 2.28, calculate the amount of capacitance in kVAR required to correct the power factor.

## Problem 2-21

A single-phase electric motor operates at 230 V, and draws 40 A and 6 390 W. Determine the kVAR required to correct the power factor to 0.950.

## Solution

First, use Equation 2.26 to determine the uncorrected power factor of the motor.

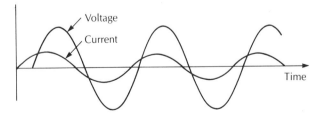

**Figure 2-34** The current leads the voltage in a circuit containing capacitance and resistance.

$$pf = \frac{W}{E \times I} = \frac{6\,390 \text{ W}}{230 \text{ V} \times 40 \text{ A}} = 0.695$$

Next, use Equation 2.28 to calculate the amount of capacitance required to correct the power factor to 0.950. Determine the tangent corresponding to each power factor.

Uncorrected:

$$pf_1 = \cos \theta_1 = 0.695$$
$$\tan \theta_1 = 1.036$$

Corrected:

$$pf_2 = \cos \theta_2 = 0.950$$
$$\tan \theta_2 = 0.325$$

$$6\,390 \text{ W} = 6.390 \text{ kW}$$
$$kVAR = 6.390 \times (1.036 - 0.325)$$
$$kVAR = 4.54$$

## EFFICIENCY

Efficiency of operation is important in the effort to manage energy wisely. *Efficiency* is the useful output of a process or machine as compared to the input required to obtain the output. There are many losses which reduce efficiency. Friction and heat produced by current flowing through the wires (Equation 2.22) are the major factors that reduce electrical machinery efficiency. Many factors affecting efficiency are controlled during the machinery design and manufacture. The electrical installer can help maintain overall efficiency by:

1. Properly sizing branch-circuit wires to reduce conductor heating;
2. Choosing and maintaining low-friction bearings; and
3. Properly adjusting belt and chain drives.

The efficiency of an electric motor-driven machine can be calculated by using Equation 2.29. Some measurements must be made on the circuit. The useful output of the machine must be known or determined.

$$\text{Efficiency (\%)} = \frac{\text{Useful output}}{\text{Input}} \times 100 \quad \textbf{Eq. 2.29}$$

The efficiency of an electric motor can be determined by using Equation 2.30.

$$\text{Efficiency (\%)} = \frac{\text{Output power}}{\text{Input power}} \times 100 \quad \textbf{Eq. 2.30}$$

## REVIEW

Refer to the *National Electrical Code* when necessary to complete the following review material. Write your answers on a separate sheet of paper.

1. ___?___ current flows in only one direction, while 60-Hz ___?___ current flows first in one direction, reverses, and then flows in the opposite direction.
2. Name the unit of measure of electrical pressure.
3. A 115-V electric motor draws 10 A in the hot wire leading to the motor. How many amperes will flow in the neutral wire leading to the motor? Explain.
4. Determine the circular mil area of a No. 12 AWG solid wire with a diameter of 0.0808 inch (in).
5. An electrical circuit of No. 14 AWG copper wire serves a load located 125 ft from a panelboard. The circuit wire has a resistance of 0.067 Ω. Determine the resistance of No. 14 AWG circuit wires serving a load located 250 ft from the panelboard.

6. Determine the amperes flowing in the circuit when a clamp-on ammeter is used to measure the current, as shown in this illustration.

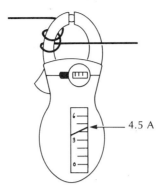

7. Identify the only meter which is properly connected to measure the voltage at the motor in this illustration.

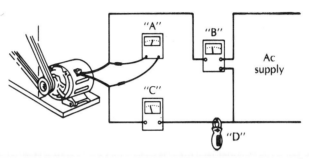

8. An electric coffeemaker has a heating element resistance of 16.4 Ω. Determine the current flowing if the appliance is plugged into a 120-V outlet.

9. In this illustration, two light bulbs are connected in series across a 115-V supply. The first bulb has a resistance of 127 Ω and the second has a resistance of 218 Ω. Determine the current flowing through the first light bulb.

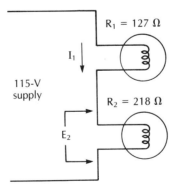

10. Referring to the series circuit of Problem 9, determine the voltage drop across the second light bulb.

11. Determine the total current flowing in the circuit with the two light bulbs connected in parallel (shown in this illustration).

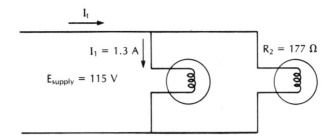

12. Determine the total load in watts of the following equipment operating at the same time: portable electric heater, 1 500 W at 120 V; coffeemaker, 800 W at 120 V; electric range oven, 4 500 W at 240 V; and an electric water heater, 3 500 W at 240 V.

13. Determine the current drawn by an electric water heater rated 5 000 W at 240 V.

14. Nine 150-W incandescent light bulbs are installed on a 15-A circuit in a barn. When all nine bulbs are burning, how much current do they draw at 120 V?

15. A 140-ft length of No. 14 AWG copper electrical cable is run from a panelboard to a poultry-house ventilation fan. Determine the resistance of the wire.

16. Which lamps in this diagram will light?

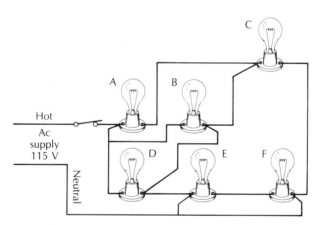

17. The wires in the following illustration are disconnected from the terminal screws of an electric water heater element. An ohmmeter is used to determine the resistance of the 2 500-W, 240-V heating element. The ohmmeter indicates infinite resistance. Is this the expected reading? What is your conclusion about the heating element?

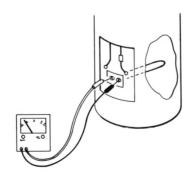

18. An electrically driven irrigation system draws 92 kW of power with a power factor of 0.72. Determine the kVAR of capacitance required to correct the power factor to 0.90.

19. Determine the efficiency of a single-phase electric motor which operates at 230 V and draws 28 A with a power factor of 0.65. The motor is developing 5 hp.

20. A farmer would like to know the power factor of a 7.5-hp, single-phase electric motor operating at 230 V and drawing 39 A. The motor is powering a silo unloader. The electrician turns off all circuits except the motor circuit and determines that the power supplier's kilowatthour meter disc makes 15.5 revolutions with a meter $K_h$ factor of 7.2. Determine the power factor of the motor.

# UNIT 3
# FARM ELECTRICAL SYSTEMS

## OBJECTIVES

After studying this unit, the student will be able to

- describe single- and 3-phase electrical systems.
- discuss nominal and actual voltages.
- explain electrical system load balancing.
- describe single-phase neutral currents.
- calculate 3-phase power.
- describe electrical system grounding.
- identify the types of electrical systems encountered in practice.

Selecting the type of electrical supply system for a farm involves careful consideration of several factors:

1. The size of the electrical service required
2. The voltages required
3. The anticipated kilowatthour usage
4. The electrical energy rates available
5. The size and number of the electrical motors powering the equipment

## SINGLE- VERSUS THREE-PHASE POWER

The size and number of electric motors generally are the determining factors when choosing 3-phase power. In most sections of the United States, the single-phase residential and farm electrical energy rates are the least costly. Three-phase power is often more expensive to obtain than single-phase power. If the 3-phase energy rate from the power supplier is higher than it is for single-phase, then the advantages of 3-phase power must be carefully considered. Often, the electrical power supplier does not allow single-phase motors larger than 10 hp to be installed on farms. A single-phase motor usually costs more than a 3-phase motor of equal horsepower, and single-phase motors require more maintenance.

Three-phase primary power lines may not be available at a farm, and most electrical power suppliers charge the customer for the cost of extending the additional wires required to provide 3-phase power. Generally, the line extension charge is based upon expected revenue over a period of several years. A farm using considerable electrical energy will be charged less for the extension of the 3-phase primary lines. A rebate may even be granted to the customer if energy usage meets or exceeds expectations.

It is important to understand line extension charges, and to weigh the charges carefully against the advantages gained from installing 3-phase power. An expanding farm enterprise utilizing many electric motors will most likely find the installation of 3-phase power an economical investment.

## SINGLE-PHASE ELECTRICAL SYSTEMS

Single-phase electrical service is available in two types: 120-V, 2-wire service, and 120/240-V, 3-wire service. The 120-V, 2-wire system is used only for small buildings on a farm with limited electrical needs. Some buildings wired many years ago may have been provided with only a 120-V, 2-wire service. A 2-wire service may be installed in a building, excluding a residence, according to the *National Electrical Code*, if the electrical load is specific and limited.

- If the installation (building or special equipment) is to contain only one circuit, the minimum size service required is 15 A, *NEC 230-79(a)*.
- If the installation is to contain two 2-wire circuits, the minimum size service required is 30 A, *NEC 230-79(b)*.

A 120-V, 2-wire, single-phase electrical system is shown in Figure 3-1.

A 2-wire service can originate at the power supplier's transformer with a grounded and an ungrounded wire leading to the service disconnect. A 2-wire service most commonly originates at the service panelboard in another building. When this is the case, one ungrounded wire (hot) and a neutral wire (grounded) feeds the 2-wire service in the outbuilding. Unless a very specific and limited load exists at the outbuilding, a 3-wire, 120/240-V service is recommended.

The 3-wire, dual-voltage, single-phase service is most common for farm buildings and rural dwellings. The voltages are generally 120/240 V. However, a 120/208-V system is occasionally encountered on a farm. The 120/208-V system is discussed later with 3-phase systems. The minimum size 120/240-V service required by the *National Electrical Code* is as follows:

- 60 A or larger for farm buildings, *NEC 230-79(d)*.
- 100 A or larger for one-family dwellings, *NEC 230-79(c)*, 1) where the initial load is computed at 10 kVA or more; 2) where the initial installation includes six or more 2-wire branch circuits.

A typical 120/240-V, single-phase, 3-wire electrical system is illustrated in Figure 3-2.

Confusion sometimes arises concerning the value of the voltages. Actual voltages measured at the service panel should be in the range of 115 V to 120 V between an ungrounded wire (hot) and the neutral (grounded wire), and from 230 V to 240 V between the two ungrounded wires. This system was once considered to be a 110/220-V system, but these lower voltages are no longer practical.

The 120/240-V, single-phase, 3-wire, alternating-current system is required to be grounded, *NEC 250-5(b) (1)*. The wire originating at the center point of the trans-

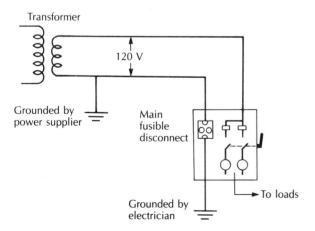

Figure 3-1  A 120-V, 2-wire, single-phase electrical system

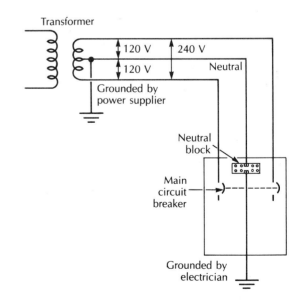

Figure 3-2  A 120/240-V, single-phase electrical system

former secondary winding is grounded at the transformer by the electric power supplier. The electrician must also ground this same wire at the service disconnect. This wire is grounded for several reasons: 1) to limit the maximum voltage which can occur on the ungrounded wires; 2) to reduce the likelihood of damage to the wiring system from lightning; and 3) to facilitate overcurrent device (fuse and circuit breaker) operation in case of a ground fault.

## BALANCING 120-VOLT LOADS

The 120-V circuits must be balanced so that half of the circuits are connected between one ungrounded wire and the neutral, while the remaining circuits are connected between the other ungrounded wire and the neutral. Tripping of the main circuit breaker sometimes is caused by too great a load on one ungrounded wire, with little load on the other. The service may be of adequate size when 120-V circuits are arranged so that the load on each ungrounded wire is nearly equal.

Assume, for example, that a farm building contains four 15-A, 120-V lighting circuits, each with a load of 12 A, and two 120-V fan circuits, each with a load of 10 A. These 120-V loads will be balanced if two lighting circuits and one fan circuit are connected to each ungrounded wire, as shown in Figure 3-3.

Equipment operating at 240 V does not unbalance the system, because both ungrounded wires carry equal current. In practice, it is not possible to balance exactly the 120-V circuits, but it is possible to keep them reasonably balanced. Some advance planning will prevent load-balancing problems later.

## NEUTRAL CURRENT

The neutral wire of a 3-wire, 120/240-V service generally carries less current than either one of the ungrounded wires. The neutral carries the difference between the currents on the two ungrounded wires. Figure 3-4 shows the neutral carrying the unbalance or difference between the loads on the two ungrounded wires. It is assumed that one ungrounded wire carries 58 A, while the other carries 42 A. When the current flows clockwise in loop A, current also flows clockwise in loop B. The two currents flowing in the neutral area traveling in opposite directions; therefore, one subtracts from the other. An ammeter measures the difference between these two currents: 58 A − 42 A = 16 A. The neutral wire of a 120/240-V, 3-wire system is often smaller than the ungrounded wires because it carries less current.

## OPEN NEUTRAL

The neutral wire of a 120/240-V, single-phase, 3-wire electrical system on occasion may become broken between the service panel and the transformer. This break can be caused by physical damage to the wire or a loose or corroded terminal connection. The electrical system on the load side of the broken neutral becomes a 240-V, 2-wire system, and 120 V are no longer present between each ungrounded wire and the neutral. The voltage across some circuits formerly at 120 V will increase, and voltage across other circuits will decrease. Failure to maintain proper voltage on 120-V circuits will result in damage to electrical equipment.

A normal 3-wire circuit with a neutral, and a 3-wire circuit with the neutral missing are shown in Figure 3-5

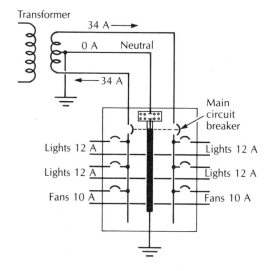

Figure 3-3 The 120-V loads are equally balanced on both ungrounded wires of this 120/240-V, single-phase system.

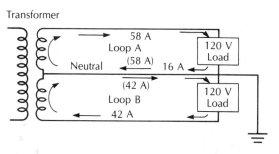

Figure 3-4 The current on the neutral of a 120/240-V, 3-wire electrical system is the difference between the current flowing on the ungrounded wires.

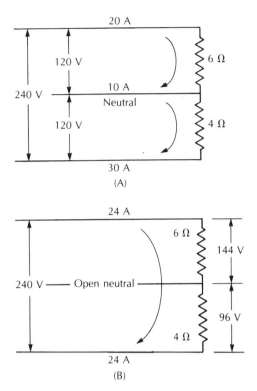

**Figure 3-5** A normal 3-wire circuit (A) maintains 120 V across each circuit. With the neutral open (B) 240 V are applied across both resistors in series with unequal voltage across each resistor.

to illustrate how the neutral maintains 120 V across each circuit. With the neutral open, Figure 3-5 (B), the total resistance of 10 Ω (6 Ω + 4 Ω) is across 240 V. Therefore, 6/10 of 240 V (144 V) are across the 6-Ω resistor, and 4/10 of 240 V (96 V) are across the 4-Ω resistor.

The resistance between each ungrounded wire and the neutral of an actual service panelboard with a broken neutral will change as different 120-V loads are turned on and off. The voltage between each ungrounded wire and the neutral will also change. Some lights will appear overly bright, while others will appear dim. This is a clue that the neutral between the service disconnect and the transformer may be broken, corroded, or loose. The main disconnect should be shut off immediately and the open neutral repaired.

> **CAUTION:** Under this open neutral condition, a person or an animal may suffer a shock by touching grounded metal equipment. This is one cause of stray voltage on farms.

## NOMINAL AND ACTUAL VOLTAGES

Confusion sometimes arises concerning the voltages of an electrical system. For example, the voltage of a general-purpose receptacle outlet is often considered to be 110 V or 115 V, and sometimes 120 V. Maintaining a constant voltage at all times is not possible and, therefore, voltages will vary within limits. The actual voltage of a general-purpose receptacle outlet should range between 115 V and 125 V. The *actual voltage* is the electrical pressure at the outlet as measured by a voltmeter.

Nominal voltages are used to help avoid confusion when referring to the voltages of an electrical system. A nominal voltage is considered the idea. It is generally available at the output terminals of the power supplier's transformer. Actual voltages are often slightly less than nominal voltages. Nominal voltages are used to describe electrical systems in this unit. Nominal and actual voltages are described in *NEC Article 100*.

A *nominal voltage* is a value assigned to a circuit or system to conveniently designate its voltage class, such as 120/240. The actual voltage at which a circuit operates may vary from the nominal voltage within a range that permits the equipment to operate satisfactorily.

The nominal voltages for a single-phase, 3-wire system were described earlier as 120/240 V, but the actual voltages encountered may be slightly lower or higher. The *National Electrical Code,* at the beginning of *Part B Examples* of *Chapter 9,* specifies that the voltages of 120 V, 208 V, and 240 V are to be used for wiring calculations. This is also specified in *NEC Section 220-1*.

## THREE-PHASE POWER

A 3-phase load, such as a motor or heater, requires three wires instead of two as is the case with single-phase loads. Generally, all 3-phase wires are ungrounded, but, for some farm applications, one of the phase wires is intentionally grounded. A single-phase load requires only two wires: one ungrounded wire and one neutral wire for a 120-V load, and two ungrounded wires for a 240-V load. Figure 3-6 shows the wires required to supply single-phase and 3-phase equipment.

The voltage, and also the current, of a 60-Hz supply of alternating current can be represented by a wave such as one of those shown in Figure 3-7. When the wave is above the horizontal line, it is assumed that the current flows in one direction; when it is below the line, current flows in the opposite direction. *Sixty hertz* means that 60 of the waves or cycles occur in one second.

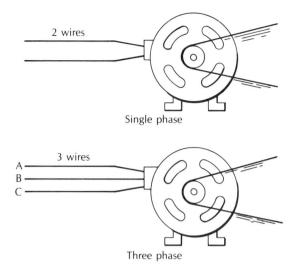

**Figure 3-6** A single-phase load requires two supply wires. A 3-phase load requires three supply wires.

A single-phase supply is represented by a single wave, such as the first wave of Figure 3-7. A 3-phase supply is represented by three separate waves, Figure 3-7. The waves are identical, except that the second wave is 120 electrical degrees behind the first, and the third wave is 120 electrical degrees behind the second. A complete cycle is 360 electrical degrees. The electrical power supplier's generators are designed to produce 3-phase power. A 3-phase supply is available by taking all three supply wires, A, B, and C. Three single-phase supplies can be derived from 3-phase by taking any two phase-wire combinations, A-B, A-C, or B-C. This gives the power supplier and the electrical installer the flexibility of supplying either single-phase loads or 3-phase loads from the same electrical system. The number of single-phase and 3-phase loads that can be supplied depends upon the type of 3-phase system and the size of service equipment.

The formula for calculating single-phase power is noted here as Equation 3.1.

---
$W = E \times I \times pf$        **Eq. 3.1**
where: W = power, in watts
            E = voltage between two wires
            I = current, in amperes flowing
               in one wire
           pf = power factor, which ranges
                from 0.0 to 1.0

---

In 3-phase equipment, three wires are involved. Inside the equipment, there are actually three different paths for the current to follow. The three individual phases also are shifted 120 electrical degrees from one another. In spite of these complications, 3-phase power can be calculated if the voltage between any two wires, the current flowing in one wire, and the power factor are known. Multiplying the power formula by the number 1.73, or $\sqrt{3}$, makes the 3-phase power calculation possible, Equation 3.2.

---
$W = 1.73 \times E \times I \times pf$        **Eq. 3-2**
where: W = power, in watts
            E = voltage between any two wires
            I = current, in amperes flowing
               in any one wire
           pf = power factor, which ranges
               from 0.0 to 1.0

---

## Problem 3-1

A 460-V, 3-phase, 20-hp electric motor draws 27 A with a power factor of 0.80. Determine the power drawn by the motor.

## Solution

---
E = 460 V
I = 27 A
pf = 0.80
W = 1.73 × 460 V × 27 A × 0.80 = 17 189 W

---

## Problem 3-2

Determine the line current drawn by a 15-kW (15 000-W), 240-V, 3-phase electric resistance water heater.

Because the water heater is a resistance device without capacitance or inductance, the power factor is 1.0.

## Solution

---
W = 15 000 W
E = 240 V
pf = 1.0

$$I = \frac{W}{1.73 \times E \times pf} = \frac{15\ 000\ W}{1.73 \times 240\ V \times 1.0}$$

I = 36.1 A

---

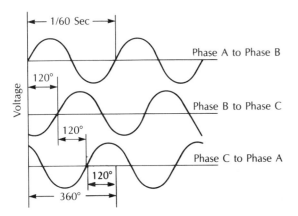

Figure 3-7 The voltages measured between each of the three combinations of phase wires (A, B, and C) of a 3-phase electrical system are represented by waves. Each wave is 120 electrical degrees out of phase with the others.

The current 36.1 A is the value used to determine wire size and circuit overcurrent protection.

## TRANSFORMER KVA RATING

The main factors considered by the power supplier in the selection of distribution transformers are the type of service to be provided, the primary voltage available, the secondary voltage desired, and the estimated load in kilovolt-amperes (kVA). The power supplier estimates the expected demand of the load in kVA and sizes the transformers. Transformers are sized according to kVA rather than kilowatts (kW) because, due to the power factor, which is generally less than 1.0, the kW will be less than the kVA. When using kVA for rating transformers, the transformers cannot be overloaded, due to a power factor value of less than 1.0.

A single-phase transformer kVA must be sufficient to supply the farm kVA-load demand. For a 3-phase kVA-load demand, three transformers forming a bank are available to carry the load, with each transformer carrying one-third of the kVA load.

## Problem 3-3

A farm 3-phase demand load is 30 kVA. Determine the kVA rating of each of the three transformers.

## Solution

$$\text{3-phase transformer kVA} = \frac{\text{3-phase load kVA}}{3}$$

$$\text{Transformer kVA} = \frac{30 \text{ kVA}}{3} = 10 \text{ kVA}$$

Three 10-kVA transformers will supply a 30-kVA load, because the total kVA rating of a 3-phase transformer bank is the sum of the kVA ratings of the three transformers. This is not true when the transformer kVA ratings differ. In this case, the maximum rating of the 3-phase transformer bank is three times the kVA rating of the smallest transformer.

## Problem 3-4

Determine the maximum kVA rating of a 3-phase transformer bank consisting of one 15-kVA transformer and two 20-kVA transformers.

## Solution

$$\text{3-phase bank kVA} = 3 \times \text{smallest transformer kVA}$$

$$\text{3-phase bank kVA} = 3 \times 15 \text{ kVA} = 45 \text{ kVA}$$

## THREE-PHASE DELTA AND WYE SYSTEMS

Three-phase power is available from two types of systems: *delta,* and *wye*. The systems are named for the shape of the transformer winding connections, as shown in Figure 3-8. Basically, the wiring of 3-phase motors and equipment is performed in the same manner whether the 3-phase electrical system is a delta or a wye. However, 3-phase voltage supplied must be the same as the voltage rating of the motors and equipment to be wired to the system.

## Three-wire Delta System

Three-phase loads, such as one or more electric motors requiring only three conductors, are often supplied with a 3-wire delta system. The customer can choose 240 V or 480 V, but remember that these are nominal values, with actual voltages being closer to 230 V and 460 V. A typical 3-phase, 3-wire delta electrical system is shown in Figure 3-9.

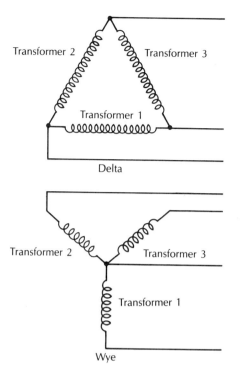

Figure 3-8 Two types of 3-phase systems are named for the shape of the transformer winding connections.

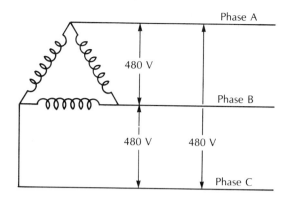

Figure 3-9 A typical 3-phase, 3-wire delta electrical system with 480 V available. Another common voltage is 240 V instead of 480 V.

A 3-phase, 3-wire electrical system often is not a grounded system but, for most farm applications, grounding is desirable and commonly required. Grounding of a 3-wire delta system can be accomplished by three methods, Figure 3-10. The first method is by grounding one phase wire, as shown in Figure 3-10 (B). This is called a *corner grounded delta*. For a 480-V corner grounded system, a voltmeter will read 480 V between each ungrounded phase wire and the metal case of the equipment.

- Grounding one phase of a multiphase system is listed in *NEC Section 250-25* as an acceptable means of grounding.
- If the grounded phase is insulated and is sized larger than No. 6, it must be identified at the time of installation with white or natural gray tape or other marking at its terminals, *NEC Section 200-6(b)*.
- The grounded-phase conductor must not be fused at the main disconnect. However, a circuit breaker which opens all ungrounded conductors simultaneously is acceptable, *NEC Section 240-22*. (Some motor circuits are an exception, *NEC 430-36*.)

A transformer of the delta system is sometimes provided with a connection at the center of the secondary winding of one of the transformers called a *center tap*. This center tap is grounded as shown in Figure 3-10 (C). The voltages obtained are identical to the 3-phase, 4-wire electrical system shown in Figure 3-11. Center tap grounding of a delta system is the method commonly used for a 240-V delta system.

The third method of grounding a 3-wire delta system is to use a special *grounding transformer*, often called a *zigzag transformer*. This method, shown in Figure 3-10 (D), is more expensive than the previous methods, but it does establish an identical voltage between each phase wire and the ground. In the case of a 480-V delta system, the voltage from the phase wires to the ground is 277 V. Voltage is ground from each phase wire of a 240-V delta system is 138 V.

## Four-wire Delta System

Single-phase, 120/240-V, 3-wire electrical service is most common on farms. However, farm expansion and the addition of large motors often make 3-phase service desirable. Single-phase service is required for the residence and some farm buildings, whereas both single-phase power and 3-phase power are required in other buildings. The 4-wire delta electrical system will supply single-phase power at 120/240 V, as well as 3-phase power at 240 V. One transformer in the 3-phase delta bank has a center tap to provide a neutral for single-phase, 120-V loads. Figure 3-11 shows a 3-phase, 4-wire delta electrical system.

Single-phase and 3-phase circuits can be taken from the same 3-phase panelboard, or each can be provided

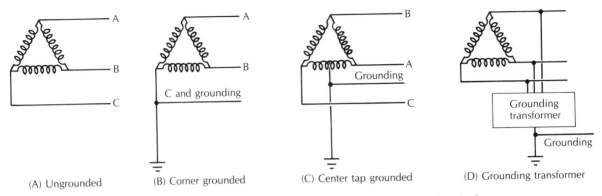

**Figure 3-10** Three methods of grounding a 3-wire, delta, 3-phase electrical system.

with a separate panelboard. A caution must be pointed out when 120-V circuits originate from a single, four-wire delta panelboard supplying both single- and 3-phase circuits.

**CAUTION:** The potential between phase wire B and the neutral in Figure 3-11 is 208 V. The B phase wire is often called the *high leg* or *wild leg*. Single-pole circuit breakers for 120-V circuits must not be placed on the B phase in the panelboard. Instead of getting 120 V, 208 V will be obtained, damaging 120-V equipment.

The installation and identification of the B phase of a 4-wire delta (wild leg, 208 V to ground) is specified by the *National Electrical Code* to help prevent accidental connection to a 120-V, single-phase circuit.

- The B phase shall be the phase having the higher voltage to ground, 208 V, *NEC Section 384-3(f)*.
- At locations where the neutral is present and connections are made (pull boxes, panelboards, disconnect switches), the B phase wire shall be identified with orange tape or paint, or by tagging, *NEC Sections 215-8, 230-56, and 384-3(e)*.
- The B phase wire shall be placed in the center lug in panelboards and disconnect switches, as shown in Figure 3-11, *NEC Section 384-3(f)*.

The B phase with 208 V to ground of a 4-wire, delta, electrical system is not always placed in the center lug of the power supplier's meter enclosure. The B phase is usually on the right of the meter socket. The meter and the meter enclosure or socket are the power supplier's equipment and do not come under the jurisdiction of the *National Electrical Code*. However, the electrician must follow the Code as closely as possible when installing service equipment, while meeting the power supplier's service equipment requirements. The meter enclosure must be wired according to the diagram inside the enclo-

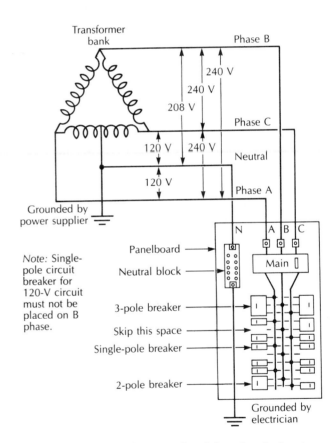

**Figure 3-11** Three-phase, 4-wire, delta, electrical system supplying single-phase, 120-V and 240-V circuits, and 3-phase 240-V circuits. Single-pole breakers are used for 120-V circuits, 2-pole breakers are used for single-phase, 240-V circuits, and 3-pole breakers are used for 3-phase, 240-V circuits.

sure or directions provided by the power supplier. The B phase must be in the center at the main service.

The 4-wire, delta electrical system is two systems in one: a 3-wire, 120/240-V, single-phase system; and a 3-wire, 240-V, 3-phase system. All three transformers provide the 3-phase power, but only one transformer provides power for the single-phase, 120-V loads. One transformer provides both single-phase, 120-V power and 3-phase, 240-V power, as shown in Figure 3-11. This transformer often has a higher kVA rating than the other two transformers.

Grounding of a 4-wire, delta, 3-phase system is accomplished by grounding the neutral wire originating at the center tap of the transformer supplying power for the single-phase loads. The power supplier grounds the neutral at the transformer location. The electrician grounds the neutral at the service disconnect.

## Open Delta Three-phase System

Three transformers are not required to provide 3-phase, 3-wire, and 4-wire delta power. One transformer can be removed from the delta bank, leaving a two-transformer bank providing the same set of voltages. This is the open delta 3-phase system common in rural areas. An open delta 4-wire system is shown in Figure 3-12.

The main advantage of the open delta 3-phase system is that 3-phase power can be provided with only two primary phase wires and the primary neutral wire. Normally, all three primary wires are required, along with the primary neutral. Many rural areas have available only single-phase power provided with two primary wires. Three-phase power can be provided with an open delta by extending one additional primary wire to the farm, as shown in Figure 3-13. Open delta is often used by the power supplier even when all three primary phase wires are available.

An open delta transformer bank has a kVA rating of 57% of the kVA rating of a closed delta, 3-phase bank, Equation 3.3

$$\text{Open delta kVA} = 3 \times \text{One transformer kVA} \times 0.57 \quad \text{Eq. 3.3}$$

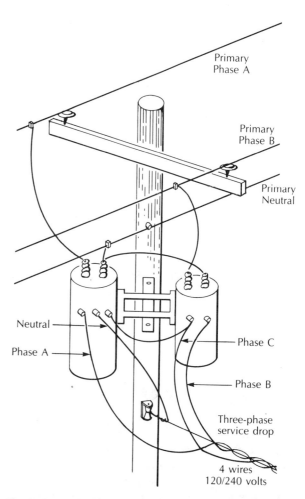

Figure 3-13 An open delta, three-phase transformer bank can provide 120/240-V single-phase service and 240-V, 3-phase service. Two primary phase wires and a primary neutral are required for an open delta system. The primary fuses have been omitted to simplify the diagram.

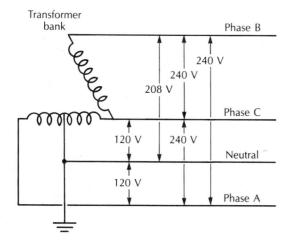

Figure 3-12 The voltages produced by an open delta, 3-phase, 4-wire electrical system are the same as those produced by the closed delta (three transformers).

Unit 3  Farm Electrical Systems  49

## Problem 3-5

A closed delta transformer bank consisting of three 20-kVA transformers has a total 3-phase rating of 60 kVA. Determine the open delta total kVA rating of the transformer bank when one transformer is removed.

## Solution

> One transformer kVA = 20
> Open delta kVA = 3 × 20 × 0.57 = 34.2 kVA

## Problem 3-6

This problem shows how open delta transformers are sized to a 3-phase load. Determine the kVA rating of each of two transformers required in an open delta bank to supply a 40-kVA, 3-phase load.

## Solution

> Open delta load kVA = 40
> One transformer kVA = $\dfrac{40}{3 \times 0.57}$ = 23.4 kVA

Two 25-kVA, single-phase transformers connected to form an open delta are required to supply the 40-kVA, 3-phase load.

An occasional disadvantage to the open delta system is a possible unequal voltage between phase wires. One of the phase voltages may be slightly lower than the other two. On occasion, this voltage unbalance causes sufficient current unbalance to trip an overload relay protecting a motor. If a voltage unbalance becomes a problem, the power supplier should be notified.

## Three-phase Wye System

The wye connection for providing 3-phase power is the electrical system least commonly encountered on farms. Another term occasionally used instead of wye is 3-phase *star* system. The wye system is frequently used in urban areas where the load consists primarily of single-phase lights and equipment, and a limited amount of 3-phase load, such as motors and electric heating. When 3-phase power for a farm is justified, compared to the single-phase load, then the load generally consists of motors and heating equipment.

An advantage of the wye system over the delta is that all three transformers in the wye bank share both the single-phase load and the 3-phase load. The wye system also provides a grounded neutral with equal voltage between each phase wire and the neutral.

Two sets of voltages are available from the 4-wire wye system: 120/208 V and 277/480 V. The 120/208-V wye system is shown in Figure 3-14. The 120/208-V system is used most commonly for small industrial plants, office buildings, stores, and schools where the circuits, and only a moderate amount of 208-V, 3-phase loads.

**CAUTION:** 230-V equipment must not be operated at 208 V unless permitted by the equipment manufacturer.

The 277/480-V wye system is used mainly for large commercial buildings and industrial plants where there is a high demand for power and lighting. The motors are

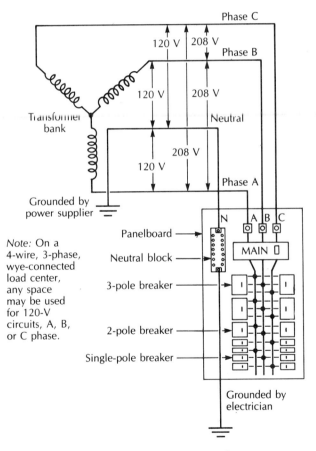

**Figure 3-14** A 3-phase, 120/208-V, 4-wire wye system provides power for 120-V and 208-V single-phase circuits, and 208-V, 3-phase circuits from a single panelboard.

operated at 480-V, 3-phase; the lighting is supplied with 277-V, single-phase power.

The 277/480-V wye system is used sometimes to provide power for a single, large electric motor or a group of motors on a farm, such as those required for an irrigation system or a field drainage, pump lift station. The center point of the wye transformer connection is grounded, and a grounding wire is provided with the 3-phase wires to the service disconnect. A certain type of processing building on a farm may contain several large motors, while having only a limited 120-V, single-phase need. A 277/480-V, 4-wire system can be used to supply the motors and lighting, and a dry-type transformer can be used to supply one or two circuits at 120 V. As the need for large horsepower motors on farms increases, this type of electrical system will become more common.

## Single-phase from a Wye System

A single-phase panelboard can be wired to a 3-phase wye electrical system to supply a load requiring only single-phase power. The single-phase power is available at 120/208 V. This type of installation is common in apartment buildings and commercial buildings where each apartment or business suite has a separate panelboard. Single-phase, 3-wire, 120/208-V power is common, but it is seldom encountered on farms.

The neutral of a 120/208-V, 3-wire, single-phase electrical system will always carry current when one or both of the ungrounded conductors is supplying a 120-V load. Figure 3-15 shows a 120/208-V system with each ungrounded wire carrying a 30-A, 120-V load.

Even though the load on both ungrounded wires is the same, the neutral also carries 30 A. If this were a 120/240-V electrical system, no current would flow in the neutral. In the 120/208-V system, it must be remembered that two transformers supply the 120-V, single-phase power, and each transformer operates 120 electrical degrees out of phase with the other. A cancellation of neutral current flowing in opposite directions does not occur with the 120/208-V system, as it does with a 120/240-V system (Figure 3-4).

If both ungrounded wires are supplying only 120-V loads, with the current on one ungrounded wire greater than on the other, the current flowing on the neutral will be only slightly less than the current flowing on the ungrounded wire carrying the greater load. For example, if one ungrounded wire is carrying an 20-A, 120-V load, while the other is carrying an 8-A, 120-V load, the neutral

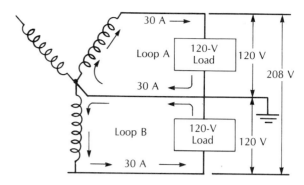

**Figure 3-15** The current flowing on the neutral of a single-phrase, 120/208-V system supplying 30-A, 120-V loads is 30 A. The current flowing in loop A is out of phase with the current flowing in loop B. A cancellation of the neutral currents does not occur.

current will be approximately 17.5 A. The *National Electrical Code* draws attention to this phenomenon in *Note 10(b)* of the *Notes to Tables 310-16* through *310-19*.

## IDENTIFYING ELECTRICAL SYSTEMS

It is important that an electrician be able to identify the type of electrical system serving a farm, residential, or commercial building. The surest method of identifying an electrical system is to call the nearest customer service office of the electrical power supplier.

Once the types of electrical systems described in this unit are understood, identification of an electrical system can be made by visual inspection and voltage measurements. Note the number of transformers on the power supplier's pole. Count the ungrounded wires at the service disconnect, and note whether there is a neutral. Determine if the system is grounded if a neutral is not present. Next, measure the voltage between each combination of ungrounded wires, and between each ungrounded wire and the neutral or service grounding wire. Remember that actual voltages are generally lower than nominal voltages.

## Problem 3-6

An electrician makes the following observations and voltage measurements at a farm service entrance, as shown in Figure 3-16.

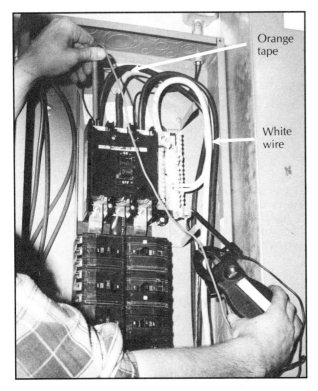

Figure 3-16 A 3-phase, 4-wire, 120/240-V electrical panelboard with wires properly installed and labeled. The electrician, however, often encounters panelboards in which the wires have not been labeled.

- Two transformers on the pole
- Three ungrounded wires
- A grounded neutral wire
- The voltages measure:
  A to B—235 V
  B to C—235 V
  C to A—235 V
  A to N—117 V
  B to N—202 V
  C to N—117 V

Identify the type of electrical system.

## Solution

The electrical system is 3-phase, because there are three ungrounded wires. A neutral is present; therefore, it is either 4-wire delta or wye. One ungrounded wire has a higher voltage to ground than the other two ungrounded wires (202 V). The electrical system is, therefore, a 4-wire delta. There are only two transformers instead of three; therefore, the system is an open delta.

## REVIEW

Refer to the *National Electrical Code* when necessary to complete the following review material. Write your answers on a separate sheet of paper.

1. What are the two most important factors influencing the choice of 3-phase electric power for use on a farm?

2. An electrician measures the current flowing on one ungrounded wire of a 3-wire, single-phase, 120/240-V electrical system to be 72 A, and 47 A on the other ungrounded wire. What amount of current can the electrician expect to measure on the neutral?

3. Describe a clue which might indicate a break in the neutral of a single-phase, 3-wire, 120/240-V electrical system.

4. An electrician measures the voltages of a single-phase, 3-wire electrical system as 116/232 V. What are the nominal voltages for this system?

5. How many current-carrying circuit wires are required for a 3-phase, 480-V electric motor?

6. Four 120-V lighting circuits each draw 12 A, two 120-V fan circuits each draw 8 A, two 120-V equipment circuits each draw 9 A, and a 240-V motor circuit draws 11-A. Determine the current flowing on each ungrounded wire if the electrical system is a 120/240-V, 3-wire, single-phase system.

7. Determine the current drawn by a 5-kW (5 000 W), 3-phase, 240-V electric resistance space heater.

8. The phase wire with the higher voltage to ground of a 4-wire, 120/240-V, delta electrical system must be tagged or labeled ___?___ in color at locations where the neutral is present and connections are made, according to the *National Electrical Code* Sections 215-___?___, 230-___?___, and 384-___?___.

9. Draw a diagram of the transformer secondary windings for a corner-grounded delta, 480-V, 3-phase, 3-wire electrical system.

10. Draw a diagram of the transformer secondary windings for a 3-phase, 4-wire, 120/240-V, open delta electrical system, and show the voltages to neutral (or ground) from each phase wire.

11. Determine the kVA rating of the maximum 3-phase load which can be connected to a bank of three transformers, two of which have a 20-kVA rating and the other a 30-kVA rating.

12. Identify, by nominal voltage and type, the electrical system which has three transformers on the pole, 3-phase wires, and a grounding wire at the service disconnect. The service supplies a 125-hp irrigation pump motor. The voltages between each phase wire are 464 V, and between each phase wire and the grounding wire 268 V.

# UNIT 4
# FARM BUILDING SERVICES

## OBJECTIVES

After studying this unit, the student will be able to

- state the purpose of an electrical service entrance.
- calculate the electrical demand load for a farm building.
- select the type of farm building electrical service equipment.
- size the farm building service-entrance wires.
- explain a procedure for installing a farm building service entrance.
- identify the sections of the *National Electrical Code* which concern the sizing and installation of a service entrance.
- determine the size of grounding wire for a farm building service entrance.
- choose an appropriate service-entrance grounding electrode.

The electrical service to a farm building must be of adequate size to handle the load, and it often must be capable of withstanding an adverse environment. One building may serve as the farm main service, or the main service may be a central distribution pole. The electric service to each farm building must be installed according to *Article 230* in the *National Electrical Code*. The individual farm building demand load must be determined before the farm main service can be sized.

## PURPOSE OF THE ELECTRICAL SERVICE ENTRANCE

Knowing the purpose of a service entrance helps to understand the *National Electrical Code* rules governing its sizing and installation. A service entrance has several important functions. It provides

- a safe means for the entrance of electrical supply wires into a building or load.

- a means of disconnecting all electrical power in a building.
- a means of limiting to a safe level the maximum amount of electrical current that can enter a building.
- a common point for grounding the electrical equipment.
- a safe means of subdividing the electrical supply to serve individual loads or groups of loads.

When properly sized and installed, the service entrance functions to provide safety from fire, electrical shock, and mechanical injury. In the event of mechanical injury or electrical shock to an animal or a human, the power can be shut down quickly. Most farm buildings contain valuable equipment, materials, or animals. Safety and reliability of a building's electrical supply equipment are essential.

**CAUTION:** An improperly sized and/or installed service entrance can be a fire hazard.

## FARM BUILDING DEMAND LOAD

The size of a farm electrical service is determined by calculating the *electrical demand load*, the maximum electrical load expected to operate for a particular length of time. The *National Electrical Code* specifies the procedure for determining the demand load for a farm building.

- According to *NEC Section 220-40*, when a farm building contains two or more branch circuits, the service-entrance equipment is sized in accordance with the demand factors in *NEC Table 220-40*. See Table 4-1.

It is necessary to make a complete list of electrical equipment for a farm building, giving full-load amperes and operating voltage. A list of all the equipment and outlets in the building must be made. The next step is to identify the equipment which is expected to operate all at the same time. The *National Electrical Code* refers to this as the load that operates *without* diversity. The remainder is the load that operates *with* diversity.

## Farm Building Loads

When listing all the electrical equipment and outlets in a farm building, the voltage and current are listed for each outlet or load. It is most convenient to calculate the load directly in amperes. Normally, a residential service-entrance calculation is done using load wattage. In most buildings, it is common to allow 1.5 A at 120 V for each lighting outlet. In farm buildings, however, many lighting outlets of low wattage are often installed. In a poultry egg-laying house, for example, the incandescent lamp wattage is usually 25 W or 40 W. Higher wattage provides too much light, which can cause problems in cage-type laying houses. A particular value of current to allow in this example is ½ A per outlet.

## CALCULATING ELECTRICAL DEMAND

The electrical loads for a livestock barn are listed in Table 4-2.

The end result of any service calculation is the electrical demand load in amperes at 240 V for a single-phase service. All 120-V loads must be divided by two to give the equivalent load at 240 V. All 120-V loads must be evenly divided between the two incoming hot wires.

The first task in determining the demand load for the service entrance is to group the loads in the following manner.

- Loads expected to operate *without* diversity. These are loads which, under normal circumstances, will operate all at the same time.
- Loads expected to operate *with* diversity. These are loads which may occur at a time other than when the loads operating without diversity are operating.

*Table 4-1 Method of computing farm loads for other than dwelling unit

| Ampere Load at 240 V | Demand Factor, Percent |
|---|---|
| Loads expected to operate without diversity, but not less than 125% of full-load current of largest motor, and not less than the first 60 A of load | 100 |
| Next 60 A of all other loads | 50 |
| Remainder of other load | 25 |

*(Adapted from *NEC Table 220-40*)

**Table 4-2  Electrical loads for a livestock barn**

| Number | Equipment Description | Voltage | Phase | Amperes At Rated Voltage | Amperes At 240 V |
|---|---|---|---|---|---|
| 10 | Lighting outlets | 120 V | 1 | 15.0 | 7.5 |
| 1 | Mercury-vapor lamp, 175 W | 120 V | 1 | 1.8 | 0.9 |
| 6 | Receptacle outlets | 120 V | 1 | 9.0 | 4.5 |
| 1 | Stock waterer, 500 W | 120 V | 1 | 4.2 | 2.1 |
| 1 | Silo unloader, 5 hp | 240 V | 1 | 28.0 | 28.0 |
| 1 | Silage bunk feeder, 1 hp | 240 V | 1 | 8.0 | 8.0 |

*Table 220-40* of the *NEC* (adapted as Table 4-1 in this text) is difficult to understand. Several examples are worked out in this unit to help make this Code table more clearly understood. Motor loads must be computed according to *NEC Sections 430-24, 430-25,* and *430-26,* as stated in *Section 220-14.* This requirement means that the motor load is computed as follows.

- Largest motor full-load current times 1.25. If two motors are of the same size, take only one of them at 125%.
- Add the full-load current of all remaining motors and other loads.

## Livestock Barn Example

All loads in a livestock barn, except for six receptacle outlets, will operate at the same time. These loads are considered to operate without diversity.

*Loads operating without diversity:*

| | |
|---|---|
| 10 lighting outlets | 7.5 A, 240 V |
| 1 mercury light | 0.9 A, 240 V |
| 1 stock waterer | 2.1 A, 240 V |
| 1 silo unloader, 5 hp, 28 A × 1.25 | 35.0 A, 240 V |
| 1 silage bunk feeder, 1 hp | 8.0 A, 240 V |
| | 53.5 A, 240 V |

*Loads operating with diversity:*

| | |
|---|---|
| 6 receptacle outlets | 4.5 A, 240 V |

The demand load is figured from Table 4-1. The load operating without diversity is taken at 100%. This load must include 125% of the full-load current of the largest motor served. For the livestock barn, the 5-hp silo unloader is the largest motor. Its full-load current times 1.25 is 35 A. Next, Table 4-1 states that the first 60 A of load in the building must be taken at 100%. The 53.5 A load without diversity is smaller than 60 A. Therefore, the first 6.5 A of the load operating with diversity must be taken at 100%. The load operating with diversity is only 4.5 A. Therefore, for the livestock barn, all of the load is taken at 100%. The livestock barn demand load calculation is summarized as follows.

**Livestock barn demand**

| | |
|---|---|
| Loads operating without diversity, but not less than the first 60 A of load | 53.5 A at 100% = 53.5 A |
| Loads operating with diversity | 4.5 A at 100% = 4.5 A |
| Next 60 A at 50% | 0.0 A at 50% = 0.0 A |
| Remaining load at 25% | 0.0 A at 25% = 0.0 A |
| Demand load | 58.0 A |

Consider the barn example again. But, this time, a special 30-hp portable single-phase motor is used twice each year to power the blower which fills the silo. The nameplate full-load current is 145 A at 240 V. When the silo is being filled it is not possible to operate the silo unloader. Usually, the silo bunk feeder is not operated either. Therefore, the silo unloader and the bunk feeder are not considered. The stock waterer also is not considered, because it is only operated during the winter. The demand load is calculated as follows.

*Loads operating without diversity:*

| | |
|---|---|
| 10 lighting outlets | 7.5 A, 240 V |
| 1 mercury light | 0.9 A, 240 V |
| 1 30-hp, portable motor, 145 A × 1.25 | 181.3 A, 240 V |
| | 189.7 A, 240 V |

*Loads operating with diversity:*

| | |
|---|---|
| 6 receptacle outlets | 4.5 A, 240 V |

**Livestock barn demand (with 30-hp motor)**

*Loads operating without diversity:*
    189.7 A at 100% = 189.7 A at 240 V

*Loads operating with diversity:*
Next 60 A at 50%  4.5 A at 50% = 2.3 A, 240 V
Demand load  192.0 A, 240 V

## Dairy Barn Example

A dairy milking barn generally contains milking and milk handling equipment, and often feed handling equipment. The following example illustrates a typical dairy milking barn demand calculation. A milkhouse is included as a part of the barn.

A dairy milking barn consisting of a milkhouse, milking parlor, and free-stall barn is to be supplied by single-phase electrical service with the following outlets and equipment.

| | |
|---|---|
| 12 fluorescent fixtures | 0.8 A, 120 V |
| 4 mercury-vapor fixtures | 1.7 A, 120 V |
| 12 incandescent lighting fixtures | 1.5 A, 120 V |
| 12 receptacle outlets | 1.5 A, 120 V |
| 2 ventilation fans, ½ hp | 4.9 A, 240 V |
| 1 vacuum pump, 5 hp | 28.0 A, 240 V |
| 1 milk tank compressor, 5 hp | 28.0 A, 240 V |
| 1 milk tank stirring motor, ⅓ hp | 3.6 A, 240 V |
| 1 milk truck pump outlet, 1 hp | 8.0 A, 240 V |
| 1 milkhouse heater, 2 kW | 8.3 A, 240 V |
| 1 water heater, 4.5 kW | 18.8 A, 240 V |
| 1 feed auger, ⅓ hp | 3.6 A, 240 V |
| 1 milk lift pump, ⅓ hp | 3.6 A, 240 V |
| 1 silo unloader, 5 hp | 28.0 A, 240 V |
| 1 silage elevator, 1 hp | 8.0 A, 240 V |
| 1 bunk feeder, 2 hp | 12.0 A, 240 V |
| 1 bulk milk tank cleaner, ½ hp | 4.9 A, 240 V |
| 2 livestock waterers, 500 W | 4.2 A, 120 V |

First, decide which loads are likely to occur at the same time; that is, to operate without diversity. For this example, it is assumed that the receptacle outlets will not be used during milking and feeding. Also, the milk truck pump outlet and the bulk milk tank cleaner will not be used during milking. It is assumed that feeding and milking may occur at the same time. Divide the loads into two groups: one group that operates without diversity, and one group that operates with diversity.

*Loads operating without diversity:*
| | |
|---|---|
| 12 fluorescent fixtures, 9.6 A at 120 V | 4.8 A, 240 V |
| 4 mercury-vapor fixtures, 6.8 A at 120 V | 3.4 A, 240 V |
| 12 incandescent fixtures, 18 A at 120 V | 9.0 A, 240 V |
| 2 ventilation fans, ½ hp | 9.8 A, 240 V |
| 1 vacuum pump, 5 hp, 28 A × 1.25 | 35.0 A, 240 V |
| 1 milk tank compressor, 5 hp | 28.0 A, 240 V |
| 1 milk tank stirring motor, ⅓ hp | 3.6 A, 240 V |
| 1 milkhouse heater, 2 kW | 8.3 A, 240 V |
| 1 water heater, 4.5 kW | 18.8 A, 240 V |
| 1 feed auger, ⅓ hp | 3.6 A, 240 V |
| 1 milk lift pump, ⅓ hp | 3.6 A, 240 V |
| 1 silo unloader, 5 hp | 28.0 A, 240 V |
| 1 silage elevator, 1 hp | 8.0 A, 240 V |
| 1 bunk feeder, 2 hp | 12.0 A, 240 V |
| 2 livestock waterers, 500 W, 8.4 A at 120 V | 4.2 A, 240 V |
| | 180.1 A, 240 V |

*Loads operating with diversity:*
| | |
|---|---|
| 12 receptacle outlets, 18 A at 120 V | 9.0 A, 240 V |
| 1 milk truck pump outlet, 1 hp | 8.0 A, 240 V |
| 1 bulk milk tank cleaner, ½ hp | 4.9 A, 240 V |
| | 21.9 A, 240 V |

Next, apply the demand factors of Table 4-1 to determine the ampere rating of the electrical service for the dairy barn.

*Loads expected to operate without diversity:*
180.1 A at 100%  180.1 A
*Loads expected to operate with diversity:*
Next 21.9 A at 50%  11.0 A
Total computed demand load  191.1 A

## SELECTING SERVICE EQUIPMENT

Selecting the type of farm building service-entrance equipment depends upon several factors.

- The type of load to be served
- The layout of the load within a building
- The environment around the service equipment

Reliability is essential for farm wiring systems, and it must be maintained while keeping equipment costs reasonable. The *National Electrical Code* allows considerable flexibility by permitting up to six separate disconnects for each service, *NEC Section 230-71(a)*. Most farm building service equipment consists of load centers, panelboards, or fusible switches. A load center is a residential-grade panelboard. Load centers corrode easily in many agricultural environments; therefore, they should be used only in dry, nonhumid areas.

The load center shown in Figure 4-1 provides both a main disconnect and branch circuit breakers. These load center panelboards are available in sizes up to 200 A. Load centers are available with a fusible main. This type is particularly useful where a large motor is supplied which may cause tripping of the main breaker. This type is also useful in locations where high levels of current could cause damage to the load center should a short circuit occur.

A fusible switch is often used as a service disconnect for farm loads. Fusible switches are commonly referred to as *safety switches*. They are particularly suited to act as a service disconnect for groups of motors such as in a grain handling system. A typical fusible safety switch is shown in Figure 4-2.

A farm building service entrance can consist of up to six disconnect switches, or a combination of load centers and disconnect switches. However, *NEC Section 230-72(a)* requires that this equipment be grouped together and labeled to indicate the load each disconnect serves. A typical application of a load center and safety-switch service entrance may be found in a dairy barn. There are two locations within a dairy barn where considerable

**Figure 4-2** A fusible safety switch can be used as a service disconnect. (*Courtesy of Square D Company*)

power is used. These areas are the milkhouse and milking parlor, and the feed room.

Heavy-duty panelboards are available in a variety of types and sizes. They are available with and without a main breaker. These panelboards may be of the fusible type or the circuit-breaker type. They are constructed of heavy gauge steel, and are often galvanized. The types shown in Figure 4-3 are available in sizes of 400 A and larger. Ordering either of these types is somewhat complicated. The installer generally must specify the:

1. Size of the enclosure box
2. Interior assembly
3. Type and size of the breakers or fuse switches
4. Style and size of the cover

Manufacturer's sales representatives are available to assist the electrical installer in ordering the necessary parts of the panelboard. After a little practice, this task becomes quite easy.

Care should be taken to avoid locating the service equipment in areas of adverse environment. Areas with high humidity and dust should be avoided. When installed in such areas, however, the panelboard should be corrosion resistant and designed to prevent the entry of dust. Frequently, there is no building in which to install

**Figure 4-1** A load center panelboard is often used for the service entrance for a farm building, but it should be used only in dry, dust-free, indoor areas. (*Courtesy of Square D Company*)

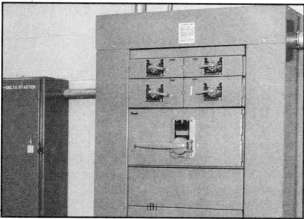

Figure 4-3 (A) Heavy-duty circuit-breaker panelboard, and (B) fusible panelboard. Both types are available in sizes of 400 A and larger.

the service equipment. A weatherproof enclosure is sometimes constructed to house the equipment. The service panelboard or disconnect switch can be installed out in the open if it is of a rainproof design.

## SIZING THE SERVICE DISCONNECT

Once the demand load for the farm building service has been determined, the disconnect can be sized. The *NEC* rules which apply to the sizing of the disconnect are found in *Section 230-79*.

- The service disconnect must be not smaller than the calculated demand load.
- If a building or load has only one circuit, the minimum size disconnect is 15 A.
- When there are only two circuits, the minimum size disconnect is 30 A.
- For a one-family dwelling served with a 3-wire system having six or more circuits, or a 10 kVA demand load, 100 A is the minimum size.
- For all other farm buildings, 60 A is the minimum size.

Determining the rating of the service equipment is based upon the calculated demand load. If additional equipment is likely to be added in the future, extra capacity should be provided. Generally, the next standard disconnect rating larger than the calculated demand load is selected, provided future expansion is not anticipated. For most farm buildings, the demand load is unlikely to last for more than a short length of time. However, if the demand load does last for a long period of time (continuous load), the demand load must be multiplied by 1.25 to select the service disconnect. This 25% increase in the rating is required in *NEC Sections 220-10(b)* and *384-16(c)*. In *NEC Article 100,* a *continuous load* is defined as a load which will last for three hours or longer. A load center loaded to 80% may even prove to be too small. Load centers, in particular, may become heated if operated continuously at 80% or more of their rating. The ratings of load centers and/or panelboards, and fusible switches are listed in Tables 4-3 and 4-4.

The following are minimum ratings of service disconnects for the farm demand loads calculated previously. To provide for future expansion, a larger service size must be installed.

|  | Demand load | Service disconnect rating |
|---|---|---|
| Livestock barn | 58 A | 100 A |
| Livestock barn plus 30-hp motor | 192 A | 200 A |
| Dairy barn | 191 A | 200 A |

## SIZING THE SERVICE-ENTRANCE WIRES

The minimum size of the service-entrance wires is dependent upon the rating of the service disconnect over-

Table 4-3  Common sizes of load centers and/or panelboards used in farm buildings (250-V rating)

| Phase | Ampere Rating | Main Lugs Only | With Main Breaker | Fusible Main[1] Pullout |
|---|---|---|---|---|
| 1 | 40 | x | | |
| 1 | 60 | | x | x |
| 1 | 70 | x | | |
| 1 | 100 | x | x | x |
| 1 | 125 | x | x | |
| 1 | 150 | x | x | |
| 1 | 200 | x | x | x |
| 1 | 225 | | x | |
| 1 | 300 | | x | |
| 1 | 400 | | x | |
| 1 | 600 | | x | |
| 3 | 60 | x | x | |
| 3 | 125 | x | x | |
| 3 | 150 | | x | |
| 3 | 200 | $x^2$ | x | |
| 3 | 225 | x | x | |
| 3 | 300 | | x | |
| 3 | 400 | $x^2$ | x | |
| 3 | 600 | $x^2$ | x | |
| 3 | 800 | | x | |

[1]Fusible safety switches or fusible circuit breakers are frequently used with sizes larger than 200 A.
[2]Fusible panelboard

current protection, and the computed demand load. *NEC Section 230-42* requires that

- the conductor be of sufficient size to handle the demand load as calculated in *Article 220*.
- where there are only two 2-wire branch circuits, the minimum wire size is No. 8 AWG copper or No. 6 AWG aluminum.
- where only a single load is served, the minimum size is No. 12 AWG copper or No. 10 AWG aluminum.
- wires are to be sized according to *NEC Table 310-16*, based upon the size of the service overcurrent protection. See *NEC Section 230-90(a)*. For the farm residence, the wire may be sized according to *Note 3* to *NEC Table 310-16*. The table in *Note 3* is adapted for this text in Table 4-5.

*NEC Section 230-41* requires that service-entrance conductors be insulated, except for the grounded neutral conductor, which may be bare. A bare neutral in a farm service raceway is not recommended. A bare neutral is permitted underground under some conditions, but it is not recommended. Soil conditions may cause the bare conductor to corrode.

Service wires may be run overhead or buried underground. If service wires are directly buried in the ground, they must be of a type identified for direct burial, according to *NEC Section 310-7*. The wire and cable used most often for direct burial on farms are Type UF and Type USE. Type USE is available as a cable or as individual wires for direct burial.

If a service entrance receives its electrical supply from an overhead service drop, the wire insulation or the service-entrance conductors must be approved for wet

Table 4-4  Common sizes of fusible safety-switch disconnects used in farm buildings (250-V rating)

| Phase | Ampere Rating |
|---|---|
| 1 | 30 plug fuse |
| 1 | 30 cartridge fuses |
| 1 | 60 |
| 1 | 100 |
| 1 | 200 |
| 1 | 400 |
| 1 | 600 |
| 3 | 30 |
| 3 | 60 |
| 3 | 100 |
| 3 | 200 |
| 3 | 400 |
| 3 | 600 |

Table 4-5 Minimum size 3-wire, single-phase service-entrance conductors for a dwelling. Insulation Types RH, RHH, RHW, THW, THWN, THHN, and XHHW

| Service Rating in Amperes | Minimum AWG Wire Size | |
|---|---|---|
| | Copper | Aluminum |
| 100 | 4 | 2 |
| 125 | 2 | 1/0 |
| 150 | 1 | 2/0 |
| 200 | 2/0 | 4/0 |

locations. These wires are exposed to the weather. The *NEC* in *Section 310-8* permits the following types of insulation for wet locations: RHW, RUW, TW, THW, THWN, and XHHW. The types of service-entrance conductor insulation used most often for farm applications are THW, THWN, and XHHW.

Either copper or aluminum conductors are permitted for use by the *NEC* in farm service entrances. Aluminum wire often is used because of its lower cost and lighter weight than copper. Aluminum conductors provide dependable, trouble-free service when they are terminated properly. The electrical installer must be sure that the lug or terminal is intended for use with aluminum wire. An oxide inhibitor or joint compound is used to coat the wire before it is installed in a service-entrance terminal lug. This is especially important in damp locations found on farms.

Once the service overcurrent device rating is known, the installer looks up the minimum size wire in *NEC Table 310-16*. Consider a typical example of a barn service. Assume the calculated demand load for the barn is 112 A, and the main overcurrent device is rated 150 A. The wire is sized to the overcurrent device rating (*NEC Section 240-3*); therefore, find a wire in *NEC Table 310-16* with a 150-A rating. If the wire is THW copper, then the minimum size is No. 0 AWG.

Consider the case where the service wire is THW aluminum. There is no wire with a rating of exactly 150 A. No. 2/0 AWG is rated 135 A, and No. 3/0 AWG is rated 155 A. The No. 2/0 AWG rated 135 A may be used because the wire ampacity is greater than the calculated demand load of 112 A, *NEC Sections 230-90(a), Exception No. 2,* and *240-3, Exception No. 1*.

Following are the rules for selecting the minimum size of service-entrance ungrounded conductors, from *NEC Table 310-16*. *Note 3* to the table is used to select the service wires for the farm dwelling.

- When a wire ampacity matches the service overcurrent device rating, that size wire is the minimum permitted.
- In the case where the wire ampacity does not match the service overcurrent device rating and the overcurrent device does not exceed 800 A, the wire with the next smaller ampere rating is the minimum permitted provided that wire is adequate to handle the load.
- Where the service consists of one to six disconnects, add together the ratings of the overcurrent devices to obtain the total equivalent overcurrent device rating, *NEC Section 230-80*.
- Consider any future loads likely to be added to the existing service.

## Problem 4-1

A dairy barn electrical service entrance consists of a 200-A, single-phase panelboard and a 100-A fusible switch (Figure 4-4). The total computed demand load is 235 A. Determine the minimum size of the ungrounded service-entrance conductors using XHHW copper wire.

## Solution

The equivalent overcurrent device rating for the service is 300 A and the demand load is 235 A. Remember, this is a wet location and the XHHW copper wire is rated 75°C. The minimum size service-entrance wire is 300 MCM copper XHHW.

## Sizing the Neutral

The neutral wire of a service entrance will almost always carry a smaller load than the ungrounded wires. This is true when there is equipment in the building that operates at the higher voltage. This voltage is 240 V for most farm buildings. The *NEC* in *Section 220-22* allows the neutral wire to be sized smaller than the ungrounded wires.

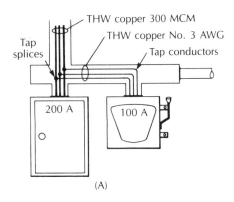
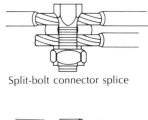
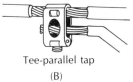

Figure 4-4 (A) The fusible disconnect is tapped from the main service-entrance conductors. (B) Many types of connectors can be used to make tap splices. Once the connection is made, the entire splice is covered with electrical tape to provide a layer of insulation.

- The feeder neutral load shall be the maximum load connected between the neutral and any one ungrounded conductor. This is called the *maximum unbalanced load*.

It is expected that the electrical installer will try to balance the 120-V loads. This is done by putting half of the load between one ungrounded service wire and the neutral. The other half is put between the other ungrounded wire and the neutral. Following this procedure minimizes the amount of current flowing on the neutral.

The unbalanced load is determined by adding up all the 120-V loads and dividing by two, Equation 4.1.

$$\text{Unbalanced load} = \frac{\text{Total 120-V load}}{2} \quad \text{Eq. 4.1}$$

## Problem 4-2

Determine the unbalanced load for the dairy farm load described earlier in this unit.

| | |
|---|---|
| 12 fluorescent fixtures | 9.6 A, 120 V |
| 4 mercury-vapor fixtures | 6.8 A, 120 V |
| 12 incandescent fixtures | 18.0 A, 120 V |
| 12 receptacle outlets | 18.0 A, 120 V |
| 2 livestock waterers | 8.4 A, 120 V |
| | 60.8 A, 120 V |

## Solution

$$\text{Unbalanced load} = \frac{60.8 \text{ A at } 120 \text{ V}}{2} = 30.4 \text{ A}$$

Recall that the total computed demand load on the ungrounded service wires was found to be 191 A. Now the question is, how small can the electrician size the neutral? Obviously, in some cases, the load on the neutral is quite small compared to the load on the ungrounded wires. The minimum size for the neutral is specified in *NEC Section 230-42(c)*.

- The grounded (neutral) conductor shall be not smaller than the minimum size grounding electrode wire required for the service.

The procedure for sizing the grounding wire for a farm electrical service entrance is discussed later in this unit. *NEC Table 250-94* (Table 4-9 in this unit) is used to determine the size of the grounding wire. The demand load for the dairy barn example was found to be 191 A. The service-entrance size chosen was 200 A. If THW copper wire is chosen for the ungrounded wires, the size according to *NEC Table 310-16* is No. 3/0 AWG. The neutral wire need only be sized to the unbalanced load, which is 30.4 A. This would mean the neutral wire would be only No. 8 AWG; but, *NEC Section 230-42(c)* sets the minimum at No. 4 AWG. Therefore, the neutral could handle an unbalance of up to 85 A.

It is difficult to know just how much additional 120-V load will be added to a building in the future. Experience has shown that 120-V loads are often not well balanced on a farm service entrance. The 120-V load connected between one ungrounded service wire and the neutral may be much greater than the load connected between the other ungrounded service wire and the neutral. The result may be loads that are more highly unbalanced than expected. It is recommended that, for a farm building containing general circuits, the neutral be not more than two sizes smaller than the ungrounded service wires. For the dairy barn example, the neutral would then be No. 1/0 AWG THW copper with a 150-A rating.

There are cases where the neutral can be sized to the minimum. The service entrance on a silo is a case in point. The only 120-V load may be a single, 15-A circuit supplying a receptacle outlet and lighting fixtures. No additional 120-V load is expected to be added in the future. A similar raintight service entrance supplying a grain-drying and storage center may have only a small 120-V load. If so, the neutral can be sized to the minimum.

## Sizing the Tap Conductors

A *tap conductor* is a short length of wire splitting off from a main conductor to supply an additional load. Figure 4-4 shows tap conductors for a barn service entrance. The fusible disconnect is connected to the service-entrance conductors by means of tap conductors.

*NEC Section 230-46* prohibits splices in service-entrance conductors. However, *Exception No. 2* of that section allows splices for the purpose of tapping to supply up to six disconnects at the service location.

The tap conductors are permitted to be smaller than the main service conductors. The rules for sizing tap conductors are given in *NEC Section 240-21. Exception No. 2* may be applied to a farm building service entrance with the wires in metal raceway.

- The tap conductors may be smaller than the service-entrance conductors provided that
1. the tap conductors are not more than 10 ft (3.05 m) long.
2. the ampere rating of the tap conductor is not smaller than the ampere rating of the switch or panelboard it supplies.
3. the ampere rating of the tap conductors is not less than the rating of the overcurrent device at their termination.
4. the tap conductors end at the switch or panelboard they supply.
5. the tap conductors are contained within a raceway, such as conduit, wireway, or auxiliary gutter.

An emergency supply of water is important on a farm in the event of fire. Water pumps should be supplied with power obtained ahead of the main service disconnect for the farm. This means that if the service disconnect is

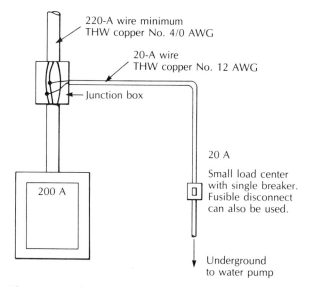

Figure 4-5 The water pump can be fed separately from the main panelboard. The tap should be made away from the main panel.

turned off during a fire, the water pump will still continue to operate. *NEC Section 230-94, Exception No. 4*, permits the water pump to be tapped ahead of the main disconnect for the purpose of fire protection. *NEC Section 230-72(a), Exception,* and *Section 230-72(b)* recommend that the disconnect switch for the water pump be located a distance from the main disconnect. If something happens to the main disconnect, the fire pump is less likely to be affected. *NEC Section 230-82, Exception No. 5*, requires that the fire pump tap comply with the service-entrance tap rules. Figure 4-5 shows an example of a service entrance in a farm building with the water pump tapped directly from the service-entrance conductors.

## Sizing Parallel Service Conductors

A farm building with a large electrical demand, such as 400 A or 600 A, requires large sized service-entrance conductors. Large sizes of wire are not always readily available to the installer. Moreover, large sized wires are difficult to handle and bend.

Consider that a farm building is to be supplied by a 400-A service. One alternative for a 400-A service entrance is to supply the service underground with two 200-A service laterals, each terminating in a 200-A disconnect or panelboard, Figure 4-6. *NEC Section 230-2, Exception No. 7*, permits this procedure.

Unit 4  Farm Building Services  63

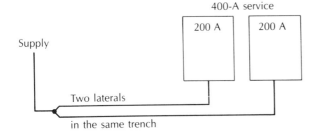

**Figure 4-6** Two underground service laterals originating with the same supply and terminating at separate disconnects are considered to be one lateral.

Parallel sets of service-entrance wires can be used to supply a single service disconnect or panelboard. *NEC Section 310-4* permits the paralleling of service-entrance conductors, and specifies the installation procedures. The rules for installing parallel service conductors must be followed *exactly* in order to prevent severe problems.

Consider that two parallel sets of THW copper service-entrance conductors are to be sized to supply a 600-A panelboard. The wires are run in a single conduit underground for a pad-mounted transformer outside to the 600-A panelboard inside the farm building, Figure 4-7.

Two or more sets of service-entrance wires may be used for a service. The sets of wires may be installed in the same conduit, or in separate conduits. If they are all placed in the same conduit, the *NEC* requires that their ampere rating be derated or lowered from the number listed in *Table 310-16*. The heat buildup from current flowing through the many wires may cause them to overheat. The derating factors are found in *Note 8 to NEC Table 310-16*, adapted for this text in Table 4-6.

The parallel sets of service-entrance wires are sized according to the following formulas, Equations 4.2 and 4.3.

Each set of conductors in a separate conduit:

$$\text{Amperes per wire} = \frac{\text{(Service ampere rating)}}{\text{(Number of parallel sets of wires)}} \quad \textbf{Eq. 4.2}$$

All wires in the same conduit:

$$\text{Amperes per wire} = \frac{\text{(Service ampere rating)}}{\text{(Number of parallel sets of wires)} \times \text{(Derating factor)}} \quad \textbf{Eq. 4.3}$$

**Table 4-6** Reduce the ampacity listed in *NEC Table 310-16* by the appropriate factor for more than three wires in conduit

| Number of Wires in the Same Conduit | Derating Factor |
|---|---|
| 4–6 | 0.8 |
| 7–24 | 0.7 |
| 25–42 | 0.6 |
| 43–more | 0.5 |

The size of wire is then found in *NEC Table 310-16*. Assume that all of the wires for the two parallel sets for the 600-A service are installed in the same conduit, as they are in Figure 4-7. For a single-phase service, there would be six wires. Therefore, the derating factor would be 0.8. The ampere rating per wire is determined using Equation 4.3

$$\text{Amperes per wire} = \frac{600 \text{ A}}{2 \text{ sets} \times 0.8} = 375 \text{ A}$$

Using THW copper wire, 400 MCM would be required as a minimum for the ungrounded conductors. If a single set had been used, the size would have been 1250 MCM. Keep in mind that *NEC Section 240-3, Exception No. 1*, applies. The actual load per wire will never reach 375 A. This is a derated ampacity.

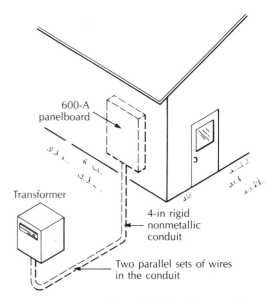

**Figure 4-7** A 600-A panelboard is fed underground from a pad-mounted transformer.

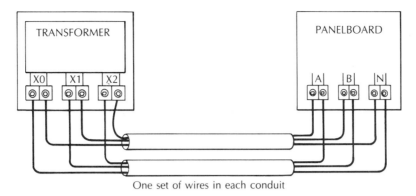

One set of wires in each conduit

**Figure 4-8** *NEC Section 310-4* must be followed exactly to avoid problems when paralleling service wires.

The neutral wire carries only the unbalanced load, as described earlier. Therefore, it is acceptable to make the neutral wire two sizes smaller than the ungrounded wires. The neutral for the example would be 300 MCM THW copper wire.

The following are the rules for installing conductors in parallel.

- Wires can be paralleled in sizes No. 1/0 AWG and larger.
- All conductors of a phase shall be exactly the same length.
- All wires must be of the same material.
- All wires must be of the same AWG or MCM size. All ungrounded (hot) wires are the same size. All neutral wires are the same size.
- All wires must have the same insulation type.
- All wires must be terminated in exactly the same way.
- When run in separate conduits, one complete set of wires must be installed in each conduit, Figure 4-8.
- When run in separate conduits, the conduits must be of the same size and material.

> **CAUTION:** The last two requirements must be followed *exactly;* otherwise, current will be induced to flow in metal piping and metal enclosures where the wires enter a panelboard. These induced currents will cause considerable damage.

## SERVICE-ENTRANCE INSTALLATION PROCEDURE

### Overhead Service Entrances

Service entrances on a farm commonly receive their electrical supply from an overhead service drop. The service drop may originate at the electrical power supplier's transformer. In this case, a kilowatthour meter socket must be installed by the electrician. Generally, the outbuildings on a farm receive their electrical supply from the central distribution pole or another building. In these cases, a revenue meter socket is not installed.

The first step in the installation of a service entrance in a farm building is to decide upon the type of materials to be used. The layout of the building, the location of the electrical service disconnect, and the direction to the nearest electrical supply must be considered. The two choices of materials for an overhead service are service-entrance cable or single wires in raceway. The main disadvantage of service-entrance cable is that, if not protected, it is subject to physical damage by animals and machines. Raceway may corrode or allow moist air to condense inside the wiring system. Clearly, the electrical installer must make a judgment and choose the most appropriate materials for the conditions.

### Locating the Service Equipment

The electrical installer must find an acceptable location for the farm building service entrance. The point of attachment must be located where the service drop will attach to the building. An acceptable location must be found to mount the load center. It is desirable to have the service drop attachment point and the load center as close together as possible.

The service drop must terminate at an insulator which is securely connected to a structural member of the building. The load center must be installed in a location accessible to farm personnel. Locating the load center in an animal pen does not meet the intent of *NEC Section 230-70(a),* which specifies that the disconnect be in a readily accessible location. However, if the enclo-

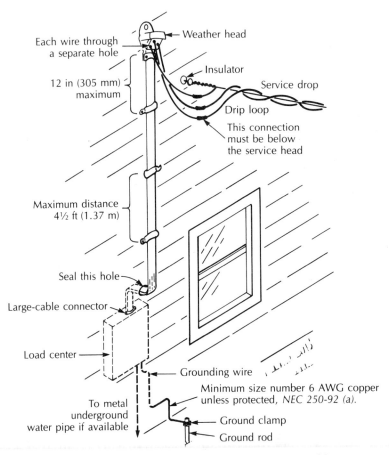

**Figure 4-9** Overhead service entrance using SE cable

sure is a large, open area for animals, the load center can be installed in the area as long as the animals contained within the area cannot endanger the equipment. In some cases, especially where dangerous animals are housed, there is no acceptable inside location for the load center. In these cases, a rainproof load center should be installed on the outside of the building.

## Service Entrance Using SE Cable

Service-entrance cable is usually easier to install than a raceway system. Ease of installation is an important factor when physical damage to the cable from machines and animals is not a problem. A typical farm building service entrance using SE cable is shown in Figure 4-9. The service drop terminates at an insulator attached securely to the building. The following are the requirements for this type of installation.

- The service head or gooseneck must be installed at the upper end of the service-entrance cable, *NEC Section 230-54(b)*.

- The service head shall be located above the point of attachment of the service drop whenever possible, *NEC Section 230-54(c)*.

- The first cable strap shall be not more than 12 in (305 mm) below the service head, *NEC Section 230-51(a)*.

- The service head or gooseneck must be at least 10 feet (3.05 m) above the ground, *NEC Section 230-26*.

- The cable shall be secured with cable straps at intervals of not more than 4½ ft (1.37 m), *NEC Section 230-51(a)*.

- A drip loop shall be formed at the point where the service drop wires attach to the service-entrance cable, and the drip loop shall be below the level of the service head, *NEC Section 230-54(f)*.

- The service-entrance cable shall be one continuous unspliced length from the drip loop to the load center. The cable can be spliced to install a meter socket, *NEC Section 230-46*.

- Do not bend the cable with a radius so small as to damage the wire covering. The rules of *NEC Article*

*336* apply in this case, where the minimum radius of a bend is set at five times the diameter of the cable, *NEC Section 336-10* and *338-4(b)*.

- The load center or disconnect must be located as close as possible to the point where the SE cable enters the building, *NEC Section 230-70(a)*.

If a kilowatthour meter is required, it is installed on the outside of the building according to the directions of the electrical power supplier. The center of the meter socket should be about 5 ft (1.52 m) above the ground. The top of the meter socket has a weatherproof threaded hub, Figure 4-10. A weatherproof SE cable connector is screwed into this hub. After installation, this connector forms a watertight seal around the cable. If the hub is too large for the SE cable connector, a reducing bushing of the proper size can be used. The reducing bushing is screwed onto the connector, then the connector is screwed into the hub. A common, large-cable connector can be used at the bottom of the meter socket where the cable leaves the socket.

Two common types of service-entrance cable are SE Type U and SE Type R, Figure 4-11. SE Type U cable has an oval cross section, and contains two insulated wires. These insulated wires are surrounded by many bare wires. When the cable covering is stripped away, these bare wires will be twisted together to form the neutral wire. SE Type U cable is the type most often used for a farm building service entrance.

SE Type R cable has a round cross section and contains four wires: two hot wires, a neutral, and a bare

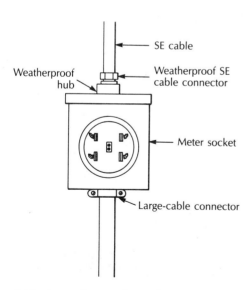

**Figure 4-10** A weatherproof SE cable connector must be used at the top of the meter socket.

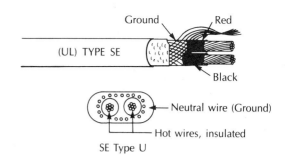

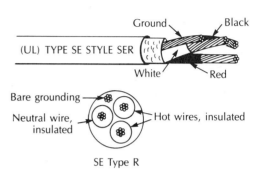

**Figure 4-11** Two types of service-entrance cable are in common use: SE Type U, and SE Type R.

grounding wire. Occasionally, there are times when the neutral wire and the grounding wire must be kept separate. In these cases, Type R cable is used.

The rules governing the use of service-entrance cable are covered in *Article 338* of the *National Electrical Code*. SE cable is permitted for use inside a building for branch circuits and feeders. When used inside, the rules for installing nonmetallic sheathed cable must be followed, *NEC Article 336*.

## Service-entrance Wires in Conduit

The service entrance can be installed using individual wires inside conduit. The conduit serves as protection for the wires. Several types of conduit may be used for a service entrance. However, if the service drop will be attached directly to the riser conduit, it must be strong enough to withstand the weight of the service drop. Common types of conduit used for a service entrance are:

- Rigid steel conduit (*NEC Article 346*)
- Rigid aluminum conduit (*NEC Article 346*)
- Intermediate metal conduit (IMC) (*NEC Article 345*)
- Electrical metallic tubing (EMT) (*NEC Article 348*)
- Rigid nonmetallic conduit (*NEC Article 347*)

Iron and steel are ferrous metals. The *NEC* usually refers to them as ferrous metals rather than calling them iron or steel.

A typical farm building service entrance using electrical metallic tubing is shown in Figure 4-12. The insulation on the wires inside the conduit is usually THW, THWN, or XHHW. The service drop is attached to an insulator securely fastened to the building. The following rules apply to the installation of this type of service entrance.

- The service raceway shall be equipped with a raintight service head, *NEC Section 230-54(a)*.
- The service head shall be at least 10 ft (3.05 m) above the ground, *NEC Section 230-26*.
- The service head shall be above the point of attachment of the service drop, *NEC Section 230-54(c)*.
- A drip loop shall be formed where the service drop connects to the service-entrance wires to prevent moisture from entering the raceway, *NEC Sections 230-54(f) and (g)*.
- Only raintight fittings can be used where the raceway is exposed to the weather, *NEC Section 230-53*.
- The lower end of the service raceway must terminate at a meter socket or the service disconnect, *NEC Section 230-55*.
- The service-entrance conductors shall not be spliced except at a meter socket. The service wires may be tapped, as discussed earlier in this unit, *NEC Section 230-46*.
- All raceway supports, bolts, straps, screws, and similar fittings shall be of a corrosion-resistant material, *NEC Sections 345-5, 346-4, and 348-4*.
- The wires must be protected from abrasion where they enter a meter socket, panelboard, fusible switch, gutter, or similar enclosure, *NEC Sections 346-8 and 373-6(c)*. If the fitting will not provide protection for the wires, a bushing must be used for wires of size No. 4 AWG and larger, Figure 4-13.
- The raceway must be supported within 3 ft (914 mm) of the service head or any enclosure it enters. It must also be supported at specified intervals along the conduit. Closely spaced supports are recommended for a service-entrance raceway, *NEC Sections 345-12, 346-12, and 348-12*. For nonmetallic conduit, support straps must be closely spaced. *NEC Table 347-8* specifies the spacing of supports.
- Conduit entering an enclosure must be securely fastened to the panel, *NEC Section 300-10*.
- The conduit entering the load center or disconnect must be effectively bonded to the grounding system. *Bonding* is the practice of electrically connecting metal parts together. A convenient and effective means of

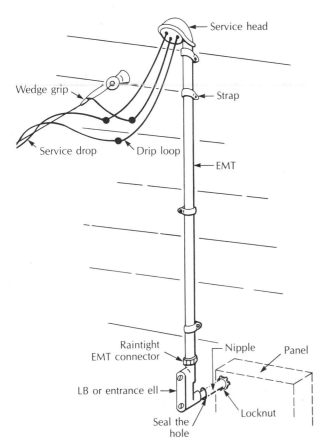

Figure 4-12 A service entrance with wires inside conduit.

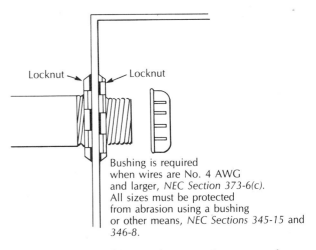

Figure 4-13 A conduit entering an enclosure must be securely fastened to the panel.

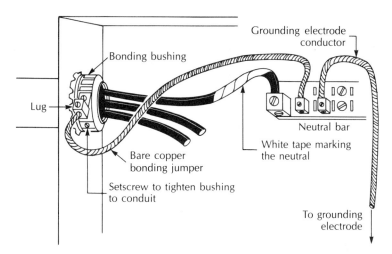

**Figure 4-14** The service conduit entering the load center must be bonded to the system grounding electrode conductor.

providing bonding is by means of a bonding bushing, *NEC Section 250-71(a)*. Figure 4-14.

The riser conduit must be of an adequate size to contain the wires. *Table 1* of *NEC Chapter 9* permits the cross-sectional area of the conductors to be up to 40% of the cross-sectional area of the conduit. *Table 3* of *NEC Chapter 9* is used to determine the size of conduit. This table gives the number of wires of a particular size and type which can be put into the various sizes of conduit. Table 4-7 in this unit gives the suggested wire and conduit sizes for common service-entrance ratings.

A service entrance using conduit is a little more difficult to install than a service using SE cable. More careful planning and measurements are required. If a meter socket does not have to be installed, the task is easier. First, determine the best location for the load center or disconnect switch. The point of attachment of the service drop to the building must be considered when locating the load center.

A surface must be prepared for mounting the load center or other service equipment. The mounting surface must be one which does not collect moisture or retain dampness. If the surface is damp, *NEC Section 373-2* requires a ¼-in (6.35 mm) air space between the enclosure and the wall. It is usually best to mount the service equipment on exterior-grade plywood or similar material. The location of the equipment is marked, and a hole is made through the wall for the conduit. Mount the load center first, and then install the conduit starting at the load center and working toward the service head. Once this equipment is securely in place, the service-entrance conductors are installed.

Measure the total length of wire needed, allowing 24 in (610 mm) of free conductor at the service head.

**Table 4-7** Suggested wire and conduit sizes for farm building service entrances. Wire insulation types are THW, THWN, and XHHW

| Service Disconnect Rating (Amperes) | Copper Wire | | Aluminum Wire | |
|---|---|---|---|---|
| | Wire Size (AWG No.) | Conduit Trade Size (Inches) | Wire Size (AWG No.) | Conduit Trade Size (Inches) |
| 30  | 8   | ¾    | 8       | ¾    |
| 40  | 8   | ¾    | 8       | ¾    |
| 60  | 6   | 1    | 4       | 1 ¼  |
| 70  | 4   | 1 ¼  | 3       | 1 ¼  |
| 100 | 3   | 1 ¼  | 1       | 1 ¼  |
| 125 | 1   | 1 ½  | 0       | 1 ½  |
| 150 | 0   | 1 ½  | 2/0     | 1 ½  |
| 200 | 3/0 | 2    | 4/0     | 2    |
| 225 | 4/0 | 2 ½  | 250 MCM | 2 ½  |

Cut all three wires to the proper length. Large size wires are best cut with a cable cutter or a hacksaw, Figure 4-15. If the neutral wire is the same size as the other wires, mark both ends of the neutral wire with white tape. This is done so that the neutral wire can be readily identified after it is installed. It is usually easiest to insert the wires from the top of the service head. The cap of the service head is removed to allow direct access straight down the conduit, Figure 4-16(A). The wires are pulled out at the LB fitting or service ell. The wires are then installed one at a time back into the LB fitting for the final run to the load center, Figure 4-16(B). This last run may range from a few inches (millimeters) to several feet (meters). A wire-pulling lubricant can be used if the wires are difficult to install. During this process, it is important to prevent damage to the insulation on the wires.

The wire should be stripped carefully with a knife so as not to gouge or nick the wire, Figure 4-17. It is best to cut the insulation with a taper. Strip only enough insulation to allow the wire to fill the lug. Do not allow insulation to enter the lug, and do not leave extra bare wire exposed outside the lug.

Aluminum wire is frequently used for service-entrance wires on farms. Some special care must be taken

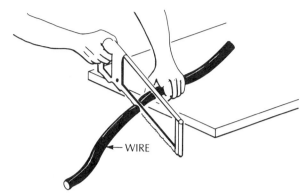

**Figure 4-15  A hacksaw with a 32-tooth-per-inch blade is an excellent method of cutting large sizes of electrical wire and cable.**

to ensure a good, long-lasting connection. First, make sure the lug is rated for aluminum wire. There will be an AL stamped on the lug. Before inserting the aluminum wire into a lug, brush the wire with a steel-bristle brush, and coat it with an oxide inhibitor. This keeps air away from the aluminum and prevents the formation of oxide which causes resistance at the connection. Make sure that the lug is tight. Aluminum tends to relax under pressure, and the connection may loosen slightly during the

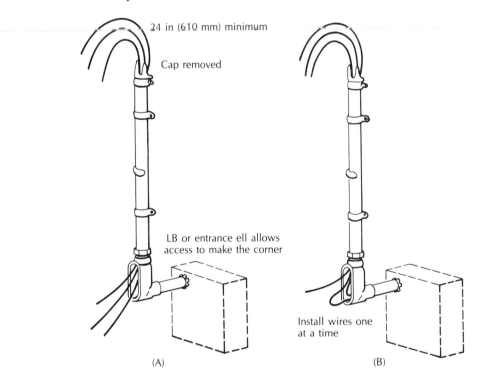

**Figure 4-16  Installing wires in a service conduit.**

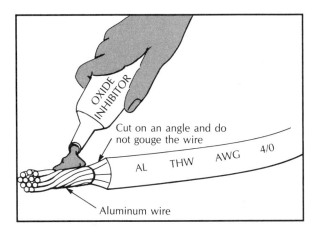

**Figure 4-17** Strip insulation away by cutting it on an angle. When installing aluminum wire, apply an oxide inhibitor before inserting the wire into the lug.

first few hours after it is installed. It is advisable to tighten the lug a second time a few hours after the aluminum wire is installed.

## Service Mast

The service-entrance conduit can be extended up through the roof to gain the needed 10 ft (3.05 m) ground clearance. This installation is called a *mast-type service*. The conduit is often referred to as a *riser*. A typical mast-type service is shown in Figure 4-18.

Strength is the most important consideration in a mast-type service. The riser conduit must be capable of supporting the weight of the service drop. In northern climates ice can build up on the wires, greatly increasing the weight of the service drop. *NEC Section 230-28* allows the use of braces or guy wires for support, if necessary. The service mast should be at least 2-in trade size rigid steel conduit, unless the service drop is quite short and of a small size wire. A service mast should be not smaller than 1¼-in trade size rigid conduit. Table 4-8

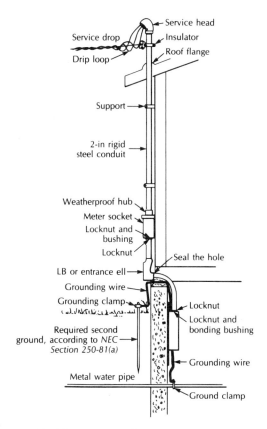

**Figure 4-18** Mast-type service

gives recommended minimum sizes of conduit for use as mast-type service risers for supporting the service drop. The mast also must be connected securely to the building to provide adequate support for the mast and service drop.

The service-drop clearance from the roof must be at least 18 in (457 mm). Therefore, the insulator attached to the riser must be at least 18 in (457 mm) above the point where the mast comes through the roof. This clearance should not be too great, however. The greater the distance between the roof and the insulator, the greater is

**Table 4-8** Recommended minimum trade size rigid steel mast to support a service drop

| | Minimum Trade Size Mast for Service Drop Length | |
|---|---|---|
| **Service Rating** | **Up to 100 ft (30.5 m)** | **More than 100 ft (30.5 m)** |
| Up to 100 | 1 ¼" | 2" |
| 125 to 200 | 2" | 2 ½" |
| 225 | 2 ½" | 3" |
| 300 and up | Structural steel support such as I beam or channel. | |

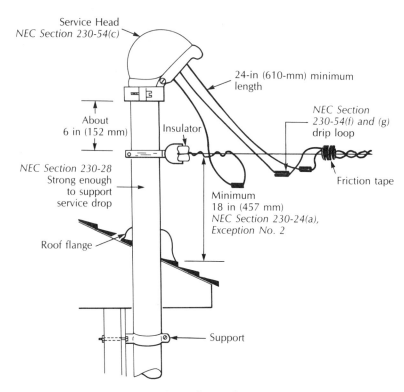

**Figure 4-19** Requirements for a mast-type service above the roof

the leverage of the service drop. This means that the weight of the service drop can bend the mast more easily when the attachment point is 3 ft (914 mm) above the roof, than when it is only 18 in (457 mm). A typical installation is shown in Figure 4-19.

A minimum clearance of 3 ft (914 mm) is required by *NEC Section 230-24(a)* when the service drop passes over more than 4 ft (1.22 m) on the roof. Figure 4-20 shows a typical service installation of this type.

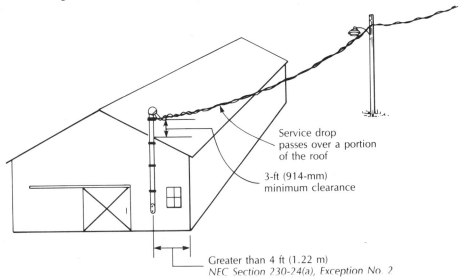

**Figure 4-20** Minimum service-drop clearances over a roof

72   Unit 4   Farm Building Services

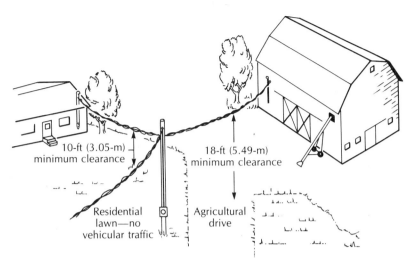

**Figure 4-21** Minimum service-drop clearances in a farmyard

## Service-drop Clearances

The clearances beneath overhead wires are of critical importance on the farm. Field machines and such portable equipment as grain augers and elevators require high clearances. Clearances for overhead wires are given in *NEC Sections 225-18* and *230-24*. In areas where agricultural equipment must pass under the wires, the required minimum clearance is 18 ft (5.49 m), Figure 4-21. An 18-ft (5.49-m) clearance also is required over public roads. Because of the sag in overhead wires, the point of attachment may have to be quite high to obtain the minimum clearance at the low part of the wire.

When many buildings on a farm require power, an overhead wire may pass over a building. A support can be attached to a building if the overhead span is quite long. If the wire simply passes over a building, a minimum clearance of 3 ft (914 mm) is required for a pitched roof, and the wire must be fully insulated. These requirements are in *NEC Sections 230-24(a)* and *230-29*, and are illustrated in Figure 4-22.

The point of attachment of an overhead wire to a building must be out of reach. *NEC Section 230-24(c)* requires that the wire be at least 3 ft (914 mm) horizontally from the edge of a window, door or porch, as shown in Figure 4-23.

The *NEC* does not cover the subject of barn openings to its fullest extent. Overhead wires should be kept much farther away than 3 ft (914 mm) to the side of some openings, and, in fact, should never be attached below certain openings. Materials which are frequently dropped from openings would hit a wire below. Elevating equipment is often maneuvered into barn openings. Wires only 3 ft (914 mm) to the side of the opening may become entangled with this equipment. It is recommended that wires be not closer than 10 ft (3.05 m) to either side of the opening, and 3 ft (914 mm) above it, and never below certain openings, as shown in Figure 4-23.

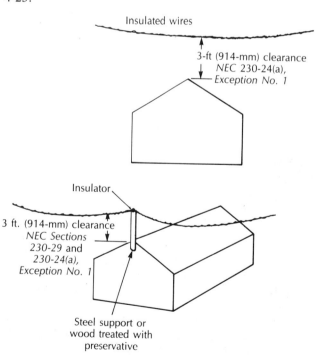

**Figure 4-22** Minimum service-drop clearance for insulated wires over a roof

Unit 4   Farm Building Services   73

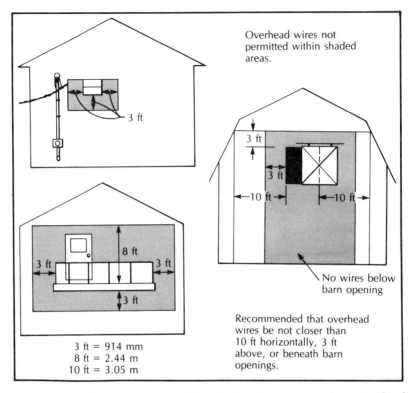

Figure 4-23 Service drops must not be attached to buildings in the shaded areas shown in the diagrams.

## Underground Services

There is a trend towards placing wires underground to avoid clearance problems with tall machines. The entire farm electrical distribution system can be placed underground. The main service can be at a building or at a central point in the farmyard. Usually, the power supplier feeds a central pole overhead, and the electrical installer runs underground feeders to the farm buildings.

Underground wiring to an individual building service entrance is called a *service lateral*. The wire permitted for direct burial to supply a service entrance is Type USE. Usually aluminum wire is used because of its light weight and lower cost as compared to copper wire. Type USE is available as a single wire for direct burial. For a typical single-phase service lateral, three individual wires would be placed in the trench. Type USE is also available as a cable. A typical triplex USE cable is shown in Figure 4-24(A). One of the triplex wires will either be white or it will have a white marker along its length. When individual USE wires are placed in the ground, the neutral wire should be marked with white tape before placing it in the trench. USE cable is described in *NEC Article 338*.

Type UF cable can be used for underground feeders, as permitted in *NEC Section 339-3*. Type UF cable is generally more expensive than Type USE in the larger

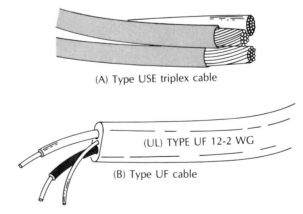

Figure 4-24 Common wires used underground are (A) Type USE triplex cable, and (B) Type UF cable. Type USE individual wires also are used.

74   Unit 4   Farm Building Services

sizes. UF cable is available with two insulated wires and a grounding wire (black, white, bare), and with three insulated wires and a grounding wire (black, red, white, and bare). UF cable is shown in Figure 4-24(B).

The requirements for the direct burial of UF cable and USE wire are given in *NEC Section 300-5*. The depth of burial without supplemental protection of the wires is 24 in (610 mm). The following are the depth requirements.

- 24 in (610 mm) with no protection
- 18 in (457 mm) if under a 2-in (50.8-mm) concrete slab
- 6 in (152 mm) if wires are in rigid conduit or IMC
- 18 in (457 mm) if wires are in nonmetallic conduit
- None if wires are in rigid conduit and under a concrete floor
- 24 in (610 mm), even if in conduit, if under an area subject to heavy traffic, *NEC Section 300-5(a), Exception No. 3*.

The minimum size wire to be used for a service lateral is set by *NEC Section 230-31*. When a limited load is served, such as for a water pump or a building with one branch circuit, the minimum size wire is No. 12 AWG copper or No. 10 AWG aluminum. Type UF cable is often used for this type of load. Normally, the minimum size wire for a service lateral is No. 8 AWG copper or No. 6 AWG aluminum.

Electrical wire installed inside a building must have an outer covering that is flame retardant. Type USE wire is not required to have flame-retardant covering, although most of the USE wire available is flame retardant. It also may carry an additional rating, such as RHW and RHH. If the USE wire is not flame retardant, it must be spliced at the point where it enters the building in an approved junction box. Normal, flame-retardant building wire is used from the junction box to the service disconnect. This splice in the service wires is permitted by *NEC Section 230-46, Exception No. 3*. Usually, a splice is not necessary because the service disconnect is located at the point where the wire enters the building.

The service disconnect must be located as close as possible to the point where the service-entrance conductors enter the building. This requirement is in *NEC Section 230-70(a)*. For an overhead service drop, the service disconnect or load center is located close to the point where the service drop connects to the building. With an underground service lateral, however, the load center can be placed at almost any convenient location within the building. This is accomplished by running the wires in rigid conduit under the floor. The conduit simply comes up through the floor and enters the bottom of the disconnect or load center. This procedure is permitted by *NEC Sections 230-44* and *300-5(c)*, which consider service wires run under a concrete floor to be outside the building.

The service lateral wires must be protected from physical damage when placed in the trench, according to *NEC Section 300-5(f)*. Clean, backfill soil must be placed in the trench adjacent to the wires. This backfill must not contain stones or other foreign material which could damage the wire insulation. Large rocks which are removed from the trench should not be put back into the trench. If there is a chance that foreign material could damage the wires, they should be run in conduit for protection. Providing a conduit sleeve for protection in areas subject to heavy vehicular traffic is recommended.

Service lateral wires must be protected by conduit at points where they emerge from the ground to enter a switch or load center. If the conduit is on a pole or subject to physical damage, then rigid conduit, IMC, or Schedule 80 PVC (rigid nonmetallic conduit) must be used, as required by *NEC Section 300-5(d)*.

A typical service entrance with a meter is shown in Figure 4-25. The power supplier provides a special meter

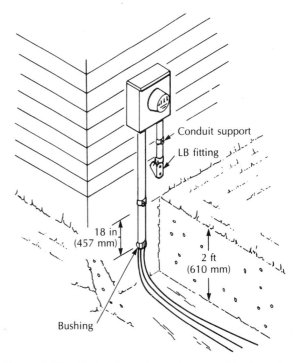

**Figure 4-25 Metered service entrance with the wires coming from underground**

Unit 4    Farm Building Services    75

*NEC Section 230-84.* If only one circuit is provided to an outbuilding, an ordinary snap switch can serve as the disconnect. However, the snap switch must be capable of disconnecting all power in the building, including receptacles. If this is not the case, then a small panelboard or a fusible switch must be provided as the disconnect.

## Fusible Load Centers

Many older farm buildings are equipped with fusible load centers using mainly Edison-base or screw-shell fuses. Typically, these load centers are rated at 60 A and 100 A. A 60-A fusible load center is shown in Figure 4-27. The internal bus connections are not always obvious. A sticker showing the internal connections is affixed to the cover of the load center. Fusible load centers are acceptable, as long as they are in good condition.

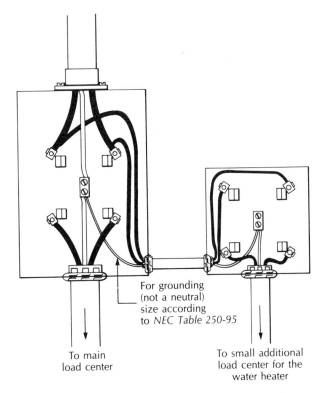

**Figure 4-26** The water heater is metered separately at a special rate. The electric utility supplies wiring instructions.

socket for underground services. The electrician installs the conduit for the service lateral. The electrical power supplier provides and installs the service lateral wires from the transformer to the meter socket. *NEC Section 300-5(h)* requires that a bushing be provided at the point where the wires leave the conduit underground.

## Water Heater Service Connections

Some power suppliers provide separate metering for electric water heaters. The water heater must then be tapped ahead of the main meter. A separate disconnect must also be provided for the water heater circuit where it enters a building. The electrical utility will supply directions for this type of installation. The wiring for a typical water heater installation is shown in Figure 4-26.

## One-circuit Service

A means of disconnecting power is required at every outbuilding, including separate garages, according to

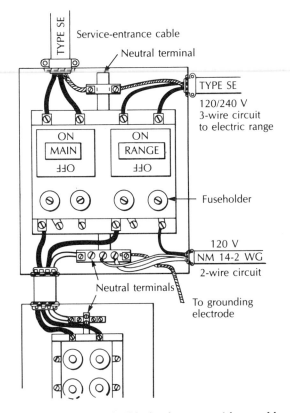

**Figure 4-27** A 60-A fusible load center with an add-on fusible panel

76    Unit 4    *Farm Building Services*

# ELECTRICAL SYSTEM GROUNDING

A wiring system on a farm must have one conductor grounded to the earth. This grounding helps to limit high voltages due to lightning surges. Grounding prevents high voltages from occurring on the wiring if a primary line should accidentally contact a secondary wire. Grounding limits the maximum voltage to ground from the hot wires. For most farm wiring systems, the maximum voltage to ground is 120 V. Grounding aids in the blowing or tripping of overcurrent devices in the event of an electrical fault.

Several decisions must be made in the grounding of an electrical system.

- Which wire should be grounded?
- How is the grounding to be accomplished?
- How is the grounding wire sized?

## Determining Which Wire To Ground

The service-entrance wire required to be grounded is described in *NEC Section 250-5(b)*. Agricultural wiring installations usually involve damp or wet locations. Therefore, only grounded wiring systems should be installed. The following wiring systems are required to be grounded, as shown in Figure 4-28. The following types of electrical services are required to be grounded.

- Single-phase, 2-wire, 120 V
- Single-phase, 3-wire, 120/240 V
- 3-phase, 4-wire, 120/208-V wye
- 3-phase, 4-wire, 120/240-V delta
- 3-phase, 4-wire, 277/480-V wye, when supplying electrically driven irrigation machines
- 3-phase, 3-wire, 240-V or 480-V delta, when supplying electrically driven irrigation machines

The grounded wire must have white insulation. If the wire is larger than No. 6 AWG, it can be labeled with white tape where the wire is visible. This is according to *NEC Section 200-6*. The corner grounded delta electrical system shown in Figure 4-28 does not have an overcurrent device installed in the grounded phase conductor. If, however, the overcurrent device serves as motor running overcurrent protection, an overcurrent device is required to be installed in all phase conductors.

## Grounding Electrode

The *grounding electrode* is the means by which electrical contact is made with the earth. The type of elec-

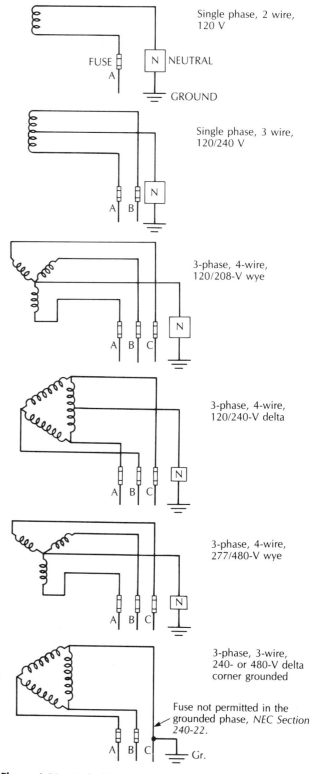

**Figure 4-28** Typical farm electrical systems showing the correct wire to be grounded

Unit 4   Farm Building Services   77

trode which can be used for grounding an electrical system is given in *NEC Sections 250-81 and 250-83*. The following electrodes must be used if available at the farm building.

- A metal underground water pipe in direct contact with the earth for at least 10 ft (3.05 m). *Note:* A metal underground water pipe must be supplemented by at least one additional electrode. (See Figure 4-18.)
- The metal frame of the building, where the metal frame is grounded.
- A bare copper wire at least 20 ft (6.1 m) long, not smaller than No. 4 AWG, encased in a concrete footing.
- A bare copper wire encircling the building, in the earth at a depth of at least 2½ ft (762 mm), No. 2 AWG or larger, at least 20 ft (6.1 m) long.

If more than one of the above electrodes are available at the building, they all must be electrically connected together with a grounding wire to form a grounding electrode system. If one of the above electrodes is not available, then one of the following shall be used.

- An underground gas pipe, if acceptable to the gas utility, or other metal pipe.
- A rod or pipe electrode driven to a depth of at least 8 ft (2.44 m) into the soil. If rock bottom is encountered, it may be driven at an angle not to exceed 45 degrees, or it may be laid in a trench at least 2½ ft (762 mm) deep. (See Figure 4-29.) The acceptable rods and pipes shall have the following minimum trade size diameters.
  1. ¾-in galvanized steel pipe
  2. ⅝-in iron or steel rod
  3. ½-in copper rod
- A plate electrode can be used in areas where soil conditions prevent the use of a pipe or rod electrode. The plate must make contact with at least 2 ft (0.186 m²) of soil. This means a 1-ft² (0.093-m²) plate, counting both the top and bottom surfaces to make 2 ft².

Rod, pipe, and plate electrodes sometimes do not make good low-resistance contact with the earth. They must be placed in areas not subject to damage, but where the soil is most likely to be moist. They should not be installed in areas protected from the weather, such as under roof overhangs. The *NEC* specifies in *Section 250-84* that a single electrode, consisting of a rod, pipe, or plate shall have a resistance to ground not exceeding 25Ω. If not, one additional rod, pipe, plate, or other electrode shall be installed. These electrodes shall be at least 6 ft (1.83 m) apart. Most electrical installers do not have an instrument capable of determining an electrode's resistance to ground. Therefore, good judgment must be used. If soil conditions are normally quite dry, an additional electrode should be provided.

System grounding electrodes for lightning protection are not permitted by the *NEC* to be used as one of the required electrodes for the electrical system. However, electrodes for the electrical and lightning systems should be electrically connected together. This bonding is required by *NEC Section 250-86*.

## The Grounding Electrode Conductor

The *grounding electrode conductor* is the wire run from the service disconnect to the grounding electrode. This wire must be corrosion resistant and suitable for the conditions. The wire can be copper or aluminum, solid or stranded, insulated, covered or bare. Copper grounding wire is preferred because it is less likely to corrode in an agricultural environment. Aluminum conductors are not permitted to be installed in masonry or within 18 in (457 mm) of the earth, *NEC Section 250-92(a)*. *NEC*

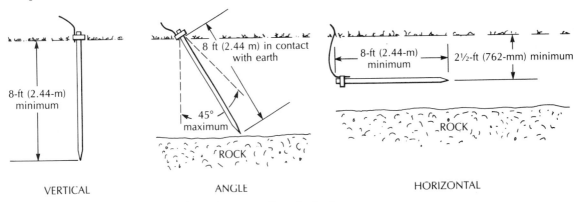

Figure 4-29 Methods of installing a rod or pipe grounding electrode

Table 4-9 (NEC Table 250-94) Grounding electrode conductor for ac systems

| Size of Largest Service-Entrance Conductor or Equivalent Area for Parallel Conductors | | Size of Grounding Electrode Conductor | |
|---|---|---|---|
| Copper | Aluminum or Copper-Clad Aluminum | Copper | *Aluminum or Copper-Clad Aluminum |
| 2 or smaller | 0 or smaller | 8 | 6 |
| 1 or 0 | 2/0 or 3/0 | 6 | 4 |
| 2/0 or 3/0 | 4/0 or 250 MCM | 4 | 2 |
| Over 3/0 thru 350 MCM | Over 250 MCM thru 500 MCM | 2 | 0 |
| Over 350 MCM thru 600 MCM | Over 500 MCM thru 900 MCM | 0 | 3/0 |
| Over 600 MCM thru 1100 MCM | Over 900 MCM thru 1750 MCM | 2/0 | 4/0 |
| Over 1100 MCM | Over 1750 MCM | 3/0 | 250 MCM |

Where there are no service-entrance conductors, the grounding electrode conductor size shall be determined by the equivalent size of the largest service-entrance conductor required for the load to be served.
*See installation restrictions in Section 250-92(a).
See Section 250-23(b).
Reprinted with permission from NFPA 70-1984, National Electrical Code®, Copyright © 1983, National Fire Protection Association, Quincy, MA. This reprinted material is not the complete and official position of the NFPA on the referenced subject which is represented only by the standard in its entirety.

*Section 250-91(a)* requires that the grounding electrode wire be one continuous length without splices.

The grounding electrode wire is sized according to *NEC Table 250-94*, reproduced in this text as Table 4-9. The grounding electrode wire size is based upon the size of the service-entrance wires. For example, a barn has a 150-A service entrance with No. 3/0 AWG THW aluminum service-entrance wires. Moving down the second column in Table 4-9, No. 3/0 aluminum is found in the second row. Moving to the right, No. 6 AWG copper wire is the minimum size permitted as the grounding electrode wire.

There is an exception to *NEC Table 250-94* where the electrical system is grounded to a made electrode such as a rod, pipe, or plate. The effectiveness of these electrodes is limited. Therefore, a maximum size grounding wire is required. *NEC Section 250-94, Exception No. 1*, sets No. 6 AWG copper wire as the maximum required size when the only grounding electrode available is a rod, pipe, or plate. If a bare copper grounding wire is attached to the outside surface of a building and is not protected, No. 6 AWG is the minimum size permitted, *NEC Section 250-92(a)*. (See Figure 4-9.)

All service disconnecting means must be grounded when there are more than one disconnect for the service.

*NEC Section 250-91(a), Exception No. 2*, allows one main grounding wire from the service to the grounding electrode. This grounding wire is sized according to the main service wires. The other separate service disconnects are allowed to be tapped to this main grounding wire. These grounding wire taps are permitted to be sized according to the tap wires feeding each disconnect. A typical example is shown in Figure 4-30.

## Installing Grounding Wire

Two problems must be considered when installing the grounding electrode conductor in an agricultural building.

1. Damage from animals and machines
2. Corrosion

The grounding electrode conductor must be protected in any area where it is exposed to animals. The grounding electrode wire can be enclosed in steel conduit, but steel conduit is likely to corrode in areas where it is exposed to animal waste or chemicals. Nonmetallic rigid conduit is a good alternative, provided the animals cannot chew on the conduit. *NEC Section 250-92(a)* requires that a grounding electrode conductor or the conduit be fastened

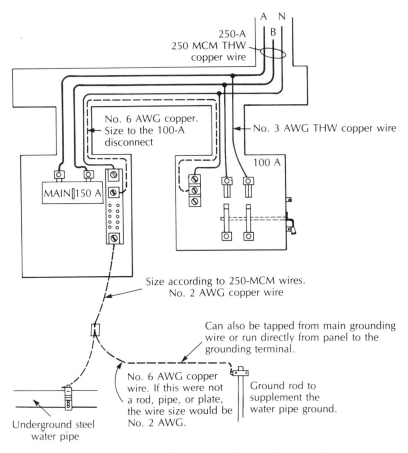

**Figure 4-30** Grounding a service entrance with more than one service enclosure, and more than one grounding electrode

securely to the building. The following materials are acceptable for supplemental protection of the grounding electrode conductor.

- Rigid metal conduit
- Intermediate metal conduit (IMC)
- Rigid nonmetallic conduit (PVC)
- Electrical metallic tubing (EMT)
- Cable armor

The grounding electrode wire shall be installed in raceway or otherwise protected for size No. 8 AWG. A No. 6 AWG grounding wire is permitted to be stapled to the building when not subject to physical damage. Otherwise, the wire must be protected by one of the materials listed earlier. A No. 4 AWG or larger grounding electrode wire must be protected when the wire is exposed to physical damage. The type of protection is not specified in the *NEC*. If not exposed to physical damage, a No. 4 AWG or larger grounding electrode wire need only be fastened securely to the structure.

A steel conduit around the grounding electrode wire affects the impedance (resistance to the flow of current) of the wire. The impedance increases, and the effectiveness of the wire decreases. This problem can be reduced by making sure that steel conduit runs continuously from the service disconnect to the grounding electrode, and is fastened securely to the grounding clamp, or is electrically connected to the grounding wire at each end of the conduit.

Grounding electrode clamps are available with a lug for the wire and a conduit fitting. However, more commonly, the grounding wire may be bonded to the conduit at the point where the conduit ends. These bonding requirements are specified in *NEC Section 250-92(a)*. Rigid nonmetallic conduit does not cause an impedance problem, and bonding conduit to the grounding electrode wire is not necessary.

*NEC Section 250-112* requires that the point of connection to the grounding electrode be accessible. However, this is impractical in the case of farm buildings where animals may be in the area with the grounding electrode. Grounding electrodes must be kept inaccessible to animals to prevent damage to the service grounding system, and to prevent injury to the animals. In the *Exception* to *Section 250-112*, the *NEC* allows for this problem by permitting concrete-encased, driven, or buried grounding electrodes to be inaccessible.

Encasing a grounding wire in a concrete footing is an effective means of providing a ground for a building service. However, it is important to prevent damage to the wire at the point where it enters the concrete. The grounding wire should be sleeved with conduit at the point where it enters the concrete.

*NEC Section 250-115* requires that the connection of the grounding wire to the grounding electrode be made with a pressure-type connecting device. Solder connections are not permitted. This connection may be buried; however, the device must be approved for direct burial.

## Bonding Service Equipment

Recall that bonding is the electrically connecting together of noncurrent-carrying metal parts of the electrical system. *NEC Section 250-71* requires that essentially every noncurrent-carrying metal part of the service entrance be bonded together to form a single conducting path. These parts include the service raceway, the meter socket, the cabinet of the load center or disconnect switch, the conduit protecting the grounding wire, and similar metal parts. The bonding together of these metal parts can be accomplished by any of the means specified in *NEC Section 250-72*.

- Bonding jumper connected at both ends with a pressure-type device, Figure 4-31
- Threaded couplings and hubs made up wrenchtight for rigid metal conduit and IMC
- Threadless couplings and connectors for rigid metal conduit, IMC, and EMT
- A bonding jumper shall be provided around concentric knockouts still in place, Figure 4-32
- Bonding bushings, bonding locknuts and similar devices. (See Figure 4-14.)

The electrical equipment grounding conductor shall be bonded to the grounded circuit conductor at the main farm disconnect. This connection shall be made on the supply side of the disconnect. This bonding is required by *NEC Section 250-50*. For most farm electrical systems, the grounded circuit conductor is the neutral. In the case of a corner grounded, 3-wire delta, 3-phase system, one of the phase conductors is the grounded circuit conductor. Without this bonding connection, any ground fault current would not have a direct path back to the transformer except through the earth. A good path back to the transformer is necessary if overcurrent devices are to operate quickly in the event of a fault.

The connection between the equipment conductor and the grounded circuit conductor is called the *main bonding jumper*. *NEC Section 250-53(b)* discusses the main bonding jumper. This jumper connects the equipment grounding conductors and the service-equipment enclosure to the grounded circuit conductor, Figure 4-33. This is generally accomplished by connecting the neutral terminal lugs to the service-equipment box. A wire, screw, or similar device is used to make this connection. Figure 4-34 shows a method of bonding the load center enclosure to the neutral bar.

Bonding a service entrance consisting of more than one enclosure, such as shown in Figure 4-35, may be

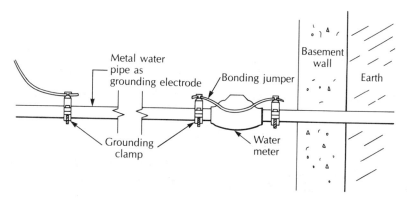

**Figure 4-31** Devices and fittings which break the electrical continuity of the grounding electrode or grounding conductor must be bonded.

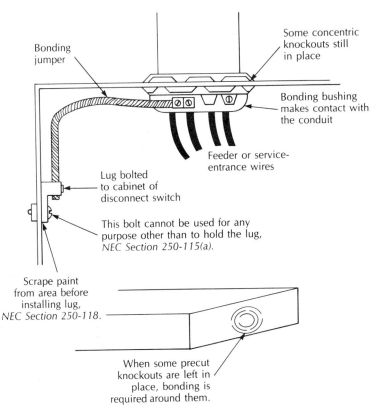

**Figure 4-32** Bonding must be provided around concentric knockouts still in place in the service enclosure case.

confusing. Bear in mind that there are several acceptable means of bonding service enclosures, in addition to the method shown in Figure 4-35. All metal service enclosures are required to be electrically connected together in a manner described in *NEC Section 250-72*.

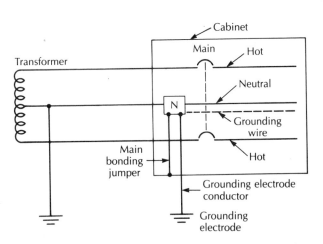

**Figure 4-33** The main bonding jumper connects the grounded circuit conductor to the service enclosure.

Ordinary locknuts, one inside and one outside, connecting a metal service conduit to an enclosure are not acceptable for bonding the service conduit to the enclosure, *NEC Section 250-76*. A threaded hub bolted to the cabinet, a bonding bushing, or a bonding-type locknut is usually used for this purpose. The installation in Figure 4-35 shows **bonding bushings**.

The purpose of the bonding bushing on the service conduit and each nipple is to bond the conduit and nipples to the grounding electrode conductor. These bonding bushings do not bond the auxiliary gutter. Therefore, a lug is bolted to the gutter. One continuous bonding wire is used as a matter of convenience. The bonding wire in this case must be the same size as the service grounding electrode conductor.

## Bonding the Grounding Wire to the Grounded Wire

Several buildings on a farm are supplied with a single electrical system. *NEC Section 250-23(a)* requires that

**Figure 4-34** A bonding screw is sometimes provided in a load center as the main bonding jumper.

the electrical service be grounded, as discussed earlier in this unit. This grounding is required to take place at the main disconnect for the farm. The grounded circuit wire and the grounding wires are connected together at the main service, as shown in Figure 4-33. *NEC Section 250-23(a)* also states that grounding connections shall not be made on the load side of the service disconnect. This means that the grounded circuit conductor, which is usually the neutral, and the grounding conductor are not permitted to be connected together after leaving the main service.

As an example of grounding at outbuildings, consider a single-phase, 120/240-V, 3-wire system. The grounded circuit wire is the neutral. Four wires would have to be run to each building if the neutral and the grounding wire were to be kept separate (two hot wires, one neutral, one grounding). However, common practice is to treat each building as though it is the main service, and run only three wires (two hot wires, and a neutral). The neutral wire also acts as the grounding wire.

The requirements for service wires from the main farm service to the individual buildings are covered in *NEC Section 250-24*. Two options are available. In the case of a single-phase, 120/240-V system, a 3-wire set of conductors is permitted. However, the neutral wire is required to be bonded to the grounding electrode at the outbuilding main service disconnect. If this system is properly sized and installed, it will provide long-term dependable service. However, if some resistance does develop in the neutral, voltage drop will occur, giving

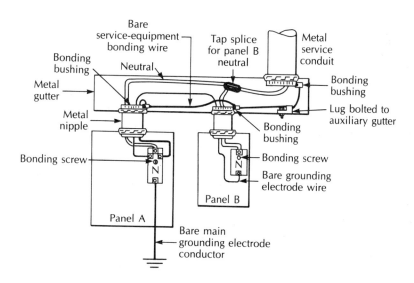

**Figure 4-35** The metal enclosures are electrically connected with a continuous bonding wire terminating at the neutral in panel A. The panel enclosures are bonded by means of the bonding screws. To simplify the diagram, the ungrounded wires are not shown.

Unit 4  Farm Building Services  83

rise to neutral-to-earth voltage (stray voltage) at one or more buildings on the farm. Figure 4-36 illustrates a 3-wire, single-phase electrical system with a small amount of resistance at the outbuilding neutral termination. A 1.7 neutral-to-earth voltage is produced at the outbuilding when a 20-A load is carried on the neutral.

*NEC Section 250-24(a), Exception No. 2,* permits a separate equipment grounding conductor in addition to the neutral conductor as an alternative to the 3-wire, single-phase system. This 4-wire, single-phase system is illustrated in Figure 4-37. The neutral is not bonded to the grounding electrode at the outbuilding. If neutral voltage drop occurs as in the case of Figure 4-36, it does not affect the equipment grounding conductors because they are not connected to the neutral. Animals making contact with grounded equipment are therefore not exposed to this source of stray voltage.

*NEC Section 250-24(a), Exception No. 2,* requires in the case of livestock buildings that the equipment grounding conductor entering the building service with the other service conductors must be insulated or covered copper wire. In the case of an overhead service drop to the building, aluminum quadruplex cable may be used with aluminum conductors entering the building. But, the grounding wire from the drip loop down to the main disconnect must be insulated or covered copper.

It is recommended that service drops and service laterals to all outbuildings be run with separated neutral and equipment grounding conductor. Simply running a 4-wire, single-phase system to the livestock barn will not guarantee the prevention of stray voltage due to voltage drop. This is illustrated in Figure 4-36. Note that a neutral-to-earth voltage occurs at the main service and transformer when there is excessive voltage drop on the neutral to the outbuilding. It is possible for voltage drop on a 3-wire feeder to the dwelling to cause a stray voltage at the barn even though the barn is served with a 4-wire system.

*NEC Section 250-24(a), Exception No. 2,* also requires that in the case where a separate equipment grounding conductor is run to an outbuilding, it must be bonded to any existing grounding electrodes at that outbuilding. If, for example, there is a metal underground water piping system, or the metal frame of the building is in good electrical contact with the earth, then these electrodes must be bonded to the equipment grounding conductor.

## Bonding Metal Piping Systems

Metal piping systems inside farm buildings must be bonded to the electrical service grounding system, *NEC Section 250-80*. This is to prevent injury to people or animals in the event a fault in some equipment or a circuit accidentally energizes the piping system. The interior metal water piping system must always be bonded to the service ground. The bonding jumper must be connected directly to the water pipe and then run to one of the following:

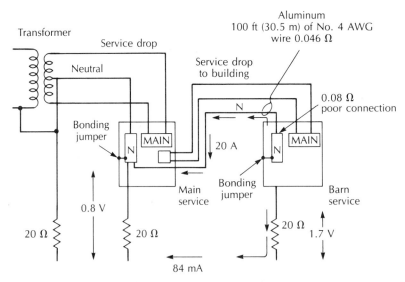

Figure 4-36  When a 3-wire feeder connects the main service to the outbuilding service, a small, harmless amount of current will flow through the ground when a load is placed on the netural.

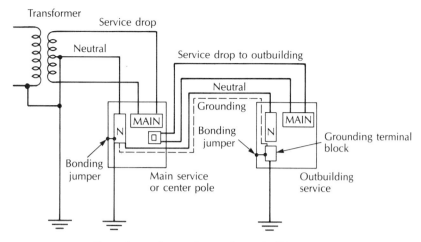

**Figure 4-37** Adding a separate grounding wire to the service to the outbuilding, and not bonding the neutral to the ground at the outbuilding panel, will prevent current from flowing through the ground.

1. The service equipment enclosure
2. The grounded conductor at the service
3. The grounding electrode conductor
4. One of the grounding electrodes

The bonding jumper shall be sized according to *NEC Table 250-94*, which means essentially that the jumper is the same size as the grounding electrode conductor.

Other interior metal piping systems, such as metal vacuum lines, shall be bonded if they could become energized. If a pipe makes good metal-to-metal contact with grounded equipment, that is considered sufficient to bond the pipe. In some cases, a bonding jumper of copper wire may be necessary. The bonding jumper is sized according to the rating of the circuit which could energize the pipe. *NEC Table 250-95* is used to size equipment grounding wires and pipe bonding jumpers.

## *Lightning Arresters*

A high-voltage surge of electricity may be set up in power lines when they are struck by lightning. These surges may even be set up when lightning strikes near the wires. The *lightning arrester* connects to each hot wire entering a building and the grounding wire. If a surge tries to enter the building on the hot wires, it will be drained directly to ground through the lightning arrester. The lightning arrester can be installed outside at the drip loop, Figure 4-38, or it can be installed on the load center or disconnect.

## *SIZING THE FARM DWELLING SERVICE ENTRANCE*

Service-entrance conductors, *NEC Sections 230-42(b), (1),* and *(2),* and the disconnecting means, *NEC Section 230-79(c),* shall be not smaller than 100 A, 3 wire.

1. for a single-family dwelling with six or more 2-wire circuits.
2. for a single-family dwelling with a computed load of 10 kVA or more.

If an electric range is installed, it will have a demand load of 8 kVA. Thus, only 2 kVA are available for remaining appliances, lighting, and special circuits. There are few single-family farm dwellings for which a service smaller than 100 A can be installed. There are two methods in the *National Electrical Code* for determining the size of service-entrance conductors for a dwelling.

The size of the service entrance for the farm dwelling described here will be determined using the two methods in *NEC Article 220*. The farm dwelling has 1 800 ft$^2$ (167.4 m$^2$) of living area and contains the following appliances:

| | |
|---|---|
| Electric range | 240 V; 14 kW |
| Microwave oven | 12 A; 120 V |
| Electric water heater | 240 V; 6 kW |
| Dishwasher | 120 V; 1.8 kW |
| Clothes dryer | 240 V; 5 kW |
| Water pump | 8 A; 240 V |
| Food waste disposer | 7.2 A; 120 V |
| Baseboard electric heat | 240 V; 15 kW |
| Air conditioners (3) | 8 A; 240 V |

**Method No. 1** *(NEC article 220, Parts A and B)*

General lighting load,
*NEC Section 220-2(b):*
   1 800 ft² × 3 VA/ft²                    = 5 400 VA
   (167.4 m² × 32.2 VA/m²)

Small appliance circuits:
*NEC Sections 220-3(b)(1)*
*and 220-16(a):*
   2 circuits × 1 500 VA                     = 3 000 VA

Laundry circuit,
*NEC Sections 220-3(c)*
*and 220-16(b)*
   1 circuit × 1 500 VA                      = 1 500 VA
                            Total:        9 900 VA

Apply demand factors from *NEC Table 220-11*.
   First 3 000 VA at 100%                    = 3 000 VA
   Remaining 6 900 VA at 35%            = 2 415 VA
         Lighting and small appliance total:           5 415 VA

Electric range, 14 kW:
   Apply demand factor from *NEC Table 220-19*.
Because size exceeds 12 kW,
see *Note 1* to *NEC Table 220-19*.
   First 12 kW (from *Column A, NEC Table 220-19*)    = 8 000 VA

14 kW exceeds 12 kW by 2 kW
2 kW × 0.05 = 0.1
0.1 × 8 kW                                  =   800 VA
                        Range demand load:                  8 800 VA

Electric space heating, *NEC Section 220-15*:
   15 kW at 100%                                       15 000 VA

Air conditioning:
   3 units × 8 A = 24 A                                   0 VA
   24 A × 240 V = 5 760 VA
This value is less than the heating load;
therefore, according to *NEC Section 220-21*,
it can be omitted.

Clothes dryer,
*NEC Section 220-18*:
   5 kW at 100%                                         5 000 VA

Other appliances,
*NEC Section 220-17:*

| | |
|---|---|
| Microwave oven; 12 A × 120 V | = 1 440 VA |
| Electric water heater | = 6 000 VA |
| Dishwasher | = 1 800 VA |
| Water pump; 8 A × 240 V × 1.25 *(NEC Section 430-22)* | = 2 400 VA |
| Food waste disposer; 7.2 A × 120 V | = 864 VA |
| Total other appliances: | 12 504 VA |

Since there are more than four appliances in this group, a demand factor of 75% can be applied:

12 504 VA × 0.75 = 9 378 VA

Total demand load: 43 593 VA

$$\text{Total feeder demand load} = \frac{43\ 593\ \text{VA}}{240\ \text{V}} = 182$$

The service equipment would be sized to 200 A. *Note 3* to *NEC Tale 310-16* indicates that No. 4/0 AWG aluminum THW wire is acceptable for this single-family dwelling.

## Sizing the Dwelling Neutral

*NEC Section 220-22* states that the neutral load shall be the maximum unbalance of the load as determined by the method of *Article 220*. The maximum unbalanced load shall be the maximum connected load between the neutral and any one ungrounded conductor.

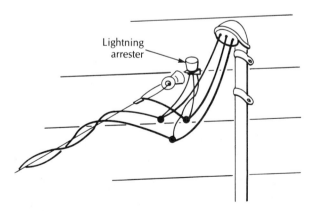

**Figure 4-38 A lightning arrester helps to prevent lightning surges from entering a building on the wiring system.**

It is evident that much of the dwelling load is due to 240-V appliances. Thus, when only these appliance are operating, no current flows on the neutral wire. The size of the neutral wire depends upon the amount of the 120-V load. The determination of the neutral size for the farm dwelling example appears on page 88.

The neutral to be installed must be capable of carrying 81 A. Referring to *NEC Table 310-16*, a No. 2 AWG aluminum wire will carry the neutral current.

Many electrical installers prefer to reduce the neutral wire one or two sizes below the size of the hot wires. This results in a neutral wire which is more than adequate to carry the load. The installer must check local and state codes, however, which may have stricter requirements on sizing the neutral than does the *National Electrical Code.*

**Method No. 2** *(NEC Section 220-30)*

A second method of determining the load for a single-family dwelling is given in *NEC Section 220-30*. This method simplifies the calculations. In most cases, the calculated demand load will be smaller than it is when using the first method. With this method, the total connected load is determined first, and then demand factors are applied.

The load center for this a single-family farm dwelling could be sized at 150 A with No. 2/0 AWG THW aluminum wires. The neutral load is determined by using the method described previously.

Minimum Size Neutral (*NEC section 220–22*)

| | | |
|---|---|---|
| General lighting and small appliance load after applying the demand factor: | | 5 415 VA |
| Range load, *NEC Section 220-22*:<br>8 800 W × 0.70 | | 6 160 VA |
| Clothes dryer, *NEC Section 220-22*:<br>5 000 W × 0.70 | | 3 500 VA |
| Other appliances:<br>  Microwave oven<br>  Dishwasher<br>  Food waste disposer,<br>  *NEC Section 430-22*:<br>  7.2 A × 120 V × 1.25 | = 1 440 VA<br>= 1 800 VA<br><br><br>= 1 080 VA | |
| Total other appliances: | | 4 320 VA |
| Total: | | 19 395 VA |

$$\text{Neutral demand load} = \frac{19\,395\text{ VA}}{240\text{ V}} = 81\text{ A}$$

**Method No. 2** (*NEC Section 220–30*)

| | | |
|---|---|---|
| Space heating: four or more separately controlled units (*Table 220-30(4)*):<br>15 000 W × 0.4 | = 6 000 VA | 6 000 VA |
| Air conditioning<br>3 units × 8 A × 240 V | = 5 760 VA | 0 VA |
| Other loads: | | |
| General lighting load:<br>1 800 ft² × 3 VA/ft²<br>(167.4 m² × 32.3 VA/m²) | = 5 400 VA | |
| Small appliance circuits:<br>2 circuits × 1 500 VA | = 3 000 VA | |
| Laundry circuit:<br>1 circuit × 1 500 VA | = 1 500 VA | |
| Electric range<br>Clothes dryer<br>Microwave oven<br>Electric water heater | = 14 000 VA<br>= 5 000 VA<br>= 1 440 VA<br>= 6 000 VA | |

                Dishwasher                    =  1 800 VA
                Water pump                    =  2 400 VA
                Food waste disposer           =    864 VA
                                                 41 404 VA

First 10 kVA at 100%                                        10 000 VA
Remaining 31 404 VA at 40%                                  12 562 VA
                           Total calculated demand load:    28 562 VA

$$\frac{28\ 562\ \text{VA}}{240\ \text{V}} = 119\ \text{A}$$

# REVIEW

Refer to the *National Electrical Code* when necessary to complete the following review material. Write your answers on a separate sheet of paper.

1. Decide whether each of the following completion statements is correct or incorrect:

    The purpose of an electrical service entrance is to provide
    a. a common point for grounding.
    b. a convenient, easy-to-see method of disconnecting power.
    c. electrical energy metering for the utility.
    d. a means of getting electrical power into a building.
    e. a method of limiting the current which can be drawn.
    f. a balanced load for the transformer.

2. The method of computing a farm building demand load for the purpose of sizing the service entrance is found in which *NEC* section?

3. A farm building has an undiversified load of 74 A and a diversified load of 88 A. What is the building total demand load?

4. Determine the demand load for a hay barn containing eight receptacle outlets, which are used only occasionally, and sixteen lighting outlets. There are a total of three 120-V circuits. Answer the following.
    a. Load operating without diversity: ___?___ A at 120 V; ___?___ A at 240 V.
    b. Load operating with diversity: ___?___ A at 120 V; ___?___ A at 240 V.
    c. Total demand load: ___?___ A at 240 V.

5. A grain-dryer system is supplied by a 120/240-V, single-phase electrical system. Determine the electrical demand load if the system contains the following electrical equipment. (Consider all equipment except e and h to operate without diversity.)
    a. Dryer fan                       15 hp; 240 V; 68 A
    b. Portable loading auger          5 hp; 240 V
    c. Stirring auger in bin           2 hp; 240 V
    d. Distributor on top of
       bin                             ⅓ hp; 240 V
    e. Unloading auger not
       operated during drying          3 hp; 240 V

f. One mercury-vapor
   lamp                         175 W; 120 V; 1.8 A
g. Six PAR-38 flood
   lamps                        150 W; 120 V
h. Circuit for tools and
   portable equipment           20 A; 120 V; 16 A

6. An electrical service entrance can consist of how many separate disconnects, according to which *NEC* section?

7. A farm building consists of a 100-A load center and two 60-A fusible disconnect switches. The electrical demand load for the building is 160 A. Determine the size of the main service-entrance conductors. Refer to the following diagram.

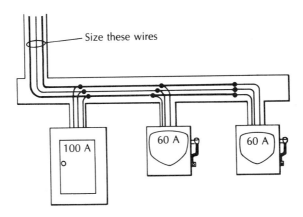

8. The electrical demand load for a farm building is determined to be 163 A. Size a load center for this service entrance.

9. A 150-A load center is to be installed in a farm residence. What is the minimum size aluminum wire for the ungrounded service-entrance conductors?

10. The minimum size THW aluminum neutral wire for a service entrance is specified in which *NEC* section?

11. A farmer has a small building serving several silos for feeding beef cattle. The silos are served by a 200-A, 120/240-V, single-phase service entrance. The ungrounded service-entrance conductors are No. 3/0 AWG XHHW copper. If this service has only two 20-A, 120-V circuits, then what is the minimum size copper neutral wire required?

12. A hog barn contains eight 20-A, 120-V circuits with a load of 1 500 W each. The building also contains a 1/2-hp, 120-V electric motor, 9.8 A; and a 1/3-hp, 120-V electric motor, 7.2 A. The service entrance is 100 A, 120/240 V, single phase with No. 3 AWG THW copper ungrounded wires. Determine the minimum size neutral wire for the service.

13. A farm building 3-phase service entrance has two No. 3/0 AWG XHHW aluminum wires for each phase (as shown in Figure 4-8). What is the ampacity for each phase? Paralleling service-entrance wires is permitted by which *NEC* section?

14. The drip loop is required to be below the level of the service head, according to which *NEC* section?

15. Service-entrance cable, SE, is available in two types. Which type of SE cable has two insulated hot wires and a bare neutral wire?

16. When making a termination with aluminum wire, the lug must be rated for aluminum, and the wire should be coated with ___?___ to prevent corrosion before inserting the wire into the lug.

17. The insulator must not be attached to a mast-type service riser closer than a minimum of ___?___ in (mm) to the roof.

18. If a service entrance is grounded to a metal underground water pipe, then a second grounding electrode is required, according to which *NEC* section?

19. What is the minimum size copper grounding electrode wire for the service entrance described in Problem 11 of this review?

20. In the following diagram, draw in (in pencil) the grounding wire for panel B. The main grounding electrode wire and bonding jumpers are already shown.

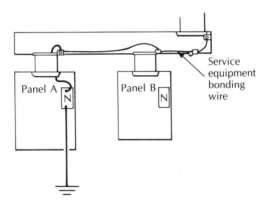

# UNIT 5
# FARM MAIN SERVICE ENTRANCE

## OBJECTIVES

After studying this unit, the student will be able to

- discuss the types of farm main electrical services.
- determine the total farm demand load when the individual building demand loads are known.
- describe the materials required for a farm central pole electrical distribution center with overhead feeders.
- describe the materials required for a farm central pole electrical distribution center with underground feeders.
- explain the advantage of current transformer metering.
- size the wires and conduit required for a farm central electrical distribution center.

The farm electrical system is unique in that one electrical service generally serves several separate buildings or loads. Each building in itself is treated as an individual electrical service. A common practice many years ago was to provide electrical service to the residence. The main panelboard or switch in the residence served as the main disconnect for the farm. The farm buildings received their electrical supply from the main service in the residence. Electrical loads in these early farm buildings were generally small; therefore, the residential service provided an economical and efficient farm electrical service. The electrical power supplier preferred this arrangement because only one transformer and one meter were required for the entire farm.

The residential service equipment serving as the entire farm service is still a good choice where the load in the outbuildings is quite small. However, since this method would be inadequate for the many large farms, an alternative farm service must be provided which is capable of supplying several hundred amperes.

## SIZING THE FARM MAIN SERVICE

The farmer must do some careful planning whenever the farm main service is to be rewired. The electric power supplier must also be involved before making decisions. The farmer should try to anticipate the number of buildings and required equipment which may be added within the first five-year period. The following are some considerations:

1. Single phase or three phase
2. One service for the entire farm, or separate service for some loads
3. Location of the various buildings and loads
4. Electrical energy rates

The first step in determining the size and type of main electrical service for the farm is to determine the demand load and size of service for each building. This procedure is covered in Unit 4. The *National Electrical Code* specifies the procedure for determining the size of the main service in *Part D* of *Article 220*.

*NEC Section 220-41* states that the total load of the farm for the purpose of sizing service-entrance conductors and service equipment shall be computed in accordance with *NEC Table 220-41*. An adaptation of this table is shown in Table 5-1 of this text.

The calculated demand load for each building is used to determine the size of the farm main service, not the size of the building service entrance. However, the size of the service should take into consideration the number of buildings or loads which may be added in the future. When two buildings have the same function, they must be added together as a single load when sizing the main service. For example, a farmer may have two identical poultry egg-laying houses, each with a demand load of 84 A. The two must be considered as a single poultry-house load at 168 A.

## Problem 5-1

A farm consists of a residence with a demand load of 62 A; a grain-drying and storage center, 58 A; a shop and machinery storage building, 38 A; a livestock barn, 30 A; and a silo and electric feeder, 42 A. Determine the total farm demand load.

## Solution

| | | |
|---|---|---|
| Grain-drying and storage center: | 58 A at 100% = | 58 A |
| Silo and feeder: | 42 A at 75% = | 32 A |
| Shop and machinery storage: | 38 A at 65% = | 25 A |
| Livestock barn: | 30 A at 50% = | 15 A |
| Dwelling: | 62 A at 100% = | 62 A |
| Total farm demand load: | | 192 A |

The farm main service must be not smaller than the calculated total demand load of 192 A. A 200-A main service would be the minimum practical size. A larger service would be more practical, because this farm load is likely to increase in the future.

## MAIN SERVICE AT A SINGLE BUILDING

The farm main service can be provided at a single building. The service-entrance equipment and service-entrance conductors must have sufficient capacity to handle the total farm demand load. The first decision is to select the type of service equipment. Some choices include the following:

1. Panelboard or load center with main fuse or circuit breaker
2. Circuit breaker or fusible switch
3. Group of switches or circuit breakers
4. Circuit breaker panelboard or fusible panelboard

The service disconnecting means may consist of up to six separate disconnecting switches or circuit breakers, *NEC Section 230-71(a)*. The disconnects must be grouped and labeled as to the load each switch serves, *NEC Section 230-72(a)*.

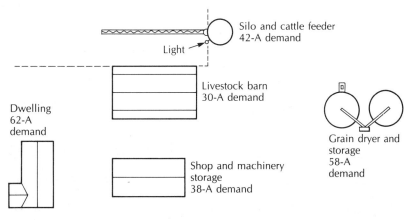

**Problem 5-1**

**Table 5-1** *(Adapted from NEC Table 220-41)* Method of determining total farm load

| Individual Farm Building Demand Loads | Demand Factor |
|---|---|
| Largest farm building load | 100% |
| Second largest farm building load | 75% |
| Third largest farm building load | 65% |
| Remaining farm building loads | 50% |
| Dwelling load | 100% |

If there are more than one separate disconnect, each disconnect may be served by separate service-entrance conductors, *NEC Section 230-45*. A single set of conductors may serve all disconnects. Where one set of service conductors supplies several disconnects, the ampacity of the service-entrance conductors shall be adequate to handle the computed load and should be equal to the combined rating of the individual disconnects.

## Problem 5-2

A farm service consists of a panelboard without a main breaker. Instead, the panelboard consists of a 100-A breaker and two 50-A breakers. Determine the ampacity and size of the service-entrance conductors.

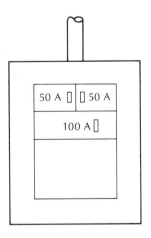

**Problem 5-2**

## Solution

Ampacity = 100 + 50 + 50 = 200 A

If the conductors are THW aluminum, the size is determined from *NEC Table 310-16*. No. 4/0 AWG aluminum with THW insulation has an ampacity of 180 A. This wire can be used as long as the demand load does not exceed 180 A. If the total farm demand load exceeds 180 A, then the next larger size wire, which is 250 MCM rated at 205 A, must be used. No. 4/0 AWG aluminum wire rated at 180 A could be used for the 200-A service because of *Exception No. 1* of *NEC Section 240-3*. When the ampacity of the wire does not correspond to a standard size overcurrent device, the next higher standard size overcurrent device is permitted. However, the wire must always have an ampacity sufficient for the load to be supplied. A 4/0 aluminum THW service-entrance conductor may be terminated at a 200-A overcurrent device, provided the load served is not greater than 180 A.

> **CAUTION:** Where a panelboard does not have a single main breaker, but, instead, has two to six overcurrent devices which act as the mains, panelboard overloading can be a problem. The panelboard must have a rating sufficient to serve the intended load, *NEC Section 384-13*. For example, a 400-A fusible panelboard used as service equipment can contain up to six switches that serve loads that do not add up to more than 400 A. To ensure protection of the panelboard, the sum of the overcurrent devices used as mains should not exceed the rating of the panelboard.

100 + 100 + 60 + 60 + 30 + 30 = 380 A

## CENTRAL POINT DISTRIBUTION

Central point distribution is a common method of distributing electrical power on farms. This distribution is usually made from a central pole, sometimes called a "maypole," located in the farmyard. The electrical power supplier's service drop provides power to the pole. The kilowatthour meter is attached to the pole, and feeders extend from the pole to each farm building. All of the electrical energy used on the farm is metered at the central pole.

Central point distribution should be considered on farms where

- the buildings to be served are widely separated.
- there is a large electrical demand load at two or more locations.
- farm buildings are on opposite sides of a public road or highway.

Central pole distribution has several advantages over a single service in a building. First, the cost of the service equipment and service-entrance wires is less. The central point distribution can be located closest to the largest loads to minimize the length of large sized feeders. Second, feeder lengths are generally shorter, thus reducing voltage drop. Third, a fire or interruption of electrical service at one building is less likely to affect the service at other buildings. Fourth, if the size of the equipment must be increased, it is less expensive to upgrade the service on the central pole than it is to upgrade a main service located in a building.

## Central Distribution Pole without Disconnect

The central distribution pole must meet the specifications of the electrical power supplier. Usually, the electrical power supplier will set the pole and, if there is a charge for this installation, the farmer will be billed directly by the power supplier. The electrician installs the equipment on the pole, according to the specifications of the electrical power supplier. Some power suppliers permit the use of service-entrance cable attached directly to the pole. Many power suppliers prefer that the installation be made with conduit. Generally, the service equipment is attached to the pole with standoff brackets, as shown in Figure 5-1. Complete central pole assemblies can be purchased, or the electrician can fabricate a central pole meter assembly.

Farm central electrical distribution pole installations are not specifically covered by the *National Electrical Code*. Therefore, a disconnect and overcurrent protection are not required at the pole, unless required by the local authority. Grounding is required because all metal enclosures for electrical wires are required to be

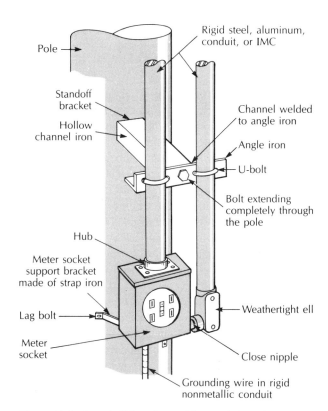

**Figure 5-1** Standoff brackets, either purchased or built, mount the metering assembly to the pole. Meter sockets may be available with two hubs on the top.

grounded. The power supplier's specifications also call for grounding at the central pole.

With regard to farm loads, the *National Electrical Code* does make reference to a central distribution pole as an *other structure*. *Section 230-21* refers to a pole upon which is mounted a disconnecting means *or* meter. It then may be assumed that a meter only, with no disconnect switch, is acceptable to the Code.

## Making Feeder Taps at the Central Pole

The load wires leaving the meter at the central pole are tapped or spliced for each feeder supplying power to a farm building. Several taps or splices are made at the central pole for a typical farm. These splices are made at the top of the pole for overhead feeders, and at the bottom of the pole for underground feeders.

Tap splices are easy to make at the top of the pole for feeding buildings with overhead wires. Splices are made in the open, because there is no danger of electrical shock or damage to the wires. However, it is important to arrange the wires so that they will not rub against one another, or against the pole or the electrical equipment on the pole. Remember that wind causes small movements of these wires.

The conduit containing the wires from the load side of the meter terminates below the conduit containing the supply wires to the meter. This is generally a requirement of the power supplier. The load wires to the various buildings are simply spliced together with split-bolt connectors or a similar pressure-type connecting device. The splices of the ungrounded (hot) wires are wrapped with tape to provide insulation and to prevent moisture from entering the splice. It is recommended that the splice be wrapped first with rubber electrical tape, and then covered with friction tape. This will provide a watertight seal which lasts indefinitely. Special electrical tape is also available for this purpose.

Tapping the load-side wires to extend to the outbuildings is more difficult when the feeders are to be run underground. An enclosure must be provided for making the tap splices. Electric power suppliers usually discourage the placing of more than one wire under a lug in the meter box. Splices can be made in a raintight junction box, using split-bolt connectors or similar connectors. The junction box can be mounted just below the meter box. Raintight junction boxes are available with lugs for making multiple taps. A typical tap box is shown in Figure 5-2.

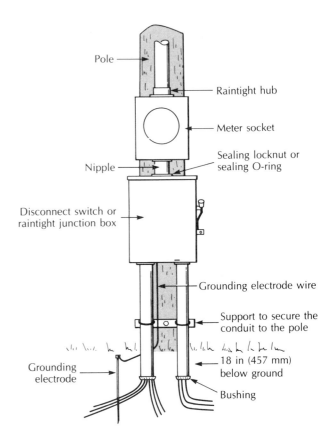

Figure 5-3 Tap splices are made in a raintight cabinet installed beneath the meter socket.

A disconnect switch can be installed below the meter socket. Double or triple lugs can be installed on the load side of the disconnect for tapping to several buildings. Figure 5-3 shows a typical central distribution pole with the supply wires coming from overhead and feeders to buildings running underground.

## Central Distribution Pole with Disconnect

A disconnect usually is not required at the central distribution pole, but it is highly desirable. Power can be easily disconnected in order to work on the wires leading to the buildings. A single disconnect for the entire farm must be rated to carry the total farm electrical demand load. The disconnect may be a raintight circuit breaker, a nonfusible switch, a fusible switch, or a standby generator transfer switch.

Individual control of feeders to buildings may be desirable. This can be accomplished with a group of disconnect switches or a circuit breaker panelboard at the

Figure 5-2 Lugs are provided inside this raintight cabinet for making feeder taps. *(Courtesy of Midwest Electric Products, Inc.)*

96   Unit 5   Farm Main Service Entrance

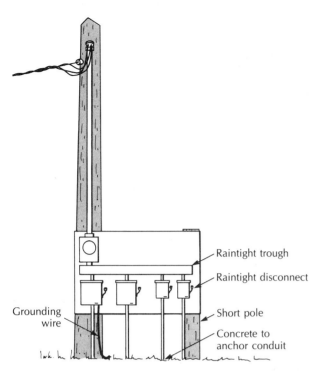

**Figure 5-4** Separate disconnects for each building feeder can be provided at the central pole.

central pole. Figure 5-4 shows a group of disconnects at the central pole. Feeders to outbuildings may be run overhead or underground.

## Raintight Installations

The equipment on the central distribution pole is exposed to the weather. The installation must be made in such a manner that water will not enter any conduit or enclosure. Only raintight equipment is acceptable. Most raintight enclosures used on farms are NEMA 3R or 4 enclosures. NEMA 7 and 9 enclosures are also acceptable for installations exposed to the weather.

An installation on a central distribution pole may involve the connecting together of two or more enclosures. A meter and a disconnect may be connected together as shown in Figures 5-3 and 5-4. A meter and a transfer switch may be connected together on the pole. These connections are usually made with conduit nipples. Where the nipples enter the side or top of an enclosure, they must be made raintight. Enclosure entering can be made raintight and dusttight with:

- sealing locknuts
- sealing O-rings
- raintight hubs

These devices are illustrated in Figure 5-5.

## Materials on the Central Pole

The wiring on the central distribution pole may be done with service-entrance cable or individual wires in conduit. Some electric power suppliers and local electrical ordinances may recommend or even require that the wiring shall consist of individual wires in conduit. Recall that the electrical installer must check local requirements before planning the central distribution pole installation.

A conduit system on the central pole is generally preferred. When the central pole electrical assembly is supported by means of standoff brackets, the conduit connections are subjected to some strain due to the weight of the equipment. Therefore, threaded connections are advisable. The most desirable materials for central pole assembly riser conduits are:

- rigid steel conduit
- rigid aluminum conduit
- intermediate metal conduit (IMC)

Conduit used to join enclosures on the pole can be made with electrical metallic tubing, or rigid nonmetallic or PVC conduit. Liquidtight flexible metal conduit can be used, as long as the enclosures are rigidly supported by a means other than the conduit. Sometimes, the conduit acts as part of the enclosure support. When this is the case, only rigid conduit or IMC should be used. Concentric knockouts left in place can break loose. If this should occur, the rigidity of the connection is lost. Therefore, concentric knockouts are not considered reliable when the raceway is to provide partial rigid support for the central pole enclosures.

## Grounding and Bonding

Metal raceway and enclosures forming the central distribution pole metering assembly must be electrically bonded together. This is required by *NEC Section 250-71*. Methods of providing adequate bonding are covered in *NEC Section 250-72*. If rigid conduit or IMC is used, the only problem with bonding occurs when concentric knockouts are left in place. A method of bonding around concentric knockouts still in place is required in the case **of service equipment.**

The central pole is treated as a service and, therefore, it must be grounded. The only means of grounding avail-

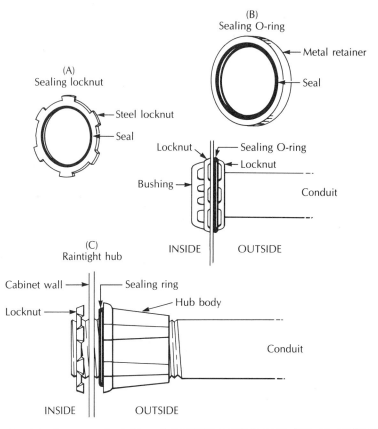

Figure 5-5 Three methods of making a watertight and dusttight conduit entry into an enclosure: (A) sealing locknut, (B) sealing O-ring, and (C) raintight hub

able at the pole is usually a made electrode such as a rod, pipe, or plate. The grounding electrode conductor is sized according to the size of the service-entrance conductors. However, when a rod, pipe, or plate electrode is used, *NEC Section 250-94, Exception No. 1,* states that the grounding electrode conductor need not be larger than No. 6 AWG copper. Aluminum grounding electrode wire is not permitted to be used at a central pole because of the restriction of *NEC Section 250-92(a).* Aluminum grounding electrode wire installed outside is not permitted to come within 18 in (457 mm) of the earth.

The grounding electrode conductor on a central distribution pole is considered to be subject to physical damage. Therefore, the wire must be provided with supplementary protection. Rigid metal conduit, IMC, or rigid nonmetallic or PVC conduit is generally used as supplementary protection for the grounding electrode wire.

The grounding electrode wire can be placed under the neutral lug in the meter socket, and then run to the grounding electrode. If a disconnect switch or standby power transfer switch is installed, the grounding wire can be connected to the neutral lug in the switch. Figure 5-6 shows the grounding of a central pole service with a standby transfer switch installed below the meter socket.

## Overhead Clearances

Adequate clearance must be provided beneath aerial feeder wires for tall farm machinery and trucks. *NEC Sections 225-18* and *230-24(b)* require a minimum clearance of 18 ft (5.49 m) where farm machinery and trucks will be traveling beneath the wires. The overhead feeders must be attached to the central pole 18 ft (5.49 m) or more to provide this minimum clearance, allowing for normal sag in the wires.

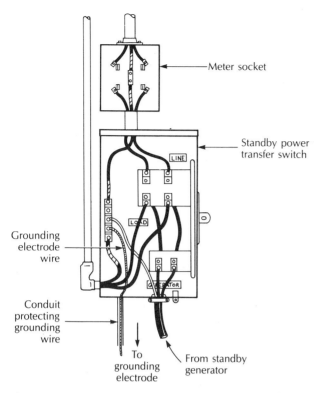

Figure 5-6 The central distribution pole electrical equipment must be grounded.

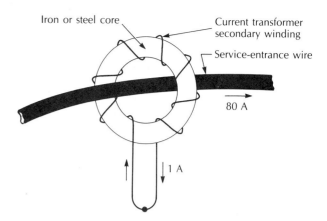

Figure 5-7 One ampere flows in the current transformer wire when 80 A flow in the service-entrance wire if the CT has a ratio of 80 to 1.

## CURRENT TRANSFORMER METERING

The purpose of metering using a current transformer is to avoid passing the entire electrical load through the kilowatthour meter. For the central poles discussed previously, it was necessary to run service wires down the pole, through the meter, and back up the pole. However, with a current transformer (often referred to as a CT), the service wires need not be run down the pole. Most electric power suppliers use current transformer metering for services larger than 200 A.

The clamp-on ammeter uses a current transformer to measure the current flowing through the wire. A current transformer consists of an iron or steel core in the shape of a doughnut. A small size wire is wrapped around the core. This wire acts as the secondary of a transformer. The service-entrance wire is passed through the center of the doughnut-shaped core, as shown in Figure 5-7. The service-entrance wire forms the primary winding of the transformer.

When the current transformer wires are shorted together, current flowing in the CT wire is proportional to the amount of current flowing in the service-entrance wire. CTs are manufactured with a specific primary to secondary current ratio. Assume that a current transformer has a ratio of 80 to 1. This means that 80 A flowing in the service-entrance wire causes 1 A to flow in the CT wire. The amount of current flowing in the service-entrance wire can then be determined by measuring the current flowing in the CT circuit. Equation 5.1 is used for determining the service-entrance current.

$$I_{\text{service wire}} = I_{CT} \times \text{Ratio} \qquad \text{Eq. 5.1}$$
$$\text{where: Ratio} = \frac{\text{Large number}}{\text{Small number}}$$

## Problem 5-3

A current transformer has a ratio of 80 to 1. The current flowing in the CT circuit, as measured with an ammeter, is 4.2 A. Determine the current flowing in the service-entrance wire.

## Solution

$$I_{\text{SE wire}} = 4.2 \times \frac{80}{1} = 336 \text{ A}$$

The current transformer of the previous example probably does not have a stated ratio of 80 to 1. Instead, the ratio is stated in such a way as to indicate the maximum current rating of the CT. The CT most likely has a ratio of 400 to 5. This means that the CT is rated for a maximum service-entrance wire current of 400 A and a maximum CT circuit current of 5 A. Dividing 400 by 5 still gives a ratio of 80 to 1. Equation 5.1 then still applies as in the case of the previous problem.

$$I_{SE\ wire} = 4.2 \times \frac{400}{5} = 336 \text{ A}$$

Current transformers used by the electric power supplier for energy metering are installed in the hot or phase wires of single-phase or 3-phase services. When this is done, the kilowatthour meter does not register the true electrical energy used. If the current transformers have a ratio of 80 to 1, the kilowatthour meter registers only 1/80th of the actual kWh used. A multiplier is printed on the face of the kilowatthour meter. The kilowatthours used in a month, as indicated by the meter, must be multiplied by the multiplier on the dial, according to Equation 5.2.

$$\text{Actual kWh} = \text{Meter kWh} \times \text{multiplier} \quad Eq.\ 5.2$$

## Problem 5-4

A potato farmer has a large potato-storage building capable of holding potatoes in excellent condition for up to 12 months. This potato-storage building is current transformer metered with a CT ratio of 400:5. Determine the actual kilowatthours used during one month if the kilowatthour meter indicates 187 kWh. The multiplier stated on the meter is 80.

## Solution

$$\text{Actual kWh} = 187 \text{ kWh} \times 80 = 14\ 960 \text{ kWh}$$

Current transformers are frequently installed at the top of the central distribution pole. Small-sized wires extend down the pole to the kilowatthour meter. Figure 5-8 shows a typical pole-top current transformer installation. The current transformers are sometimes contained within a weathertight box, and sometimes they are left in the open. A pole-top disconnect or a pole-top standby power transfer switch can also be added. A handle extends down the pole so that the disconnect or transfer switch can be operated from ground level.

Often it is desirable to put feeders to the farm buildings underground from the central distribution pole. In this case, the current transformers are installed at the

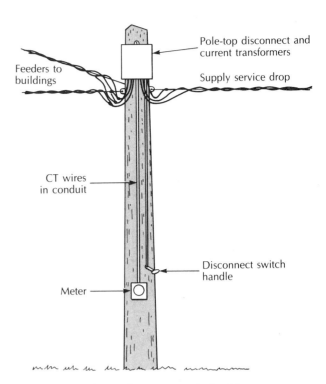

**Figure 5-8  Pole-top disconnect switch with current transformers inside**

bottom of the pole in a raintight cabinet. The kilowatt hour meter is attached to the side of the CT cabinet, Figure 5-9. Usually, there are several lugs available on the load side of the CTs for tapping feeders to several buildings.

## LOCATING THE CENTRAL DISTRIBUTION POLE

Ideally, the central distribution pole should be located in the midst of the group of farm buildings, unless the buildings are too close together. The pole should be located closest to the buildings with the greatest demand loads. This location minimizes the length of large size feeders, and reduces installation cost and the problem of voltage drop. The actual location depends upon the position of trees and obstacles in the area. The power supplier can provide assistance in determining a practical location. A suggested location for the central distribution pole for the farmstead discussed in Problem 5-1 is shown in Figure 5-10. Usually, the pole must be placed a reasonable distance from the power supplier's transformer, without obstacles between the transformer and the pole.

**Figure 5-9** The current transformer cabinet is usually installed at the bottom of the pole when feeders to farm buildings are run underground.

The electric power supplier generally allows the central metering pole to be placed at any reasonable location within the vicinity of the farm buildings. The pole must be located so that power supplier personnel have easy access to the pole. If the power supplier has to set additional poles to support the service drop or primary wires, there may be an extra charge for setting the poles and extending the lines. The extra expense, if any, may be

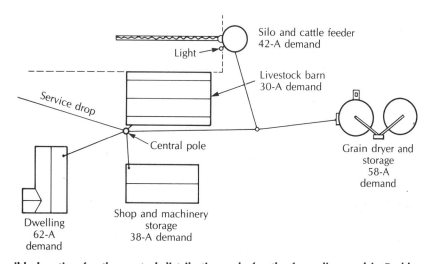

**Figure 5-10** A possible location for the central distribution pole for the farm discussed in Problem 5-1

Unit 5 Farm Main Service Entrance 101

justified in order to locate the meter pole closest to the largest farm loads.

It is desirable for the transformer to be as close to the central metering pole as possible, especially if a long service drop is required. A long service drop from the transformer to the central farm service location may result in neutral wire voltage drop that is measured on the farm as neutral-to-earth voltage. With the transformer at or near the central farm service, this neutral voltage drop is generally reduced to insignificant levels.

## REVIEW

Refer to the *National Electrical Code* when necessary to complete the following review material. Write your answers on a separate sheet of paper.

1. The service entrance in a single building can act as the farm main service, or a ___?___ distribution ___?___ can be the farm main service.

2. A farm has several buildings with the following electrical demand loads:

    | Poultry house | 126 A, 240 V |
    | Shop | 34 A, 240 V |
    | Livestock barn | 28 A, 240 V |
    | Machinery storage barn | 18 A, 240 V |
    | Dwelling | 64 A, 240 V |

    The total farm demand load for the purpose of sizing the farm main service entrance is which of the following?
    a. 185 A
    b. 219 A
    c. 243 A
    d. 270 A

3. The total single-phase demand load for a farm is determined to be 190 A. Determine the minimum size THW aluminum service-entrance wires for the central electrical distribution pole.
    a. 3/0
    b. 4/0
    c. 250 MCM
    d. 300 MCM

4. Demand factors for determining a farm total demand load are found in which *NEC* Section?

Refer to the following diagram to complete the statements in Problems Nos. 5, 6, 7, and 8.

5. The pole standoff bracket in the diagram is lettered ___?___.

6. The conduit containing the load side wires to the farm buildings is lettered ___?___ in the diagram.

7. The farm main disconnect in the diagram is lettered ___?___.

8. The place in the diagram where a raintight hub, sealing O-ring, or sealing locknut is required is lettered ___?___.

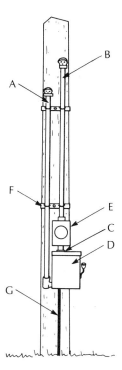

9. Describe a method of making tap connections at the central pole for underground feeders to several buildings.

10. A 400-A, single-phase farm service is metered using current transformers with a stated ratio of 40 to 1. If the kilowatthour meter indicates that the meter reading for one month was 124 kWh, the actual electrical energy used was which of the following?
    a. 2 480 kWh
    b. 3 720 kWh
    c. 4 464 kWh
    d. 4 960 kWh

# UNIT 6
# OVERCURRENT PROTECTION: FUSES AND CIRCUIT BREAKERS

## OBJECTIVES

After studying this unit, the student will be able to

- explain the problems that can cause an overcurrent condition in a circuit.
- explain the protective functions of overcurrent protective devices.
- identify common circuit fuses according to Underwriters' Laboratories class designations.
- explain how a circuit breaker operates.
- determine the interrupting rating of a fuse or circuit breaker.

## PURPOSE OF OVERCURRENT PROTECTION

*Overcurrent devices* (fuses and circuit breakers) are the safety valves of an electrical system. Overcurrent protective devices are some of the most important components in the electrical system on a farm. An overcurrent device must safely open an electrical circuit whenever an overload, short circuit, or ground fault occurs.

Heat energy is damaging to electrical wiring and equipment. Heat is produced as electrical current flows through the wires and electrical parts of the circuit. The amount of heat produced is proportional to the amount of current squared and the length of time the current flows. Damage to a circuit occurs when there is a current overload on the circuit for a long period of time. Damage also may occur in just a few seconds when a high level of current flows during a short circuit.

Arcing is another producer of heat. *Arcing* occurs when two electrical conductors are separated a short distance with air in between. At lower voltages, the electrical conductors must first touch and then become separated by a short distance. The electrical current flows the short distance from one conductor to the other through the air, causing an arc. A large amount of heat is concentrated in a small space during an arc. This heat often melts some of the conductor metal if it is allowed to continue for more than a fraction of a second.

*Magnetic force* may also damage electrical equipment and wiring during high levels of overcurrent. Magnetic forces may be strong enough to bend or move electrical conductors. They may even bend or break parts in electrical equipment. Electrical conductors of different voltages are sometimes forced to touch one another by magnetic forces produced during a high overcurrent level. The arcing that results can totally destroy the equipment. An explosion may even occur, endangering personnel in the area.

Overcurrent in circuits and equipment can be controlled to prevent damage to the wiring or equipment, to avoid fire, and to protect humans or animals from danger. Fuses and circuit breakers are selected to provide overcurrent protection. Unfortunately, no single type of fuse or circuit breaker can meet the requirements of every circuit. Different types are required. A purpose of this unit is to explain how to select the correct overcurrent device for agricultural applications. The *National Electrical Code* provides rules for the selection of overcurrent devices to provide minimum safety for humans and property. *Article 240* of the Code deals with overcurrent protection. However, many other important requirements are covered in other articles of the Code. Many of these requirements are discussed in this unit.

## TYPES OF OVERCURRENT CONDITIONS

Overcurrent devices must provide protection for overcurrent levels that are only slightly above the rated continuous current rating of the device, to levels that are many times the continuous current rating of the device. The cause of the overcurrent condition may be an overload, a short circuit, or a ground fault.

### *Overloads*

An *overload* is a condition where too much current flows through the wires, an appliance, a motor, a transformer, or other load being served. The current stays *within* the intended path, such as the wires and connected load. An overload condition is illustrated in Figure 6-1. A load of 20 A is flowing in the circuit, but the wires are rated for a maximum load of 15 A. If this overload continues, the wires will get hot and could cause a fire.

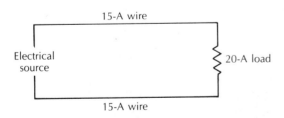

**Figure 6-1** The circuit is overloaded with 20 A flowing through conductors rated to safely carry 15 A.

### *Short Circuit*

A *short circuit* is a condition where too much current flows through the wires because the wires of the circuit are touching one another. When this happens, the current flow can greatly exceed the normal, safe value, causing overheating of the conductors. In addition, the energy released at the point where the conductors are touching causes much heat and sparking, possibly contributing to a fire. In a short circuit, the current flows *outside* of its intended path. The effect of a short circuit on a circuit is illustrated in Figures 6-2 and 6-3. The normal current flow of the circuit is illustrated in Figure 6-2.

$$I = \frac{E}{R} = \frac{240}{0.05 + 0.05 + 12} = \frac{240}{12.1} = 19.83 \text{ A}$$

The abnormal current flow (overcurrent) is illustrated in Figure 6-3. The short circuit is assumed to have zero resistance for this example, although this is not always the case.

$$I = \frac{E}{R} = \frac{240}{0.05 + 0.05} = \frac{240}{0.1} = 2\,400 \text{ A}$$

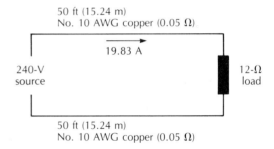

**Figure 6-2** A normally loaded electrical circuit

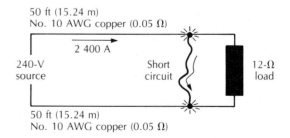

**Figure 6-3** A circuit that is short circuited

Note that the 12-Ω load is no longer in the circuit. It has been shorted out, resulting in a flow of short-circuit current of approximately 2 400 A.

## *Ground Faults*

A *ground fault* is a condition where the hot wires come into contact with a grounded metal conduit, water pipe, gas pipe, or any other grounded conductive object or surface. This condition may result in excessive current flow through the wire that is touching the grounded surface, causing insulation damage.

> **CAUTION:** The sparking at the point of ground can cause a fire. If the object with which the hot wire comes into contact is not properly grounded, then the object itself becomes live. This condition can cause injury or death by electrical shock both to humans and livestock.

It is possible that the ground fault may result in a very low value flow of current due to the high resistance of the total path through which the current flows to ground. This condition may not cause the insulation to become damaged on the hot conductor itself, but the resultant arcing and sparking at the point of the ground fault can cause severe damage to the equipment.

This is a serious problem nationally. Ground faults are often a common source of voltage on metal equipment around farms. "Stray" voltages on equipment may result. These stray voltages may cause shocks to animals, possibly resulting in death or loss of production.

The resistance at the point of the ground fault may be very low (large value of current flow), or it may be very high (low value of current flow). The arcing damage, electrical shock hazard, or other damage to the electrical equipment varies considerably from one situation to another. During a ground fault, the current is flowing *outside* of its intended path. A typical ground fault to a metal pipe is illustrated in Figure 6-4.

The hot wire solidly touches the grounded metal pipe. For the case of this ground-fault example, it is assumed that the resistance of the grounding path along the metal pipe is 0.01 Ω. The circuit is derived from a 3-wire, 120/240-V system. The voltage to ground is 120 V and the current is trying to return to the center tap point on the transformer.

$$I = \frac{E}{R} = \frac{120}{0.05 + 0.01} = \frac{120}{0.06}$$
$$= 2\ 000 \text{ A to the metal pipe}$$

Note in these examples that the *overload*, the *short circuit*, and the *ground fault* are all specific types of overcurrent. Thus, the need for proper and adequate overcurrent protection for each situation becomes apparent.

Properly sized fuses or circuit breakers should provide good overcurrent protection against those currents that exceed the circuit's normal capabilities. Ground-fault circuit interrupters (GFCI) are available to protect against the hazards of low value ground-fault currents.

A ground fault to poorly grounded metal equipment is illustrated in Figure 6-5. The hot wire touches the ungrounded or very poorly grounded metal surface of stanchions in a barn. Resistance of the total circuit is very high, so current flow is very low. Arcing at the point of the ground fault can cause damage. The stanchion becomes live with nearly 120 V to ground. Electrical shock to livestock which touch the stanchions may be fatal.

$$I = \frac{E}{R} = \frac{120}{0.05 + 100} = \frac{120}{100.05}$$
$$= 1.199 \text{ A flowing to ground}$$

This low-level current is not sufficient enough to open an overcurrent device and, therefore, it may persist for a long time until arcing is noticed or a person or animal suffers a shock.

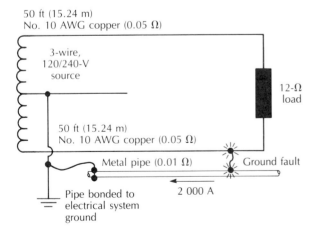

**Figure 6-4** This circuit has a fault to a grounded metal pipe.

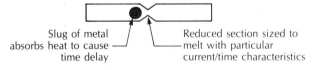

**Figure 6-5** A hot wire is making contact with the metal surface of the stanchion.

## HOW A FUSE FUNCTIONS

A fuse is an extremely simple and reliable overcurrent device. It operates on the principle that heat is generated when current flows through the fuse link. As the current flow increases, the heat buildup within the link increases. At some specific link temperature the link melts, thereby opening the circuit, Figure 6-6. These actions are really a simple application of the formula:

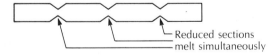

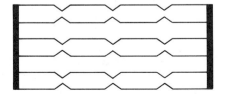

**Figure 6-6** Typical types of melting links used in fuses

$Q = I^2 \times R \times t$, where Q is the heat generated in the link, I is the current in amperes flowing through the link, R is the resistance of the link in ohms, and t is the length of time the current was flowing in seconds.

In some fuses, a small slug of metal is added to the link adjacent to the reduced section of the link. This slug absorbs heat and gives the fuse some time delay. It can also reduce the melting point of the link as the slug melts into the actual fuse link and changes the characteristics of the metal. Fuse links are made of zinc, copper, and sometimes silver. Most fuses contain a sand or powder filler to help quench the arc that occurs when the fuse link melts. The single-section fuse link with a metal slug for time delay is shown in Figure 6-6(A).

Some fuse links are designed with a number of reduced sections, so that the melting occurs at more than one point within the fuse, Figure 6-6(B). Thus, the arcing voltage appears across each open section that has been melted away. This permits the fuse to perform better, since voltage divides in a series circuit. The arcing after melting is extinguished quickly with this type of fuse.

Some fuses contain a number of fuse links, each containing a number of reduced sections, Figure 6-6(C). In this case, the current flow is divided equally through each link. These links are considered to be connected in parallel. The reduced sections of each link can be considered as being in series in each of the fuse links.

One can readily see that a fuse operates on the basic principles of electricity in the generating of sufficient heat to melt the link, in dividing the current flow according to parallel circuits, and by spreading out the arcing across several reduced sections of the link, according to the principle of series circuits.

## TYPES OF FUSES

Many types of fuses are used in today's modern electrical systems. Only those types commonly found in farm wiring installations are discussed in this unit. These fuse types are listed and described in Table 6-1.

A fuse must be labeled to show its *ampere rating*, *voltage rating*, and *interrupting rating* when it is rated at more than 10 000 A; whether it is of the *current-limiting* type; and the name of the *manufacturer*.

The *National Electrical Code* covers fuses in *Sections 240-50* through *240-61*. These sections contain the general requirements. When desiring specific information on how to size overcurrent protective devices for motors, transformers, branch circuits, or for other appli-

**Table 6-1 Fuse types, ratings, UL classes, and their applications** (Photos courtesy of Bussmann Manufacturing Division, McGraw-Edison Co.)

| Type | Amperes | Voltage | Interrupting Rating | UL Class | Application |
|---|---|---|---|---|---|
| Ordinary nonrenewable cartridge fuses. | 0-600 | 250 and 600 | 10 000 A*  *Some manufacturers provide K5 one-time fuses having 50 000-A interrupting rating. These fuses are superior to those having 10 000-A interrupting rating. | H* | Least expensive of all types of cartridge fuses. For general-purpose circuits and normal loading. For general-purpose circuits, do not load to over 80% of fuse ampere rating. For motors, size at 300% to 400% of motor full-load ampere rating. Because of this large sizing, they are not recommended for motor circuits as a larger and more expensive than necessary switch may be required. Fuse has little time delay. Will not fit into any switch that has Class R rejection fuse clips. |
| Dual-element, time-delay fuses, Class RK5.  Dual-element, time-delay fuses, Class RK1.  Do not be confused by the notch at one end of the Class RK1 and RK5 fuses. These are the most modern types of fuses. Fuses of similar charac- | 0-600 | 250 and 600 | 200 000 A | RK5 and RK1  Some of the older fuses of this category carry the marking K1 or K5. These fuses function in the same way as the newer ones, but they do not have the rejection Class R slot. | Most popular and common fuse of all. Recommended for general-purpose circuits, mains, feeders, services, motors, welders, transformers. Load to not over 80% of fuse ampere rating for general-purpose circuits. For motor circuits, may be sized at 110% to 175% of motor rating. Most motor circuits may be fused at 125% for good running overcurrent protection. Where close protection is required, size at 110% of motor current. Fuses of this type hold five times their ampere rating for at least 10 seconds. Class RK5 and Class RK1 are UL designations. The dual-element, time-delay types have the same time delay. The RK1 is much faster than the RK5 when a short circuit occurs, thereby offering superior protection. The RK1 is more expensive than the RK5. Class RK1 and Class RK5 fuses fit all existing switches, as well as the newer types of switches that may have Class R rejection fuse clips. These |

**Table 6-1**  (*cont.*)

| Type | Amperes | Voltage | Interrupting Rating | UL Class | Application |
|---|---|---|---|---|---|
| teristics without the notch are of the older variety. The important thing to remember is to look at the label to make sure the fuses are rated at 200 000-A interrupting rating. This is the best available. If the fuse label does *not* show any interrupting rating, then it has a maximum interrupting rating of 10 000 A. | | | | | clips make it impossible to insert the less expensive, low-interrupting capacity fuses of Class H type. Class RK1 and Class RK5 fuses are used ahead of circuit breakers to protect the breakers where the available short-circuit current exceeds the interrupting rating of the circuit breakers. |
| Fast-acting, current-limiting cartridge fuses, Class RK1. | 0–600 | 250 and 600 | 200 000 A | RK1  Some of the older fuses of this category carry the marking K1. The K1 fuse functions in the same way as the RK1, but does not have the rejection Class R slot. | Moderately expensive. Extremely fast-acting on overloads, ground faults, and shorts. Load to not over 80% of fuse ampere rating for general-purpose circuits. For motors, size at 300% to 400% of motor full-load ampere rating. Excellent for protection of electric heating circuits where little or no current surge or inrush exists.  Momentary overloads, power outages, and other situations that cause widely fluctuating currents may cause nuisance blowing of this type of fast-acting fuse.  This straight current-limiting fuse is very sensitive to surges of current.  For loads having varying degrees of inrush current surges, use the dual-element, time-delay fuses of the RK1 type when it is also necessary to |

Unit 6  Overcurrent Protection: Fuses and Circuit Breakers   109

**Table 6-1** (*cont.*)

| Type | Amperes | Voltage | Interrupting Rating | UL Class | Application |
|---|---|---|---|---|---|
| | | | | | have current limitation. This fuse is fast when it should be fast, and slow when it should be slow. |
| | | | | | Used ahead of circuit breakers to protect the breakers where the available short-circuit current exceeds the interrupting rating of the circuit breakers. |
| Renewable link cartridge fuses | 0-600 | 250 and 600 | 10 000 A | H | Initially, this fuse is expensive but, when the fuse blows, only the inexpensive link need be replaced. Do not load to more than 80% of fuse rating on general-purpose circuits for motors, size at 300% to 400% of motor full-load ampere rating. Because of this large sizing, they are not recommended for motor circuits as a larger and more expensive than necessary switch may be required. |
| | | | | | Fuse has a small degree of time delay. |
| | | | | | Never install more than one link in the fuse, unless specifically recommended by the fuse manufacturer. *To double up on links is hazardous.* Make sure links and end ferrules are tight. Looseness can cause excessive heat. |
| | | | | | These fuses will not fit into any switch that has Class R rejection fuse clips. |
| | | | | | The renewable link has a long history. They are *never* installed in new installations. They are a carry-over from years ago. Many old industrial plants are changing over their fuses to the time-delay RK5 and the time-delay, current-limiting RK1 when updating their electrical systems. |

**Table 6-1** (*cont.*)

| Type | Amperes | Voltage | Interrupting Rating | UL Class | Application |
|---|---|---|---|---|---|
| Ordinary plug fuse, Edison base. Ordinary plug fuse blown on an overload. Ordinary plug fuse blown on a short circuit. | 0-30 | 125* <br> * May be used in 240-volt circuits if the voltage to ground does not exceed 150 volts. For instance, on a 120/240-volt, single-phase system, the voltage to ground is 120 volts. | 10 000 A | None. Listed as Edison-base plug fuse. | This is the original plug fuse. Good for general-purpose circuits. Does not have much time delay, so circuits supplying motors, motor-operated appliances, and such, may experience nuisance blowing of this ordinary plug fuse. <br><br> If this problem exists, install dual-element, time-delay Type T or Type S fuses. <br><br> NEC Section 240-51(b) requires that Edison-base plug fuses be used only to replace blown fuses on existing installations where there has been no evidence of overfusing or tampering. <br><br> These fuses are the least expensive of the Edison-base types. In most cases, repeated nuisance blowing can be eliminated or reduced by installing dual-element, time-delay Type S or Type T fuses. |
| Plug fuse, Edison-base, Fusetron time-delay type. Cutaway view of plug-type Fusetron fuse. Note the spring-loaded element. | 0-30 | 125* <br> *May be used in 240-volt circuits if the voltage to ground does not exceed 150 volts. For instance, on a 120/240-volt, single-phase system, the voltage to ground is 120 volts. | 10 000 A | None. Listed as Edison-base plug fuse. | Fits any Edison-base fuseholder. <br><br> Has time-delay features that make this fuse ideal for motor loads and motor-driven appliance loads. <br><br> Two types of time-delay plug fuses are on the market. The better type is the one having a spring-loaded overload element. This type offers excellent overcurrent protection, and provides thermal protection to the fuseholder as well. The other type is recognized by its loaded-link construction. A heavy slug is squeezed onto the fuse element, which results in time delay. <br><br> NEC Section 240-51(b) requires that Edison-base plug fuses be used only to replace blown fuses on existing installations where there has been no evidence of overfusing or |

**Table 6-1** (*cont.*)

| Type | Amperes | Voltage | Interrupting Rating | UL Class | Application |
|---|---|---|---|---|---|
| Time-delay plug fuse with loaded-link element. | | | | | tampering. |
| | | | | | The spring-loaded type is available in sizes 3/10 ampere through 30 amperes. All sizes 14 amperes and less are designed specifically for motors or other loads having high-inrush characteristics. The 15-, 20-, 25-, and 30-ampere sizes are typical branch-circuit sizes for No. 14, No. 12, and No. 10 wire sizes. |
| | | | | | If nuisance blowing occurs with regular plug fuses, use this time-delay type. |
| Plug fuse, Type S Fustat fuse with spring-loaded element. Adapter for Type S fuses. Time-delay Type S fuse of the loaded-link design. | 0-30 | 125 | 10 000 A | | Has the same features and characteristics as the Edison-base, Fusetron time-delay fuse. |
| | | | | | The Type S fuse requires the insertion of an adapter before the Type S fuse can be screwed in. The adapter prevents inserting a fuse of a larger ampere rating. |
| | | | | | *NEC Section 240-52* requires that fuses installed in a new installation must be of the Type S design. |
| | | | | | Two types of Type S fuses are available. The type that has the spring-loaded element is the better fuse. It offers overload protection, and provides thermal protection to the fuseholder as well. The other type is recognized by its loaded-link design. A heavy slug is squeezed onto the fuse element, which results in time delay. |
| Current-limiting, high-interrupting capacity, cartridge fuses. Very small dimension as compared to | 1-600 | 300 and 600 | 200 000 A | Class T | This type is the newest fuse available. It is approximately one-third the size of ordinary cartridge fuses. |
| | | | | | The fuses are fast acting; thus, they should be sized at 125% for continuous loads. They may be sized at 100% |

**Table 6-1** (*cont.*)

| Type | Amperes | Voltage | Interrupting Rating | UL Class | Application |
|---|---|---|---|---|---|
| ordinary cartridge fuses. | | | | | for noncontinuous loads.<br><br>Size at 300% for motor loads, unless the motor has extreme starting current inrushes or if it is started and stopped frequently; in which case, the fuse may be sized at 400%.<br><br>Class T fuses will fit only into equipment having fuse clips suitable for Class T fuses.<br><br>Many equipment manufacturers are installing Class T fuses in their equipment because of their space-saving feature and superior protection.<br><br>Electrical equipment manufacturers are in the process of manufacturing panelboards and disconnect switches that will accept the Class T fuse.<br><br>Class T fuses are *not* dual-element fuses. |
| High capacity, current-limiting fuses, Class L type.<br><br>Time-delay Class L fuse.<br><br>Fast-acting Class L fuse. | 601-6 000 | 600 volts | 200 000 A | L | These fuses are large in physical size. They are installed in main switches that are rated at more than 600 amperes.<br><br>They are used on systems of 600 volts or less.<br><br>Available in fast-acting and time-delay types.<br><br>Both types are UL listed as Class L current-limiting fuses.<br><br>The time-delay type is ideal for general-purpose mains and feeders, as well as for large feeders supplying motors.<br><br>These fuses are bolted into the switches designed for Class L fuses. The fuse links in these fuses are made of silver strips. The fuse is filled with a high-grade arc-quenching filler of silica sand. Hence, it is sometimes called a "silver-sand fuse." |

**Table 6-1** *(cont.)*

| Type | Amperes | Voltage | Interrupting Rating | UL Class | Application |
|---|---|---|---|---|---|
| Miscellaneous fuses | 0-30 | 32-600 | Varies | UL listed as miscellaneous fuses | Many varieties of miscellaneous fuses are available. They are the smaller types, such as ¼″ × 1¼″ or ¼″ × 1½″, and ¹³⁄₃₂″. Some have glass tubes; others have fiber or ceramic tubes. These fuses are often found as an integral part of appliances, such as fence chargers, tractors, automobiles, and so forth. They are available in normal-speed, slow-blowing, and fast-acting types. Always replace a blown fuse with one that has the same voltage and ampere rating, or one you are certain will provide proper protection. |

cations, refer to the section of the Code that pertains to the specific topic.

## CIRCUIT BREAKERS

Circuit breakers most commonly installed in residential and farm wiring are of the *molded-case* types. That is, their internal parts are contained within a plasticlike enclosure. Their trip settings are generally not adjustable, although more expensive molded-case circuit breakers are available with adjustable trip.

A thermal-magnetic circuit breaker has a bimetallic element that will cause the breaker to trip when subjected to continuous overload conditions. The greater the overload, the sooner the circuit breaker trips. Momentary small overloads do not damage the wiring or equipment, and they do not cause the circuit breaker to trip. But, when subjected to heavy overloads, short circuits, or ground faults, a magnetic coil in the circuit breaker will trip the mechanism to the off position in a fraction of a second.

*NEC Sections 240-80* through *240-83* give the requirements for circuit breakers. The following important points are found in the Code.

- Circuit breakers shall be trip free so that even if the handle is held in the on position, the internal mechanism can trip to the off position.
- Circuit breakers shall clearly indicate whether they are in the on or off position.
- A circuit breaker shall be nontamperable so that it cannot be altered (trip point changed) without dismantling the circuit breaker or breaking the seal.
- The rating shall be durably marked on the circuit breaker. For circuit breakers rated at 100 A or less and 600 V or less, the rating must be labeled, stamped, or etched on the handle or on another part of the circuit breaker that will be visible after the cover of the panel is installed.
- Every circuit breaker with an interrupting rating other than 5 000 A shall have its rating marked on the circuit breaker.
- Circuit breakers for fluorescent lighting loads at 120 V and 227 V shall not be used as switches unless marked "SWD."

Panels for circuit breakers should be installed in locations where there will be no corrosion or moisture problems; otherwise, the circuit-breaker mechanism is likely to become inoperative, resulting in an unsafe situation. It is a good practice to turn the circuit breaker on and off periodically to "exercise" or test its moving parts.

In most cases, circuit breakers go for many years without ever tripping. To gain confidence in the reliability of the circuit breaker, connect or plug in enough electrical equipment or appliances to cause an overload on the circuit. Then time how long it takes for the circuit breaker to trip the circuit off.

**CAUTION:** Do not do this test by causing a short circuit. Testing in this way is *dangerous*. Test equipment is available for the purpose of testing circuit breakers. Some electrical contractors perform a service of testing circuit breakers and overload elements for motor controllers.

All equipment and materials for electrical wiring have a maximum safe operating temperature. The lowest maximum operating temperature is found in the wire ampacity tables of *NEC Article 310*. This lowest maximum temperature is 60°C (140°F). Circuit breakers have a maximum operating temperature. If this maximum operating temperature is greater than 60°C, it is so marked on the circuit breaker. Many common circuit breakers have a maximum operating temperature of only 60°C. This is important, because even if a wire is used for a circuit with a higher insulation rating, such as THWN (75°C), if it is connected to a circuit breaker rated at only 60°C, then the wire ampacity must be based upon the 60°C temperature and not 75°C.

## STANDARD RATINGS OF OVERCURRENT DEVICES

The standard ampere ratings of fuses and circuit breakers are given in *NEC Section 240-6* as follows: 15, 20, 25, 30, 35, 40, 45, 50, 60, 70, 80, 90, 100, 110, 125, 150, 175, 200, 225, 250, 300, 350, 400, 450, 500, 600, 700, 800, 1 000, 1 200, 1 600, 2 000, 2 500, 3 000, 4 000, 5 000, and 6 000 amperes. Not all of these sizes are available in circuit breakers. Additional standard ampere ratings for fuses *only* are 1, 3, 6, 10, and 601 amperes. Underwriters' Laboratories recognizes many other sizes of fuses for closer protection of electrical equipment. Refer to Table 6-2 for time-delay 250-V and 600-V fuses rated at less than 15 A.

The fuse sizes listed in the table are but a few of the many sizes that are applied to motor circuits for added running overcurrent protection. Such fuses are used to protect against burnout of the motor due to overloaded conditions, worn bearings, V-belts jumping out of the

**Table 6-2 Time-delay 250-V and 600-V fuses rated at less than 15 A**

| 1/10, 1/8, 15/100, 2/10, 1/4, 3/10, 4/10, 1/2, 6/10, 8/10, 1, 1 1/8, 1 1/4, 1 4/10, 1 1/2, 1 6/10, 1 8/10, 2, 2 1/4, 2 1/2, 2 8/10, 3, 3 2/10, 3 1/2, 4, 4 1/2, 5, 5 6/10, 6, 6 1/4, 7, 7 1/2, 8, 9, 10, 12 |
|---|

pulley and becoming jammed or dragging, and similar situations that could result in a motor burnout.

Dual-element, time-delay fuses are used to protect small transformers; solenoids; sump pumps; electric overhead doors; hoists; ventilation fans; and such power tools as saws and drills, among others. These fuses are also used for food waste disposers, dishwashers and many other applications. The cost is minimal for the properly sized fuse when compared to the cost of repairing or replacing the electrical equipment.

## BASIC REQUIREMENTS FOR FUSES AND CIRCUIT BREAKERS

Four extremely important factors must always be considered when selecting and installing fuses or circuit breakers. These factors are:

1. Voltage rating
2. Continuous current rating
3. Interrupting rating
4. Speed of response

These four factors are so important that each is discussed in depth in this section.

### Voltage Rating

The *National Electrical Code* requires that the voltage rating of a fuse or circuit breaker must be equal to or greater than the voltage of the circuit in which the fuse or circuit breaker is to be installed, Figure 6-7.

A fuse or circuit breaker rated at 600 V may be used

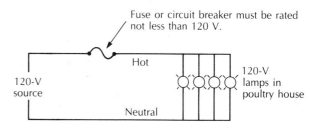

**Figure 6-7** Voltage rating requirement for fuses and circuit breakers

where the system voltage is less than 600 V. A fuse or circuit breaker rated at 250 V may not be used on a 480 V circuit. Similarly, a 125 V fuse may be used in a 32 V circuit, but a 32 V fuse may not be used in a 125 V circuit.

## Continuous Current Rating

The continuous current rating of a fuse or circuit breaker generally must be equal to or greater than the ampacity of the circuit or the equipment which it protects, although there are exceptions. For example, for a No. 8 AWG copper wire Type TW rated at 40-A, a 40-A fuse or circuit breaker would be selected as the protective device. Circuits for motors are frequently exceptions to this rule. Proper selection of an overcurrent device for a 240 V, 32-A continuous load is illustrated in Figure 6-8.

The ampere rating of the circuit is determined by the continuous current rating of the overcurrent device protecting the circuit. The ampere rating of the load or loads to be served and the operating characteristics of the load must be known before the ampere rating of the circuit overcurrent device can be determined. Is the load made up of a single piece of equipment, or does it also serve outlets or lighting fixtures? Add up the ampere rating of the circuit loads. If the load consists of a single piece of equipment or appliance, the load usually may not exceed 80% of the overcurrent device rating. This is true for cord- and plug-connected equipment used on 15-A and 20-A circuits, *NEC Section 210-23(a)*.

A *continuous load* is one that under normal operation will operate for 3 hours or more. A continuous load is not permitted to exceed 80% of the circuit overcurrent device rating, *NEC Sections 210-22(c)* and *384-16(c)*. Water heaters, ventilation fans, and most nonresidential lighting circuits are considered to be continuous loads. This means that a circuit protected with a 15-A overcurrent device is not permitted to be loaded continuously beyond 12-A. The maximum continuous load is 16-A for a 20-A circuit.

$$\text{Maximum continuous load} = 15 \text{ A} \times 0.80 = 12 \text{ A}$$
$$\text{Maximum continuous load} = 20 \text{ A} \times 0.80 = 16 \text{ A}$$

An exception to this rule is stated in *NEC Section 384-16(c)*. The exception states that some overcurrent device and panel combinations are rated for operation at 100% load. But, generally, this is not the case.

When the conductor ampacity does not match a standard rating of a fuse or circuit breaker, the *NEC* in *Section 240-3 Exception No 1* permits the use of the next larger size fuse or nonadjustable trip-circuit circuit breaker. An example of the use of this exception is illustrated in Figure 6-9. A simple one-line circuit diagram is used to represent the circuit. Note that the load served is less than the ampere rating of the wire. This must always be the case.

The next larger size overcurrent device is not permitted where the overcurrent device rating exceeds 800 A. Consider an example where a 900-A noncontinuous load is supplied with three parallel sets of 500 MCM copper wires, Type THW. The total ampere rating of the parallel sets of wires is 1 140 A. However, it is not permitted to use a 1 200-A fuse or circuit breaker in this case. The maximum size permitted is 1 000 A, Figure 6-9(B). Note that the load served does not exceed 1 000 A.

Some situations exist where the fuse or circuit breaker may be rated larger than the load it serves. An

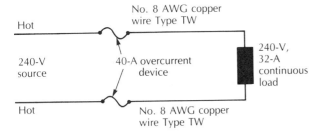

Figure 6-8 Example of an overcurrent device selected for a load

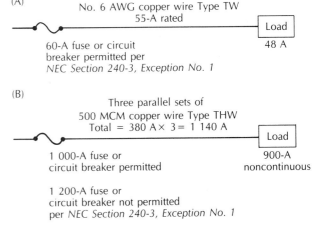

Figure 6-9 Examples of selecting overcurrent devices for the wire ampere rating and the load to be supplied

example of this is an electric motor. When an electric motor is started, the motor draws a very high current for a few seconds. As the motor comes up to speed, the momentary starting inrush current begins to decrease until the motor reaches its normal full-load speed, at which time the motor is now drawing its normal full-load current.

For this application, the *National Electrical Code* permits the installation of a fuse or circuit breaker having an ampere rating greater than the actual current rating of the motor and/or the conductor feeding the motor. This is permitted because the conductors are protected against overload by means of the motor running overcurrent protective devices, such as the thermal overloads in the starter. This situation is covered in *NEC Article 430*.

A grain-dryer motor circuit is illustrated in Figure 6-10 with the branch-circuit fuses having a higher ampere rating than the branch-circuit wires. This is permitted by *NEC Section 430-52* for the specific application of a motor circuit.

It is possible to use ordinary nontime-delay fuses for a motor circuit, although they are not recommended. Size for size, when compared with dual-element, time-delay fuses, the ordinary nontime-delay fuse is the least expensive. But, nontime-delay fuses must be sized at 300% to 400% of a motor's full-load current rating, whereas dual-element, time-delay fuses are sized at 100% to 125% of the motor's, full-load current rating.

Using the grain-dryer motor of Figure 6-10 as an example, the ordinary nontime-delay fuses would be sized at about 45 A.

$$15.2 \text{ A} \times 3 = 45.6 \text{ A (Use 45 A fuses)}$$

The 45-A fuses would require the installation of a 60-A disconnect switch. Using dual-element, time-delay fuses, the switch size is 30 A? Thus, by attempting to use lower-priced fuses, a larger and more expensive switch would have to be used.

For motor circuits, it is much better to install dual-element, time-delay fuses in a properly sized disconnect switch. Also, when sizing the dual-elements, time-delay fuses at 125% of the motor full-load current rating, additional running overcurrent protection is provided by the fuses, in addition to the thermal overloads in the motor starter. Installation money is saved by installing a properly sized switch rather than by installing an oversized switch. All manufacturers of switches and fuses provide literature and catalog information which show the information needed to select switches and fuses for motor circuits.

## Interrupting Rating

The term *interrupting rating*, also called *interrupting capacity*, is a measure of the ability of the fuse or circuit breaker to open an electrical circuit under fault conditions. *Fault conditions* include 1) overloads, 2) short circuits, and 3) ground faults. The interrupting rating of an overcurrent device, as marked on its label, is the maximum current that it can safely interrupt at rated voltage, Figure 6-11.

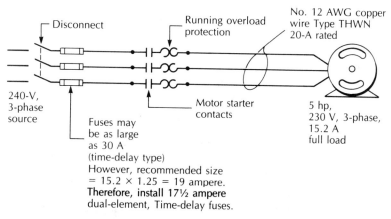

Figure 6-10  In the case of a motor circuit, the ampere rating of the circuit overcurrent device may be larger than the ampere rating of the wire.

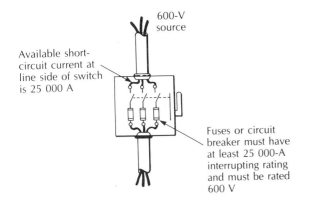

**Figure 6-11** Example of how fuses and circuit breakers must have adequate interrupting rating

*Section 110-10* of the Code states that the overcurrent devices, the total impedance, the component short-circuit withstand ratings, and other characteristics of the circuit to be protected, shall be selected and coordinated in such a way that the circuit protective devices used must clear a fault without the occurrence of extensive damage to the electrical components of the circuit.

*NEC Section 110-9* is an all-encompassing section that covers the interrupting ratings of service mains, feeders, subfeeders, and branch circuits. This section states that the interrupting rating of the equipment used to break the current at fault levels shall be sufficient for the system voltage and the current that is available at the line terminal of the equipment.

*NEC Section 230-98* refers specifically to services, stating that the service equipment must be appropriate for the short-circuit current available at the supply terminals.

The interrupting rating of a fuse is found on the label or body of the fuse if the interrupting rating is other than 10 000 A. If the interrupting rating is not printed on the fuse label, it has an interrupting rating of 10 000 A. Small dimension fuses (¼ by 1¼ inch and 13/32 by 1½ inch) can have even lower interrupting ratings, such as 1 000 A. The interrupting rating of these types is available from the manufacturer of the fuse.

The interrupting rating of a circuit breaker is found on the label or body of the breaker if the interrupting rating is other than 5 000 A. Therefore, if the interrupting rating does not appear on the breaker, it may be assumed that the interrupting rating is 5 000 A maximum. As with fuses, small types of circuit breakers may have interrupting ratings less than 5 000 A. This information is available from manufacturers.

An example of a violation of interrupting rating requirements is given in Figure 6-12. The main circuit breaker has sufficient interrupting rating to withstand the 50 000 A that are available from the source if a fault should occur at the line terminal of the main breaker. The circuit breaker does not open quickly enough, and fault current in excess of 10 000 A may pass through the main to the branch circuit where the fault occurs. The branch-circuit breaker cannot withstand this high-fault current, and consequently it explodes.

It is usual practice to first determine the size of the service main disconnect, the size of the subpanels, the number of branch circuits required, the conductor sizes, and all other basic load requirements. Next, make sure that all of the overcurrent devices have proper interrupting ratings.

**CAUTION:** When a fuse or circuit breaker of inadequate interrupting rating is installed in switches and other equipment where the available short-circuit current exceeds the interrupting rating of the fuse or breaker, the result is the same as a bomb just waiting for a short circuit to trigger it into an explosion. Personal injury may result as the equipment blows up and destroys itself. Damage to the equipment can be costly, but personal injury or death is impossible to measure in terms of dollars.

Recently, manufacturers of circuit breakers have submitted some of their circuit breakers to Underwriters' Laboratories for testing in series with each other. For instance, a main breaker rated 22 000 A interrupting capacity and a branch breaker rated at 10 000 A inter-

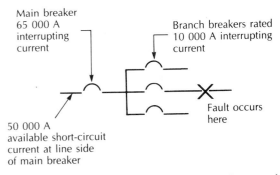

**Figure 6-12** If a fault occurs on a branch circuit, excessive fault current may pass through the main breaker and destroy the branch breaker.

rupting capacity are connected in series. A 22 000-A fault is then established on the load side of the breaker rated at 10 000 A interrupting capacity, allowing both breakers to be in the circuit. The added impedance of both breakers connected in series reduces the fault levels seen by each breaker. Both breakers trip open and the entire electrical system to the panelboard or load center is shut off. This installation, while not especially desirable, does meet UL standards. These series-tested breakers must be used only with other series-tested breakers. The equipment will be clearly marked with the UL markings identifying it as "series tested."

Although the interrupting rating of an overcurrent device is its maximum rating, it is good practice to install the device so as not to subject it to fault current of more than 80% of its interrupting rating. For example, for a circuit breaker that has an interrupting rating of 18 000 A, the breaker should not be used in an electrical system having more than 14 400 A of available short-circuit current.

$$18\ 000\ A \times 0.80 = 14\ 400\ A$$

An overloaded condition resulting from a miscalculation of connected loads, or by simply plugging in too many electrical appliances, will cause a fuse to blow or breaker to trip in a safe, normal manner. However, an incorrect guess, a miscalculation, or plain ignorance regarding the amount of short-circuit current available can result in the selection and installation of fuses or circuit breakers that do not have adequate interrupting ratings. The result is a hazardous and unsafe installation which is in violation of the *National Electrical Code*.

**CAUTION:** Do not attempt to use a standard ammeter to get a reading of short-circuit currents, as this will damage the meter, and possibly result in personal injury.

Various easy methods are available for making estimates of the available short-circuit current at any point on an electrical system. These methods include technical data and impedance factors for wires provided by fuse manufacturers. Such short-circuit current calculations should be made for farm services of large capacity, and for farms close to a power supplier substation.

## Speed of Response

The speed of response is found by referring to the time-current curves for a fuse or circuit breaker. Manufacturers' literature provides this information. The *speed of response* is an indication of how fast or how slowly the fuse or circuit breaker reacts when subjected to some value of current. The label may also give some indication as to the speed of opening by stating something like, "fast acting," "slow blow," "current limiting" (usually these are fast-acting), "time delay" (usually slow-acting), and "dual element" (usually slow-acting).

A fuse or circuit breaker should not open needlessly on a momentary surge of current, yet it should open fast when a true fault occurs. The ideal response is to be fast, but not too fast, and to be slow, but not too slow.

The time required for the fusible element (link) of a fuse to open varies inversely with the amount of current flowing through the fuse. That is, the time required to melt the fuse element decreases as the amount of current flowing through it increases.

The time required for the mechanism of a breaker to trip varies inversely with the amount of current flowing through the breaker. In other words, the time required to trip a circuit breaker decreases as the amount of current flowing through it increases. Above some value of current, however, a circuit breaker cannot open any faster due to the inertia of the moving parts contained within the breaker. This phenomenon is called the *irreducible tripping time* of the breaker.

The circuit breaker for the motor in Figure 6-13(A) is not properly sized for the load. The circuit breaker may not have sufficient time delay to pass through the period of high inrush current during motor starting. Therefore, according to *NEC Section 430-52*, the circuit breaker for

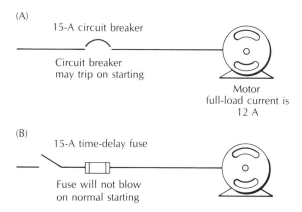

**Figure 6-13** A motor with a 12-A full-load current is shown protected with a 15-A circuit breaker and with a 15-A time-delay fuse. During normal starting the breaker may trip, but the fuse will not blow.

branch-circuit protection is usually permitted to be sized at 250% of the motor full-load current.

Time-delay fuses are designed with an element for time delay, and another for short circuits. A proper match of motor and time-delay fuse is shown in Figure 6-13(B). The time delay of the fuse is sufficient to allow the motor inrush current to pass without opening either the time-delay or short-circuit element. In most cases, a time-delay fuse can be sized only slightly above the motor full-load current.

A typical inrush current curve for a motor is shown in Figure 6-14. Motors powering such farm loads as silo unloaders and gutter cleaners typically draw high inrush currents for several seconds, especially when starting these loads in winter. If a circuit breaker is not sized much larger than the motor full-load current it may trip before the starting current inrush peak passes. A time-delay fuse can usually be sized small enough to provide running overcurrent protection and still not nuisance blow during motor start-up.

Time-current curves clearly show the speed of response of a fuse or circuit breaker. Consider a motor that has a full-load current rating of 48 A and is protected with time-delay fuses sized at 125% of the full-load current rating.

$$48 \text{ A} \times 1.25 = 60 \text{ A fuses}$$

Assume that the starting inrush current for this motor is approximately 200 A. This will vary with the type of motor, supply voltage, and load. The higher the code letter on the nameplate of the motor, the higher is the starting inrush current.

Here is how to determine the time it will take a fuse or circuit breaker to trip at a particular level of current. Draw a vertical line from the 200-A value at the bottom of the time-current curve, moving upward until it intersects with the 60-A fuse curve of Figure 6-15. Then draw a horizontal line to the left edge of the curve and find that the 60-A fuse will hold 200-A for approximately 60 seconds. This is enough time to let the motor come up to speed.

Should a short circuit or ground fault occur in the magnitude of 600 A, the fuses will open the circuit in approximately 0.03 second. A cycle for 60-Hz power is 0.016 second; so the 60-A fuses with this example will open the circuit in just slightly less than two cycles. This is very fast and will minimize damage to the motor and the equipment.

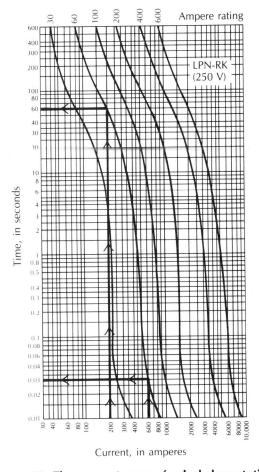

Figure 6-15 Time-current curve of a dual-element, time-delay fuse. Manufacturers of fuses and breakers provide time-current curves for all sizes and types of their fuses and breakers.

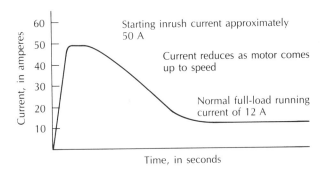

Figure 6-14 Typical motor starting current curve. A circuit breaker sized only slightly above the motor full-load current will likely trip during the starting inrush current portion of the curve.

120    Unit 6    Overcurrent Protection: Fuses and Circuit Breakers

Circuit breaker curves are read in much the same manner as fuse curves. The one exception involves the breaker irreducible tripping time mentioned previously. Just as with fuses, the higher the current, the faster the breaker will trip. But at some value of current, no matter how much more current flows, the breaker trip unit will unlatch, but the contacts will still take a given amount of time to finally open. This is illustrated in Figure 6-16, where a 100-A circuit breaker is compared to a 100-A time-delay fuse. It is very likely that more than one breaker in the system might trip on high values of short-circuit or ground-fault currents.

It is useful to obtain time-current curves of fuses and circuit breakers from a manufacturer. Such information is contained in manufacturers' literature and technical publications.

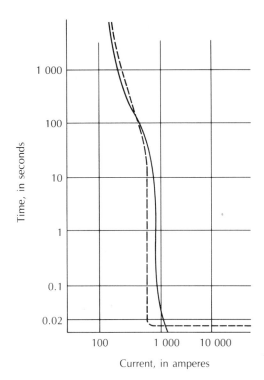

**Figure 6-16** Approximate time-current curve of a 100-A circuit breaker (dotted line) and a 100-A time-delay fuse (solid line)

## REVIEW

Refer to the *National Electrical Code* when necessary to complete the following review material. Write your answers on a separate sheet of paper.

1. What is the purpose of overcurrent protection?
2. Define *overload, short circuit,* and *ground fault.*
3. What value of current will flow if the hot conductor of a circuit comes into contact with a metal pipe having a resistance to the system ground of $5\Omega$?
4. Name two common types of overcurrent devices.
5. State briefly how a fuse operates.
6. List the voltage, range of ampere ratings available, and interrupting rating of each of the following types of fuses.
    a. Class H fuses
    b. Class RK1 fuses
    c. Class RK5 fuses
    d. Plug fuses
    e. Class T fuses
    f. Class L fuses
7. Is it a good idea to periodically exercise breakers by manually turning them off and on?

8. List the standard ampere ratings of fuses and circuit breakers up to and including 200 A.

9. What type of fuse is the best choice of fuses to use for motor circuits?

10. Name the four basic considerations when selecting fuses or circuit breakers.

11. The voltage rating of a fuse or circuit breaker must be equal to or ___?___ than the voltage of the circuit.

12. It is permitted to install a fuse or circuit breaker the next size larger than the ampacity of the conductor up to and including ___?___ amperes.

13. When an electric motor is started, does it draw less than, the same as, or more than its normal full-load ampere current draw?

14. Explain in your own words the meaning of the term *interrupting rating*.

15. What sections of the 1984 *National Electrical Code* pertain to interrupting rating?

# UNIT 7
# VOLTAGE DROP

> ## OBJECTIVES
>
> After studying this unit, the student will be able to
>
> - discuss the effects of voltage drop.
> - calculate the voltage drop in a simple 2-wire, single-phase circuit.
> - determine the wire size in a simple 2-wire circuit, based upon voltage drop calculations.
> - calculate the voltage drop in a simple 3-wire, 3-phase circuit.
> - calculate the proper conductor size required for a 3-phase, 3-wire circuit, based upon voltage drop calculations.
> - measure the voltage drop in a circuit.
> - determine when line reactance and power factor must be considered in voltage drop.

## EFFECTS OF VOLTAGE DROP

Excessive voltage drop in a circuit can cause lights to flicker and/or burn dimly, heaters to heat poorly, and motors to run hot to the extent of motor burnout. What is really happening when we refer to *voltage drop* or *voltage loss* is that the voltage is literally used up while pushing the current through the conductors. This condition is called *IR drop*.

## Solving for Voltage Drop

The basic formula for calculating voltage drop is:

```
E = I × R
where: E = voltage
       I = current, in amperes
       R = resistance, in ohms
```

## Problem 7-1

A simple 2-wire circuit is shown in Figure 7-1. Solve for the voltage drop.

## Solution

The voltage drop, $E_d$, in conductor A equals:

```
E_d = I × R
    = 10 A × 1 Ω
    = 10 V
```

The voltage drop in conductor C also equals 10 V. Thus, there is a total of 20-V drop or loss across the two conductors supplying the 10-A load B, resulting in 100 V across the load terminals.

What happens when such a voltage drop occurs? In

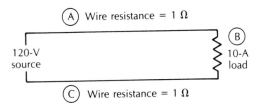

**Figure 7-1** A simple 2-wire circuit

the case of an electric motor, a voltage to 90% of the motor's rated voltage will

- decrease the full-load speed by 1½%.
- decrease the full-load efficiency by 2%.
- decrease the starting current by 10% to 12%.
- decrease the maximum overload capacity by 19%.
- increase the temperature rise at full load by 6°C to 7°C.
- increase the full-load current by 11%.

What happens when low voltage occurs on a circuit supplying incandescent lights, or on electric resistive heating elements such as the heater that might be installed in a chicken brooder?

## Problem 7-2

An electric heating unit is rated at 3 000 W, 236 V. Solve first for its rated current draw and resistance, then solve for the current draw at 215 V.

## Solution

$$I = \frac{W}{E} = \frac{3\,000 \text{ W}}{236 \text{ V}} = 12.7 \text{ A at rated voltage}$$

$$R = \frac{E^2}{W} = \frac{236 \times 236}{3\,000} = 18.56 \text{ }\Omega$$

Another way to solve for rated current draw is:

$$I = \frac{E}{R} = \frac{236 \text{ V}}{18.56 \text{ }\Omega} = 12.7 \text{ A at rated voltage}$$

If a different voltage is experienced, such as a low-voltage condition of 215 V across the heater terminals, the new wattage rating and current draw would be calculated as follows:

$$W = \frac{E^2}{R} = \frac{215 \times 215 \text{ V}}{18.56 \text{ }\Omega} = 2\,491 \text{ W}$$

$$I = \frac{E}{R} = \frac{215 \text{ V}}{18.56 \text{ }\Omega} = 11.58 \text{ A current draw}$$

Another way to solve for the new lower current draw is:

$$I = \frac{W}{E} = \frac{2\,491 \text{ W}}{215 \text{ V}} = 11.58 \text{ A}$$

At 215 V, the electric resistance heater produces only 2 491 W. Remember the original rating of 3 000 W at 236 V. With a 9% reduction in supply voltage, the heater output is reduced 17%.

What happens when the same electric heater is connected to 118 V? This voltage is only one-half of the rated voltage of the heater.

$$W = \frac{E^2}{R} = \frac{118 \times 118 \text{ V}}{18.56 \text{ }\Omega} = 750 \text{ W}$$

$$I = \frac{E}{R} = \frac{118 \text{ V}}{18.56 \text{ }\Omega} = 6.36 \text{ A}$$

Another way to solve for the new lower current draw is:

$$I = \frac{W}{E} = \frac{750 \text{ W}}{118 \text{ V}} = 6.36 \text{ A}$$

The voltage was reduced by 50%, but the output of the heater dropped to 25%. The example shows that as the voltage is reduced in a resistance load, such as incandescent lights or an electric heating element, the current flow will also be reduced. The output of a resistance heating element will drop about twice as fast as the voltage and current.

An electric motor operates on a different principle than a light bulb or an electric heating element. Current flowing through a motor experiences the resistance of the wire in the motor windings and a reverse electromotive force (reverse voltage) caused by the magnetic fields within the motor. As the motor voltage is reduced, the motor slows down. This causes a reduction of reverse electromotive force, resulting in an increase in motor current, shown in Table 7-1. This increased current can lead to premature motor burnout.

Table 7-1  The effects on an electric motor due to a voltage drop or a voltage rise

|  | VOLTAGE DROP (Low-voltage Condition) 90% of Rated Voltage | VOLTAGE RISE (High-voltage Condition) 110% of Rated Voltage |
|---|---|---|
| Full-load speed | Decrease ........ 1½% | Increase ........... 1% |
| Full-load efficiency | Decrease ......... 2% | Increase ........ ½–1% |
| Starting current | Decrease .... 10–12% | Increase ...... 10–12% |
| Maximum overload capacity | Decrease ........ 19% | Increase .......... 21% |
| Temperature rise, full Load | Increase ....... 6–7°C | Decrease ....... 3–4°C |
| Full-load current | Increase ........ 11% | Decrease ......... 7% |

## PERMISSIBLE VOLTAGE DROP LEVELS

Excessive voltage drop on feeder and branch-circuit wires is a problem on many farms. The electrical installer must understand the effects of excessive voltage drop. Wires must be selected properly to avoid excessive voltage drop. The electrical installer must keep in mind that farm electrical loads will often be increased in the future. As more load is added, the voltage drop could become excessively high.

The *National Electrical Code*, in the fine-print note of *Section 210-19(a)*, recommends limiting the voltage drop to 3% on a branch circuit to the farthest output for power, heating, or lighting. The fine-print note to *NEC Section 215-2(b)* recommends limiting voltage drop on feeder conductors to a maximum of 3%. Voltage over the combined length of both the feeder conductors and the branch circuit to the farthest outlet should not exceed 5%. Both fine-print notes make this suggestion. Figure 7-2 illustrates these voltage drop recommendations.

## EXCESSIVE VOLTAGE

Now, examine what would happen to the circuit current if the electric heater previously discussed is connected to an overvoltage. The heater is rated for 236 V, according to the nameplate. If the heater is connected to a 480-V supply, the heater will produce 12 414 W. Recall that the electric heater resistance was determined to be 18.56 Ω.

$$W = \frac{E^2}{R} = \frac{480 \times 480 \text{ V}}{18.56 \text{ }\Omega} = 12\ 414 \text{ W}$$

$$I = \frac{E}{R} = \frac{480 \text{ V}}{18.56 \text{ }\Omega} = 25.86 \text{ A}$$

A resistive load, such as an electric heating element, experiences an increase in current when the supply voltage is increased. When the supply voltage is doubled, the load current of the heater is doubled. The wattage output and, therefore, the rate of heat production, is four

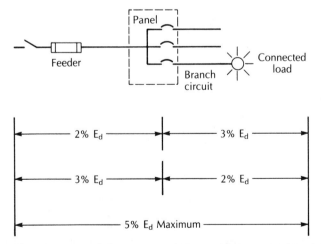

Figure 7-2  For light, power, and heating loads, the combined voltage drop should not exceed 5% on the feeder and branch-circuit wiring. As shown, if a feeder has a 2% voltage drop, then the voltage drop in the branch circuit should not exceed 3%.

times as great with a 480-V supply as compared to a 236-V supply.

**CAUTION:** This excessive overvoltage condition is dangerous. Damage will result to the heater, and a fire or shock hazard may be created.

## *MEASURING VOLTAGE DROP*

Conductors for farm applications are selected using *NEC Tables 310-16, 310-17,* and *310-19* as the minimum size permitted for the load to be served. These tables do not take voltage drop into consideration. When voltage drop is suspected to be excessive for an existing farm feeder or circuit, the electrical installer can determine the extent of the problem by using a voltmeter.

Voltage drop cannot be measured when there is no load on the feeders and branch circuit. The extent of the voltage drop is determined only when all of the equipment involved is operating. For example, voltage drop may be a problem on a dairy farm only when electrically powered feeding equipment is operated during the evening milking. The electrical load in the residence often reaches a maximum in the evening. This load may compound the problem of voltage drop. In the fall, the farmer may have a crop dryer operating. Thus, voltage drop may become excessive only during certain times of the day or, perhaps, in certain seasons.

## *Procedure*

The following procedure can be followed to measure the extent of voltage drop.

1. Turn on all electrical equipment which is normally in operation at the time voltage drop is suspected to be a problem.
2. Measure the voltage at the farm main service. It should be 235 V or more between hot conductors of a 120/240-V, single-phase system.
3. Measure the voltage at the service panelboard of the building in which voltage drop is suspected to be a problem. It should be 228 V or greater between hot conductors (maximum of 3% voltage drop on feeders).
4. Measure the voltage at such utilization equipment as motors, electric heaters, or lighting outlets farthest from the service panelboard. It should be 223 V or greater for a 240-V circuit, or 112-V for a 120-V circuit (maximum of 5% voltage drop back to main service).

Voltage measurements substantially below the suggested values are to be considered excessive. Steps should be taken by the electrical installer to find the source of the problem.

## *CAUSES OF VOLTAGE DROP*

It would be ideal if there were no resistance in the conductors leading to lights, motors, and heaters. There would be no voltage drop along the wires, and full voltage would be available at the lights, motor, or heater. But, wires do have resistance, and it is this resistance that uses up a small portion of the electrical power and reduces the voltage. The longer the path, the greater is the resistance and, therefore, the greater is the voltage drop. Increasing the cross-sectional area of the wire reduces the resistance. In the case of a long conductor, it is necessary to increase the size of the wire to limit the resistance and thus the voltage drop.

Voltage drop is caused by resistance in the conductor leading to the electrical load. There are many causes of resistance in the conductor path. The following are some of the common sources of resistance.

1. The wire is too small for the current to be carried the distance from the power source to the load.
2. A loose or corroded connection can cause high resistance.
3. Nicking a wire with a stripping device, as shown in Figure 7-3, reduces the cross-sectional area of the conductor. Voltage drop may occur at this nick when the wire is carrying current near its capacity.
4. Broken or cut strands of a stranded wire reduce the cross-sectional area of a wire, Figure 7-4. Voltage drop may occur when the wire is carrying current near its normal capacity.
5. Conductors, especially in large sizes, must be terminated properly. An improper termination can result in insufficient contact between the wire and the lug.

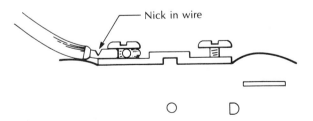

**Figure 7-3** If the wire stripping device is not properly adjusted, the wire may be nicked.

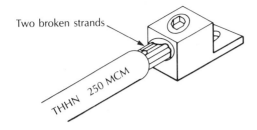

**Figure 7-4** Care must be taken to prevent strands of wire from being broken or cut.

The result may be reduced cross-section contact area, and voltage drop when the conductor is carrying current near its capacity. There may be evidence of heating damage to the wire if the problem is severe. A proper and an improper wire termination at a lug are shown in Figure 7-5.

6. Electrical equipment may not have a current rating sufficient for the load. The maximum ampere rating is marked on all electrical equipment and this rating must never be exceeded. Manufacturers generally design equipment so that the equipment would have to be altered in order to exceed its rating. With farm wiring, however, care must be taken not to keep adding new loads without making sure that the central distribution equipment is sufficient to handle the loads.
7. Plugs and receptacles must have a current rating and a voltage rating sufficient for the load to be supplied. A cord and plug too small for the current requirement of the load will cause excessive voltage drop and may be a *fire hazard*.

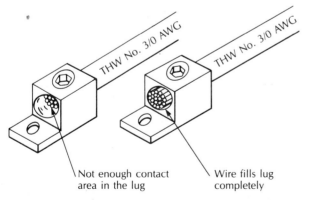

**Figure 7-5** Terminations must be made properly, especially with aluminum wire, to prevent terminal heating and voltage drop.

8. Circuit breakers and fuses should be checked to make sure that they have been installed properly. Make sure that there has been no attempt to tamper with the overcurrent device. Improper installation or tampering often results in increased resistance. Make sure that fuses fit tightly into their holders. Poor contact often causes fuses to blow needlessly, and can cause the fiber fuse tube to burn up. Tighteners (clip-clamps) are available to correct a loose connection that has resulted from the "temper" or "spring" being taken out of the fuse clips because of excessive heat. See Figure 7-6.
9. If the farm electrical load and the farm service equipment have been increased without notifying the electrical power supplier, the transformer may not be sized adequately to supply the farm load. An overloaded transformer may not be able to maintain the desired secondary voltage.
10. High resistance can be the result of control device failure. Contacts in switches, relays, motor starters and contactors break down due to wear and arcing. Sometimes it is possible to replace the contacts; at other times, the entire device must be replaced.

## SIZING WIRES TO LIMIT VOLTAGE DROP

An adequate branch-circuit or feeder wire goes beyond just selecting the minimum size wire from *NEC Tables 310-16, 310-17,* and *310-19*. The ampacity values in the tables are the maximum values allowed in order to prevent heating damage to the insulation. Remember that if heat is being produced in the wires, voltage is being used up or lost. Therefore, it is necessary to size the wire to minimize voltage drop. The voltage drop occurring on single-phase circuit wires can be determined with Equation 7.1.

**Figure 7-6** Fuse clip-clamps (tighteners) used to ensure good contact between fuse clips and the fuse itself

*Single Phase.* To find voltage drop if wire size is known:

$$E_d = \frac{K \times I \times L \times 2}{CSA} \qquad \text{Eq. 7-1}$$

where: $E_d$ = permissible voltage drop, in volts
  $K$ = 12.0 ohms-circular mils/foot for copper, 19.5 ohms-circular mils/foot for aluminum
  $I$ = current drawn by the load, in amperes
  $L$ = length or distance from power source to the load, in feet
  $CSA$ = cross-sectional area of the conductor, in circular mils. (See Table 7-3.)

The resistances of copper and aluminum wire, given in *Table 8* of *Chapter 9* of the *National Electrical Code*, are based on a wire temperature of 167°F (75°C). The values for K are 12.9 ohms-circular mils/foot for bare stranded copper and 21.2 for stranded aluminum.

The actual temperature of the wire is usually lower than 167°F. This means that the actual resistance of the wire will be lower than the values given in the *NEC*. Electrical current flowing through the wire produces heat, and this generally raises the temperature a few degrees above ambient temperature.

A temperature typical of summer should be chosen to make voltage drop calculations. It is suggested that 130°F (55°C) be taken as the operating temperature for electrical wire in agricultural installations. The values for resistivity of copper and aluminum wire are given in Table 7-2. These values for K are used for calculations in this text.

The electrical installer is most interested in determining the size of wire required to limit voltage drop to a certain amount. Equation 7.2 is used to determine the minimum wire size to limit voltage drop.

*Single Phase.* To find wire size when voltage drop is known:

$$CSA = \frac{K \times I \times L \times 2}{E_d} \qquad \text{Eq. 7.2}$$

Current is drawn on three wires in the case of a 3-phase load. A slight change in the previous equations allows the electrical installer to make an approximate determination of 3-phase voltage drop. Equation 7.3 will work as long as the currents in all of the three wires are nearly the same.

*Three Phase.* To find voltage drop if wire size is known:

$$E_d = \frac{K \times I \times L \times 1.73}{CSA} \qquad \text{Eq. 7.3}$$

When the maximum allowable voltage drop is known, wire size for a 3-phase load can be determined with Equation 7.4.

*Three Phase.* To find wire size if voltage drop is known:

$$CSA = \frac{K \times I \times L \times 1.73}{E_d} \qquad \text{Eq. 7.4}$$

Problem 7-3 shows how to determine the voltage drop on the wires of a 2-wire circuit when the conductor size is known.

## Problem 7-3

A 120-V, 24-A, single-phase load is located 90 ft (27 m) from the supply panelboard, Figure 7-7. The circuit wire is No. 10 AWG copper stranded with THW insulation. The CSA or circular mil area is found in Table 7-3. Determine the voltage drop in the wires, and the voltage across the load.

Table 7-2  Values of resistivity for copper and aluminum wire (resistivity is in ohms-circular mils/foot)

| Wire | K at 77°F (25°C) | K at 130°F (55°C) | K at 167°F (75°C) |
|---|---|---|---|
| Copper solid | 10.6 | 11.7 | 12.6 |
| Copper stranded | 10.8 | 12.0 | 12.9 |
| Aluminum solid | 17.4 | 19.1 | 20.8 |
| Aluminum stranded | 17.7 | 19.5 | 21.2 |

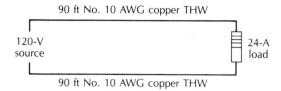

**Figure 7-7** Determining voltage drop on a 2-wire circuit when conductor size is known

## Solution

$$E_d = \frac{K \times I \times L \times 2}{CSA}$$
$$= \frac{12.0 \times 24\ A \times 90\ ft \times 2}{10\ 380\ CM} = 5.0\ V$$

The voltage across the load terminals will be approximately:

$$120\ V - 5.0 = 115.0\ V$$

**Table 7-3** Circular mil areas of conductors

| AWG Size of Conductor | Circular Mil Area of Conductor |
|---|---|
| 18 | 1 620 |
| 16 | 2 580 |
| 14 | 4 110 |
| 12 | 6 530 |
| 10 | 10 380 |
| 8 | 16 510 |
| 6 | 26 240 |
| 4 | 41 740 |
| 3 | 52 620 |
| 2 | 66 360 |
| 1 | 83 690 |
| 0 | 105 600 |
| 00 | 133 100 |
| 000 | 167 800 |
| 0000 | 211 600 |
| 250 MCM | 250 000 |
| 300 | 300 000 |
| 350 | 350 000 |
| 400 | 400 000 |
| 500 | 500 000 |
| 600 | 600 000 |
| 700 | 700 000 |
| 750 | 750 000 |
| 800 | 800 000 |
| 900 | 900 000 |
| 1 000 | 1 000 000 |

Voltage drop is generally expressed in terms of the percent of supply voltage rather than the actual drop in volts. Equation 7.5 is used to determine the percent drop if the actual drop in volts is known.

$$\text{Percent } E_d = \frac{E_d \times 100}{E_s} \qquad \text{Eq. 7.5}$$
where: $E_d$ = drop, in volts
$E_s$ = supply voltage

In order to calculate the size of wire needed to limit voltage drop to within a specified percent, the actual drop in volts ($E_d$) must be determined. Equation 7.6 is used to determine the actual drop in volts when the percent drop in volts is known.

$$E_d = \frac{\text{Percent } E_d \times E_s}{100} \qquad \text{Eq. 7.6}$$

## Problem 7-4

Determine the percent drop in volts caused by the wires in Problem 7-3.

## Solution

$$\text{Percent } E_d = \frac{E_d \times 100}{E_s} = \frac{5.0 \times 100}{120} = 4.2\%$$

## Problem 7-5

How much is the allowable actual drop in volts if the maximum voltage drop is to be 2% of 240 V?

$$E_d = \frac{\text{Percent } E_d \times E_s}{100} = \frac{2\% \times 240\ V}{100} = 4.8\ V$$

The typical problem encountered by an electrical installer is to determine the wire size required to limit voltage drop to within a maximum allowable percent of the supply voltage. The installer must determine the minimum wire size to satisfy the voltage drop limitation as well as the ampacity *Tables 310-16, 310-17,* and *310-19* in the *National Electrical Code*. Problem 7-6 illustrates this procedure.

## Problem 7-6

Determine the minimum THW copper wire size required to prevent the voltage drop on the wires from exceeding 3%. The electrical supply is 230 V, single phase. A 40-A load is located 65 ft from the electrical supply, Figure 7-8. The circuit consists of the wires in metal conduit.

## Solution

First determine the voltage drop permitted.

$$E_d = \frac{3\% \times 230 \text{ V}}{100} = 6.9 \text{ V}$$

Next, determine the circular mil area of the wire.

$$CSA = \frac{K \times I \times L \times 2}{E_d}$$
$$= \frac{12.0 \times 40 \text{ A} \times 65 \text{ ft} \times 2}{6.9 \text{ V}} = 9\,043 \text{ CM}$$

According to Table 7-3, a No. 12 AWG wire is 6 530 circular mils in area. A No. 12 wire is too small; therefore, a No. 10 wire is required. Choose No. 10 AWG copper based upon 3% voltage drop.

The electrical installer must then check *Table 310-16* of the *National Electrical Code* to determine the minimum size copper THW wire permitted to carry 40 A. According to *NEC Table 310-16*, a No. 8 AWG THW copper wire is the smallest size permitted to carry 40 A. Choose No. 8 AWG copper based upon the ampacity table. The conclusion is that voltage drop is less than 3% when a No. 8 wire is carrying 40 A.

As a general rule, voltage drop is not a problem when a 120-V load is located less than 50 ft from the electrical supply, and a 240-V load is located less than 100 ft from the supply.

Voltage drop calculations for 3-phase loads are made using Equations 7.3 and 7.4. These formulas work when the 3-phase load is balanced. A *balanced load* is one in which all three of the wires are carrying about an equal amount of current. Problem 7-7 illustrates how to calculate 3-phase voltage drop for a farm situation.

## Problem 7-7

A 74-A, 3-phase load is supplied with three No. 3 AWG aluminum wires, Figure 7-9. The load is located 80 ft from the electrical supply panelboard. Determine the voltage drop caused by the current flowing in these wires.

## Solution

For this 3-phase balanced load, 74 A flow on each wire.

$$E_d = \frac{K \times I \times L \times 1.73}{CSA}$$
$$= \frac{19.5 \times 74 \text{ A} \times 80 \text{ ft} \times 1.73}{52\,620 \text{ CM}} = 3.8 \text{ V}$$

(The 19.5 value for resistivity of aluminum wire is determined from Table 7-2).

Inadequate sizing of overhead feeder wires between buildings is a common source of excessive voltage drop on farms. The *National Electrical Code* allows a single conductor suspended in the air to carry more current than a conductor of the same size in cable or conduit. The reason is that the air flowing over the wire in free air takes the heat away from the wire. The single wire in free air will not heat up as quickly as the same wire would in cable or conduit. Voltage drop, however, depends mainly upon wire type, wire size, length of wire, and the operating temperature.

As an example, assume that a 100-A, single-phase service entrance is installed in a farm building (not a dwelling). If THW aluminum wire is used, *NEC Table*

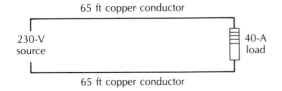

**Figure 7-8** Determining the minimum THW copper wire size required to limit voltage drop to 3%

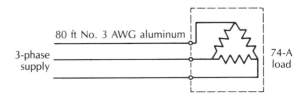

**Figure 7-9** Calculating 3-phase voltage drop for a farm situation

*310-16* requires size No. 1 AWG wire. *Table 310-17* allows No. 4 AWG THW single wires in free air to carry 100 A. If the overhead feeder is carrying 70 A and its length is 200 ft, the voltage drop will be 5.5%.

$$E_d = \frac{19.5 \times 70 \text{ A} \times 200 \text{ ft} \times 2}{41\,740 \text{ CM}} = 13.1 \text{ V}$$

$$\text{Percent } E_d = \frac{13.1 \text{ V} \times 100}{240 \text{ V}} = 5.5\%$$

Increasing the size of the overhead feeder to No. 1 AWG will reduce the voltage drop to 2.7%.

The use of the higher-temperature conductor insulation, such as THHN and XHHW, may appear to permit smaller size wire for a given load, but could result in an excessive voltage drop in the circuit conductors.

Always check the ampacity tables in the *NEC*, and then compare with the wire size determined by voltage drop calculations. The wire size is not permitted to be smaller than the *NEC* tables allow, no matter how low the voltage drop.

It is important to remember that voltage drop is unaffected by the type of insulation on the conductors. It is affected, however, by the size of the conductor, the length of the run, the spacing of the conductors, the power factor of the circuit, and the temperature of the conductors. Inductive reactance and power factor add to voltage drop when wires are larger than No. 1 AWG. This effect is considered in the next section.

## NONRESISTANCE EFFECTS ON VOLTAGE DROP

Voltage drop is more complex than just the resistance of the wire. Alternating current encounters several effects not encountered with direct current. One of these factors is the *skin effect*. When current flows through a wire of large diameter, such as 250 MCM and larger, more current flows near the outer portion of the wire than in the center. The problem is worsened when the wires are contained within steel conduit.

Another significant factor for wires larger than No. 1 AWG is inductive reactance due to line inductance. The current flowing through the wire builds up a magnetic field around the wire. In the case of large wires, this magnetic field produces a reverse voltage that opposes the flow of current similar to resistance. (Reverse voltage is explained in a later unit.) The problem becomes worse the farther apart from one another the wires are placed. Most agricultural wires, however, are placed next to one another. Line inductance is of little effect when the power factor is 1.0. Where much of the load is electric motors or electric discharge lighting, however, the power factor will be less than 1.0, and the effect of line inductance can be significant. In the case of a 700-MCM wire and an electrical load with a power factor of 0.85, the line inductance is about equal to the wire resistance. The line inductance effect is more severe when the wires are contained within steel conduit than when in earth, air, or nonsteel conduit.

The actual power factor of the load on an agricultural feeder or branch-circuit wire is difficult to predict. It will actually change as the load changes. A good assumption is to consider farm electrical load power factor at 0.85 when a feeder supplies loads consisting of electric-discharge lighting and electric motors, as well as such resistance-type loads as incandescent lights and electric heaters.

The tables in the next section can be used for sizing branch-circuit and feeder wires on farms. If the load consists mainly of such resistance-type loads as incandescent lights and electric heating equipment, the 100% power factor column is used. Otherwise, the 85% column is chosen. Skin effect and line inductance are included in the tables.

## Voltage Drop Tables

Four unique voltage drop tables have been developed for this text which take into consideration all of the significant factors which cause voltage drop on farm electrical feeders and branch circuits. They are Tables 7-4, 7-5, 7-6, and 7-7. The maximum permitted ampere rating for the wires has been summarized in each table in columns 2, 3, and 4. The numbers in the table columns 5, 6, 7, and 8 are the amounts of current which, if flowing on the wire supplying a load 100 ft (30.5 m) away from the electrical supply, are required to cause a drop of 1 V. These numbers are called *voltage drop factors*.

The tables are used for any length of circuit, and any desired percent voltage drop. The following steps provide the correct size wire for any farm feeder or branch circuit.

1. Determine if power factor should be considered. If the load consists mostly of electric discharge lighting or electric motors, then use the column for 85% power factor. If the load is mainly incandescent lamps or resistance electric heaters, then use the

Table 7-4 Voltage drop factors for copper wire in steel raceway.

$$\text{Voltage drop factor} = \frac{(\text{Load amperes}) \times (\text{One-way circuit length})}{(\% \text{ Voltage drop}) \times (\text{Circuit voltage})}$$

| Wire Size AWG or MCM | Maximum Load Amperes Rating | | | Voltage Drop Factor | | | |
|---|---|---|---|---|---|---|---|
| | 60°C (140°F) TW | 75°C (167°F) THW THWN | 90°C (194°F) THHN XHHW | Single-phase | | Three-phase | |
| | | | | Power Factor 100% | Power Factor 85% | Power Factor 100% | Power Factor 85% |
| 14 | 15 | 15 | 15 | 1.75 | 1.75 | 2.01 | 2.01 |
| 12 | 20 | 20 | 20 | 2.77 | 2.77 | 3.20 | 3.20 |
| 10 | 30 | 30 | 30 | 4.41 | 4.41 | 5.08 | 5.08 |
| 8 | 40 | 50 | 55 | 6.90 | 6.90 | 7.95 | 7.95 |
| 6 | 55 | 65 | 75 | 11.0 | 11.0 | 12.6 | 12.6 |
| 4 | 70 | 85 | 95 | 17.4 | 17.4 | 20.1 | 20.1 |
| 3 | 85 | 100 | 110 | 21.9 | 21.9 | 25.3 | 25.3 |
| 2 | 95 | 115 | 130 | 27.6 | 27.6 | 32.0 | 32.0 |
| 1 | 110 | 130 | 150 | 34.8 | 34.8 | 40.3 | 40.3 |
| 0 | 125 | 150 | 170 | 43.1 | 40.7 | 49.9 | 46.2 |
| 2/0 | 145 | 175 | 195 | 54.0 | 48.3 | 62.2 | 54.1 |
| 3/0 | 165 | 200 | 225 | 67.2 | 58.2 | 77.7 | 64.8 |
| 4/0 | 195 | 230 | 260 | 84.8 | 68.9 | 97.1 | 75.8 |
| 250 | 215 | 255 | 290 | 98.1 | 73.2 | 114 | 84.8 |
| 300 | 240 | 285 | 320 | 117 | 84.2 | 135 | 97.2 |
| 350 | 260 | 310 | 350 | 135 | 89.3 | 156 | 103 |
| 400 | 280 | 335 | 380 | 152 | 95.2 | 175 | 110 |
| 500 | 320 | 380 | 430 | 184 | 107 | 213 | 125 |
| 600 | 355 | 420 | 475 | 216 | 118 | 249 | 135 |
| 700 | 385 | 460 | 520 | 245 | 127 | 283 | 145 |
| 750 | 400 | 475 | 535 | 258 | 130 | 299 | 149 |
| 800 | 410 | 490 | 555 | 273 | 135 | 316 | 152 |
| 900 | 435 | 520 | 585 | 298 | 141 | 344 | 161 |
| 1 000 | 455 | 545 | 615 | 321 | 145 | 371 | 167 |

100% power factor column. For wire sizes No. 1 AWG and smaller, the power factor will not affect the wire size required.

2. Choose the maximum permitted voltage drop in percent (not decimal).
3. Choose the circuit voltage.
4. Determine the load to be carried by the wires. If the load will be increased in the future, then be sure to take this into consideration.
5. Choose the type of conductor for the circuit, either copper or aluminum. Consider also the type of insulation on the wires.
6. Is the load single phase or 3 phase?
7. Are the wires to be run in steel conduit, such as rigid steel, IMC, or EMT? Or, will the wires be run in rigid aluminum, PVC (rigid nonmetallic) conduit, multiplex cable in air, cable, or underground in a trench?
8. Determine the one-way length of the circuit. This is the distance from the source to the load. For example, this length may be the distance from the central distribution pole to the barn.
9. Determine the voltage drop factor for the feeder or branch circuit by using Equation 7.7.

$$\text{Voltage drop factor} = \frac{(\text{Load amperes}) \times (\text{One-way length})}{(\% \text{ Voltage drop}) \times (\text{Circuit voltage})} \quad \text{Eq. 7.7}$$

10. Choose the correct table and column and find the minimum size wire. Move down the column until a number is found as large or larger than the voltage drop factor calculated.

The voltage drop factors in the tables are based upon

**Table 7-5** Voltage drop factors for copper wire in PVC (rigid nonmetallic) or aluminum conduit, cable, multiplex cable in air, or direct-burial wire or cable.

$$\text{Voltage drop factor} = \frac{(\text{Load amperes}) \times (\text{One-way circuit length})}{(\% \text{ Voltage drop}) \times (\text{Circuit voltage})}$$

| Wire Size AWG or MCM | Maximum Load Amperes Rating | | | Voltage Drop Factor | | | |
|---|---|---|---|---|---|---|---|
| | 60°C (140°F) TW | 75°C (167°F) THW THWN | *Multiplex Aerial Cable | Single-phase Power Factor 100% | Single-phase Power Factor 85% | Three-phase Power Factor 100% | Three-phase Power Factor 85% |
| 14 | 15 | 15 | — | 1.75 | 1.75 | 2.01 | 2.01 |
| 12 | 20 | 20 | — | 2.77 | 2.77 | 3.20 | 3.20 |
| 10 | 30 | 30 | 55 | 4.41 | 4.41 | 5.08 | 5.08 |
| 8 | 40 | 50 | 70 | 6.90 | 6.90 | 7.95 | 7.95 |
| 6 | 55 | 65 | 100 | 11.0 | 11.0 | 12.6 | 12.6 |
| 4 | 70 | 85 | 130 | 17.4 | 17.4 | 20.1 | 20.1 |
| 3 | 85 | 100 | 150 | 21.9 | 21.9 | 25.3 | 25.3 |
| 2 | 95 | 115 | 175 | 27.6 | 27.6 | 32.0 | 32.0 |
| 1 | 110 | 130 | 205 | 34.8 | 34.8 | 40.3 | 40.3 |
| 0 | 125 | 150 | 235 | 44.0 | 41.2 | 50.9 | 47.6 |
| 2/0 | 145 | 175 | 275 | 55.5 | 50.0 | 64.1 | 57.8 |
| 3/0 | 165 | 200 | 320 | 69.9 | 60.6 | 80.8 | 69.9 |
| 4/0 | 195 | 230 | 370 | 88.2 | 73.5 | 101 | 84.0 |
| 250 | 215 | 255 | 410 | 104 | 82.6 | 120 | 95.2 |
| 300 | 240 | 285 | — | 124 | 93.5 | 144 | 105 |
| 350 | 260 | 310 | — | 145 | 104 | 167 | 120 |
| 400 | 280 | 335 | — | 165 | 114 | 191 | 132 |
| 500 | 320 | 380 | — | 205 | 130 | 237 | 152 |
| 600 | 355 | 420 | — | 244 | 145 | 282 | 167 |
| 700 | 385 | 460 | — | 282 | 156 | 326 | 182 |
| 750 | 400 | 475 | — | 301 | 161 | 348 | 189 |
| 800 | 410 | 490 | — | 319 | 167 | 369 | 192 |
| 900 | 435 | 520 | — | 355 | 179 | 411 | 204 |
| 1 000 | 455 | 545 | — | 391 | 185 | 451 | 217 |

*Refer to NEC Section 230-22 and Table 310-19.

copper and aluminum wire K values for 130°F (55°C). Skin effect and line inductance are also considered.

## Problem 7-8

An overhead aluminum multiplex feeder supplies a 150-A, single-phase, 120/240-V service in a barn from the central distribution pole. If the distance from the pole to the barn is 185 ft, size the wire to limit the voltage drop to 2½%.

## Solution

The barn contains electric motors and fluorescent lights; therefore, a power factor of 85% will be assumed.

Next, determine the voltage drop factor for the overhead feeder, using Equation 7.7.

$$\text{Voltage drop factor} = \frac{150 \text{ A} \times 185 \text{ ft}}{2.5\% \times 240 \text{ V}} = 46.3$$

The aluminum overhead wire size is determined from the 85% power factor, single-phase column of Table 7-7. Move down the column of the table until a drop factor is encountered which is equal to or greater than the voltage drop factor calculated. For this example, the minimum recommended size overhead aluminum multiplex cable is No. 4/0 AWG. Check the multiplex aerial cable column of the table to make sure that an aluminum multiplex cable is allowed to carry 150 A.

Table 7-6  Voltage drop factors for aluminum wire in steel raceway.

$$\text{Voltage drop factor} = \frac{(\text{Load amperes}) \times (\text{One-way circuit length})}{(\% \text{ Voltage drop}) \times (\text{Circuit voltage})}$$

| Wire Size AWG or MCM | Maximum Load Amperes Rating | | | Voltage Drop Factor | | | |
|---|---|---|---|---|---|---|---|
| | 60°C (140°F) TW | 75°C (167°F) THW THWN | 90°C (194°F) THHN XHHW | Single-phase | | Three-phase | |
| | | | | Power Factor 100% | Power Factor 85% | Power Factor 100% | Power Factor 85% |
| 12 | 15 | 15 | 15 | 1.72 | 1.72 | 1.99 | 1.99 |
| 10 | 25 | 25 | 25 | 2.73 | 2.73 | 3.16 | 3.16 |
| 8 | 30 | 40 | 45 | 4.34 | 4.34 | 5.02 | 5.02 |
| 6 | 40 | 50 | 60 | 6.91 | 6.91 | 7.98 | 7.98 |
| 4 | 55 | 65 | 75 | 11.0 | 11.0 | 12.7 | 12.7 |
| 3 | 65 | 75 | 85 | 13.8 | 13.8 | 16.0 | 16.0 |
| 2 | 75 | 90 | 100 | 17.5 | 17.5 | 20.2 | 20.2 |
| 1 | 85 | 100 | 115 | 22.0 | 22.0 | 25.5 | 25.5 |
| 0 | 100 | 120 | 135 | 27.8 | 27.8 | 32.1 | 32.1 |
| 2/0 | 115 | 135 | 150 | 35.0 | 32.6 | 40.5 | 37.6 |
| 3/0 | 130 | 155 | 175 | 43.7 | 39.2 | 50.5 | 45.2 |
| 4/0 | 150 | 180 | 205 | 55.1 | 47.2 | 63.7 | 54.3 |
| 250 | 170 | 205 | 230 | 64.5 | 53.2 | 74.6 | 61.3 |
| 300 | 190 | 230 | 255 | 77.4 | 61.3 | 89.5 | 70.9 |
| 350 | 210 | 250 | 280 | 89.4 | 68.0 | 103 | 78.7 |
| 400 | 225 | 270 | 305 | 101 | 74.1 | 117 | 85.5 |
| 500 | 260 | 310 | 350 | 124 | 85.5 | 144 | 98.0 |
| 600 | 285 | 340 | 385 | 146 | 94.3 | 169 | 110 |
| 700 | 310 | 375 | 420 | 166 | 102 | 192 | 118 |
| 750 | 320 | 385 | 435 | 176 | 106 | 196 | 122 |
| 800 | 330 | 395 | 450 | 185 | 109 | 213 | 127 |
| 900 | 355 | 425 | 480 | 204 | 115 | 236 | 132 |
| 1 000 | 375 | 445 | 500 | 221 | 120 | 256 | 139 |

## VOLTAGE DROP ON SERIES FEEDERS

Feeder wires serving buildings on a farm are frequently connected in series. More than one building may be served. Figure 7-10 shows a typical example of two loads in series. The feeder between the central pole and the barn carries both the barn load and the shop load. The feeder between the barn and the shop carries only the shop load. The following procedure is used to determine the wire size to limit the voltage drop to the shop to only 3%.

First, the actual load on each feeder must be determined. The 110-ft feeder in Figure 7-10 will carry 300 A, and the 70-ft feeder will carry 100 A.

Next, the amount of voltage drop to be allowed on

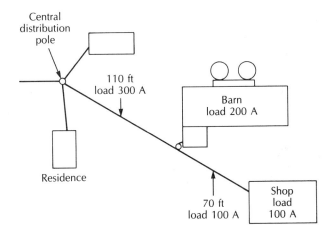

Figure 7-10  Two farm loads occurring at the same time and sharing a portion of the same feeder wire

Table 7-7  Voltage drop factors for aluminum wire in PVC (rigid nonmetallic) or aluminum conduit, cable, multiplex cable in air, or direct-burial wire or cable.

$$\text{Voltage drop factor} = \frac{(\text{Load amperes}) \times (\text{One-way circuit length})}{(\% \text{ Voltage drop}) \times (\text{Circuit voltage})}$$

| Wire Size AWG or MCM | Maximum Load Amperes Rating | | | Voltage Drop Factor | | | |
|---|---|---|---|---|---|---|---|
| | 60°C (140°F) TW | 75°C (167°F) THW THWN | *Multiplex Aerial Cable | Single-phase | | Three-phase | |
| | | | | Power Factor 100% | Power Factor 85% | Power Factor 100% | Power Factor 85% |
| 12 | 15 | 15 | — | 1.72 | 1.72 | 1.99 | 1.99 |
| 10 | 25 | 25 | — | 2.73 | 2.73 | 3.16 | 3.16 |
| 8 | 30 | 40 | 55 | 4.34 | 4.34 | 5.02 | 5.02 |
| 6 | 40 | 50 | 80 | 6.91 | 6.91 | 7.98 | 7.98 |
| 4 | 55 | 65 | 100 | 11.0 | 11.0 | 12.7 | 12.7 |
| 3 | 65 | 75 | 115 | 13.8 | 13.8 | 16.0 | 16.0 |
| 2 | 75 | 90 | 135 | 17.5 | 17.5 | 20.2 | 20.2 |
| 1 | 85 | 100 | 160 | 22.0 | 22.0 | 25.5 | 25.5 |
| 0 | 100 | 120 | 185 | 27.8 | 27.8 | 32.1 | 32.1 |
| 2/0 | 115 | 135 | 215 | 35.0 | 35.0 | 40.5 | 40.5 |
| 3/0 | 130 | 155 | 250 | 43.7 | 41.0 | 50.5 | 47.6 |
| 4/0 | 150 | 180 | 290 | 55.1 | 50.0 | 63.7 | 57.5 |
| 250 | 170 | 205 | 320 | 65.7 | 56.8 | 75.9 | 65.8 |
| 300 | 190 | 230 | 360 | 78.7 | 65.4 | 91.0 | 75.2 |
| 350 | 210 | 250 | 400 | 91.7 | 74.1 | 106 | 85.5 |
| 400 | 225 | 270 | 435 | 105 | 81.3 | 121 | 94.3 |
| 500 | 260 | 310 | 490 | 131 | 96.2 | 151 | 111 |
| 600 | 285 | 340 | — | 156 | 108 | 181 | 123 |
| 700 | 310 | 375 | — | 182 | 118 | 210 | 137 |
| 750 | 320 | 385 | — | 194 | 123 | 224 | 143 |
| 800 | 330 | 395 | — | 207 | 127 | 239 | 147 |
| 900 | 355 | 425 | — | 232 | 137 | 268 | 156 |
| 1 000 | 375 | 445 | — | 256 | 145 | 297 | 167 |

*Refer to NEC Section 230-22 and Table 310-19.

each wire must be decided. This is a matter of judgment on the part of the electrical installer. Usually, more voltage drop is allowed on the feeder carrying the larger load. For this example, allow 2% on the 110-ft wire and 1% on the 70-ft wire. This adds up to a total of 3% from the central pole to the shop.

The size of wire is then determined using the voltage drop factor tables. If the loads are 120/240-V, single-phase, the calculations for aluminum multiplex wire in free air are as follows.

110-ft wire with a 300-A load:

$$\text{Voltage drop factor} = \frac{300 \text{ A} \times 100 \text{ ft}}{2\% \times 240 \text{ V}} = 68.8$$

The power factor is assumed to be 85%. Therefore, from Table 7-7, the minimum size wire which will limit the voltage drop to 2% is 350 MCM.

Now consider the 70-ft (21.34-m) length of wire carrying the 100-A load.

$$\text{Voltage drop factor} = \frac{100 \text{ A} \times 70 \text{ ft}}{1\% \times 240 \text{ V}} = 29.2$$

The minimum wire size for this feeder to limit voltage drop to 1% is No. 2/0 AWG.

A check of the multiplex aerial cable column of Table 7-7 shows that these wires are rated to carry the intended load.

If 350-MCM aluminum multiplex cable is not available, then two smaller sizes of multiplex cable may be run in parallel, each carrying half of the load. (Refer to Unit 4 for proper installation of feeder wires in parallel.) The minimum size wire permitted to be run in parallel is No. 0 AWG, *NEC Section 310-4*.

The solution to the previous problem using two sets of parallel multiplex aerial cables for the 110-ft run begins by determining the load for each wire.

$$\frac{300 \text{ A}}{2 \text{ sets}} = 150 \text{ A per cable set}$$

$$\text{Voltage drop factor} = \frac{150 \text{ A} \times 110 \text{ ft}}{2\% \times 240 \text{ V}} = 34.4$$

Remember that the problem assumes 85% power factor. This time, it is found that two sets of No. 2/0 aluminum cables will carry 300 A a distance of 110 ft without exceeding a voltage drop of 2%.

## SERVING DISTANT LOADS

Occasionally, a farm electrical load is located a considerable distance from the electrical supply, Figure 7-11. When practical, the power supplier's primary wires should be brought as close to the load as possible. If the load is quite small, or operated only seasonally, then the cost of bringing primary power close to the load is often not economically justified.

Voltage drop must be considered carefully when serving a distant load with secondary power. The example in Problem 7-9 illustrates how judgment must be applied.

## Problem 7-9

A beef farmer intends to install two stock waterers in adjacent feed lots located 800 ft from the main panelboard in the barn. These stock waterers contain electric heating elements to prevent freezing in the winter. The waterers each draw 4.3 A, and are to be served by a 15-A, 120-V circuit using copper UF cable installed underground. The barn is close to the central distribution pole. Assume that the feeder between the pole and the barn experiences only a 1% voltage drop. Determine the size copper wire for this 800-ft branch circuit.

## Solution

A 5% voltage drop can be allowed to the most distant outlet. If 1% is lost on the feeder from the central pole to the barn, then 4% can be allowed on the 800-ft circuit from the barn to the waterers. Determine the voltage drop factor for the circuit.

$$\text{Voltage drop factor} = \frac{8.6 \text{ A} \times 800 \text{ ft}}{4\% \times 120 \text{ V}} = 14.3$$

The stock waterers are resistance loads; therefore, the 100% column of Table 7-5 is used. The minimum wire size to limit the voltage drop to 4% is No. 4 AWG.

An alternative is to supply the stock waterers with 240 V. Then, at the load, a small transformer could be used to step down the voltage to 120 V to match the load, Figure 7-12. The cost of installing the transformer would have to be compared to the savings from using a reduced wire size. At 240 V, the load will be only 4.3 A.

$$\text{Voltage drop factor} = \frac{4.3 \text{ A} \times 800 \text{ ft}}{4\% \times 240 \text{ V}} = 3.6$$

From Table 7-5, the minimum copper wire size is now No. 10 AWG.

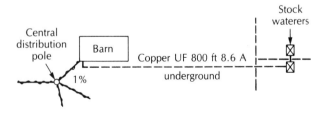

Figure 7-11 Serving distant loads

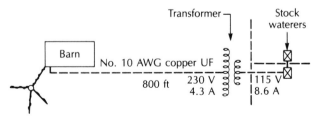

Figure 7-12 A small transformer steps down the voltage to match the load.

# REVIEW

Refer to the *National Electrical Code* when necessary to complete the following review material. Write your answers on a separate sheet of paper.

1. Is voltage drop commonly referred to as *IR drop,* or is it referred to as *watts loss?*
2. To determine the approximate voltage drop in a conductor, multiply the __?__ passing through the conductor by the __?__ of the conductor.
3. When a low-voltage condition occurs in a circuit supplying lights or electric heaters, will the lights burn dimly, normally, or brightly? Will the electric heaters heat normally, below normal, or above normal?
4. Will a low-voltage condition supplying a fully loaded electric motor result in the motor running hot, normally, or cool? Is this because the full-load running current is normal, below normal, or above normal?
5. Can voltage drop be determined only when the load is turned off, or only when the load is turned on? Is voltage drop determined by measuring the current at various locations along the circuit, or is it determined by measuring the voltage at various locations along the circuit?
6. A single-phase, 2-wire electric water heater is rated at 240 V, 5 500 W. Calculate the normal current draw.
7. An electric water heater is rated at 4 500 W, 240 V, single phase. The minimum size wire required for the water heater is No. 10 AWG THW copper. The water heater is located 200 ft (61 m) from the service panelboard. If 240 V are available at the panelboard, determine the actual voltage drop on this branch-circuit conductor.
8. A 4 000-W, 230-V, single-phase heater draws 17.4 A, but the heater was connected by error to a 115-V source. Determine the current that the heater will draw at 115 V, and the actual operating wattage of the heater.
9. A 100-A, 3-phase feeder is run in steel conduit from the main service equipment to a panelboard 135 ft (41.15 m) away. The power factor of the load at the panelboard is assumed to be 85%. Determine the minimum size THHN copper wire required to limit the voltage drop to not more than 2% if the voltage at the main service is 240 V.
10. Study the following illustration. Three buildings are served with a 230-V, 3-phase power with a common overhead aluminum multiplex feeder. Size the wires to limit the voltage drop to 3% to the buildings with the 100-A load if the power factor is assumed to be 85%.

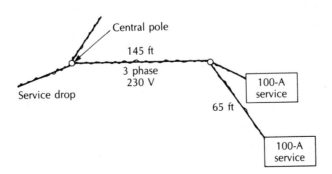

# UNIT 8
# FEEDERS

## OBJECTIVES

After studying this unit, the student will be able to

- choose an appropriate wire for an inside, overhead, or underground feeder.
- select the proper size wire for a feeder.
- size the overcurrent protection for a feeder.
- provide proper grounding and bonding to panelboards supplied by feeders.
- list clearance requirements for overhead feeders.
- list depth of burial requirements for underground feeders.
- size feeder taps and describe the procedure for protecting taps from overcurrent.

An *electrical feeder* provides power to a point where branch circuits originate. It derives its name from the function it performs: feeding power to circuits. The fuse or circuit breaker protecting the feeder, therefore, is not the last overcurrent device before the electricity is used to produce light, power, or heat. *NEC Article 220* refers to the service-entrance conductors as feeders.

This unit is concerned with the sizing and installation of feeder conductors on farms between buildings and inside buildings. The situations discussed include overhead and underground feeders from the main service or central distribution pole to farm buildings, and feeders connecting the building service entrance to subpanels within the building.

## FEEDER WIRE

Many materials and methods are available to the electrician for installing electrical supply feeders between and in farm buildings. Most situations involve only a few common techniques and materials:

- Individual overhead wires on insulators
- Messenger supported multiplex overhead cables, Figure 8-1
- Individual underground wires
- Underground cables
- Interior cables
- Individual wires in raceway

Figure 8-1 Messenger supported triplex overhead feeder supplying power to a dairy barn

Feeder conductors for agricultural applications are of either copper or aluminum, *NEC Section 310-2(b)*. Conductors are required to be insulated, except grounding conductors and the neutral wire for overhead feeders and service entrances, *NEC Section 310-2(a)*. Wire manufactured for overhead applications is available as a solid wire in some large sizes, but usually it is stranded for sizes No. 8 AWG and larger. When electrical wire is installed in raceway, sizes No. 8 AWG and larger are required to be stranded, *NEC Section 310-3*.

**Feeder Wire and Cable for Damp Locations**

| *Individual wires in raceway* | *Cables* |
|---|---|
| RHW | UF |
| RUW | SE Type U |
| TW | SE Type R |
| THW | MI |
| THWN | MC |
| XHHW | USE |
|  | SNM |
|  | Messenger supported multiplex cable |

The environmental conditions around the feeder wire must be considered carefully for agricultural installations. Moisture is found frequently on farms, and, sometimes, corrosive atmospheres are present. Wire and cable installed in wet locations must be listed for use in wet locations, *NEC Section 310-8*. The following feeder wire and cable are suitable for interior use as farm building feeders in damp or wet locations.

Feeders are commonly placed underground on farms, and several types of wire or cable are used for this application. Types UF cable, USE cable, USE individual wire, and MC cable are approved for direct burial in the earth. Underground service-entrance cable and its permitted use is described in *NEC Article 338*.

In the past, overhead feeder wires were generally single copper wires connected to insulators. Due to the cost and weight associated with copper, most overhead feeders on farms are now aluminum. Occasionally, a feeder is run as three or four single wires on insulators. However, most overhead aluminum feeders consist of a set of two insulated wires and a bare neutral wire. The neutral wire is called a *messenger*. The messenger contains one steel strand for strength. This 3-wire set is referred to as *triplex*. Another type consists of three insulated wires and a messenger. This type is called *quadruplex*. The insulation on these wires is generally cross-linked polyethylene, XHHW.

In the past, triplex and quadruplex were not recognized by the *National Electrical Code;* therefore, they were not approved for use in a building interior. This type of cable falls within the category of messenger supported wiring. Its installation and use is covered in *NEC Article 321*. For other cables that may be supported with a messenger wire, see *NEC Section 321-3(a)*.

## SIZING FEEDER CONDUCTORS

The size of a feeder conductor is based upon the load as determined by *NEC Section 220-B*. This requirement applies both to outside feeders, *NEC Section 225-5*, and interior feeders, *NEC Section 215-2*. Several factors must be considered when determining the ampere rating of the feeder:

- Calculated load or demand load
- Voltage drop
- Future loads

The voltage drop on a feeder should not exceed 3%, according to the fine-print note of *NEC Section 215-2*. Usually voltage drop on a feeder is not a problem unless the feeder is more than 100 ft (30.5 m) in length.

Farm operators frequently add new equipment to their enterprise to increase efficiency and productivity. Electrical feeders should be of adequate size to handle some additional load beyond the need at the time of installation. This is a situation requiring judgment and input from the farm operator. The electrical installer should question the farmer as to the possibility of adding future loads, and explain the merits of providing some extra capacity. The farmer, however, is the one who will make the final decision after considering the information provided by the installer.

Once the ampacity of the feeder has been determined, the size of the wire is found from one of the *NEC* wire tables. The size of wire is based upon several factors:

- Copper or aluminum wire
- Insulation temperature rating
- Single conductor in air, multiplex cable in air, or wire in raceway, cable, or direct burial

Aluminum wire has a higher resistance than copper, so a larger size is required when aluminum is selected.

Aluminum wire is used frequently for feeders on farms because it is less costly than copper. Aluminum is generally used for aerial feeders because of its lighter weight. However, under certain circumstances, some electrical installers may prefer to use copper wires overhead.

A limiting factor for the ampacity of a wire is the type of insulation covering on the wire. All wire has some resistance and, as current flows through the wire, heat is produced. This heat raises the temperature of the wire. The maximum safe ampacity of the wire is the maximum continuous current which can flow without heating the insulation beyond its maximum temperature limit. If the temperature rating of the insulation is exceeded repeatedly, the insulation will break down, causing a *dangerous* situation. The factors affecting temperature rating are listed in *NEC Section 310-10*. Electrical wire has a minimum insulation temperature rating of 60°C (140°F). This temperature rating applies to RUW and TW wire, and UF cable. The wires inside UF cable are rated TW. An "H" in the insulation type indicates an additional 15°C (59°F) maximum temperature limit. Common wires in the 75°C (167°F) category are RHW, THW, THWN, USE, and XHHW. XHHW is rated at 75°C (167°F) in damp or wet locations, and at 90°C (194°F) in dry locations.

The size of conductor for a feeder is determined from *NEC Table 310-16* for wires in raceway, cable, and direct burial. Aerial triplex and quadruplex cable is sized as directed by the fine-print note of *NEC Section 230-22*. The columns of *NEC Table 310-19* for bare conductors are used for multiplex aerial cable run outside between buildings. *NEC Table 310-19* is represented in this text as Table 8-1.

Overhead feeder wires may be installed as individual wires mounted on insulators. If the wire is insulated, then the size is determined from *NEC Table 310-17* (represented in this text as Table 8-2). If bare wire is mounted outside in free air on insulators, then the bare conductors columns of Table 8-1 are used to determine minimum size.

An example will help to illustrate how farm electrical feeders are selected. Assume that a farm shop has a demand load of 65 A, and an electrician is to install a 100-A load center in the shop. The feeder to the shop must have a minimum ampacity of 65 A, as determined by the method in *NEC Article 220*. In this case, the electrician must size the feeder based upon the 100-A rating of the load center, because the feeder is protected by a 100-A overcurrent device. The feeder will be overhead aluminum triplex cable.

**Table 8-1** (*NEC Table 310-19*) Ampacities for insulated conductors rated 0–2000 V, 110°C to 250°C, and for bare or covered conductors. Single conductors in free air, based on ambient temperature of 30°C (86°F)

| Size AWG MCM | 110°C (230°F) Types AVA, AVL | 125°C (257°F) Types AI, AIA | 150°C (302°F) Type Z | 200°C (392°F) Types A, AA, FEP, FEPB, PFA | Bare or covered conductors | 250°C (482°F) Types PFAH, TFE | 110°C (230°F) Types AVA, AVL | 125°C (257°F) Types AI, AIA | 200°C (392°F) Types A, AA | Bare or covered conductors | Size AWG MCM |
|---|---|---|---|---|---|---|---|---|---|---|---|
| | Copper | | | | | Nickel or Nickel-Coated Copper | Aluminum or Copper-Clad Aluminum | | | | |
| 14 | 40 | 40 | 40 | 45 | 30 | 60 | .... | .... | .... | .... | |
| 12 | 50 | 50 | 50 | 55 | 40 | 80 | 40 | 40 | 45 | 30 | 12 |
| 10 | 65 | 70 | 70 | 75 | 55 | 110 | 50 | 55 | 60 | 45 | 10 |
| 8 | 85 | 90 | 95 | 100 | 70 | 145 | 65 | 70 | 80 | 55 | 8 |
| 6 | 120 | 125 | 130 | 135 | 100 | 210 | 95 | 100 | 105 | 80 | 6 |
| 4 | 160 | 170 | 175 | 180 | 130 | 285 | 125 | 135 | 140 | 100 | 4 |
| 3 | 180 | 195 | 200 | 210 | 150 | 335 | 140 | 150 | 165 | 115 | 3 |
| 2 | 210 | 225 | 230 | 240 | 175 | 390 | 165 | 175 | 185 | 135 | 2 |
| 1 | 245 | 265 | 270 | 280 | 205 | 450 | 190 | 205 | 220 | 160 | 1 |
| 0 | 285 | 305 | 310 | 325 | 235 | 545 | 220 | 240 | 255 | 185 | 0 |
| 00 | 330 | 355 | 360 | 370 | 275 | 605 | 255 | 275 | 290 | 215 | 00 |
| 000 | 385 | 410 | 415 | 430 | 320 | 725 | 300 | 320 | 335 | 250 | 000 |
| 0000 | 445 | 475 | 490 | 510 | 370 | 850 | 345 | 370 | 400 | 290 | 0000 |
| 250 | 495 | 530 | .... | .... | 410 | .... | 385 | 415 | 320 | 320 | 250 |
| 300 | 555 | 590 | .... | .... | 460 | .... | 435 | 460 | 360 | 360 | 300 |
| 350 | 610 | 655 | .... | .... | 510 | .... | 475 | 510 | 400 | 400 | 350 |
| 400 | 665 | 710 | .... | .... | 555 | .... | 520 | 555 | 435 | 435 | 400 |
| 500 | 765 | 815 | .... | .... | 630 | .... | 595 | 635 | 490 | 490 | 500 |
| 600 | 855 | 910 | .... | .... | 710 | .... | 675 | 720 | .... | 560 | 600 |
| 700 | 940 | 1 005 | .... | .... | 780 | .... | 745 | 795 | .... | 615 | 700 |
| 750 | 980 | 1 045 | .... | .... | 810 | .... | 775 | 825 | .... | 640 | 750 |
| 800 | 1 020 | 1 085 | .... | .... | 845 | .... | 805 | 855 | .... | 670 | 800 |
| 900 | .... | .... | .... | .... | 905 | .... | .... | .... | .... | 725 | 900 |
| 1 000 | 1 165 | 1 240 | .... | .... | 965 | .... | 930 | 990 | .... | 770 | 1 000 |
| 1 500 | 1 450 | .... | .... | .... | 1 215 | .... | 1 175 | .... | .... | 985 | 1 500 |
| 2 000 | 1 715 | .... | .... | .... | 1 405 | .... | 1 425 | .... | .... | 1 165 | 2 000 |

**Ampacity Correction Factors**

| Ambient Temp. °C | For ambient temperatures other than 30°C, multiply the ampacities shown above by the appropriate factor shown below. | | | | | | | | | | Ambient Temp. °F |
|---|---|---|---|---|---|---|---|---|---|---|---|
| 31–40 | .94 | .95 | .96 | .... | | .... | .94 | .95 | .... | | 87–104 |
| 41–45 | .90 | .92 | .94 | .... | | .... | .90 | .92 | .... | | 105–113 |
| 46–50 | .87 | .89 | .91 | .... | | .... | .87 | .89 | .... | | 114–122 |
| 51–55 | .83 | .86 | .89 | .... | | .... | .83 | .86 | .... | | 123–131 |
| 56–60 | .79 | .83 | .87 | .91 | | .95 | .79 | .83 | .91 | | 132–141 |
| 61–70 | .71 | .76 | .82 | .87 | | .91 | .71 | .76 | .87 | | 142–158 |
| 71–75 | .66 | .72 | .79 | .86 | | .89 | .66 | .72 | .86 | | 159–167 |
| 76–80 | .61 | .68 | .76 | .84 | | .87 | .61 | .69 | .84 | | 168–176 |
| 81–90 | .50 | .61 | .71 | .80 | | .83 | .50 | .61 | .80 | | 177–194 |
| 91–100 | .... | .51 | .65 | .77 | | .80 | .... | .51 | .77 | | 195–212 |
| 101–120 | .... | .... | .50 | .69 | | .72 | .... | .... | .69 | | 213–248 |
| 121–140 | .... | .... | .29 | .59 | | .59 | .... | .... | .59 | | 249–284 |
| 141–160 | .... | .... | .... | .... | | .54 | .... | .... | .... | | 285–320 |
| 161–180 | .... | .... | .... | .... | | .50 | .... | .... | .... | | 321–356 |
| 181–200 | .... | .... | .... | .... | | .43 | .... | .... | .... | | 357–392 |
| 201–225 | .... | .... | .... | .... | | .30 | .... | .... | .... | | 393–437 |

Reprinted with permission from NFPA 70-1984, *National Electrical Code*®, Copyright © 1983, National Fire Protection Association, Quincy, Massachusetts 02269. This reprinted material is not the complete and official position of the NFPA on the referenced subject, which is represented only by the standard in its entirety.

**Table 8-2** (*NEC Table 310-17*) Ampacities for insulated conductors rater 0–2000 Volts, 60°C to 90°C. Single conductors in free air, based on ambient temperature of 30°C (86°F)

| Size AWG MCM | Temperature Rating of Conductor, See Table NEC 310-13. | | | | | | | | Size AWG MCM |
|---|---|---|---|---|---|---|---|---|---|
| | 60°C (140°F) | 75°C (167°F) | 85°C (185°F) | 90°C (194°F) | 60°C (140°F) | 75°C (167°F) | 85°C (185°F) | 90°C (194°F) | |
| | Types †RUW, †T, †TW | Types †FEPW, †RH, †RHW, †RUH, †THW, †THWN, †XHHW, †ZW | Types V, MI | Types TA, TBS, SA, AVB, SIS, †FEP, †FEPB, †RHH †THHN, †XHHW* | Types †RUW, †T, †TW | Types †RH, †RHW, †RUH, †THW, †THWN, †XHHW | Types V, MI | Types TA, TBS, SA, AVB, SIS, †RHH, †THHN, †XHHW* | |
| | Copper | | | | Aluminum or Copper-Clad Aluminum | | | | |
| 18 | .... | .... | .... | 18 | .... | .... | .... | .... | .... |
| 16 | .... | .... | 23 | 24 | .... | .... | .... | .... | .... |
| 14 | 25† | 30† | 30 | 35† | .... | .... | .... | .... | .... |
| 12 | 30† | 35† | 40 | 40† | 25† | 30† | 30 | 35† | 12 |
| 10 | 40† | 50† | 55 | 55† | 35† | 40† | 40 | 40† | 10 |
| 8 | 60 | 70 | 75 | 80 | 45 | 55 | 60 | 60 | 8 |
| 6 | 80 | 95 | 100 | 105 | 60 | 75 | 80 | 80 | 6 |
| 4 | 105 | 125 | 135 | 140 | 80 | 100 | 105 | 110 | 4 |
| 3 | 120 | 145 | 160 | 165 | 95 | 115 | 125 | 130 | 3 |
| 2 | 140 | 170 | 185 | 190 | 110 | 135 | 145 | 150 | 2 |
| 1 | 165 | 195 | 215 | 220 | 130 | 155 | 165 | 175 | 1 |
| 0 | 195 | 230 | 250 | 260 | 150 | 180 | 195 | 205 | 0 |
| 00 | 225 | 265 | 290 | 300 | 175 | 210 | 225 | 235 | 00 |
| 000 | 260 | 310 | 335 | 350 | 200 | 240 | 265 | 275 | 000 |
| 0000 | 300 | 360 | 390 | 405 | 235 | 280 | 305 | 315 | 0000 |
| 250 | 340 | 405 | 440 | 455 | 265 | 315 | 345 | 355 | 250 |
| 300 | 375 | 445 | 485 | 505 | 290 | 350 | 380 | 395 | 300 |
| 350 | 420 | 505 | 550 | 570 | 330 | 395 | 430 | 445 | 350 |
| 400 | 455 | 545 | 595 | 615 | 355 | 425 | 465 | 480 | 400 |
| 500 | 515 | 620 | 675 | 700 | 405 | 485 | 525 | 545 | 500 |
| 600 | 575 | 690 | 750 | 780 | 455 | 540 | 595 | 615 | 600 |
| 700 | 630 | 755 | 825 | 855 | 500 | 595 | 650 | 675 | 700 |
| 750 | 655 | 785 | 855 | 885 | 515 | 620 | 675 | 700 | 750 |
| 800 | 680 | 815 | 885 | 920 | 535 | 645 | 700 | 725 | 800 |
| 900 | 730 | 870 | 950 | 985 | 580 | 700 | 760 | 785 | 900 |
| 1 000 | 780 | 935 | 1 020 | 1 055 | 625 | 750 | 815 | 845 | 1 000 |
| 1 250 | 890 | 1 065 | 1 160 | 1 200 | 710 | 855 | 930 | 960 | 1 250 |
| 1 500 | 980 | 1 175 | 1 275 | 1 325 | 795 | 950 | 1 035 | 1 075 | 1 500 |
| 1 750 | 1 070 | 1 280 | 1 385 | 1 445 | 875 | 1 050 | 1 145 | 1 185 | 1 750 |
| 2 000 | 1 155 | 1 385 | 1 505 | 1 560 | 960 | 1 150 | 1 250 | 1 335 | 2 000 |
| **Ampacity Correction Factors** | | | | | | | | | |
| Ambient Temp. °C | For ambient temperatures other than 30°C, multiply the ampacities shown above by the appropriate factor shown below. | | | | | | | | Ambient Temp. °F |
| 31–40 | .82 | .88 | .90 | .91 | .82 | .88 | .90 | .91 | 87–104 |
| 41–45 | .71 | .82 | .85 | .87 | .71 | .82 | .85 | .87 | 105–113 |
| 46–50 | .58 | .75 | .80 | .82 | .58 | .75 | .80 | .82 | 114–122 |
| 51–60 | .... | .58 | .67 | .71 | .... | .58 | .67 | .71 | 123–141 |
| 60–70 | .... | .35 | .52 | .58 | .... | .35 | .52 | .58 | 142–158 |
| 71–80 | .... | .... | .30 | .41 | .... | .... | .30 | .41 | 159–176 |

†The overcurrent protection for conductor types marked with an obelisk (†) shall not exceed 20 amperes for 14 AWG, 25 amperes for 12 AWG, and 40 amperes for 10 AWG copper, or 20 amperes for 12 AWG and 30 amperes for 10 AWG aluminum and copper-clad aluminum after any correction factor for ambient has been applied.

*For dry locations only. See 75°C column for wet locations.

Reprinted with permission from NFPA 70-1984, *National Electrical Code®*, Copyright © 1983, National Fire Protection Association, Quincy, Massachusetts 02269. This reprinted material is not the complete and official position of the NFPA on the referenced subject, which is represented only by the standard in its entirety.

The minimum size aluminum triplex wire is found in the columns for bare aluminum in *NEC Table 310-19*. The electrician will install aluminum triplex size No. 4 AWG with a rating of 100 A. It may be necessary to check voltage drop.

Assume, instead, that the electrician decided to install individual XHHW copper wires overhead on insulators (75°C rated because of wet conditions) for the farm shop in the previous example. The size of wire is determined from *NEC Table 310-17*. No. 4 AWG copper wire is permitted to carry 125 A as a single conductor in free air. No. 6 AWG wire is permitted to carry 95 A. The electrician may choose to install the No. 6 AWG wire, provided the voltage drop does not exceed 3%.

If the 100-A feeder wire is installed as direct burial-type USE wire or cable, then the wire is sized according to *NEC Table 310-16*. For direct burial, the minimum size aluminum feeder wire is No. 1 AWG.

A minimum ampere rating for feeder conductors is stated in *NEC Section 215-2*. The rating applies to feeders inside buildings as well as to overhead and underground feeders supplying buildings. The minimum rating is 30 A. This minimum rating applies when

- a 2-wire feeder supplies two or more 2-wire circuits.
- a 3-wire feeder supplies three or more 2-wire circuits.
- a 3-wire feeder supplies two or more 3-wire circuits.

The neutral wire of a single-phase, 120/240-V, 3-wire feeder carries only the unbalanced load. This is the difference in amperes flowing on each hot wire. The neutral wire is sized according to *NEC Section 220-22*. Usually, the neutral wire selected is two AWG numbers smaller than the hot conductors. For the previous farm shop example, if the hot wires were No. 6 AWG copper, the neutral wire would most likely be size No. 8 AWG. The manufacturer automatically sizes the neutral or messenger smaller for multiplex cables.

## Problem 8-1

A dairy barn with a demand load of 128 A will be supplied with an underground aluminum feeder. The load center in the barn is 150 A, single phase, 3 wire. The ampacity of the feeder must be sufficient to handle the demand load and minimize voltage drop, and must be sized to the feeder overcurrent protection. Determine the type and size of underground feeder wire.

## Solution

The underground feeder will be individual aluminum wire Type USE. The minimum size is found in *NEC Table 310-16*: No. 2/0 AWG with a maximum continuous rating of 135 A. The neutral wire need not be sized as large as the two ungrounded wires. The neutral could be No. 1 AWG with a rating of 100 A. This wire does not allow for future loads, and it may result in excessive voltage drop. The electrician may prefer to install No. 3/0 AWG wire with a rating of 155 A. Note that *NEC Section 240-3, Exception No. 1*, was applied to this problem.

## OVERCURRENT PROTECTION

A feeder wire from the central distribution pole to a building is considered a service drop, *NEC Section 230-21*. The service drop is part of the service-entrance conductors. Therefore, the disconnect and overcurrent protection is permitted to be located at the end of the feeder. When a feeder originates at a load center, panelboard, or other service equipment at a building, the feeder is required to be protected from overcurrent at the source of supply, *NEC Section 240-21*. This even includes feeders to another building. Figure 8-2 shows the location of the disconnect and overcurrent protection for several types of farm feeders.

Feeder wires are required to be protected from overcurrent by fuses or circuit breakers based upon the ampacity listed for the wire in *NEC Tables 310-16, 310-17, or 310-19*. However, *Exception No. 1* to *NEC Section 240-3* allows the next standard size larger overcurrent device when the ampacity does not correspond to a standard size. The following is a procedure for choosing the size of wire and overcurrent protection for feeder wires on farms.

1. Determine the load to be served. Usually, the load is determined with one of the following methods: *Part B, NEC Article 220*, or *Part B, NEC Article 430*.
2. Choose and size a disconnect switch, load center, or panelboard large enough to handle the load and provide for future expansion if desirable.
3. Select the feeder wire size based upon the size of main fuse or circuit breaker of the load center, panelboard, or disconnect switch.

The feeder wire must be large enough to serve the load. However, the rating of the feeder is determined by the ampere rating of the overcurrent device protecting the feeder. The wire size, therefore, must be selected from *NEC Tables 310-16, 310-17, or 310-19*, based upon the ampere rating of the overcurrent device.

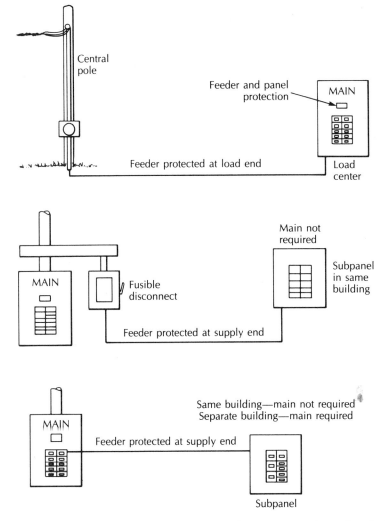

Figure 8-2 Location of overcurrent protection for three common farm electrical feeders

## FEEDER DISCONNECT

A method must be provided for disconnecting a feeder wire from the source of power. A disconnect installed at the central distribution pole is preferred. Many electrical power suppliers and local inspection authorities do not require a disconnect at the central pole. The power supply to a feeder seldom needs to be shut off. When power must be disconnected, however, the power supplier or the electrician usually removes the meter. A disconnect at the central pole can be a safety switch, a circuit breaker, or a standby power transfer switch. If the disconnect is installed outside, it must be in a raintight enclosure, or it must be mounted within a protective enclosure. *NEC Section 230-21* implies that a meter and no disconnect is acceptable at a central distribution pole.

Disconnects for feeders are required to be clearly and durably labeled, *NEC Section 110-22*. This is especially important if two or more disconnects are grouped together. Inside a building, the disconnect for a feeder must be labeled so that power can be shut down quickly in case of an emergency. A label should also be placed at the load end of the feeder, indicating at which location in the building the disconnect can be found. Figure 8-3 shows typical farm electrical feeders, feeder disconnects, and proper labeling.

## SUBPANELS

Subpanels are supplied by feeders within a building. Subpanels often are not required to have main fuses or a

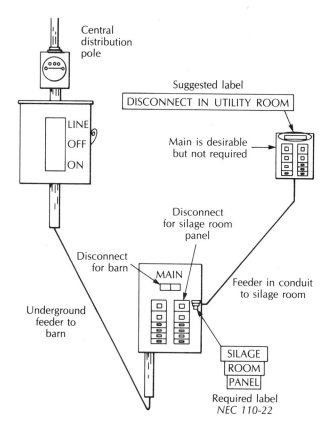

Figure 8-3 Outside and inside electrical feeders showing disconnects, overcurrent protection, and labeling

Figure 8-4 Panelboard with main lugs only. A main circuit breaker can be added by the electrical installer, if desired.

main circuit breaker. Panelboards and load centers are available with main lugs only, Figure 8-4. The cost of a panelboard is lower if a main is not provided.

Load centers and panelboards have a maximum ampere rating for which they have been constructed and tested by the manufacturer and a testing agency, such as UL or CSA. The load center or panelboard must be protected so that the current cannot exceed the maximum rating. *NEC Section 384-16* requires that the panelboard be protected in such a way that its rating cannot be exceeded. *Exception No. 1* to *NEC Section 384-16(a)* permits the overcurrent protection for the feeder to also protect the panelboard. This means that main fuses or a main circuit breaker in the panelboard are not required.

The ampere rating of a panelboard can be less than the ampere rating of the feeder. The only restriction is that the ampere rating of the panelboard cannot be exceeded by applying too much load. The panelboard is easily protected with main fuses or a main circuit breaker.

## GROUNDED FEEDER WIRE

Virtually all farm electrical systems are grounded systems. This means that one current-carrying wire is intentionally grounded. A grounding wire for the purpose of safety may also be present in addition to the grounded circuit wire. Usually, the grounded wire is the neutral of a single-phase, 120-V, 2-wire system, or the neutral of a 120/240-V, 3-wire system. The grounded wire may be opened by the service disconnect switch, but usually this is not the case. An insulated grounded conductor is required by *NEC Section 200-6* to have white or gray insulation. For sizes larger than No. 6 AWG, the wire is permitted to have a white marking at terminal points. Usually, white tape is used to mark the grounded wire. An overcurrent device is not permitted to be installed in the grounded wire, except for a circuit breaker which simultaneously opens all ungrounded wires, *NEC Sections 230-90(b)* and *240-22*.

Large 3-phase farm motors are sometimes supplied with electrical power by a corner grounded, 3-phase delta, 3-wire electrical system. One of the phase wires is intentionally grounded. This grounded wire is *not* a neutral; it is a grounded phase wire. However, the rules for grounded wires still apply. An overcurrent devices must not be placed in the grounded wire, except for a circuit breaker which opens all ungrounded wires at the same time. A 3-pole circuit breaker can be used as the disconnect to a corner grounded delta feeder.

If the fuses are sized to provide motor overload protection, then a fuse is required in the grounded phase wire, *NEC Section 430-36*. Experience has shown that when time-delay fuses are *properly* sized and serve as motor running overcurrent protection, all fuses will open on an overload, fault, short circuit, or loss of voltage on one phase.

A 3-pole fusible disconnect switch is most frequently used on farms for corner grounded, 3-phase delta feeders. Fuses are placed in the two hot phases, and a properly sized copper shorting bar is placed in the fuseholder of the grounded phase. These copper shorting bars are more commonly known as *solid neutrals*. Fuse manufacturers also manufacture solid neutrals, using the same size (length—width—thickness) of copper as is used for a fuse of the same ampere rating. For instance, to order a copper shorting bar (solid neutral) for a 3-pole fusible disconnect switch rated at 200 A, 250 V, the electrician would order one 200-A, 250-V solid neutral. The cost of a solid neutral is the same as that of an ordinary nonrenewable, one-time fuse having the same ampere rating and voltage rating.

It is important that the cross-sectional area of a shorting bar be adequate to handle the current. It is also very important that the surface of the copper shorting bar, in contact with the fuseholder clip, be clean and free from corrosion and oxide.

A properly installed disconnect switch for a corner grounded, 3-phase delta feeder is shown in Figure 8-5. The grounded phase is located on the left side of the disconnect. The grounded phase wire is marked with white tape, *NEC Section 200-6*. The switch of Figure 8-5 is the service disconnect. Therefore, it must be grounded. In this case, it is grounded to a ground rod and the metal underground water pipe for an irrigation system. The ground connection is made on the supply side of the disconnect with an add-on lug, *NEC Sections 250-23(a), 250-61,* and *384-3(c)*. An alternative is to use a disconnect switch with a solid neutral terminal for the grounded phase.

## WIRE COLOR CODE

Ungrounded feeder conductors are not permitted to have an outer covering which is white, gray, or green, *NEC Section 310-12(c)*. A wire with a white or gray outer covering may be used as an ungrounded wire if the wire is permanently marked with paint or tape at all accessible location, *NEC Section 200-7, Exception No. 1*.

A 4-wire, 3-phase delta feeder presents a special case when one of the ungrounded wires requires special identification. The fourth wire of this system is a neutral. The ungrounded wire, which is called the *high leg* or *wild leg*, has a higher voltage to the neutral than the other two ungrounded wires, usually 208 V. This high-leg wire is required to be marked or labeled orange in color at access points and terminal locations when the neutral wire is also present, *NEC Section 215-8*. This high leg must be marked or labeled orange at all switchboards and panelboards, *NEC Sections 230-56* and *384-3(e)*.

As discussed earlier in this unit, a grounded wire is required to have an outer covering of white or gray along its entire length for wires No. 6 AWG and smaller. For wires larger than No. 6 AWG, a marking of white at locations where the wire is visible is acceptable, *NEC Sections 310-12(a)* and *200-6*.

A copper neutral wire of a service entrance is not required to be insulated in a service raceway, *NEC Section 230-41*. Both copper and aluminum neutral wires may be bare, and usually are, when they are part of a

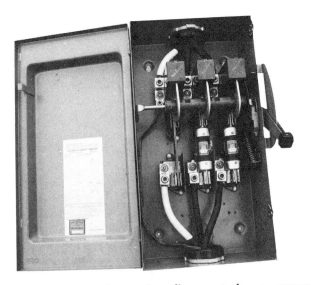

Figure 8-5  Main service disconnect for a corner grounded, 3-phase electrical system.

service-entrance cable assembly. A bare copper neutral wire in a service raceway is not recommended for a farm building service.

The neutral wire of an overhead cable (multiplex cable) is not required to be insulated, *NEC Section 230-22, Exception*. The copper neutral wire of an underground feeder may, under proper soil conditions, be permitted to be bare, *NEC Sections 230-30(b) and (c)*. This, however, is not recommended on farms.

A wire used for the purpose of grounding is permitted to be insulated or bare. If insulated, grounding wires must have a green outer covering, or a green outer covering with one or more yellow stripes, *NEC Section 250-57*. Aluminum wire, however, is not permitted to be bare when in direct contact with masonry or the earth, or where subject to corrosive conditions, *NEC Section 250-92(a)*. Where used outside, bare aluminum grounding wire is not permitted within 18 in (457 mm) of the earth.

Insulated wire other than green in sizes larger than No. 6 AWG can be used as a grounding wire if properly identified, *NEC Section 250-57(b), Exception No. 1*. The insulation can be stripped away, leaving the entire exposed grounding wire bare. The entire exposed wire can be painted green or covered with green tape.

## FEEDER TAPS

A *tap* is a wire of limited length connected to a feeder. Feeder taps are permitted by *NEC Section 215-7*. The rules for feeder taps are found in *NEC Section 240-21*. Several tap situations occur on farms.

1. The tap conductor ampacity is not smaller than the overcurrent device protecting the feeder, Figure 8-6(A) (*NEC Section 240-21, Exception No. 1*).
2. The ampacity of the tap conductor may be less than the overcurrent device protecting the feeder if the following restrictions are satisfied (*NEC Section 240-21, Exception No. 2*):
    a. The total length of the tap is not more than 10 ft (3.05 m).
    b. The ampere rating of the tap is sufficient to supply the tap load.
    c. The tap ends at an overcurrent device sized not greater than the ampere rating of the tap wire.
    d. The tap conductor is enclosed within raceway for its entire length, Figure 8.6(B).
3. The ampacity of the tap conductor must be at least one-third the ampacity of the feeder wire or feeder overcurrent protection, and the following restrictions are satisfied (*NEC Section 240-21, Exception No. 3*):
    a. The total length of the tap is not more than 25 ft (7.62 m).
    b. The tap conductor ends at a single circuit breaker or set of fuses.
    c. The tap conductor is enclosed in raceway, and is protected from physical damage, Figure 8-6(C).

## Problem 8-2

A tap is made to a No. 1 AWG copper THW feeder wire protected at 125 A. The tap load is 16 A and is located so that the tap conductor will be 18 ft (5.49 m) long. The tap will end at a fusible switch containing 20-A fuses. Determine the minimum size THW copper tap wire.

## Solution

The tap wire ends at a 20-A fuse; therefore, to serve the load, the wire must be rated at not less than 20 A.

The tap is more than 10 ft (3.05 m) long, but less than 25 ft (7.62 m). Therefore, the tap wire must have an ampacity not less than one-third the ampacity of the feeder wire or the feeder overcurrent device, whichever is larger. In this case, the ampere rating of the wire is greater.

Feeder wire, THW copper No. 1 AWG, 130 A:

$$\boxed{\frac{130 \text{ A}}{3} = 43 \text{ A}}$$

Find a THW copper wire in *NEC Table 310-16* which is rated at not less than 43 A. The minimum size THW copper tap wire is No. 8 AWG, which has a 50-A rating.

## GROUNDING AT OUTBUILDINGS

Farm wiring is unique in that it usually involves several buildings supplied from a single electrical supply. Each building is usually treated as a separate service. This means that grounding is required at each building. Further, the grounded wire, usually the neutral, is generally bonded to a grounding electrode, *NEC Section 250-24*. There are exceptions, however. There is only one circuit in the building, and there is no equipment requiring grounding. This practice of grounding the neutral at each building can sometimes lead to a difference in electrical potential between grounding electrodes at the various buildings. A feeder neutral wire that is too small, or a poor connection in the neutral conductor, will create voltage drop, which results in the grounding electrodes

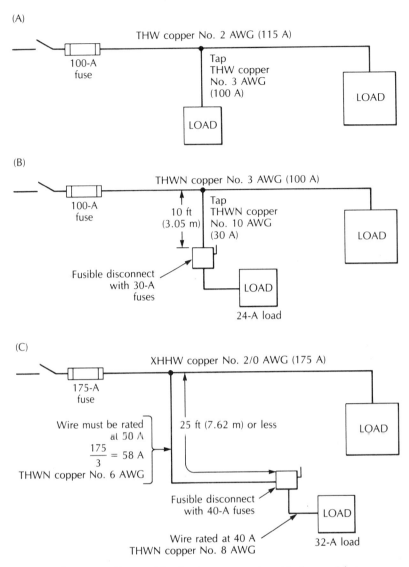

**Figure 8-6   Types of feeder taps encountered in farm wiring**

at buildings being at different voltages. This causes a small amount of current to flow through the ground. To avoid this problem, the following steps are recommended.

1. Size the feeder wires to handle the load and hold voltage drop to not more than 3%.
2. Make sure that all splices are made properly.
3. Make sure that 120-V loads in buildings are balanced between the two hot wires to reduce the load on the neutral.
4. Run a neutral wire and a separate grounding conductor with all feeders. The grounding conductor bonds all of the grounding electrodes together on the farm to form one grounding electrode system.

Figure 8-7 illustrates the bonding together of the grounding electrodes on a farm to form a grounding electrode system. This bonding is best accomplished by adding a separate grounding wire. Instead of using triplex cable overhead for a 120/240-V, 3-wire feeder, use quadruplex. The bare messenger wire serves as the grounding wire. One of the insulated wires is marked with white tape and is used as a neutral wire. For an underground feeder, a fourth wire is installed in the trench. This grounding wire may be insulated copper or

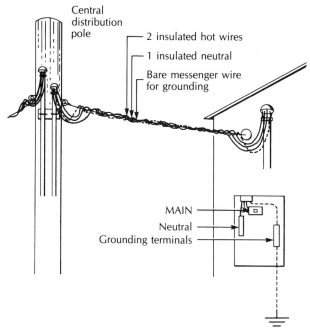

Figure 8-7 A 4-wire, single-phase overhead feeder with the neutral insulated and the messenger used for grounding

aluminum wire. It may also be bare copper wire. However, this is not recommended because of the possibility of corrosion.

## SIZING THE FEEDER GROUNDING WIRE

Two separate situations must be considered when sizing the equipment grounding for feeders. The first situation involves a feeder to a separate building. This feeder originates at the central distribution pole or at another building. The feeder may be protected from overcurrent at the supply end, but often it is not protected. This grounding conductor is sized according to *NEC Table 250-94*, per *Section 250-79(c)* when the feeder is not protected at the supply end. The second situation involves a feeder supplying a subpanel or other load in the same building. This feeder originates on the load side of the service disconnect; therefore, *NEC Section 250-79(d)* applies. The grounding wire run with a feeder in the same building is sized according to *NEC Table 250-95*.

## Problem 8-3

A livestock barn has a 100-A load center fed underground with No. 1 AWG aluminum USE wire. The feeder is not protected at the supply end. The electrician decides to run a copper wire to bond the grounding electrode at the livestock barn to the ground electrode at the central pole. Determine the minimum size copper grounding wire.

## Solution

This is a service in a separate building. Therefore, the grounding wire is sized according to *NEC Table 250-94*. Look up the size of copper grounding wire required for a No. 1 AWG aluminum service-entrance wire. The correct size is No. 8 AWG copper.

## Problem 8-4

A potato storage and packaging building is supplied with a 600-A, 120/208-V, 3-phase electrical system. A 225-A feeder from the main service supplies a subpanel in the packaging area. The feeder consists of No. 4/0 AWG THHN copper wire in PVC (rigid nonmetallic) conduit. The wire is considered to be in a wet location and is rated at 75°C (167°F). Determine the minimum size bare copper grounding wire to be run in the conduit.

## Solution

The wire is sized according to *NEC Table 250-95*. The first column gives the largest size overcurrent device which can be used to protect the feeder wires for the size grounding wire given. The fuses on the feeder in the problem are rated at 225 A. No. 6 AWG copper wire cannot be used because the largest fuse permitted is 200 A. Therefore, the minimum size grounding wire permitted is No. 4 AWG bare copper.

The *National Electrical Code*, in *Section 250-95*, requires that when a feeder wire is increased in size to compensate for voltage drop, the equipment grounding wire must also be increased in size according to the ratio of change in circular mil area of the ungrounded wires. Determine the circular mil area of the wire from *NEC Table 8* of *Chapter 9* (this table is found in Unit 2 of this text).

## OVERHEAD FEEDER CLEARANCES

Maintaining adequate clearances for overhead wires is important around the farm. The major problem is contact with tractors, trucks, and farm machinery. It should be remembered that tall machinery may pass through such areas as residential yards. Therefore, high clearances should be maintained as much as is practical. Minimum clearance requirements for overhead wires are given in *NEC Section 225-18,* as follows.

- 10 ft (3.05 m)—above areas where there will be no vehicular traffic, and the wires do not exceed 150 V to ground
- 12 ft (3.66 m)—above areas not subject to truck traffic, and wires do not exceed 300 V to ground
- 15 ft (4.57 m)—above areas accessible to pedestrians only, and wires exceed 300 V to ground
- 18 ft (5.49 m)—above public roads, agricultural drives, and commercial drives

These clearances are minimums, and the points of attachment to poles and buildings may actually be higher to allow for normal sag in the wires.

It may be necessary for overhead feeder wires to pass directly over low buildings. This is acceptable as long as minimum clearances are maintained. In the case of a roof not accessible to pedestrians, and with the wires fully insulated, the minimum clearance is 3 ft (914 mm), *NEC Section 225-19(a), Exception No. 1.* If the wires are not fully insulated, the minimum clearance is 10 ft (3.05 m). However, if the roof is too steep for walking, the clearance may be reduced to 3 ft (914 mm), *Exception No. 4.* If the roof is accessible to pedestrians and the wires are fully insulated, the minimum clearance is 8 ft (2.44 m), *Exception No. 2.* Feeder wires are not permitted to be installed above swimming pools or in areas adjacent to pools. Exact limits and clearances for pools are given in *NEC Section 680-8.*

Sometimes, tall buildings stand between the central distribution pole and the building or structure to be supplied with an overhead feeder. At the beginning of this unit, Figure 8-1 shows a pole that is set to allow the electrical supply feeder access to the desired location at the barn. Wires are permitted to be attached to the outside surface of a building, *NEC Section 225-10,* although this should be avoided if possible.

A multiplex cable may be run along one side of a tall building, with a support at each corner of the barn. A minimum horizontal clearance of 3 ft (914 mm) is permitted between the cable and the barn, *NEC Section 225-19(c).* Individual wires may be attached to the side of a building on insulators with insulator spacing as required in *NEC Article 320.* Raceway or cable may also be attached to the side of a building for this purpose.

## SPLICING AND TERMINATING OVERHEAD FEEDERS

Individual overhead feeder wires are terminated at an insulator, *NEC Section 225-12,* and wrapped to hold them in place, as shown in Figure 8-8. Multiplex cable may be terminated in the same manner. Multiplex cable can also be terminated with a messenger grip. The messenger grip need not be terminated at an insulator. It need only be terminated in a secure manner.

Wires can be supported at poles by several methods. It is important to provide adequate support for the wire or cable. It is also important that the hot insulated wires be arranged so that in the wind they will not rub against any surface, causing damage to the insulation.

Feeder wire splices made outdoors must be made in a manner which will protect the wires from the weather. One effective method of making a weathertight splice covering is to first cover the splice completely with rubber tape. When applying the rubber tape, it must be stretched slightly to make a tight seal. The tape should overlap the previous layer about halfway. This will provide a double layer of tape when the splice is completed. The insulation on the splice must be equivalent to the insulation on the wires, *NEC Section 110-14(b).* A layer of friction tape or plastic tape on the outside protects the rubber tape from damage. This procedure is shown in Figure 8-9. Aluminum wires must be coated with an oxide inhibitor before the splice is covered, to ensure a good, long-lasting electrical connection.

## OVERHEAD SPANS

Overhead wires will sag to some extent when they are installed. The greater the span between supports, the greater is the sag. The wire contracts or becomes shorter when it gets cold. This means that the wire will become shorter and tighter in the winter. To this tension, add the

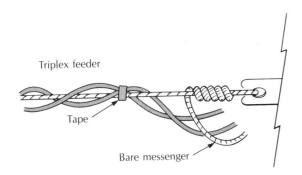

**Figure 8-8** Multiplex cable messenger wire and single insulated wire can be terminated by wrapping the wire at an insulator.

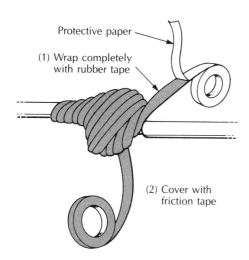

**Figure 8-9** A splice in an overhead wire exposed to the weather must be weatherproof for a long-lasting, no-resistance connection. Plastic tapes are available for making a weatherproof seal.

weight of ice in northern climates, and, along with the wind, damage to insulators and mast risers at the ends of the wire may occur. When an overhead wire is installed in the summer, some sag must be provided to allow for contraction in the winter. With some sag allowed in the wire, the wind is less likely to exert stress at the termina-tions and supports for the wire. Little sag need be allowed when the wire is installed in the winter.

A maximum span of 50 ft (15.2 m) for No. 10 AWG copper and No. 8 aluminum overhead wire is specified in *NEC Section 225-6(a)*. The *NEC* does not restrict the span for larger wires; however, large sized wire is heavy, thus placing great strain on wire supports. Table 8-3 lists sizes of wire and the recommended maximum length of span between supports.

## UNDERGROUND FEEDERS

The types of wire permitted for direct burial are UF cable and USE wire and cable. The installation of underground wire is discussed in Unit 4. Adequate depth of burial is important on a farm to prevent damage due to heavy equipment traveling above the wire. It is particularly important that the fill earth adjacent to the wire be free from stones and foreign material. If clean fill is not practical to obtain, the wire must be protected by a suitable conduit sleeve, *NEC Section 300-5(f)*.

Underground wires which become damaged, or are not long enough for the required length at the time of installation, are permitted to be spliced, *NEC Section 300-5(e)*. Underground wire splicing kits are available for direct burial with complete and easy-to-follow installation instructions.

**Table 8-3** Maximum recommended length of span between supports for overhead wire

| Wire Size AWG No. | | Maximum Span | |
|---|---|---|---|
| Copper | Aluminum | Feet | Meters |
| 10 | 8 | 50 | 15.2 |
| 8 | 6 | 100 | 30.5 |
| 6—2 | 4—1 | 150 | 45.7 |
| 1 and larger | 0 and larger | 100 | 30.5 |

## REVIEW

Refer to the *National Electrical Code* when necessary to complete the following review material. Write your answers on a separate sheet of paper.

1. A farm building service receives the electrical supply from an overhead service drop. If the service is wired with aluminum XHHW service-entrance conductors, what is the temperature rating of the wire insulation, in degrees celsius and Fahrenheit?

2. Determine the type of conductor and minimum size aluminum wire for an underground feeder to a farm building requiring a feeder with 100-A capacity.

3. A farm storage building will be wired with one 20-A circuit. The building will be fed underground from a nearby building, using Type UF copper cable. Determine the minimum size wire.

4. A farm central electrical distribution pole is not supplied with a main disconnect or individual overcurrent protection for the feeders to the buildings. Does the *NEC* permit the use of UF cable for an underground feeder under these conditions? Explain.

5. Determine the maximum size fuse for the feeder disconnect switch feeding a load center having no main and rated at 125 A, single-phase, 120/240 V. The feeder wire is THWN copper No. 0 AWG.

6. Four ventilation fans for a poultry building are tapped from a single-phase, 240-V feeder of THWN copper wire No. 4 AWG. The feeder is protected by 60-A time-delay fuses. Each tap is 12 ft (3.66 m) long, in conduit, and ends at a disconnect switch containing 20-A time-delay fuses. Each fan is rated at 1 hp, 240 V, and draws 8 A. Determine the minimum size THWN copper tap wire. (The footnote to *NEC Table 310-16* indicates that the wire ampere rating is as shown in the table for the purpose of making tap calculations, but the actual load placed on the wire may not exceed the values stated in the footnote.)

7. A bare copper grounding wire is placed in a trench with a 150-A underground single-phase feeder to a barn. The feeder wire is USE aluminum No. 3/0 AWG. Determine the minimum size grounding wire for this feeder.

8. What is the minimum clearance for an aerial triplex feeder over a farm drive, in feet and meters?

9. An underground feeder is placed in PVC (rigid nonmetallic) conduit under a farm drive. What is the minimum depth of burial of the feeder, in inches and centimeters?

10. Requirements for messenger supported wiring are included in which *NEC Article*?

11. A disconnect for a feeder must be marked as to its purpose, according to which *NEC Section*?

12. The grounded phase wire of a 3-phase, corner grounded delta must be identified with what color, according to which *NEC* Section?

13. A messenger supported wire passing over a building with a roof not accessible to pedestrians must maintain a minimum clearance of how many feet (meters)? (Wires are assumed to be fully insulated.)

14. The grounded phase wire of a corner grounded, 3-phase feeder not serving a motor load is not permitted to contain a fuse, according to which *NEC Section*?

15. A 200-A feeder wire, THW copper No. 3/0 AWG, is run in PVC (rigid nonmetallic) conduit from the main service in a barn to the silage room. What is the minimum size bare copper equipment grounding wire to be run in the conduit? The feeder is not protected for overcurrent at the supply end.

# UNIT 9
# FARM WIRING METHODS AND MATERIALS

## OBJECTIVES

After studying this unit, the student will be able to

- describe the wiring methods suitable for agriculture.
- identify the types of cable and conduit used in agricultural wiring.
- explain the procedures for installing aluminum wire.
- select the proper type and size of wire for a branch circuit.
- identify and diagram the internal circuit of a single-pole switch, a 2-pole switch, a 3-way switch, and a 4-way switch.
- diagram the wiring of a circuit, given the outlets, devices, and switches in the circuit.
- identify the wire colors of a circuit.
- determine the proper electrical boxes used in a circuit.
- determine the minimum size of conduit for the number, type, and size of wires to be placed in the conduit.
- identify common electrical tools.
- find *NEC* references stating requirements for supporting conduit and cable.
- describe the wiring materials used in damp, wet, or dusty agricultural locations.

Agricultural wiring is subjected to a variety of conditions and environments. In some areas, the wires must be protected from physical damage caused by animals and machines. The electrician frequently must contend with problems of corrosion, dust, and moisture, particularly in buildings housing animals. Some buildings on farms do not pose problems for the wiring. *NEC Article 547* deals directly with wiring in agricultural buildings.

Therefore, before starting interior wiring of a farm building, the electrician must first determine whether the building falls within one or both of the categories described in *NEC Section 547-1*.

Totally enclosed and environmentally controlled buildings, such as those for poultry and livestock require a dusttight and moisturetight wiring system where excessive dust and dust combined with water may accumulate. This is also true of similar buildings where corrosive atmospheres exist. The following areas often produce corrosive atmospheres.

1. Animal confinement areas where excrement may accumulate
2. Areas in which such corrosive particles as chemicals and fertilizer may combine with water
3. An area, such as a milking parlor, where there is periodic washing or spraying of the room with cleaning agents
4. Other areas where corrosive liquids or vapors may be produced

Buildings on the farm which are dry and do not have conditions where there is excessive dust or dust in combination with moisture are wired in accordance with other appropriate sections of the *National Electrical Code*.

## AMPERE RATING OF WIRES

The limiting factor which determines the amount of current a particular wire can carry is the temperature limitation of the type of insulation used on the wire. All wires have resistance, and current flowing through the wire produces heat. The heat produced in the wire is dissipated to adjacent materials by conduction, and then by air convection and radiation to the surrounding environment. Heat always travels from a warm body to a cold body. Heat, therefore, can flow to the wire from the surrounding environment. Wires inside a grain dryer will heat up if they are exposed to hot air. A wire may heat up from radiation of the sun or from an adjacent hot surface. If many wires are contained in a cable or a conduit, the wires will heat one another due to the heat each wire produces as current flows through it. It is important to remember that the maximum temperature rating of the insulation must not be exceeded, *NEC Section 310-10*.

Under normal conditions, the maximum ampere rating of a wire for agricultural wiring is found in *NEC Tables 310-16* or *310-17*. However, there are three situations where the ampere rating shown in the tables must be reduced.

1. A continuous load. This is a load which under normal operation may continue for more than three hours. (See the definition in *NEC Article 100*.)
2. More than three current-carrying wires in a cable or conduit. (*Note 8* to the *NEC* ampacity tables.)
3. The normal temperature of the area around the wires is unusually high. (Correction factors at the bottom of *NEC* ampacity tables.)

Electrical wires, except for a few specific applications, must be insulated, *NEC Section 310-2(a)*. The wire material must be copper, aluminum, or copper-clad aluminum, *NEC Section 310-2(b)*. Copper-clad aluminum is aluminum wire with a thin skin of copper on the outside surface of the wire. It looks like copper wire at first glance, but examination of the cut crosssection of the wire reveals that it is actually aluminum. The use of this type of wire is discussed later in this unit. The minimum size of wire used for an interior wiring system is No. 14 AWG copper or No. 12 aluminum, *NEC Section 310-5*.

## TYPES OF WIRE

Wire or cable installed in wet locations must have insulation which is suitable for wet locations. These types are indicated with the letter W in the type designation, such as THW. Wires exposed to corrosive conditions must also be suitable for the particular conditions. *NEC Table 310-13* states the insulation type, temperature limitation, and the suitability of the insulation for various conditions. Common wires used in agricultural wiring are shown in Table 9-1.

Many wires carry more than one rating. For example, a wire may be marked THW and MTW. Or, a wire may be marked THWN, THHN, and MTW, Figure 9-1. These markings mean that the wire is rated THWN 75°C (167°F) in wet locations, and THHN 90°C (194°F) in dry locations. The designation also means that the wire can be used on machine tools or in locations where oil is present. If a wire is rated MTW, the words "oil resistant" also are often printed on the wire.

## WIRE AMPERE RATING

It is essential to refer to the notes to the table when consulting the wire ampere ratings given in *NEC Table 310-16*, represented in this text as Table 9-2. *Table 310-*

Table 9-1 Common electrical wires used in agriculture

| Type | Maximum Operating Temperature | Application |
|---|---|---|
| TW | 60°C(140°F) | Dry and wet locations |
| THW | 75°C(167°F) 90°C(194°F) | Dry and wet locations. When inside a fluorescent fixture |
| THHN | 90°C(194°F) | Dry locations |
| THWN | 75°C(167°F) | Dry and wet locations |
| XHHW | 75°C(167°F) 90°C(194°F) | Wet locations Dry locations |
| MTW | 60°C(140°F) 90°C(194°F) | Machine tool wiring in wet locations (oil resistant) Machine tool wiring in dry locations (oil resistant) |
| RHW | 75°C(167°F) | Dry and wet locations |

16 applies where there are not more than three current-carrying wires in cable, conduit, or direct burial in the earth. Grounding wires are not counted, *Note 11*. A neutral wire may be counted as a current-carrying wire, but not always. Figure 9-2 illustrates the situations where the neutral is or is not counted.

## DERATING FACTORS

The wire ampere rating in *NEC Table 310-16* is limited to not more than three wires, because each wire produces heat when current flows through it. With more than three wires enclosed by a cable jacket or conduit, heat is trapped and the wires do not cool as quickly as if they were in open air. Therefore, the ampere rating of the wire must be reduced or derated when there are four or more wires in cable or conduit. The *NEC* gives the derating factors for more than three wires in cable and conduit in the table in *Note 8* to the ampacity tables, represented in this text as Table 9-3.

An example will help to illustrate how conductor derating works. The wires to three electric motors are run in the same conduit. The motors are 3-phase, and the wire is No. 8 AWG copper with THWN insulation. The maximum ampere rating of the wire for these conditions is determined as follows.

A 3-phase motor requires three wires; therefore, with three motors there will be nine wires in the conduit. Table 9-3 indicates that the ampere rating shown in *NEC Table 310-16* must be reduced to 70%.

The value from *Table 310-16* for No. 8 AWG THWN copper is 50 A. For nine No. 8 AWG copper THWN wires in conduit:

$$50 \text{ A} \times 0.70 = 35 \text{ A}$$

The maximum load which can be placed on the nine No. 8 AWG THWN copper wires in the same conduit is 35 A.

An important footnote at the bottom of *NEC Table 310-16* must be considered concerning wire sizes No. 14, No. 12, and No. 10 AWG. This footnote applies to common wire and cable types used for agricultural

Figure 9-1 Electrical wire is required to have the insulation type and size in either AWG number or MCM legibly shown along the entire length of the wire. Other information to aid the electrician may also be present.

**Table 9-2** (*NEC Table 310-16*) Allowable ampacities of insulated conductors rated 0–2000 V, 60°C to 90°C (140°F to 194°F). Not more than three conductors in raceway or cable or earth (directly buried), based on ambient temperature of 30°C (86°F).

| Size AWG MCM | Temperature Rating of Conductor. See Table 310-13 ||||||||  Size AWG MCM |
|---|---|---|---|---|---|---|---|---|---|
| | 60°C (140°F) | 75°C (167°F) | 85°C (185°F) | 90°C (194°F) | 60°C (140°F) | 75°C (167°F) | 85°C (185°F) | 90°C (194°F) | |
| | Types *RUW, †T, †TW, †UF | Types †FEPW, †RH, †RHW, †RUH, †THW, †THWN, †XHHW, †USE, †ZW | Types V, MI | Types TA, TBS, SA, AVB, SIS, †FEP, †FEPB, †RHH, †THHN, †XHHW* | Types †RUW, †T, †TW, †UF | Types †RH, †RHW, †RUH, †THW, †THWN, †XHHW, †USE | Types V, MI | Types TA, TBS, SA, AVB, SIS, †RHH, †THHN, †XHHW* | |
| | Copper |||| Aluminum or Copper-Clad Aluminum |||| |
| 18 | .... | .... | .... | 14 | .... | .... | .... | .... | .... |
| 16 | .... | .... | 18 | 18 | .... | .... | .... | .... | .... |
| 14 | 20† | 20† | 25 | 25† | .... | .... | .... | .... | .... |
| 12 | 25† | 25† | 30 | 30† | 20† | 20† | 25 | 25† | 12 |
| 10 | 30 | 35† | 40 | 40† | 25† | 30† | 30 | 35† | 10 |
| 8 | 40 | 50 | 55 | 55 | 30 | 40 | 40 | 45 | 8 |
| 6 | 55 | 65 | 70 | 75 | 40 | 50 | 55 | 60 | 6 |
| 4 | 70 | 85 | 95 | 95 | 55 | 65 | 75 | 75 | 4 |
| 3 | 85 | 100 | 110 | 110 | 65 | 75 | 85 | 85 | 3 |
| 2 | 95 | 115 | 125 | 130 | 75 | 90 | 100 | 100 | 2 |
| 1 | 110 | 130 | 145 | 150 | 85 | 100 | 110 | 115 | 1 |
| 0 | 125 | 150 | 165 | 170 | 100 | 120 | 130 | 135 | 0 |
| 00 | 145 | 175 | 190 | 195 | 115 | 135 | 145 | 150 | 00 |
| 000 | 165 | 200 | 215 | 225 | 130 | 155 | 170 | 175 | 000 |
| 0 000 | 195 | 230 | 250 | 260 | 150 | 180 | 195 | 205 | 0 000 |
| 250 | 215 | 255 | 275 | 290 | 170 | 205 | 220 | 230 | 250 |
| 300 | 240 | 285 | 310 | 320 | 190 | 230 | 250 | 255 | 300 |
| 350 | 260 | 310 | 340 | 350 | 210 | 250 | 270 | 280 | 350 |
| 400 | 280 | 335 | 365 | 380 | 225 | 270 | 295 | 305 | 400 |
| 500 | 320 | 380 | 415 | 430 | 260 | 310 | 335 | 350 | 500 |
| 600 | 355 | 420 | 460 | 475 | 285 | 340 | 370 | 385 | 600 |
| 700 | 385 | 460 | 500 | 520 | 310 | 375 | 405 | 420 | 700 |
| 750 | 400 | 475 | 515 | 535 | 320 | 385 | 420 | 435 | 750 |
| 800 | 410 | 490 | 535 | 555 | 330 | 395 | 430 | 450 | 800 |
| 900 | 435 | 520 | 565 | 585 | 355 | 425 | 465 | 480 | 900 |
| 1 000 | 455 | 545 | 590 | 615 | 375 | 445 | 485 | 500 | 1 000 |
| 1 250 | 495 | 590 | 640 | 665 | 405 | 485 | 525 | 545 | 1 250 |
| 1 500 | 520 | 625 | 680 | 705 | 435 | 520 | 565 | 585 | 1 500 |
| 1 750 | 545 | 650 | 705 | 735 | 455 | 545 | 595 | 615 | 1 750 |
| 2 000 | 560 | 665 | 725 | 750 | 470 | 560 | 610 | 630 | 2 000 |
| **Correction Factors** |||||||||| |
| Ambient Temp. °C | For ambient temperatures over 30°C, multiply the ampacities shown above by the appropriate correction factor to determine the maximum allowable load current. |||||||| Ambient Temp. °F |
| 31–40 | .82 | .88 | .90 | .91 | .82 | .88 | .90 | .91 | 86–104 |
| 41–45 | .71 | .82 | .85 | .87 | .71 | .82 | .85 | .87 | 105–113 |
| 46–50 | .58 | .75 | .80 | .82 | .58 | .75 | .80 | .82 | 114–122 |
| 51–60 | .... | .58 | .67 | .71 | .... | .58 | .67 | .71 | 123–141 |
| 61–70 | .... | .35 | .52 | .58 | .... | .35 | .52 | .58 | 142–158 |
| 71–80 | .... | .... | .30 | .41 | .... | .... | .30 | .41 | 159–176 |

†The overcurrent protection for conductor types marked with an obelisk (†) shall not exceed 15 amperes for 14 AWG, 20 amperes for 12 AWG, and 30 amperes for 10 AWG copper; or 15 amperes for 12 AWG and 25 amperes for 10 AWG aluminum and copper-clad aluminum after any correction factors for ambient temperature and number of conductors have been applied.

*For dry locations only. See 75°C column for wet locations.

Reprinted with permission from NFPA 70-1984, *National Electrical Code®*, Copyright © 1983, National Fire Protection Association, Quincy, Massachusetts 02269. This reprinted material is not the complete and official position of the NFPA on the referenced subject, which is represented only by the standard in its entirety.

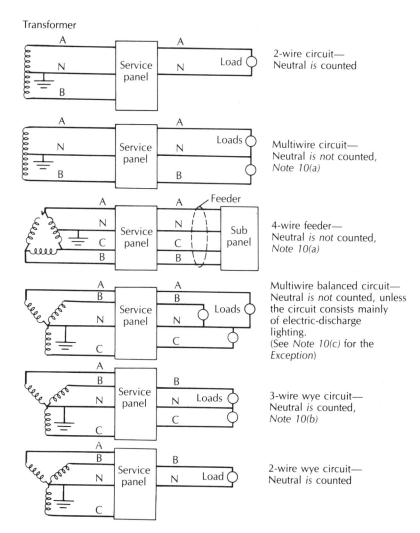

**Figure 9-2** Situations where the neutral wire is or is not counted as a current-carrying wire for the purposes of conductor derating.

wiring, Types TW, THW, THWN, THHN, XHHW, UF, USE, and others. The footnote sets the maximum fuse or circuit-breaker size for these wires. A continuous load current or the load current of specific equipment must not exceed 80% of the overcurrent device rating, *NEC 384-16(c)*.

Consider another example, where there are eight THHN copper wires in conduit. Two of the wires supply a milk pipeline washer that draws 17 A. Determine the minimum size copper THHN wire which can be used to supply the load.

There are more than six wires in the conduit. Therefore, according to Table 9-3, the current values in *NEC Table 310-16* must be reduced to 70% of the table value.

**Table 9-3** Derating factors for more than three wires in cable or conduit

| Number of Conductors | Percent of Value in NEC Table 310-16 |
|---|---|
| 4 thru 6 | 80 |
| 7 thru 24 | 70 |
| 25 thru 42 | 60 |
| 43 and above | 50 |

No. 14 AWG THHN copper—25 A
25 A × 0.70 = 17.5 A

Unit 9    *Farm Wiring Methods and Materials*    157

This value is larger than the load; however, the footnote to *NEC Table 310-16* sets the maximum overcurrent device rating for No. 14 AWG copper wire at 15 A. Therefore, size No. 14 AWG is too small for this example. Try, then the next larger size, No. 12 AWG.

> No. 12 AWG THHN copper—30 A
> 30 A × 0.70 = 21 A

The footnote at the bottom of *NEC Table 310-16* limits the overcurrent protection of No. 12 AWG wire to 20 A. A single equipment load or a continuous load is not permitted to exceed 80% of the overcurrent device rating.

> 20 A × 0.80 = 16 A

It is clear that the 17-A load is too great for the No. 12 AWG copper wire. As a result, the minimum size wire permitted for this circuit is No. 10 AWG copper.

Electrical wire ampere ratings are required to be derated if the load supplied is considered a continuous load. Recall that continuous load is any load which normally operates for more than three hours. Most farm lighting circuits are considered to be continuous loads. Electric space heaters and electric water heaters are considered to be continuous loads, and, therefore, branch-circuit wires are sized at 125% of the load current, *NEC Sections 422-14(b)* and *424-3(b)*. Wires used to supply continuous loads are required to be derated to 80% of their current rating shown in *NEC Table 310-16*. The overcurrent device must also be derated in the case of a continuous load, *NEC Section 384-16(c)*. This means that a 15-A circuit breaker or fuse can be loaded only to 12 A, a 20-A breaker can be loaded only to 16 A and so forth.

## Problem 9-1

A lighting circuit in a livestock barn will supply nine fixtures which draw 1.5 A each. Determine the minimum size circuit breaker, and minimum size copper UF cable permitted for this circuit.

## Solution

The total load on the lighting circuit is:

> 9 fixtures × 1.5 A per fixture = 13.5 A

This is a continuous load; therefore, the circuit breaker and wire must be derated to 80% of their ampere rating.

> 15-A circuit breaker × 0.80 = 12 A
> 20-A circuit breaker × 0.80 = 16 A

The smallest size circuit breaker permitted for this circuit is 20 A. The minimum size copper wire for this 20-A circuit breaker is No. 12 AWG. The ampere rating of No. 12 AWG UF cable is 25 A.

> 25 A × 0.80 = 20 A

No. 12 AWG copper UF cable is the minimum size permitted for this lighting circuit.

The question often arises concerning what to do when more than three circuit wires in conduit supply continuous loads. *Exception No. 2* to *Note 8* answers this question. When the derating factors of Table 9-3 are used, it is not necessary to derate because the circuit supplies a continuous load.

Under some circumstances, the temperature of the environment around the wires can make it necessary to lower the ampere rating of a wire below the values given in *Table 310-16*. Under *no* circumstances is a wire permitted to be placed in an environment where the ambient

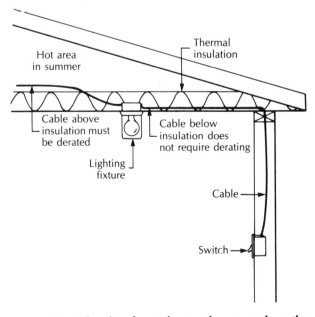

**Figure 9-3** Wires in a hot attic, or other area where the temperature is abnormally high, may have to be derated.

temperature exceeds the temperature rating of the insulation, *NEC Section 310-10*. The term *ambient* is used in the *National Electrical Code*, and means the surrounding environment.

An unventilated attic space in the summer is a typical example of an area where the ampere rating of a wire may have to be reduced because of high ambient temperature Figure 9-3.

Assume that the area reaches a temperature of 120°F (49°C). Determine the maximum load which is permitted to be placed on No. 12 AWG copper UF cable above the insulation.

> No. 12 AWG copper UF cable—25 A

The correction factors for 120°F (49°C) ambient temperature are found in the third row of the table at the bottom of *NEC Table 310-16*. For copper UF cable, the factor is 0.58. The maximum permitted load current on copper UF cable in 120°F (49°C) ambient temperature is:

> 25 A × 0.58 = 14.5 A

The temperature derating factors are part of the ampacity tables. Therefore, the derating factors for a continuous load, or more than three wires in cable or conduit, also apply when relevant.

## HIGH-TEMPERATURE WIRE

Heating-type appliances and equipment used on farms, such as electric water heaters, cooking appliances, space heaters, lighting fixtures, booster heaters, grain dryers, and similar equipment, may produce high temperatures in areas where electrical wires are contained. Wire rated with high-temperature insulation may be required. Wires of this type are described in *NEC Table 310-13*. The ampere rating of these wires is given in *NEC Table 310-18*, represented in this text as Table 9-4. The derating factors previously discussed also apply. The rural electrician performing repairs on a grain dryer and similar heating equipment should keep a small quantity of high-temperature wire on hand.

## ALUMINUM WIRE

Aluminum electrical wire is frequently used in homes, farms, businesses, and factories because it is lighter in weight and less expensive than copper. Aluminum metal is much more abundant than copper and, therefore, it is less expensive. Compared with copper, aluminum is the next best conductor of electricity among the long list of metals available. Silver is the best conductor of electricity; but, obviously, it is too expensive except for special applications.

Aluminum and copper wire are each installed frequently by electricians and, when installed properly, both provide safe, dependable, trouble-free service. But, aluminum wire has a few undesirable characteristics which require special attention.

Bare aluminum wire in the presence of air forms a thin layer of aluminum oxide on the outside surface. This is comparable to rust forming on steel, only it happens more quickly with aluminum. The problem is that this oxide acts as an insulator. An insulator on the surface of a bare wire is a bad situation; it causes heat to build up when current flows. The problem of oxide buildup is minimized with copper-clad aluminum wire. However, the problem of thermal expansion still exists.

Aluminum's rate of thermal expansion presents another problem. All materials, especially metals, expand when they are heated. Aluminum expands about 40% faster than copper. The problem occurs when aluminum wire is terminated at a connector, a screw on a switch or receptacle, or a wire lug.

A wire heats up a little when electrical current flows. This heat causes the wire and the screw or lug to expand. But, aluminum wire may expand faster than the screw or lug. When this is so, the aluminum wire squeezes out from under the screw. When the circuit is turned off, current stops flowing. The screw contracts back to its size when it cools, but the aluminum does not. Now the connection becomes loose. Air gets into the connection, building up more oxide; this causes more resistance, which produces even more heat the next time current flows. The problem is solved with screw and lug terminals specially designed for aluminum wires.

Aluminum wire must be terminated at proper connectors and lugs rated for aluminum wire. Large sizes of wire, such as those feeding power to the circuit-breaker panel in a building, must be terminated in lugs marked AL. The wire must also be coated with a pastelike material, called an *oxide inhibitor*, which keeps air away from the aluminum wire.

Aluminum wire is permitted to be directly connected only to switches and receptacles marked CO/ALR on the metal strap. The *NEC* requires that switches rated at 20 A or less directly connected to aluminum conductors must be marked CO/ALR, *NEC Section 380-14(c)*. The *NEC* requires that receptacles rated at 20 A or less di-

**Table 9-4** (*NEC Table 310-18*) Ampacities for insulated conductors rated 0–2000 V, 110°C to 250°C (230°F to 482°F). Not more than three conductors in raceway or cable based on ambient temperature of 30°C (86°F)

| Size AWG MCM | Temperature Rating of Conductor. See Table 310-13 | | | | | | | | Size AWG MCM |
|---|---|---|---|---|---|---|---|---|---|
| | 110°C (230°F) | 125°C (257°F) | 150°C (302°F) | 200°C (392°F) | 250°C (482°F) | 110°C (230°F) | 125°C (257°F) | 200°C (392°F) | |
| | Types AVA, AVL | Types AI, AIA | Types Z | Types A, AA, FEP, FEPB, PFA | Types PFAH, TFE | Types AVA, AVL | Types AI, AIA | Types A, AA | |
| | Copper | | | | Nickel or Nickel-Coated Copper | Aluminum or Copper-Clad Aluminum | | | |
| 14 | 30 | 30 | 30 | 30 | 40 | .... | .... | .... | .... |
| 12 | 35 | 40 | 40 | 40 | 55 | 25 | 30 | 30 | 12 |
| 10 | 45 | 50 | 50 | 55 | 75 | 35 | 40 | 45 | 10 |
| 8 | 60 | 65 | 65 | 70 | 95 | 45 | 50 | 55 | 8 |
| 6 | 80 | 85 | 90 | 95 | 120 | 60 | 65 | 75 | 6 |
| 4 | 105 | 115 | 115 | 120 | 145 | 80 | 90 | 95 | 4 |
| 3 | 120 | 130 | 135 | 145 | 170 | 95 | 100 | 115 | 3 |
| 2 | 135 | 145 | 150 | 165 | 195 | 105 | 115 | 130 | 2 |
| 1 | 160 | 170 | 180 | 190 | 220 | 125 | 135 | 150 | 1 |
| 0 | 190 | 200 | 210 | 225 | 250 | 150 | 160 | 180 | 0 |
| 00 | 215 | 230 | 240 | 250 | 280 | 170 | 180 | 200 | 00 |
| 000 | 245 | 265 | 275 | 285 | 315 | 195 | 210 | 225 | 000 |
| 0 000 | 275 | 310 | 325 | 340 | 370 | 215 | 245 | 270 | 0 000 |
| 250 | 315 | 335 | .... | .... | .... | 250 | 270 | .... | 250 |
| 300 | 345 | 380 | .... | .... | .... | 275 | 305 | .... | 300 |
| 350 | 390 | 420 | .... | .... | .... | 310 | 335 | .... | 350 |
| 400 | 420 | 450 | .... | .... | .... | 335 | 360 | .... | 400 |
| 500 | 470 | 500 | .... | .... | .... | 380 | 405 | .... | 500 |
| 600 | 525 | 545 | .... | .... | .... | 425 | 440 | .... | 600 |
| 700 | 560 | 600 | .... | .... | .... | 455 | 485 | .... | 700 |
| 750 | 580 | 620 | .... | .... | .... | 470 | 500 | .... | 750 |
| 800 | 600 | 640 | .... | .... | .... | 485 | 520 | .... | 800 |
| 1 000 | 680 | 730 | .... | .... | .... | 560 | 600 | .... | 1 000 |
| 1 500 | 785 | .... | .... | .... | .... | 650 | .... | .... | 1 500 |
| 2 000 | 840 | .... | .... | .... | .... | 705 | .... | .... | 2 000 |

### Ampacity Correction Factors

| Ambient Temp. °C | For ambient temperatures other than 300°C, multiply the ampacities shown above by the appropriate factor shown below. | | | | | | | | Ambient Temp. °F |
|---|---|---|---|---|---|---|---|---|---|
| 31–40 | .94 | .95 | .96 | .... | .... | .94 | .95 | .... | 87–104 |
| 41–45 | .90 | .92 | .94 | .... | .... | .90 | .92 | .... | 105–113 |
| 46–50 | .87 | .89 | .91 | .... | .... | .87 | .89 | .... | 114–122 |
| 51–55 | .83 | .86 | .89 | .... | .... | .83 | .86 | .... | 123–131 |
| 56–60 | .79 | .83 | .87 | .91 | .95 | .79 | .83 | .91 | 132–141 |
| 61–70 | .71 | .76 | .82 | .87 | .91 | .71 | .76 | .87 | 142–158 |
| 71–75 | .66 | .72 | .79 | .86 | .89 | .66 | .72 | .86 | 159–167 |
| 76–80 | .61 | .68 | .76 | .84 | .87 | .61 | .69 | .84 | 168–176 |
| 81–90 | .50 | .61 | .71 | .80 | .83 | .50 | .61 | .80 | 177–194 |
| 91–100 | .... | .51 | .65 | .77 | .80 | .... | .51 | .77 | 195–212 |
| 101–120 | .... | .... | .50 | .69 | .72 | .... | .... | .69 | 213–248 |
| 121–140 | .... | .... | .29 | .59 | .59 | .... | .... | .59 | 249–284 |
| 141–160 | .... | .... | .... | .... | .54 | .... | .... | .... | 285–320 |
| 161–180 | .... | .... | .... | .... | .50 | .... | .... | .... | 321–356 |
| 181–200 | .... | .... | .... | .... | .43 | .... | .... | .... | 357–392 |
| 201–225 | .... | .... | .... | .... | .30 | .... | .... | .... | 393–437 |

Reprinted with permission from NFPA 70-1984, *National Electrical Code®*, Copyright © 1983, National Fire Protection Association, Quincy, Massachusetts 02269. This reprinted material is not the complete and official position of the NFPA on the referenced subject, which is represented only by the standard in its entirety.

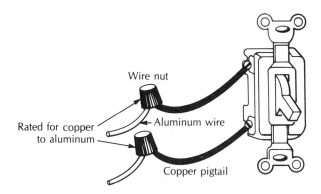

**Figure 9-4** A copper pigtail can be used to connect aluminum wire to a device not rated for aluminum. The wire nut must be rated for aluminum.

rectly connected to aluminum conductors must be marked CO/ALR, *NEC Section 410-56(b)*.

Another method of terminating aluminum wire to switches and receptacles may be used which does not require devices to be rated CO/ALR. A short length of copper wire, often referred to as a *pigtail,* is connected to the aluminum wire and the device, as shown in Figure 9-4. The wire nut or connector joining the aluminum and copper wire *must* be rated for aluminum wire. This technique is often used to solve aluminum wire termination problems in existing buildings.

## *WIRE PROTECTION*

Single wires are required to be protected from physical damage. Methods of protecting wire are 1) cable, 2) open wiring on insulators, 3) flexible cords, 4) raceway, such as conduit, and 5) other wiring methods described in *NEC Article 300*. Essentially, this means that individual wires in buildings are not permitted to be run unprotected, *NEC Section 300-1(c)*.

## *CABLE WIRING*

Cable wiring is the most common wiring method for farms. In cable wiring, electrical wires are contained within a protective outer sheath. The individual wires must be insulated, except for grounding wires which are permitted to be bare, *NEC Section 310-2*. Common types of cable used in agricultural buildings are:

- Type NM (*NEC Article 336*)—normally dry locations only
- Type NMC (*NEC Article 336*)—dry, damp, wet, or corrosive locations
- Type UF (*NEC Article 339*)—dry, damp, wet, or corrosive locations
- Type SE (*NEC Article 338*)—dry, damp, or wet locations

*Nonmetallic-sheathed cable* is defined as a factory assembly of two or more insulated conductors enclosed in an outer sheath of moisture-resistant, flame-retardant, nonmetallic material. This type of cable is available with two or three current-carrying conductors. The conductors range in size from No. 14 through No. 2 for copper conductors, and from No. 12 through No. 2 for aluminum or copper-clad aluminum conductors. Two-wire cable contains a black conductor and a white conductor. Three-wire cable contains a black, a white, and a red conductor. The current-carrying conductors are now required to have 90°C (194°F) rated insulation. Type NM cable ampacity, however, is based upon the 60°C (140°F) column of *Table 310-16, NEC Section 336-2*.

Nonmetallic-sheathed cable is also available with an uninsulated copper conductor which is used for grounding purposes only, Figure 9-5. This grounding conductor is not intended for use as a current-carrying circuit wire.

Two types of nonmetallic-sheathed cable are discussed in *NEC article 336:* Type NM, and Type NMC. Type NM cable may be installed so that the cable is on an exposed surface or concealed in a wall or ceiling space. It is permitted to be used only for normally dry locations. It may be placed in the void space of masonry or tile block walls as long as the walls are not subject to excessive moisture or dampness. Type NM cable is not permitted for use where it is exposed to corrosive fumes or vapors. It is not permitted to be embedded in concrete, masonry, or adobe.

Type NMC cable differs from Type NM in that it is permitted to be installed in dry, damp, wet, or corrosive locations. However, it is not permitted to be embedded

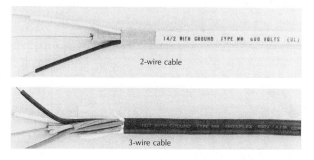

**Figure 9-5** Nonmetallic-sheathed cable is available with two or three insulated wires, plus a bare grounding wire.

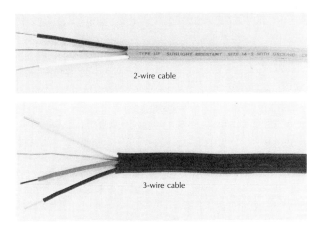

**Figure 9-6** Underground feeder and branch-circuit cable, Type UF, is available with two or three insulated wires, plus a bare grounding wire.

in poured cement or concrete. Type NMC cable is not readily available from electrical supply houses. Type UF cable, described next, is usually used instead of NMC where the area may be damp or wet or where corrosive vapors may exist.

Type UF underground feeder and branch-circuit cable contains two or three insulated wires and a bare grounding wire in an outer sheath which completely surrounds each wire, Figure 9-6. The outer covering is required to be flame retardant, and resistant to moisture, fungus, and corrosion. The insulation on the wires must be moisture resistant. Type TW insulation is generally used. UF cable is available in sizes No. 14 to No. 4/0 for copper, and in sizes No. 12 to No. 4/0 for aluminum and copper-clad aluminum. The ampere rating of UF cable is found in *NEC Table 310-16* for wire with a 60°C (140°F) rating.

Type UF cable is permitted to be used for surface and concealed wiring in buildings, and for direct burial in the earth. Type UF cable is one of the permissible wiring materials for farm buildings which are moist and/or dusty. Type UF cable is not permitted to be embedded in poured concrete.

Type SE cable is permitted to be used for interior wiring, Figure 9-7. Type SE cable is available with two or three insulated wires and a bare grounding wire. The insulation on the wires is usually THW or XHHW, and carries a 75°C (167°F) rating. The outer covering is flame retardant. Type SE cable must be protected from physical abuse, as is true of the other types of cable.

## FLEXIBLE CORDS

Some agricultural machines and equipment, though permanently installed, may need to be free to move about in normal operation, such as a silo unloader. Some equipment must be free to be moved for periodic servicing, such as a milking system automatic washer, which involves the coupling together of several components with liquid hoses and electrical wires, Figure 9-8. Flexible electrical cords often provide the most satisfactory means of wiring for some agricultural equipment. Electrical cords used for such purposes must be approved for hard usage, *NEC Section 547-3(b)*.

Flexible cords for agricultural applications must be suitable for damp locations, and resistant to oil and oil-based compounds. *NEC Table 400-4* lists types of cords and uses. Suitable hard-usage cords commonly used for agricultural equipment applications are Type SJO and Type SJTO. Insulated copper wires inside the protective jacket are available in sizes from No. 18 AWG to No. 12 AWG. Flexible cords supplying such equipment as silo unloaders are required to be rated for extra-hard usage.

**Figure 9-7** Type SE cable may be used for interior wiring.

**Figure 9-8** Flexible cords supply power and control to a milking system automatic washer. *(Courtesy of Baas Agrisystems, Inc.)*

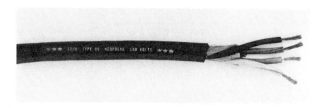

Figure 9-9 Moisture-resistant flexible cords, rated for extra-hard service, are required for some agricultural applications.

Type SO and Type STO are commonly used. These cords are available with insulated copper wires in sizes from No. 18 to No. 2 AWG. A typical flexible cord used with agricultural equipment is shown in Figure 9-9.

## OPEN WIRING ON INSULATORS

Open wiring on insulators, NEC Article 320, is used to a minor extent for animal housing in some areas. This wiring method is permitted for areas where there is dust or dust with moisture, but not in areas where there is a corrosive atmosphere, NEC Section 547-3. The wire is supported on insulating knobs, cleats, tubes, and flexible tubing. The wiring is not permitted to be concealed inside walls. It must be mounted on an exposed surface. This type of wiring is not recommended, unless warranted by the circumstances and the desires of the farm operator.

## WIRING WITH CONDUIT

Electrical wires are often contained within conduit for many farm wiring applications. Conduit has the advantage of providing the wires with protection from physical damage. Many wires can be placed in a single conduit. Wires with the desired color insulation can be installed. Wires with high-temperature insulation can be installed where necessary. New wires can be added or old wires can be replaced quite easily. Disadvantages are that conduit installation often is more difficult and time consuming than cable installation, which results in higher installation cost. However, this is not always true. Sometimes, installation is faster and less expensive with conduit rather than cable.

Various conditions for wiring in farm buildings require a variety of types of conduit. Moist and extremely corrosive conditions may destroy metal conduit in a year or two. Animals may chew nonmetallic conduit or it may become damaged by impact from animals or machines.

NEC Section 547-3 states that raceway shall be suitable for the conditions. Clearly, the electrical installer must judge the conditions and select appropriate materials. The types of conduit used in farm buildings are:

- Rigid Metal Conduit (NEC Article 346)
  galvanized steel or aluminum
- Intermediate Metallic Conduit, IMC (NEC Article 345)
- Electrical Metallic Tubing, EMT (NEC Article 348)
  thin-wall conduit
- Rigid Nonmetallic Conduit (NEC Article 347)
  Schedule 40—standard wall
  Schedule 80—thick wall
- Flexible Metal Conduit (NEC Article 350)
- Liquidtight Flexible Conduit (NEC Article 351)
  metallic and nonmetallic

Electrical conduits other than flexible come in 10-ft (3.05-m) lengths and are available in nominal trade sizes according to inside diameter. One-half inch (in) trade size conduit is the minimum size permitted, except for special applications of 3/8-in flexible conduits. Common trade sizes of conduit available are shown in Table 9-5.

## Rigid Metal Conduit

Rigid metal conduit provides mechanical protection to the wires where severe physical abuse may occur. This type of conduit is suitable for all wiring conditions, except in areas where corrosive conditions exist. Rigid metal conduit is available with a corrosion-resistant nonmetallic coating. Because of cost and corrosion problems, rigid metal conduit should only be used where physical abuse is a problem.

Rigid metal conduit is joined by means of threads. Each length of conduit is supplied with one threaded coupling. Rigid conduit is fastened to boxes and cabinets using two locknuts, one on the inside and one on the outside. Numerous fittings and boxes are available with threaded hubs for rigid conduit. Steel and aluminum boxes and fittings are available. Aluminum fittings may be used with steel conduit, NEC Section 346-1(b). A type of fitting used to make an abrupt change of direction is called a *conduit body* or *condulet*. These fittings have removable covers to aid in installing wires.

Common types of conduit bodies are shown in Figure 9-10. A special box made for surface mounting has threaded hubs, and is designed to accept devices such as switches and receptacle outlets. These boxes are commonly referred to as *FS boxes*. The common types of FS boxes are shown in Figure 9-11.

Table 9-5  Common trade sizes of conduit (in inches)

|  | 3/8 | 1/2 | 3/4 | 1 | 1 1/4 | 1 1/2 | 2 | 2 1/2 | 3 | 3 1/2 | 4 | 5 | 6 |
|---|---|---|---|---|---|---|---|---|---|---|---|---|---|
| Rigid Metal | — | x | x | x | x | x | x | x | x | x | x | x | x |
| IMC | — | x | x | x | x | x | x | x | x | x | — | — | |
| EMT | — | x | x | x | x | x | x | x | x | x | — | — | |
| Rigid Nonmetallic | — | x | x | x | x | x | x | x | x | x | x | x | |
| Flexible Metal | x | x | x | x | x | x | x | x | x | — | — | — | |
| Liquidtight Flexible | | | | | | | | | | | | | |
| • Metallic | x | x | x | x | x | x | x | x | x | x | — | — | |
| • Nonmetallic | x | x | x | x | x | x | x | — | — | — | — | — | |

## Intermediate Metal Conduit (IMC)

Intermediate metal conduit (IMC) is a metal alloy which can be used in place of rigid metal conduit for most applications. The conduit is threaded the same as rigid metal conduit. The primary advantages of IMC over rigid metal conduit is its lower cost. The installation of IMC is covered in *NEC Article 345*.

## Electrical Metallic Tubing

Electrical metallic tubing (EMT) is often called *thinwall conduit*. EMT is never threaded. It is not permitted to be installed in locations where it is subject to physical abuse or a corrosive environment. It is recommended for use only in farm buildings that are dry. The advantages of EMT are its low cost, light weight, and ease of bending and installation.

Electrical metallic tubing is joined and terminated with pressure-type couplings and connectors. The most common types are termed *setscrew* and *raintight*. Raintight fittings only are permitted to be used outdoors. Typical fittings for use with EMT are shown in Figure 9-12.

## Rigid Nonmetallic Conduit

Rigid nonmetallic conduit is the most ideal raceway for farm wiring, except in those areas where the conduit is subjected to extreme physical abuse. The most common nonmetallic material used for this conduit is polyvinyl chloride (PVC). It is available with two wall thicknesses: Schedule 40, which is the standard thickness; and Schedule 80, which has an extra-thick wall. Schedule 80 is permitted to be used in areas where physical abuse is judged to be not excessive.

PVC conduit is corrosion resistant, and forms a watertight and dusttight system after installation. The installation of PVC conduit is covered in *NEC Article 347*.

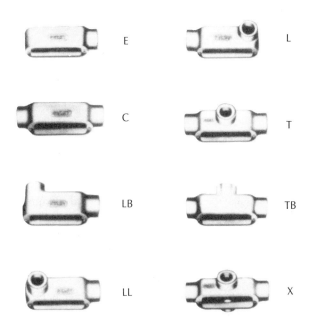

Figure 9-10  Conduit bodies with threaded hubs are used to provide access to the conduit system, and to make abrupt changes in direction.

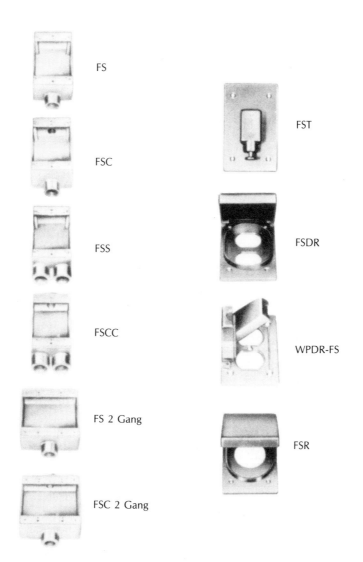

Figure 9-11 FS boxes with threaded hubs are available to accept one or more devices.

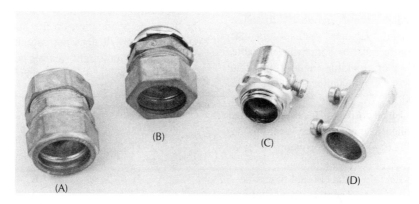

Figure 9-12 EMT couplings and connectors. (A) raintight coupling, (B) raintight connector, (C) setscrew connector, and (D) setscrew coupling

Unit 9  Farm Wiring Methods and Materials

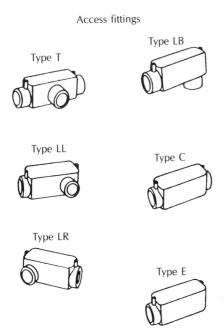

**Figure 9-13** Nonmetallic conduit access fittings

**Figure 9-14** Flexible metal conduit fittings

PVC is not threaded. It is cut with a saw or special cutter. An adhesive is applied to the end of the conduit, and then the conduit is pressed into the fitting.

Couplings, connectors, adapters, and boxes manufactured from PVC are available for use with PVC conduit. Using only PVC, it is possible to install a completely nonmetallic wiring system, except for the fasteners. PVC is often the most satisfactory and least costly wiring system for new farm building construction. PVC access fittings are shown in Figure 9-13.

## Flexible Metal Conduit

Flexible metal conduit is used frequently to make connections to equipment, machines, and motors that either produce vibration or must be somewhat movable for maintenance and repair. Flexible metal conduit is used frequently on machines by manufacturers. Flexible metal conduit, covered in *NEC Article 350,* is not permitted to be used in wet or corrosive locations. In the past, it was often installed in inappropriate locations on the farm, which resulted in severe corrosion in only weeks or months.

The minimum size flexible metal conduit that is permitted to be installed is ½ in. There is an exception, however, in the case of leads to lighting fixtures and motors, where the minimum size permitted is ⅜ in. Special fittings are available for terminating flexible metal conduit, as shown in Figure 9-14. The fittings are not dusttight, and must not be used in areas where dust is present.

## Liquidtight Flexible Conduit

Liquidtight flexible conduit provides a liquidtight and dusttight flexible wiring system. This conduit is available in two types: metallic and nonmetallic. Liquidtight flexible metal conduit is most commonly used. It is a flexible metal conduit with a nonmetallic outer covering which resists water and petroleum products, and the dusty, corrosive materials, and vapors encountered on the farm. This type of flexible conduit is shown in Figure 9-15. Fittings are designed to prevent moisture from entering the wiring system. Typical fittings are shown in Figure 9-16.

Liquidtight flexible nonmetallic conduit can be used for the same application on the farm as the type previously discussed. It does not contain a metal inner core. Both types of liquidtight flexible conduit will collapse if subjected to physical abuse; therefore, this limitation

**Figure 9-15** Liquidtight flexible metal conduit has a nonmetallic outer covering.

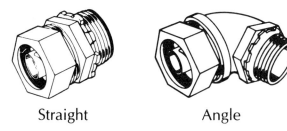

Straight　　　　　Angle

**Figure 9-16** Connectors for liquidtight flexible metal conduit

must be considered during installation. The installation and use of liquidtight flexible conduit is covered in *NEC Article 351*.

## NUMBER OF WIRES IN CONDUIT

Tables are provided in *NEC Chapter 9* to be used in selecting the size conduit required for electrical wires. If the conduit is too small, installation of the wires becomes difficult. Difficult installation may result in damage to the insulation on the wires, and it can be too time consuming. The following factors determine the size of conduit required for electrical wires.

1. Size of wires
2. Type of insulation on wires
3. Number of wires
4. Are all wires of the same size and type?

If all the wires are of the same size and have the same type of insulation, then conduit sizing is simple. The minimum size conduit is given in *NEC Table 3, Chapter 9*, represented in this text as Table 9-6. As an example, assume that an electric furnace in a milkhouse is supplied with two No. 4 AWG copper wires with THW insulation. Referring to *NEC Table 3A*, find the type of insulation in the left column. Next find the size of wire in the second column. Then move horizontally to the right until you find the correct number of wires. The number in the table can be larger, but not smaller. The minimum trade conduit is shown at the top of the table. The minimum size conduit for this example is 1 in.

## Problem 9-2

A conduit contains four No. 8 AWG copper wires with THWN insulation. Determine the minimum trade size conduit.

## Solution

Wire with THWN insulation is found in *NEC Table 3B* (text Table 9-6). The minimum trade size conduit is ¾ in.

Sizing the conduit requires some calculations when the wires are of different sizes or different insulation types. The exact method of determining the minimum size conduit requires some calculation. First the total cross-sectional area of all the wires in the conduit must be determined. The area of the wire is found in *NEC Table 5, Chapter 9*, represented in this text as Table 9-7. Be sure to consider the footnotes to the table. As an example, assure that a conduit contains two No. 6 AWG wires with THWN insulation, and four No. 10 AWG wires with THW insulation. Determine the minimum conduit size.

> No. 6 AWG THWN:
> $\quad 0.051\ 9\text{ in}^2 \times 2 = 0.103\ 8\text{ in}^2$
> No. 10 AWG THW:
> $\quad 0.031\ 1\text{ in}^2 \times 4 = \underline{0.124\ 4\text{ in}^2}$
> Total wire area $\quad\quad\quad = 0.228\ 2\text{ in}^2$

The total cross-sectional areas of the wire is not permitted to exceed a specified percent of the area of the conduit. This information is given in *NEC Table 1, Chapter 9*, summarized in this text as Table 9-8.

For the example, there are six wires; therefore, they are not permitted to exceed 40% of the total area of the conduit. The minimum size conduit can be found in *NEC Table 4, Chapter 9*, represented in the text as Table 9-9. Find the correct column for the number of wires. For this example, use the 40% column headed "Not Lead Covered." Go down the column until an area is found which is equal to or greater than the total area of the wire calculated previously. The minimum size conduit, found in the left-hand column, is 1 in.

## Problem 9-3

A conduit contains three No. 4/0 AWG THWN aluminum wires and four No. 6 AWG THW copper wires. Determine the minimum trade size conduit.

## Solution

First, determine the total cross-sectional area of the wires. Start by finding the cross-sectional area of each wire in *NEC Table 5* (text Table 9-7).

Table 9-6 (*NEC Table* 3A) Maximum number of conductors in trade size conduit or tubing (based on *NEC Table 1, Chapter 9*)

| Type Letters | Conductor Size AWG, MCM | ½ | ¾ | 1 | 1¼ | 1½ | 2 | 2½ | 3 | 3½ | 4 | 5 | 6 |
|---|---|---|---|---|---|---|---|---|---|---|---|---|---|
| TW, T, RUH, RUW, XHHW (14 thru 8) | 14 | 9 | 15 | 25 | 44 | 60 | 99 | 142 | | | | | |
| | 12 | 7 | 12 | 19 | 35 | 47 | 78 | 111 | 171 | 176 | | | |
| | 10 | 5 | 9 | 15 | 26 | 36 | 60 | 85 | 131 | | 108 | | |
| | 8 | 2 | 4 | 7 | 12 | 17 | 28 | 40 | 62 | 84 | | | |
| RHW and RHH (without outer covering), THW | 14 | 6 | 10 | 16 | 29 | 40 | 65 | 93 | 143 | 192 | | | |
| | 12 | 4 | 8 | 13 | 24 | 32 | 53 | 76 | 117 | 157 | 163 | 133 | |
| | 10 | 4 | 6 | 11 | 19 | 26 | 43 | 61 | 95 | 127 | 85 | | |
| | 8 | 1 | 3 | 5 | 10 | 13 | 22 | 32 | 49 | 66 | | | |
| TW, T, THW, RUH (6 thru 2), RUW (6 thru 2), | 6 | 1 | 2 | 4 | 7 | 10 | 16 | 23 | 36 | 48 | 62 | 97 | 141 |
| | 4 | 1 | 1 | 3 | 5 | 7 | 12 | 17 | 27 | 36 | 47 | 73 | 106 |
| | 3 | 1 | 1 | 2 | 4 | 6 | 10 | 15 | 23 | 31 | 40 | 63 | 91 |
| | 2 | 1 | 1 | 2 | 4 | 5 | 9 | 13 | 20 | 27 | 34 | 54 | 78 |
| | 1 | | 1 | 1 | 3 | 4 | 6 | 9 | 14 | 19 | 25 | 39 | 57 |
| FEPB (6 thru 2), RHW and RHH (without outer covering) | 0 | | 1 | 1 | 2 | 3 | 5 | 8 | 12 | 16 | 21 | 33 | 49 |
| | 00 | | 1 | 1 | 1 | 3 | 5 | 7 | 10 | 14 | 18 | 29 | 41 |
| | 000 | | 1 | 1 | 1 | 2 | 4 | 6 | 9 | 12 | 15 | 24 | 35 |
| | 0 000 | | | 1 | 1 | 1 | 3 | 5 | 7 | 10 | 13 | 20 | 29 |
| | 250 | | | 1 | 1 | 1 | 2 | 4 | 6 | 8 | 10 | 16 | 23 |
| | 300 | | | 1 | 1 | 1 | 2 | 3 | 5 | 7 | 9 | 14 | 20 |
| | 350 | | | | 1 | 1 | 1 | 3 | 4 | 6 | 8 | 12 | 18 |
| | 400 | | | | 1 | 1 | 1 | 3 | 4 | 5 | 7 | 11 | 16 |
| | 500 | | | | 1 | 1 | 1 | 2 | 3 | 4 | 6 | 9 | 14 |
| | 600 | | | | | 1 | 1 | 1 | 3 | 4 | 5 | 7 | 11 |
| | 700 | | | | | 1 | 1 | 1 | 2 | 3 | 4 | 7 | 10 |
| | 750 | | | | | 1 | 1 | 1 | 2 | 3 | 4 | 6 | 9 |

(Table 9-6 continued)                                                                                                    (NEC Table 3B, Chapter 9)

| Type Letters | Conductor Size AWG, MCM | ½ | ¾ | 1 | 1¼ | 1½ | 2 | 2½ | 3 | 3½ | 4 | 5 | 6 |
|---|---|---|---|---|---|---|---|---|---|---|---|---|---|
| THWN, THHN, FEP (14 thru 2), FEPB (14 thru 8), PFA (14 thru 4/0) | 14 | 13 | 24 | 39 | 69 | 94 | 154 | | | | | | |
| | 12 | 10 | 18 | 29 | 51 | 70 | 114 | | | | | | |
| | 10 | 6 | 11 | 18 | 32 | 44 | 73 | 164 | 160 | | | | |
| | 8 | 3 | 5 | 9 | 16 | 22 | 36 | 104 | 79 | 106 | 136 | | |
| | | | | | | | | 51 | | | | | |
| PFAH (14 thru 4/0) Z (14 thru 4/0) | 6 | 1 | 4 | 6 | 11 | 15 | 26 | 37 | 57 | 76 | 98 | 154 | 137 |
| | 4 | 1 | 2 | 4 | 7 | 9 | 16 | 22 | 35 | 47 | 60 | 94 | 116 |
| | 3 | 1 | 1 | 3 | 6 | 8 | 13 | 19 | 29 | 39 | 51 | 80 | 97 |
| XHHW (4 thru 500MCM) | 2 | 1 | 1 | 3 | 5 | 7 | 11 | 16 | 25 | 33 | 43 | 67 | 72 |
| | 1 | | 1 | 1 | 3 | 5 | 8 | 12 | 18 | 25 | 32 | 50 | |
| | 0 | | 1 | 1 | 3 | 4 | 7 | 10 | 15 | 21 | 27 | 42 | 61 |
| | 00 | | 1 | 1 | 2 | 3 | 6 | 8 | 13 | 17 | 22 | 35 | 51 |
| | 000 | | 1 | 1 | 1 | 3 | 5 | 7 | 11 | 14 | 18 | 29 | 42 |
| | 0 000 | | 1 | 1 | 1 | 2 | 4 | 6 | 9 | 12 | 15 | 24 | 35 |
| | 250 | | | 1 | 1 | 1 | 3 | 4 | 7 | 10 | 12 | 20 | 28 |
| | 300 | | | 1 | 1 | 1 | 3 | 4 | 6 | 8 | 11 | 17 | 24 |
| | 350 | | | 1 | 1 | 1 | 2 | 3 | 5 | 7 | 9 | 15 | 21 |
| | 400 | | | | 1 | 1 | 1 | 3 | 5 | 6 | 8 | 13 | 19 |
| | 500 | | | | 1 | 1 | 1 | 2 | 4 | 5 | 7 | 11 | 16 |
| | 600 | | | | 1 | 1 | 1 | 1 | 3 | 4 | 5 | 9 | 13 |
| | 700 | | | | | 1 | 1 | 1 | 3 | 4 | 5 | 8 | 11 |
| | 750 | | | | | 1 | 1 | 1 | 2 | 3 | 4 | 7 | 11 |
| XHHW | 6 | 1 | 3 | 5 | 9 | 13 | 21 | 30 | 47 | 63 | 81 | 128 | 185 |
| | 600 | | | | 1 | 1 | 1 | 1 | 3 | 4 | 5 | 9 | 13 |
| | 700 | | | | | 1 | 1 | 1 | 3 | 4 | 5 | 7 | 11 |
| | 750 | | | | | 1 | 1 | 1 | 2 | 3 | 4 | 7 | 10 |

(Table 9-6 continued)

| Conduit Trade Size (Inches) | | ½ | ¾ | 1 | 1¼ | 1½ | 2 | 2½ | 3 | 3½ | 4 | 5 | 6 |
|---|---|---|---|---|---|---|---|---|---|---|---|---|---|
| Type Letters | Conductor Size AWG, MCM | | | | | | | | | | | | |
| RHW, RHH (with outer covering) | 14 | 3 | 6 | 10 | 18 | 25 | 41 | 58 | 90 | 121 | 155 | | |
| | 12 | 3 | 5 | 9 | 15 | 21 | 35 | 50 | 77 | 103 | 132 | | |
| | 10 | 2 | 4 | 7 | 13 | 18 | 29 | 41 | 64 | 86 | 110 | | |
| | 8 | 1 | 2 | 4 | 7 | 9 | 16 | 22 | 35 | 47 | 60 | 94 | 137 |
| | 6 | 1 | 1 | 2 | 5 | 6 | 11 | 15 | 24 | 32 | 41 | 64 | 93 |
| | 4 | 1 | 1 | 1 | 3 | 5 | 8 | 12 | 18 | 24 | 31 | 50 | 72 |
| | 3 | 1 | 1 | 1 | 3 | 4 | 7 | 10 | 16 | 22 | 28 | 44 | 63 |
| | 2 | | 1 | 1 | 3 | 4 | 6 | 9 | 14 | 19 | 24 | 38 | 56 |
| | 1 | | 1 | 1 | 1 | 3 | 5 | 7 | 11 | 14 | 18 | 29 | 42 |
| | 0 | | 1 | 1 | 1 | 2 | 4 | 6 | 9 | 12 | 16 | 25 | 37 |
| | 00 | | | 1 | 1 | 1 | 3 | 5 | 8 | 11 | 14 | 22 | 32 |
| | 000 | | | 1 | 1 | 1 | 3 | 4 | 7 | 9 | 12 | 19 | 28 |
| | 0 000 | | | 1 | 1 | 1 | 2 | 4 | 6 | 8 | 10 | 16 | 24 |
| | 250 | | | | 1 | 1 | 1 | 3 | 5 | 6 | 8 | 13 | 19 |
| | 300 | | | | 1 | 1 | 1 | 3 | 4 | 5 | 7 | 11 | 17 |
| | 350 | | | | 1 | 1 | 1 | 2 | 4 | 5 | 6 | 10 | 15 |
| | 400 | | | | 1 | 1 | 1 | 1 | 3 | 4 | 6 | 9 | 14 |
| | 500 | | | | 1 | 1 | 1 | 1 | 3 | 4 | 5 | 8 | 11 |
| | 600 | | | | | 1 | 1 | 1 | 2 | 3 | 4 | 6 | 9 |
| | 700 | | | | | 1 | 1 | 1 | 1 | 3 | 3 | 6 | 8 |
| | 750 | | | | | 1 | 1 | 1 | 1 | 3 | 3 | 5 | 8 |

(NEC Table 3C, Chapter 9)

Reprinted with permission from NFPA 70-1984, *National Electrical Code*®, Copyright © 1983, National Fire Protection Association, Quincy, MA. This reprinted material is not the complete and official position of the NFPA on the referenced subject which is represented only by the standard in its entirety.

Table 9-7 (NEC Table 5, Chapter 9) Dimensions of rubber-covered and thermoplastic-covered conductors

| Size AWG MCM | Types RFH-2, RH, RHH,*** RHW,*** SF-2 | | Types TF, T, THW,† TW, RUH,** RUW** | | Types THHN, THWN | | Types**** FEP, FEPB, FEPW, TFE, PF, PFA, PFAH, PGF, PTF, Z, ZF, ZFF | | Type XHHW, ZW†† | | Types KF-1, KF-2, KFF-1, KFF-2 | |
|---|---|---|---|---|---|---|---|---|---|---|---|---|
| | Approx. Diam. Inches | Approx. Area Sq. In. | Approx. Diam. Inches | Approx. Area Sq. In. | Approx. Diam. Inches | Approx. Area Sq. In. | Approx. Diam. Inches | Approx. Area Sq. Inches | Approx. Diam. Inches | Approx. Area Sq. In. | Approx. Diam. Sq. In. | Approx. Area Sq. In. |
| Col. 1 | Col. 2 | Col. 3 | Col. 4 | Col. 5 | Col. 6 | Col. 7 | Col. 8 | Col. 9 | Col. 10 | Col. 11 | Col. 12 | Col. 13 |
| 18 | .146 | .0167 | .106 | .0088 | .089 | .0062 | .081 | .0052 | ... | ... | .065 | .0033 |
| 16 | .158 | .0196 | .118 | .0109 | .100 | .0079 | .092 | .0066 | ... | ... | .070 | .0038 |
| 14 | 30 mils .171 | .0230 | .131 | .0135 | .105 | .0087 | .105 .105 | .0087 .0087 | ... | ... | .083 | .0054 |
| 14 | 45 mils .204* | .0327* | ... | ... | ... | ... | ... | ... | ... | ... | ... | ... |
| 14 | ... | ... | .162† | .0206† | ... | ... | ... | ... | .129 | .0131 | ... | ... |
| 12 | 30 mils .188 | .0278 | .148 | .0172 | .122 | .0117 | .121 .121 | .0115 .0115 | ... | ... | .102 | .0082 |
| 12 | 45 mils .221* | .0384* | ... | ... | ... | ... | ... | ... | ... | ... | ... | ... |
| 12 | ... | ... | .179† | .0252† | ... | ... | ... | ... | .146 | .0167 | ... | ... |
| 10 | .242 | .0460 | .168 | .0222 | .153 | .0184 | .142 .142 | .0158 .0158 | ... | ... | .124 | .0121 |
| 10 | ... | ... | .199† | .0311† | ... | ... | ... | ... | .166 | .0216 | ... | ... |
| 8 | .328 | .0845 | .245 | .0471 | .218 | .0373 | .206 .186 | .0333 .0272 | .241 | .0456 | ... | ... |
| 8 | ... | ... | .276† | .0598† | ... | ... | ... | ... | ... | ... | ... | ... |
| 6 | .397 | .1238 | .323 | .0819 | .257 | .0519 | .244 | .0468 .0716 | .282 | .0625 | ... | ... |
| 4 | .452 | .1605 | .372 | .1087 | .328 | .0845 | .292 | .0670 .0962 | .328 | .0845 | ... | ... |
| 3 | .481 | .1817 | .401 | .1263 | .356 | .0995 | .320 | .0804 .1122 | .356 | .0995 | ... | ... |
| 2 | .513 | .2067 | .433 | .1473 | .388 | .1182 | .352 | .0973 .1320 | .388 | .1182 | ... | ... |
| 1 | .588 | .2715 | .508 | .2027 | .450 | .1590 | .420 | .1385 | .450 | .1590 | ... | ... |
| 0 | .629 | .3107 | .549 | .2367 | .491 | .1893 | .462 | .1676 | .491 | .1893 | ... | ... |
| 00 | .675 | .3578 | .595 | .2781 | .537 | .2265 | .498 | .1948 | .537 | .2265 | ... | ... |
| 000 | .727 | .4151 | .647 | .3288 | .588 | .2715 | .560 | .2463 | .588 | .2715 | ... | ... |
| 0 000 | .785 | .4840 | .705 | .3904 | .646 | .3278 | .618 | .3000 | .646 | .3278 | ... | ... |

(Table 9-7 continued) (NEC Table 5, Chapter 9)

| Size AWG MCM | Types RFH-2, RH, RHH,*** RHW,*** SF-2 | | Types TF, T, THW,† TW, RUH,** RUW*** | | Types THHN, THWN | | Types**** FEP, FEPB, FEPW, TFE, PF, PFA, PFAH, PGF, PTF, Z, ZF, ZFF | | Type XHHW, ZW†† | |
|---|---|---|---|---|---|---|---|---|---|---|
| | Approx. Diam. Inches | Approx. Area Sq. In. | Approx. Diam. Inches | Approx. Area Sq. In. | Approx. Diam. Inches | Approx. Area Sq. In. | Approx. Diam. Inches | Approx. Area Sq. Inches | Approx. Diam. Inches | Approx. Area Sq. In. |
| Col. 1 | Col. 2 | Col. 3 | Col. 4 | Col. 5 | Col. 6 | Col. 7 | Col. 8 | Col. 9 | Col. 10 | Col. 11 |
| 250 | .868 | .5917 | .788 | .4877 | .716 | .4026 | ... | ... | .716 | .4026 |
| 300 | .933 | .6837 | .843 | .5581 | .771 | .4669 | ... | ... | .771 | .4669 |
| 350 | .985 | .7620 | .895 | .6291 | .822 | .5307 | ... | ... | .822 | .5307 |
| 400 | 1.032 | .8365 | .942 | .6969 | .869 | .5931 | ... | ... | .869 | .5931 |
| 500 | 1.119 | .9834 | 1.029 | .8316 | .955 | .7163 | ... | ... | .955 | .7163 |
| 600 | 1.233 | 1.1940 | 1.143 | 1.0261 | 1.058 | .8791 | ... | ... | 1.073 | .9043 |
| 700 | 1.304 | 1.3355 | 1.214 | 1.1575 | 1.129 | 1.0011 | ... | ... | 1.145 | 1.0297 |
| 750 | 1.339 | 1.4082 | 1.249 | 1.2252 | 1.163 | 1.0623 | ... | ... | 1.180 | 1.0936 |
| 800 | 1.372 | 1.4784 | 1.282 | 1.2908 | 1.196 | 1.1234 | ... | ... | 1.210 | 1.1499 |
| 900 | 1.435 | 1.6173 | 1.345 | 1.4208 | 1.259 | 1.2449 | ... | ... | 1.270 | 1.2668 |
| 1 000 | 1.494 | 1.7530 | 1.404 | 1.5482 | 1.317 | 1.3623 | ... | ... | 1.330 | 1.3893 |
| 1 250 | 1.676 | 2.2062 | 1.577 | 1.9532 | ... | ... | ... | ... | 1.500 | 1.7671 |
| 1 500 | 1.801 | 2.5475 | 1.702 | 2.2751 | ... | ... | ... | ... | 1.620 | 2.0612 |
| 1 750 | 1.916 | 2.8832 | 1.817 | 2.5930 | ... | ... | ... | ... | 1.740 | 2.3779 |
| 2 000 | 2.021 | 3.2079 | 1.922 | 2.9013 | ... | ... | ... | ... | 1.840 | 2.6590 |

*The dimensions of Types RHH and RHW.

**No. 14 to No. 2.

†Dimensions of THW in sizes No. 14 to No. 8. No. 6 THW and larger is the same dimension as T.

***Dimensions of RHH and RHW without outer covering are the same as THW No. 18 to No. 10, solid; No. 8 and larger stranded.

****In Columns 8 and 9 the values shown for sizes No. 1 thru 0000 are for TFE and Z only. The right-hand values in Columns 8 and 9 are for FEPB, Z, ZF, and ZFF only.

††No. 14 to No. 2.

Reprinted with permission from NFPA 70-1984, *National Electrical Code®*, Copyright © 1983, National Fire Protection Association, Quincy, MA. This reprinted material is not the complete and official position of the NFPA on the referenced subject which is represented only by the standard in its entirety.

**Table 9-8  Total area of the wire cannot exceed this percent area of the conduit. (From *NEC Table 1, Chapter 9*.)**

| | |
|---|---|
| One wire or cable in conduit | 53% |
| Two wires in conduit | 31% |
| Three or more wires in conduit | 40% |

**Table 9-9  (*NEC Table 4, Chapter 9*) Dimensions and percent area of conduit and tubing. (Areas of conduit or tubing for the combinations of wires permitted in *NEC Table 1, Chapter 9*)**

| Trade Size | Internal Diameter Inches | Area—Square Inches | | | | | | | | |
|---|---|---|---|---|---|---|---|---|---|---|
| | | | Not Lead Covered | | | Lead Covered | | | | |
| | | Total 100% | 2 Cond. 31% | Over 2 Cond. 40% | 1 Cond. 53% | 1 Cond. 55% | 2 Cond. 30% | 3 Cond. 40% | 4 Cond. 38% | Over 4 Cond. 35% |
| ½ | .622 | .30 | .09 | .12 | .16 | .17 | .09 | .12 | .11 | .11 |
| ¾ | .824 | .53 | .16 | .21 | .28 | .29 | .16 | .21 | .20 | .19 |
| 1 | 1.049 | .86 | .27 | .34 | .46 | .47 | .26 | .34 | .33 | .30 |
| 1 ¼ | 1.380 | 1.50 | .47 | .60 | .80 | .83 | .45 | .60 | .57 | .53 |
| 1 ½ | 1.610 | 2.04 | .63 | .82 | 1.08 | 1.12 | .61 | .82 | .78 | .71 |
| 2 | 2.067 | 3.36 | 1.04 | 1.34 | 1.78 | 1.85 | 1.01 | 1.34 | 1.28 | 1.18 |
| 2 ½ | 2.469 | 4.79 | 1.48 | 1.92 | 2.54 | 2.63 | 1.44 | 1.92 | 1.82 | 1.68 |
| 3 | 3.068 | 7.38 | 2.29 | 2.95 | 3.91 | 4.06 | 2.21 | 2.95 | 2.80 | 2.58 |
| 3 ½ | 3.548 | 9.90 | 3.07 | 3.96 | 5.25 | 5.44 | 2.97 | 3.96 | 3.76 | 3.47 |
| 4 | 4.026 | 12.72 | 3.94 | 5.09 | 6.74 | 7.00 | 3.82 | 5.09 | 4.83 | 4.45 |
| 5 | 5.047 | 20.00 | 6.20 | 8.00 | 10.60 | 11.00 | 6.00 | 8.00 | 7.60 | 7.00 |
| 6 | 6.065 | 28.89 | 8.96 | 11.56 | 15.31 | 15.89 | 8.67 | 11.56 | 10.98 | 10.11 |

Reprinted with permission from NFPA 70-1984, *National Electrical Code®*. Copyright © 1983, National Fire Protection Association, Quincy, MA. This reprinted material is not the complete and official position of the NFPA on the referenced subject which is represented only by the standard in its entirety.

> No. 4/0 AWG THWN:
> $\quad 0.327\ 8\ \text{in}^2 \times 3 = 0.983\ 4\ \text{in}^2$
> No. 6 AWG THW:
> $\quad 0.081\ 9\ \text{in}^2 \times 4 = 0.327\ 6\ \text{in}^2$
> Total wire area $\quad\quad\quad = 1.311\ 0\ \text{in}^2$

Next, under the column headed "Not Lead Covered," find the minimum trade size conduit in the 40% column of *NEC Table 4,* text Table 9-9. The minimum trade size conduit is 2 in.

## GENERAL WIRING PROCEDURES

A wiring system must be installed in such a manner that it will not create a shock or fire hazard. This is the purpose of the *National Electrical Code*. The wiring system must be installed so that an electrician or qualified maintenance personnel can gain access to necessary parts of the wiring system. The following are some important general wiring rules and procedures.

- The completed wiring system must be free from short circuits and ground faults, *NEC Section 110-7*. This means that splices must be properly insulated, and any damaged insulation on wires must be repaired.
- Wires, materials, and equipment must be identified for use in the type of environment in which they will be installed, *NEC section 110-11*. For example, wires used in damp or wet locations must have a letter W in their type designation, such as THW.
- The electrical system must be installed in a neat and workmanlike manner, *NEC Section 110-12*. Seldom does an electrical inspector find a problem in a wiring system installed by an electrician who practices good workmanship.
- Electrical equipment must be securely fastened to the mounting surface, *NEC Section 110-13(a)*.
- Wires shall be spliced with an appropriate splicing device or the wires shall be soldered, *NEC Section 110-14(b)*.
- An electrical box or approved fitting must be installed at every location when wires are spliced, or when switches, outlets, or receptacles are installed, *NEC Section 300-15(a)*.
- At every outlet or switch, at least 6 in (152 mm) of free conductor must be left for easy connecting to the device and for making splices, *NEC Section 300-14,* Figure 9-17. For deep boxes, an extra inch or two (25.4 mm or 50.8 mm) of free conductor should be provided.

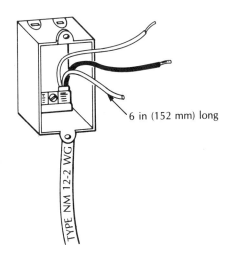

**Figure 9-17** Allow 6 in (152 mm) of free conductor at each outlet box. For deep boxes, allow a little more.

- Wires size No. 8 AWG and larger installed in conduit must be stranded, not solid, *NEC Section 310-3*. It is recommended that all sizes of wire installed in conduit be stranded for ease of installation.

## ELECTRICAL HAND TOOLS

Each electrician develops a preference for particular hand tools and techniques. However several basic types of hand tools are most frequently used by electricians. The following is considered a starter set of tools. Some of the tools are shown in Figure 9-18.

Knife with hook blade
Screwdrivers; several sizes
Wire stripper
Cable ripper
Diagonal cutting plier
Needlenose plier
Side cutting or lineman's plier
Pump pliers; two are needed
Awl
Hammer
Measure; folding wood-type is preferred
Crimping tool
Tool pouch
Hacksaw

Cable jacket can be stripped open with the hooked blade on the knife or with a cable stripper. With 2-wire cable, the wires lie parallel inside the jacket and there is little danger of damaging the insulation on the wires.

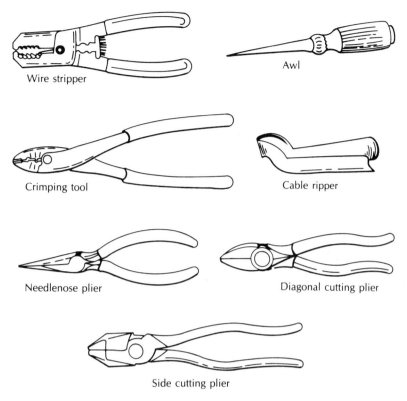

**Figure 9-18** Some electricians' handtools

Type NM 3-wire cable has the wires twisted inside the jacket. The jacket must be stripped carefully to avoid damaging the insulation. Type UF cable is much more difficult to strip than NM cable because the wires are completely molded in the thermoplastic jacket. The diagonal cutting plier is used to cut away the cable jacket and paper filler in NM cable. The wire stripper is used to remove insulation from the wire. Care must be taken not to nick the wire. A nick will reduce the ampere rating of the wire and may cause the wire to break.

Needlenose pliers are most frequently used for making hooks on the end of wires which are to be placed under a screw. The wire must curve around the screw in a clockwise direction. This tends to draw the wire in tightly under the screw as the screw is tightened, Figure 9-19. Some screw terminals are designed for the wire to be inserted straight. These terminals are often designed for two wires.

Boxes and equipment are usually attached to wood surfaces in farm buildings. A pan-head, ¾-in No. 10 self-tapping screw can be used for mounting equipment to most wood surfaces. The location of the screw is marked, and then a guide hole for the screw is formed with a small drill or a hammer and awl, Figure 9-20. Screws with plastic anchors are used to mount equipment to masonry surfaces. Masonry kits are available containing screws, anchors, and masonry bit. Self-tapping metal screws are convenient for mounting equipment to metal surfaces. Lag bolts, toggle bolts, and other heavy-duty fasteners are required for mounting heavy equipment.

**Figure 9-19** The wire is curved around the screw in the direction the screw turns to tighten.

Unit 9   Farm Wiring Methods and Materials   175

Figure 9-20 A hole can be started for a self-tapping wood screw using a hammer and an awl.

Electricians and electrical apprentices are required to learn the safe and proper use of other hand and power tools. Power tools and specialized hand tools are generally supplied by the electrical contractor.

## Safety Equipment

The electrician should always wear appropriate and durable work clothing. All areas of the body, except possibly the arms, should be covered to provide a shield against possible flying objects. A hard hat must be worn at all times when in the construction area. Eye protection, such as safety glasses with side shields, or goggles, should be worn when working in areas where tools may produce flying objects, Figure 9-21

> **CAUTION:** Electricians working with live equipment should always wear eye protection to prevent injury from flying sparks.

Only those power tools that are double insulated or have a grounding wire plugged into a grounded outlet should be used. Portable ground-fault circuit interrupters (GFCI) are available to help prevent electrical shock to the electrician working in damp locations.

All electricians should have training in first aid and emergency care for the protection of fellow workers. It is highly recommended that the electrician also receive training in cardiopulmonary resuscitation (CPR). Classes in first aid and CPR are available to the public in nearly every community.

## SPLICING WIRES

Branch-circuit wires size No. 10 AWG and smaller frequently require splicing. Splices are most commonly made with twist-type devices, often called wire nuts. An electrician commonly uses several sizes of wire nuts. It is important that the completed splice covers exposed bare wire, and that all wires are held securely. Poor splices are among the main causes of short circuits, ground faults, and electrical malfunctions.

The following procedure produces a good splice using wire nuts.

1. Read the manufacturer's directions for the type of wire nut to be used.
2. Choose the correct size wire nut for the wire size and number of wires to be spliced.
3. Strip the wire insulation to expose the correct length of bare conductor. This is usually ½ in (13 mm) to ¾ in (19 mm).
4. Hold all of the wires to be spliced with one hand so the ends of the bare wires are even.
5. Holding the wires firmly, twist on the wire nut, pushing the wires into the wire nut and pushing the wire nut onto the wires, Figure 9-22. Some wire nuts will stop turning when tight. Others will slip or click.
6. Test the splice by holding the wire nut in one hand and pulling each wire individually to make sure it is secure. If the splice is made properly, you will not be able to pull out any of the wires.
7. Check the wires to make sure there is no exposed bare wire.

Figure 9-21 Electricians should wear hard hats and eye protection.

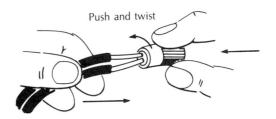

Figure 9-22 Push and twist to install a wire nut.

**Figure 9-23** Ring and fork terminals, insulated and bare, are crimped to the wire.

Crimping-type splices are also commonly used. One type of crimping splice uses a copper sleeve or barrel. The wires are inserted and crimped with a hand crimping tool. Lineman's pliers often have a crimping die as well as side cutter. Many electricians use crimp sleeves for splicing grounding wires in boxes. Insulating caps are available for crimp sleeves to be used when splicing insulated circuit wires.

Wires are frequently terminated with ring or fork terminals, Figure 9-23. The wire is stripped, inserted into the terminal, and then crimped. Splices using ring terminals are often made, especially at a motor wire box where several wires must be spliced together. Ring terminals are crimped to the wires. The wires are then joined using a short bolt and nut. The splice is then covered with tape, Figure 9-24.

Solder splices are used when there is a chance that other methods may not prove satisfactory. Soldering is preferred for example, where the available space or length of wire may be too small to use an alternative method, where extreme vibration could cause wires to become loose, or when wire sizes to be joined differ enough to make a proper joining questionable.

The following is a proper method for making a good solder splice.

1. Use a solder iron or solder gun that produces an ample amount of heat. A soldering tool rated at 125 W to 250 W works well on branch circuit wires No. 10 AWG and smaller.
2. Choose an appropriate solder for electrical wire. A better solder connection is obtained when flux is applied to the wires. The flux must be of a type appropriate for electrical wire.
3. Make the splice mechanically secure before soldering *NEC Section 110-14(b)*.
4. Put the heated soldering gun against the wires and heat them, but do not overheat the wire insulation.
5. Touch the solder to the wire, and it will melt and flow completely around each wire, fusing them together.
6. Tape the splice to apply an adequate layer of insulation.

## ELECTRICAL BOXES

Electrical boxes are available in many types and sizes. Boxes are constructed of a variety of materials. Steel, nonmetallic, and corrosion-resistant tight metal boxes are commonly used in farm buildings.

Dust can be a serious problem if it enters boxes. Figure 9-25 shows a serious dust accumulation on a porcelain lamp receptacle. Moisture in the air can make the dust conductive, causing electrical current to short circuit from the hot terminal to the neutral. Boxes with holes in them must be arranged to reduce the possibility of dust from entering. Tight boxes and sealed boxes are required in areas where there are excessive dust and excessive dust with moisture.

Device boxes, octagonal boxes, and square boxes are

**Figure 9-24** Splices in a motor terminal box may be made by crimping on ring terminals and then bolting the wires together.

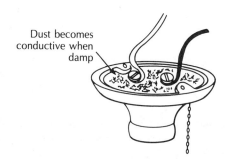

**Figure 9-25** Dust can enter through holes in boxes and settle on the inside surface of lampholders.

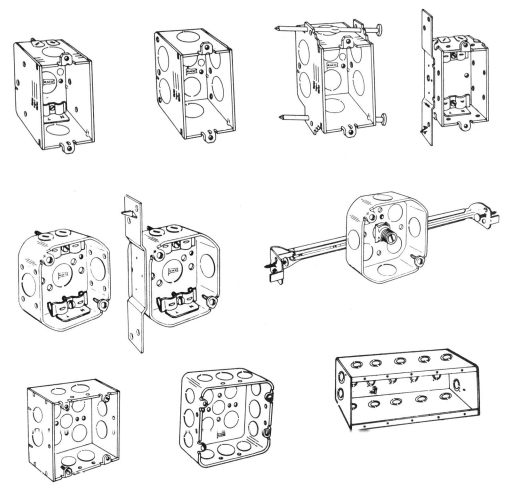

**Figure 9-26** Steel box types commonly used in agricultural buildings where dust and moisture are not a problem.

the most commonly used steel boxes for farm wiring. Boxes are available with cable clamps or with knockouts to accept conduit or cable connectors. Steel boxes are made for concealed wiring in hollow walls, setting into masonry, and surface mounting, Figure 9-26. Old work boxes are made for wiring in concealed existing walls, Figure 9-27.

Conduit outlet boxes, known as *FS boxes*, are commonly used in farm wiring. These boxes are available in a nonmetallic type with smooth hubs to accept PVC conduit, or metallic-type (steel or aluminum) boxes with threaded hubs. Hub sizes available are ½-in, ¾-in, and 1-in trade sizes. Most are single-gang boxes for one device, but 2-gang FS boxes are also available. A variety of gasketed covers are available for switches and receptacles. These boxes are intended for surface mounting, particularly in areas where there is dust or moisture. The nonmetallic boxes are recommended for use in areas where constant moisture is a problem or corrosive conditions are present. (Common FS box configurations are shown in Figure 9-11.)

Large junction boxes are sometimes needed in farm wiring where metallic and nonmetallic conduit systems are exposed to the weather. Sealed molded PVC boxes and sealed metallic JIC boxes are ideal for these applications. The PVC boxes usually come without hubs. Entry holes in the desired size and location are easily made with conduit trade size hole saws, Figure 9-28. JIC boxes come with and without hubs. For farm wiring applications, holes in the desired size and location can be made easily using conduit trade size knockouts, Figure 9-29.

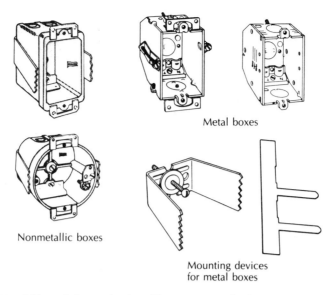

Figure 9-27　Old work boxes for installing a new outlet in an existing hollow wall

Nonmetallic boxes designed for use with NM cable are used extensively for residential wiring, Figure 9-30. Single-gang boxes usually do not contain a means of securing the cable to the box. This is permitted as long as the cable is secured (usually stapled) within 8 in (203 mm) of the box, measured along the cable, *NEC Section 370-7(c)*. Multigang boxes are required to have cable clamps to secure the cable to the box, *NEC Section 370-7(c), exception*. These boxes are not recommended for use in farm buildings, unless the conditions are similar to those of a residential occupancy.

Dusttight and watertight nonmetallic boxes are recommended for nonmetallic conduit and UF cable wiring in farm buildings. However, the variety of such boxes is limited. The use of a metal box, such as the FS type, may be required in order to construct an economical, watertight, and dusttight wiring system. Some watertight, surface-mount metal boxes which are frequently used are shown in Figure 9-31.

## Sizing Outlet and Junction Boxes

A box must contain adequate free space for devices and conductors, *NEC Section 370-6*. Inadequate box size

Figure 9-28　Hole saws can be used to cut wood or metal.

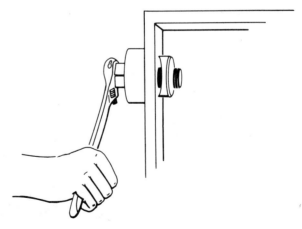

Figure 9-29　Knockouts are used for making holes in metal boxes and cabinets.

Unit 9　Farm Wiring Methods and Materials　179

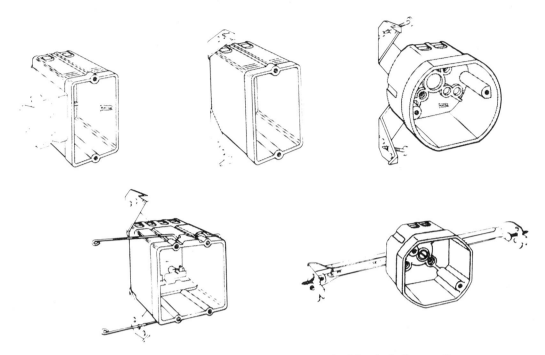

Figure 9-30　Nonmetallic boxes for concealed wiring in hollow walls

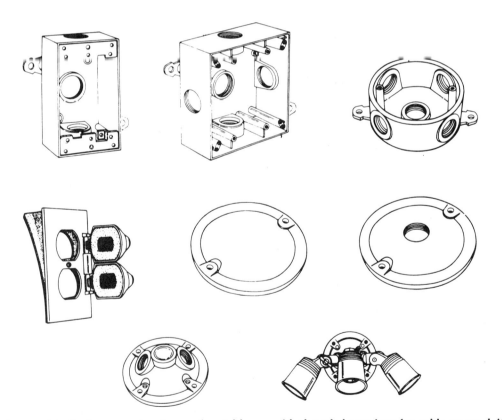

Figure 9-31　Surface-mount waterproof metal boxes with threaded openings for cable or conduit

**Table 9-10** (*NEC Table 370-6[b]*) **Volume required per conductor**

| Size of Conductor | Free Space Within Box for Each Conductor |
|---|---|
| No. 14 | 2. cubic inches |
| No. 12 | 2.25 cubic inches |
| No. 10 | 2.5 cubic inches |
| No. 8 | 3. cubic inches |
| No. 6 | 5. cubic inches |

*This is *NEC Table 370-6(b)*.

Reproduced by permission from NFPA 70-1984, *National Electrical Code®*, Copyright © 1983, National Fire Protection Association, Quincy, MA. This reprinted material is not the complete and official position of the NFPA on the referenced subject which is represented only by the standard in its entirety.

is a problem area in electrical wiring. With experience, an electrician learns to choose boxes of the proper type and size for the wiring requirements and conditions.

A method for determining the proper box size is given in *NEC Section 370-6*. This method is based upon a minimum volume required for each wire in the box. These minimum volume requirements are listed in *NEC Table 370-6(b)*, represented in this text as Table 9-10.

The number of wires in a box is counted to determine the minimum size box required. But, counting the wires is not as easy as it may seem at first. The rules for counting the wires in a box are stated in *NEC Section 370-6(a)(1)*. Refer to Figure 9-32 for a summary of the rules for counting current-carrying wires in boxes.

1. A wire passing through a box, not spliced or connected to a device, is counted as one wire. For Figure 9-32(A), the wire count is two.
2. Each wire that terminates within the box is counted as one wire. A spliced wire is considered to be terminated; therefore, each wire is counted. A wire connected to a device is considered terminated. The boxes of Figure 9-32(B) and (C) illustrate this rule.
3. A wire contained completely within the box is not counted at all. A pigtail connecting to a device is such a wire, as illustrated in Figure 9-32(D).

Next, grounding wires are considered. For a nonmetallic conduit system, a grounding wire is run with the circuit wires in order to ground any receptacles or metal equipment requiring grounding. A bare grounding wire is run with UF and NM cable. These grounding wires must be spliced together at each box. The rule for counting grounding wires is simple.

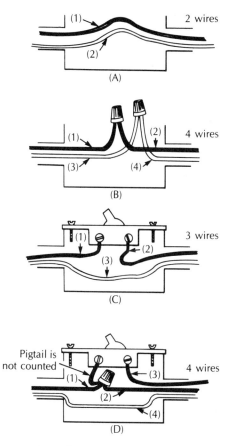

**Figure 9-32** Counting current-carrying wires in boxes. The wires inside the boxes of these figures are shown shorter than is normal to simplify the diagrams.

4. All of the grounding wires entering a box are counted as only one wire, Figure 9-33.

Devices that take up space inside the box, such as switches and receptacles, must be considered somehow in the sizing of a box. Therefore, devices are considered by counting them as wires for the purpose of sizing boxes.

5. Each strap containing one or more devices is counted as one wire.
   a. A switch counts as one wire.
   b. A single receptacle counts as one wire.
   c. A duplex receptacle counts as one wire.

Other devices that take up space within the box must be considered when sizing boxes. The *National Electrical Code* requires that several other devices if present within the box must be counted. The common devices are fixture studs, cable clamps, and hickeys. Cable con-

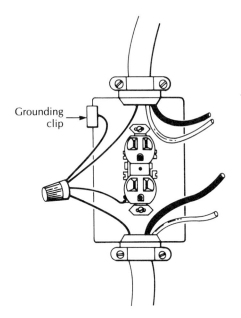

**Figure 9-33** All of the grounding wires count as only one wire.

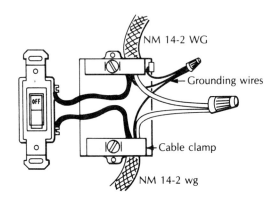

**Figure 9-35** This device box has an equivalent conductor count of seven wires.

nectors which have only a locknut on the inside of a box are not counted.

6. Each type of device, such as cable clamps, fixture studs, and similar devices taking up space within the box, must be counted as one wire, Figure 9-34.
   a. A fixture stud counts as one wire.
   b. One cable clamp, or more, counts as one wire.
   c. A hickey counts as one wire.

Consider an example to see how the required volume of a box is determined. The device box of Figure 9-35 contains a switch and cable clamps. The two cables entering the box are both Type NM 14-2 with ground. The conductor count for the box is:

| | |
|---|---:|
| Circuit wires | 4 |
| Grounding wires | 1 |
| Device, switch | 1 |
| Cable clamps | 1 |
| Total equivalent wire count | 7 |

The wire size is No. 14 AWG; therefore, from *NEC Table 370-6(b)* (text Table 9-10), there must be at least 2 cubic inches (2 in$^3$) of space in the box for each equivalent wire counted.

$$7 \text{ wires} \times 2 \text{ in}^3/\text{wire} = 14 \text{ in}^3$$

The box must have a minimum capacity of 14 in$^3$. *NEC Table 370-6(a)*, represented in this text as Table 9-11,

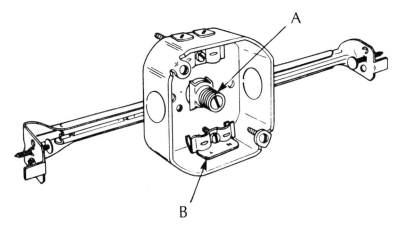

**Figure 9-34** The fixture stud (A) and cable clamp (B) each count as one wire when sizing boxes.

**Table 9-11** (*NEC Table 370-6[a]*) Metal boxes

| Box Dimension, Inches Trade Size or Type | Min. Cu. In. Cap. | Maximum Number of Conductors | | | | |
|---|---|---|---|---|---|---|
| | | No. 14 | No. 12 | No. 10 | No. 8 | No. 6 |
| 4 × 1 ¼ Round or Octagonal | 12.5 | 6 | 5 | 5 | 4 | 0 |
| 4 × 1 ½ Round or Octagonal | 15.5 | 7 | 6 | 6 | 5 | 0 |
| 4 × 2 ⅛ Round or Octagonal | 21.5 | 10 | 9 | 8 | 7 | 0 |
| 4 × 1 ¼ Square | 18.0 | 9 | 8 | 7 | 6 | 0 |
| 4 × 1 ½ Square | 21.0 | 10 | 9 | 8 | 7 | 0 |
| 4 × 2 ⅛ Square | 30.3 | 15 | 13 | 12 | 10 | 6* |
| 4 ¹¹⁄₁₆ × 1 ¼ Square | 25.5 | 12 | 11 | 10 | 8 | 0 |
| 4 ¹¹⁄₁₆ × 1 ½ Square | 29.5 | 14 | 13 | 11 | 9 | 0 |
| 4 ¹¹⁄₁₆ × 2 ⅛ Square | 42.0 | 21 | 18 | 16 | 14 | 6 |
| 3 × 2 × 1 ½ Device | 7.5 | 3 | 3 | 3 | 2 | 0 |
| 3 × 2 × 2 Device | 10.0 | 5 | 4 | 4 | 3 | 0 |
| 3 × 2 × 2 ¼ Device | 10.5 | 5 | 4 | 4 | 3 | 0 |
| 3 × 2 × 2 ½ Device | 12.5 | 6 | 5 | 5 | 4 | 0 |
| 3 × 2 × 2 ¾ Device | 14.0 | 7 | 6 | 5 | 4 | 0 |
| 3 × 2 × 3 ½ Device | 18.0 | 9 | 8 | 7 | 6 | 0 |
| 4 × 2 ⅛ × 1 ½ Device | 10.3 | 5 | 4 | 4 | 3 | 0 |
| 4 × 2 ⅛ × 1 ⅞ Device | 13.0 | 6 | 5 | 5 | 4 | 0 |
| 4 × 2 ⅛ × 2 ⅛ Device | 14.5 | 7 | 6 | 5 | 4 | 0 |
| 3 ¾ × 2 × 2 ½ Masonry Box/Gang | 14.0 | 7 | 6 | 5 | 4 | 0 |
| 3 ¾ × 2 × 3 ½ Masonry Box/Gang | 21.0 | 10 | 9 | 8 | 7 | 0 |
| FS—Minimum Internal Depth 1 ¾ Single Cover/Gang | 13.5 | 6 | 6 | 5 | 4 | 0 |
| FD—Minimum Internal Depth 2 ⅜ Single Cover/Gang | 18.0 | 9 | 8 | 7 | 6 | 3 |
| FS—Minimum Internal Depth 1 ¾ Multiple Cover/Gang | 18.0 | 9 | 8 | 7 | 6 | 0 |
| FD—Minimum Internal Depth 2 ⅜ Multiple Cover/Gang | 24.0 | 12 | 10 | 9 | 8 | 4 |

*Not to be used as a pull box. For termination only.

Reprinted with permission from NFPA 70-1984, *National Electrical Code®*, Copyright © 1983, National Fire Protection Association, Quincy, MA. This reprinted material is not the complete and official position of the NFPA on the referenced subject which is represented only by the standard in its entirety.

lists the commonly used metal boxes, giving their cubic-inch capacity and maximum number of wires permitted. A single-gang metal device box is 3 in by 2 in by the desired depth. For the example, the minimum size device box is 3 × 2 × 2¾ in.

A manufacturers' catalog of boxes may be obtained from an electrical wholesale distributor which lists the cubic inch-capacity of available boxes. Nonmetallic boxes have the cubic-inch capacity stamped inside the box.

## Problem 9-4

Determine the dimensions of a ceiling-type metal octagonal box with cable clamps. The wire entering the box is Type NM 14-2 cable with ground. The lamp receptacle does not take up space within the box. See Figure 9-36.

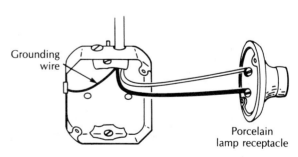

**Figure 9-36** This ceiling-type octagonal box has an equivalent conductor count of four wires.

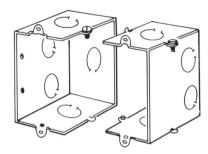

Figure 9-37 Metal device boxes can be ganged together.

## Solution

The first task is to make an equivalent conductor count:

| | |
|---|---:|
| Circuit wires | 2 |
| Grounding wires | 1 |
| Cable clamps | 1 |
| Total equivalent wire count | 4 |

The 4-in octagonal box listed in *NEC Table 370-6(a)* (text Table 9-11) is selected. The wires are No. 14; therefore, find an octagonal box which is permitted to contain four wires. A box 1¼ in deep is the minimum size permitted.

Metal device boxes for concealed wiring in hollow walls have removable sides so that boxes can be ganged (connected) together. Two device boxes can be connected together to accept two devices, as shown in Figure 9-37. Box dimensions are standard, so boxes by different manufacturers are interchangeable.

Masonry boxes are required to have smooth sides and square corners. A masonry box of the desired size must be selected to accept the number of devices required. Masonry boxes are available in lengths for one to ten devices.

Square boxes are used frequently where wiring is exposed on the wall surface. Raised covers to permit the attachment of devices are available for square boxes, Figure 9-38. These covers can add several cubic inches of capacity to the box. Device covers are available for square boxes that are recessed into the wall, Figure 9-39. Some are flat, and others are raised. The cubic-inch capacity of raised covers is added to the volume of the box.

## BRANCH-CIRCUIT WIRE COLOR CODE

The color code for wires is simple and clear-cut when the wires are run in raceway.

- Equipment grounding wire—green, bare, or green with yellow stripes, *NEC Sections 210-5(b), 250-57(b),* and *310-12(b).*
- Neutral grounded wire—white or gray, *NEC Sections 200-6(a), 210-5(a),* and *310-12(a).*
- Hot ungrounded wire—color other than white, gray, or green, *NEC Section 310-12(c).*

But, when wiring is installed using NM or UF cable, this color code is not always possible to follow. Two-wire

Figure 9-38 Raised covers for attaching devices to surface-mounted square boxes.

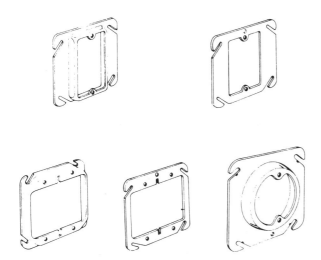

**Figure 9-39** Flat covers and plaster rings for square boxes recessed into hollow or masonry walls.

cable has a black, a white, and a bare wire. Three-wire cable has a black, a red, a white, and a bare wire. The following are the color code rules using cable.

- Equipment grounding wire—green or bare, *NEC Sections 210-5(b), 250-57(b),* and *310-12(b)*.
- Neutral grounded wire—white or gray, *NEC Sections 200-6(a), 210-5(a),* and *310-12(a)*. White wires are not necessarily neutral wires. *NEC Section 200-7, Exception No. 2.*
- Hot ungrounded wires—usually black or red, or any color other than green, *NEC Section 310-12(c)*.

Often with cable wiring, a hot wire is required to be the white wire. An example is the switch loop shown later in this unit (in Figure 9-42). The traveler wires of Figures 9-46 and 9-47 (shown later in this unit) have one wire which is white. Using the following rules will avoid confusion when a white wire is used as a hot wire, *NEC Section 200-7*.

- The wire, where it is visible in a box, can be painted or covered with tape in a color other than white, gray, or green.
- A white wire connected to a switch is considered to be a hot wire.
- If a white or gray wire in a switch loop is spliced to a wire of a color other than white, gray, or green, the white wire is considered to be a hot wire, Figure 9-40, *NEC Section 200-7, Exception No. 2.*

One additional rule for white wires is important. For a switch loop, the wire from the switch to the lamp must

**Figure 9-40** A white wire connected to a wire of another color in a switch loop is considered to be a hot wire, *NEC Section 200-7, Exception No. 2.*

not be white or gray. *NEC Section 200-7, Exception No. 1.* However, if a white wire to the lamp is used, it must be identified black, or another acceptable color, using tape or paint.

Keep in mind that wire colors on appliances and machines are not required to follow the same color code used for a building wiring system. This is true with wires for electric equipment such as a motor, *NEC Section 210-5(b), Exception No. 2.* A diagram is always included with motors and appliances to identify the wires.

## DEVICES

Switches and receptacle outlets are the devices most frequently used in general-purpose circuits. Switches control the flow of electricity, and receptacle outlets provide a convenient method of connecting portable equipment to the wiring system. Several different types of switches are required to conveniently control electrical circuits.

It must be pointed out that a switch is permitted to control the hot wire only, *NEC Section 380-2*. See Figure 9-41. However, a switch may be used to simultaneously open both the hot and the neutral wires, *NEC Section 380-2, Exception No. 1.*

> **CAUTION:** Installing the switch in the neutral wire can create a very dangerous safety hazard.

## TOGGLE (SNAP) SWITCH

The flush-mounted toggle switch (often called a snap switch) is the switch most frequently used in general-purpose circuits. When mounted in a box, the switch is concealed in the wall with only the insulated handle or toggle protruding. Switches are given a voltage and current rating. These ratings are stamped into the metal yoke or mounting strap. Underwriters' Laboratories classifies toggle switches used for general-purpose circuits as general-use snap switches.

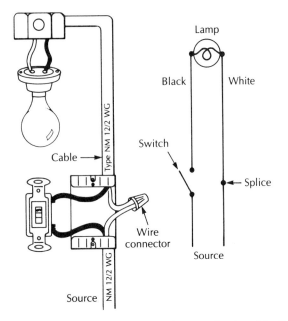

**Figure 9-41** A single-pole switch in a circuit with the source wire at the switch. Grounding wires are not shown in the illustration, but must be connected in an actual circuit.

The *National Electrical Code* separates general-use snap (toggle) switches into two categories in *Article 380*.

Category 1, *NEC Section 380-14(b)*, contains those ac/dc general-use snap switches which are used to control

- alternating-current or direct-current circuits.
- resistive loads not to exceed the ampere rating of the switch at rated voltage.
- inductive loads, such as motors and electric discharge lighting, not to exceed one-half the ampere rating of the switch at rated voltage.
- tungsten-filament incandescent lamp loads not to exceed the ampere rating of the switch at 125 V when marked with the letter T. (A tungsten-filament lamp draws a very high current at the instant the circuit is closed. As a result, the switch is subjected to a current surge.)

The ac/dc general-use snap switch normally is not marked ac/dc. However, it is always marked with the current and voltage rating, such as 10 A—125 V, or 5 A—250 V. The letter T (for tungsten) is also marked on an ac/dc snap switch that is T-rated.

Category 2, *NEC Section 380-14(a)*, contains those ac general-use snap switches which are used to control

- alternating-current circuits only.
- resistive, inductive, and tungsten-filament lamp loads not to exceed the ampere rating of the switch at 125 V.
- motor loads not to exceed 80% of the ampere rating of the switch at rated voltage, but not exceeding 2 hp.

Ac general-use snap switches may be marked "ac only," or they may also be marked with the current and voltage ratings. A typical switch marking is 15 A, 120—277 V ac. The 277-V rating is required on 277/480-V systems.

The different types of common snap (toggle) switches are shown in Figure 9-42, along with the internal connections for the two positions of the switch. The single-pole and the double-pole switches are either open or closed. The double-pole switch is actually two switches in one. At first, the double-pole looks like a 4-way switch, except that the double-pole is marked on and off whereas the 4-way is not.

The 3-way and 4-way switches are used when multiple switches are to be used to control outlets or equipment. It is important to note that in either position there is a circuit through the 3-way and 4-way switch.

## Single-pole Switch

A single-pole switch is used to interrupt the ungrounded wire to outlets or equipment. A light controlled by a single-pole switch is shown in Figure 9-41, with the electrical supply wires entering the switch box.

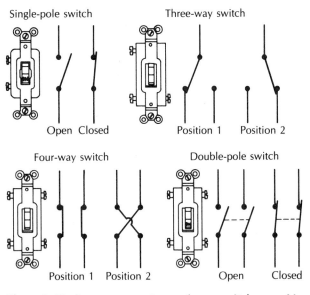

**Figure 9-42** Four common types of snap switches used in agricultural wiring

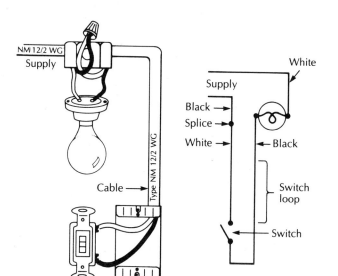

The electrical power wires frequently enter the outlet box where power is used, such as in the light in Figure 9-43. The neutral wire is connected directly to the light. However, the ungrounded wire must go to the switch before returning to the light. The wires from the light to the switch are called a *switch loop*. NEC Section 200-7, Exception No. 2, covers the case of the switch loop, and states that when cable is used, the white wire must be the feed to the switch, and the black wire must be the return to the load.

Circuits frequently are not as simple as a switch controlling a light. Figure 9-44 is an example of how an

**Figure 9-43** A single-pole switch in a circuit with the supply at the lighting outlet. Grounding wires are not shown in the illustration, but must be connected in an actual circuit.

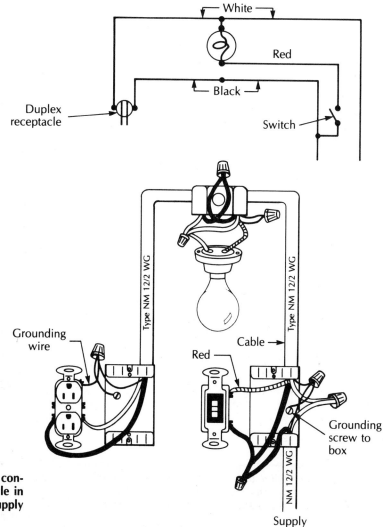

**Figure 9-44** A ceiling light fixture is controlled with a switch, and a receptacle in the circuit is live all of the time. The supply is at the switch.

Unit 9   Farm Wiring Methods and Materials   187

ungrounded supply conductor must be carried through the circuit to provide power to the receptacle outlet. In this circuit, a 3-wire cable is required between the switch and the light. The black wire is power, and the red wire is controlled by the switch and goes to the light.

## Three-way Switch

If a light or an outlet is to be controlled from two locations, then 3-way switches are used. The 3-way switch has a terminal called a *common* and two terminals called *travelers*. The traveler terminals of one switch are connected directly to the traveler terminals of the other switch. Thus, only the two common terminals remain. The ungrounded supply wire is connected to one switch common, and the other switch common goes to the light, as shown in Figure 9-45.

It can be seen from Figure 9-45 that the light may be turned on or off at either switch. A common error in wiring 3-way switches is to have one of the travelers and the common interchanged. The common terminal is usually identified by a screw which is different in shade of color from the travelers. The common is not always in the same location on switches produced by different manufacturers.

A switch loop with 3-way switches is shown in Figure 9-46. The traveler terminals of one switch are connected to the traveler terminals of the other switch. A black wire is used as the return to the light.

## Four-way Switch

A 4-way switch is used when a light or outlet is to be controlled from more than two locations. Figure 9-47 is a diagram of a light controlled by three switches. The 4-way switch is wired in series with the two 3-way switches. Figure 9-47 shows how the light can be turned on or off from any one of the three locations.

The 4-way switch has only traveler terminals. The traveler terminals of one 3-way switch run to the traveler terminals of the 4-way switch. The other traveler terminals of the 4-way switch run to the second 3-way switch. Now the only terminals remaining are the common terminals, one at each 3-way switch. It is important to connect the traveler wires correctly to the 4-way switch.

The most common type of 4-way switch is shown in Figure 9-47. One set of travelers connects to the top of the switch, and the other set connects to the bottom.

## Double-pole Switch

A double-pole switch is actually two switches in one. Two separate circuits can be controlled with one switch, as shown in Figure 9-48. This type of switch is especially handy when some lights in a building are to be controlled by an emergency power source. With this switch, lights with emergency power backup are always on when the main lights are on.

## RECEPTACLE OUTLETS

Receptacle outlets provide a convenient means of connecting electrical equipment to a wiring system.

## Single and Duplex

The most common receptacle outlets are the 15-A, 125-V, single and duplex types. Receptacle outlets are also available in a locking type which prevents the plug

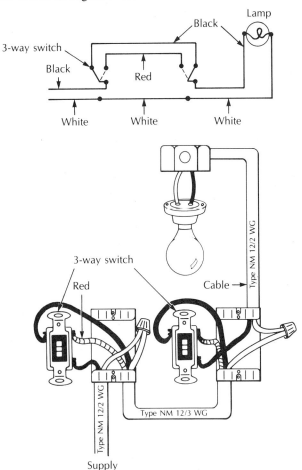

Figure 9-45 A circuit with 3-way switch control with the supply at the first switch. Grounding wires are not shown in the illustration, but must be connected in an actual circuit.

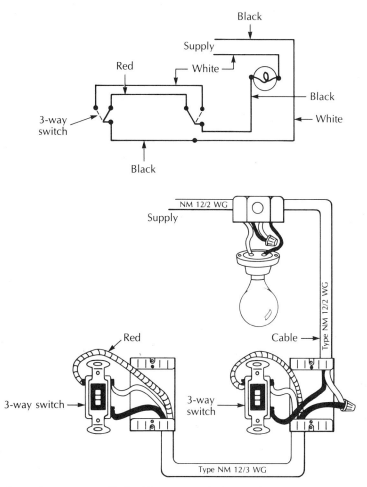

Figure 9-46  A lighting fixture controlled with 3-way switches with the supply at the lighting outlet

from coming loose. These are frequently used for portable or movable agricultural machines where vibration can cause the plug to separate from the receptacle.

The choice of receptacle outlets for a specific load must be based upon the number of phases, circuit voltage, and load current. This information is provided on each receptacle outlet. A receptacle must never be connected to a power source different from the receptacle rating, *NEC Section 410-56(g)*. The National Electrical Manufacturers Association (NEMA) has developed a set of standard receptacle and attachment plug configurations. Typical NEMA standard plug configurations used in agriculture are shown in Figure 9-49. Other plug configurations encountered in farm wiring are standard configurations used prior to the development of the NEMA standards. Replacements for these older styles are manufactured, but they are not always readily available.

The *NEC* requires that receptacles installed on 15-A and 20-A branch circuits be of the grounding type, *Section 210-7(a)*. When an old receptacle is replaced with a new one, a receptacle of the grounding type shall be used. However, this is not a hard-and-fast rule. If a grounding means is not available, the receptacle is not required to be of the grounding type, *NEC Section 210-7(d)*. In farm wiring, however, replacement receptacles should always be of the grounding type, even if it is necessary to replace the wire feeding the receptacle. A GFCI receptacle may be used as a replacement for a nongrounding type.

If a receptacle is not grounded, then a grounding-type replacement is not permitted to be used. *NEC Section 210-7(b)* requires that the grounding contact of a grounding-type receptacle must always be properly grounded.

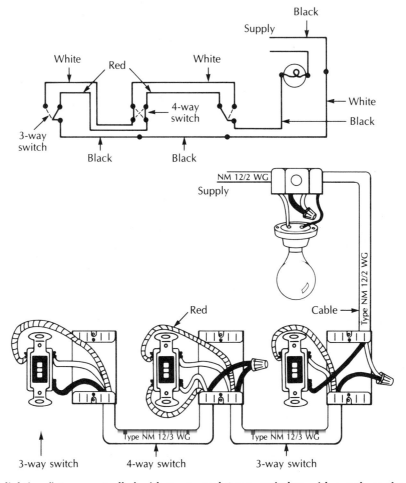

**Figure 9-47** A lighting fixture controlled with 3-way and 4-way switches with supply at the lighting outlet

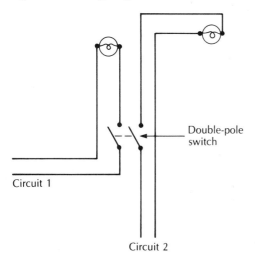

**Figure 9-48** A double-pole switch controlling two separate circuits such as one normal power circuit and an emergency circuit

## Polarity

Most electrical wiring systems are polarized for electrical safety. The placement of the neutral wire is *not* a matter of choice. The neutral wire is *always* connected to the screw or terminal which is more silvery in color than the others. If there is not a distinctive difference in color of the terminal, the letter W or the word white appears beside the neutral terminal, *NEC Sections 200-9* and *200-10(b)*.

Look carefully at the duplex receptacle outlet of Figure 9-50, and notice that the plug slot on the right is larger than the one on the left. If the receptacle has been wired according to correct polarity, the large slot is the neutral wire. Small appliances sometimes have a polarized 2-prong attachment plug so that the plug can be inserted only one way into the receptacle. The neutral blade on the attachment plug is slightly larger than the

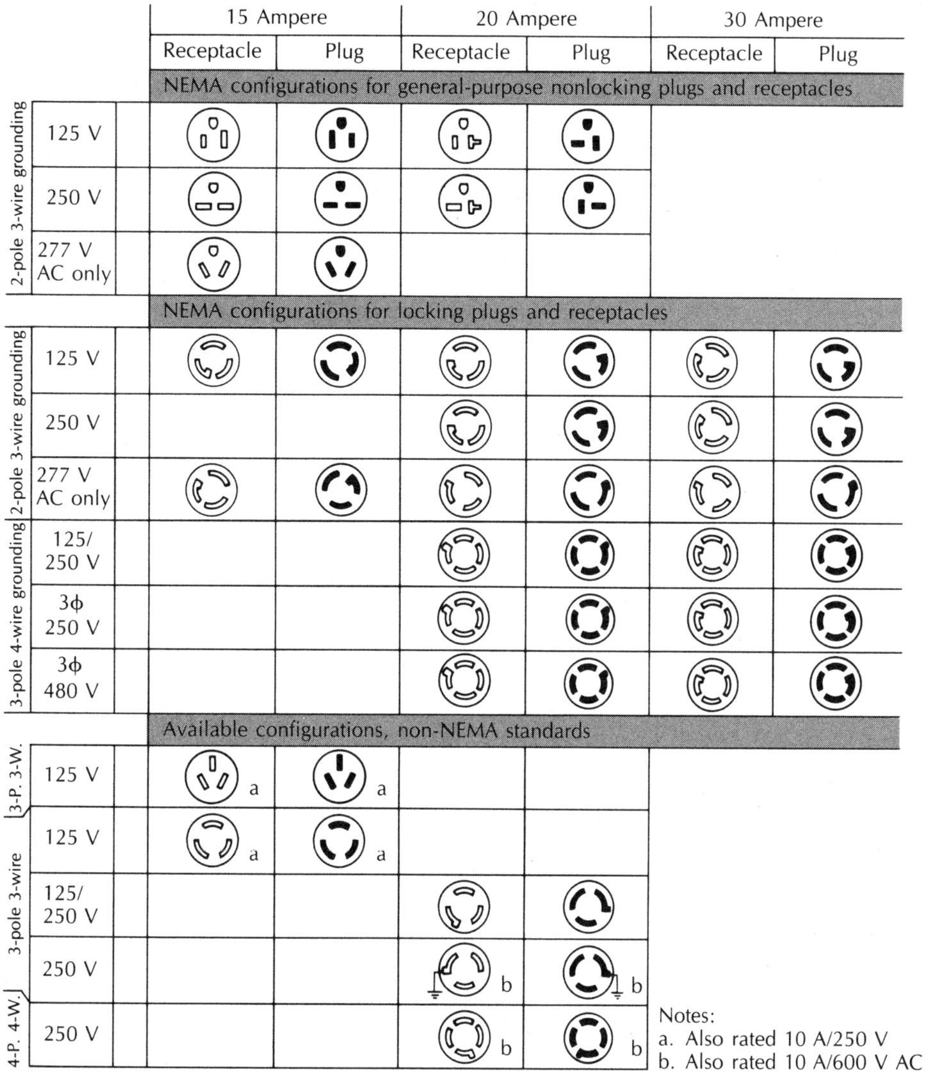

Figure 9-49 Plug and receptacle configurations most frequently used for agricultural applications

hot blade. The neutral blade of the attachment plug will not fit into the receptacle slot intended for the hot conductor. Television sets have a polarized attachment plug. Three-prong plugs are automatically polarized, because this type of plug usually can be plugged only one way into the receptacle.

## Split Duplex

A duplex receptacle outlet is constructed with a removable tab between the wire terminal screws, Figure 9-50. The purpose of this tab is to allow the two receptacles to function in parallel (with the tab in place), or to allow the receptacles to be controlled separately (with the tab removed). A typical installation is a split receptacle with the top receptacle live all the time and the bottom receptacle controlled by a switch, Figure 9-50. Only the tab on the hot side (brass colored) is removed. Do not remove the tab on the neutral side (silver colored).

## Multiwire

A split-wired receptacle circuit, or a multiwire circuit, as it is referred to in the *NEC* definition, *Article 100*, has two hot wires and a neutral wire. The potential between each hot wire and the neutral is 120 V. The

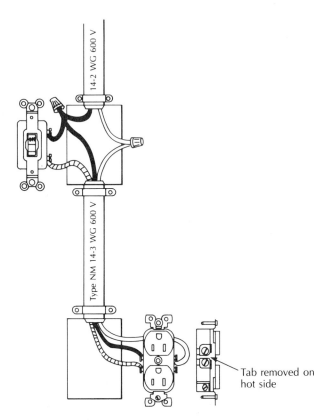

Figure 9-50 Split duplex receptacle with the top live and the bottom controlled by the switch

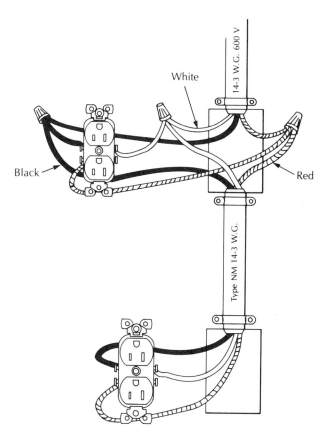

Figure 9-51 A multiwire duplex receptacle circuit with two circuits available at each device

potential between the hot wires is either 208 V or 240 V, depending upon the type of electrical supply. The advantage of this device is that two circuits using a common neutral requires only three wires, Figure 9-51.

Special attention must be given to proper installation of the neutral wire of a multiwire circuit. If the neutral should become broken due to a loose connection, the appliances plugged into the receptacles would then be subjected to an improper voltage. This situation is discussed in an earlier unit as an open neutral. To reduce the chance of a loose connection on a receptacle causing an open neutral, *NEC Section 300-13(b)* requires that the neutral be connected to the receptacle in such a way that removal of the receptacle will not break the neutral.

It is preferred that a multiwire circuit consisting of three wires be protected by a 2-pole circuit breaker. This procedure is required for multiwire circuits in a dwelling, *NEC Section 210-4*.

## OTHER DEVICES

### Combination Devices

Combination devices are available on one yoke to provide two devices in the space of one. The most common are two switches, or one switch and one receptacle outlet, Figure 9-52.

### Lampholders

Porcelain lampholders are frequently used in areas around the farm where dust and moisture are not a problem. Three types are in common use: keyless, pull chain, and pull chain with receptacle. Porcelain lampholders must never be installed in areas where excessive dust can accumulate on them or where the hot incandescent lamp can contact flammable material.

A hidden hazard can develop when lampholders are installed in dusty areas. Dust may enter the box and set-

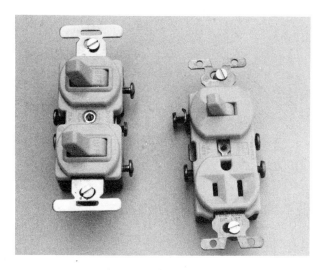

Figure 9-52 Combination devices on a single yoke

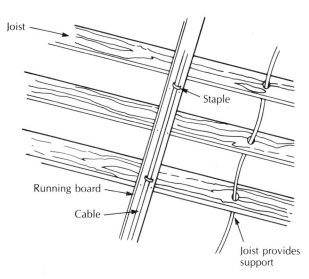

Figure 9-53 Cable may be run through bored holes in joists or run on running boards attached to the bottom of the joists.

tle on the inside face of the lampholder. A layer of dust may make contact between the hot screw and the neutral screw on the lampholder. If this dust becomes moist, it can become conductive, causing a short circuit between the two contact screws (see Figure 9-25).

## INSTALLING CABLE

The installation requirements for nonmetallic-sheathed cable Types NM and NMC, and underground feeder and branch-circuit cable are found in *NEC Articles 300* and *336*. Type NM cable is permitted only in dry locations; Type UF is used in damp, wet, or corrosive locations. The primary advantage of using cable as compared to conduit is the ease of installation. A major disadvantage is that cable must be installed so that it cannot be easily damaged.

Exposed electrical cable must closely follow the surface of the building and be fastened securely, *NEC Section 336-6(a)*. If a cable is required to span an open area, it must be supported by a running board or contained within raceway, Figure 9-53. Cable without a running board is not permitted to be fastened directly to the bottom of joists in basements, unless the cable contains two No. 6 or larger wires or three No. 8 or larger wires, *NEC Section 336-8*.

Electrical cable is required to be fastened with staples or similar devices within 12 in (305mm) of every cabinet, box, or fitting. The cable must also be fastened at intervals not to exceed 4½ ft (1.37 m). Closer spacing is generally required for a good installation.

In an accessible attic, cable run across joists must be protected with guard strips. Guard strips must also be provided for cables attached in a similar manner to studs and rafters if the cable is within 7 ft (2.13 m) of the floor, *NEC Sections 336-9* and *333-12*. An attic is considered accessible only if a permanent stairway or ladder is provided. If the attic is only accessible through a scuttle hole, guard strips must be provided for cable installed within 6 ft (1.83 m) of the hole. Usually, the cable can be attached to the sides of joists and rafters so that guard strips are not necessary near the scuttle hole. See Figure 9-54.

Cables must be securely fastened to boxes, *NEC Section 370-7*. Usually, this is accomplished with a cable clamp supplied with the box or a box connector, Figure 9-55. There is an exception in the case of single-gang nonmetallic boxes. The cable is not required to be secured to the box if the sheath extends into the box at least ¼ in (6.35 mm), and is stapled within 8 in (203 mm) of the box, Figure 9-56, *NEC Section 370-7(c)*. All cables entering cabinets, such as circuit-breaker panels, must be secured to the cabinet, *NEC Section 373-5(c)*.

Cables are permitted to be run through bored holes in structural materials, such as studs and joists. When this is done, the cable need only be secured to the structure at each end of the run. Sometimes, there is a chance that a nail driven into a wood frame member, such as a stud, may hit the cable. In the case of a stud, the bored hole is

Unit 9   Farm Wiring Methods and Materials   193

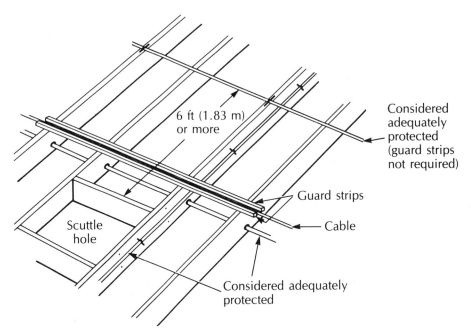

**Figure 9-54** Protecting cable in an attic

not permitted to be closer than 1 1/4 in (31.8 mm) from the nearest edge, *NEC Section 300-4(a)(1)*. If this distance cannot be maintained, the cable must be protected with a 1/16-in (1.59-mm) thick metal plate or some other acceptable means.

Bored holes in wood framing members are required to be as close as possible to the center of the member. A hole in the center will weaken the member less than a hole near the edge. Actually, holes should not be bored in farm building members, unless there is no other practical alternative. Holes must never be bored in roof or floor trusses or in barn rafters or joists required to support heavy weight. Cables run through holes in metal framing members must be protected from abrasion, *NEC Section 300-4(b)*.

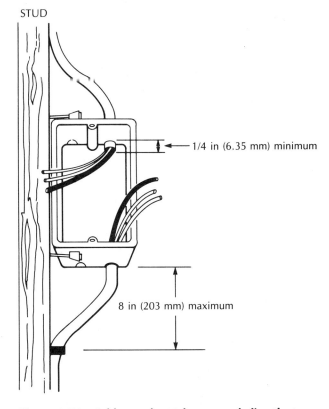

**Figure 9-55** A box connector can be used to secure cable to a box or cabinet if a cable clamp is not used.

**Figure 9-56** Cable need not be secured directly to a single-gang nonmetallic box if the cable is secured within 8 in (203 mm) of the box.

## CONNECTING WIRES TO DEVICES

An electrical wire can be connected to a switch or receptacle by placing the wire under a screw; by inserting the wire into an opening in the back of the device, and tightening the screw; or by inserting the wire into a locking-type opening in the back of the device, Figure 9-57. A gauge is provided as a guide for stripping away the proper amount of insulation. More than one wire is not permitted to be placed under one screw unless the screw is designed for more than one, *NEC Section 110-14(a)*. Several methods of connecting multiple wires to a device are shown in Figure 9-58. Many electricians prefer the method of Figure 9-58 because removal of the device does not affect the remainder of the circuit.

## Grounding of Devices and Boxes

An effective grounding system is essential to prevent electrical shocks to humans and animals on a farm. Good workmanship will ensure a dependable grounding system.

A metal conduit system serves as the grounding system as long as the metal is continuous all the way back to the electrical supply panel. A receptacle grounding terminal is grounded simply by connecting the grounding screw of the device to the box by means of a grounding or bonding jumper, Figure 9-59.

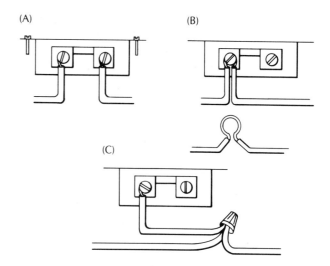

**Figure 9-58** Connecting multiple wires to a device

A grounding wire is used for grounding cable wiring, or nonmetallic conduit. This grounding wire must be joined at each box, *NEC Section 250-14*. If the box is metallic, it must be bonded to the grounding wire. If there is a receptacle at the box, its grounding screw must also be bonded to the grounding conductor, Figure 9-60. Manufacturers provide a receptacle with a bonding clip at the mounting screw which ensures electrical continuity from the yoke to the screw and then to the box, Figure 9-61. With this receptacle outlet, the grounding screw on the receptacle is bonded to the grounding wire through

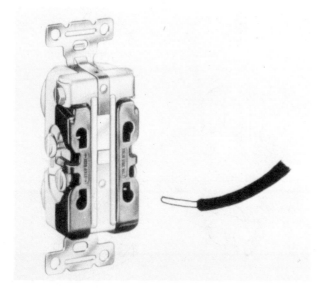

**Figure 9-57** Wires may be pushed into slots on the back of a switch or receptacle outlet.

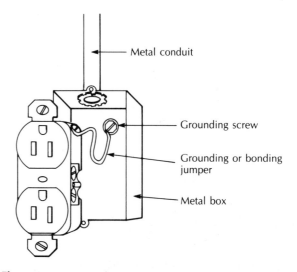

**Figure 9-59** Grounding a receptacle outlet to a metal box within a metal conduit wiring system

Unit 9   Farm Wiring Methods and Materials   195

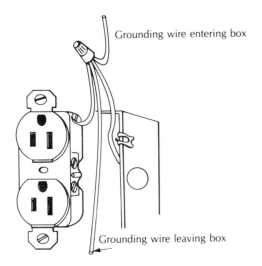

**Figure 9-60** Proper grounding of a metal box and receptacle outlet with cable or nonmetallic conduit wiring

the grounding clip to the box. Continuity of the grounding conductor may not depend upon the device, *NEC Section 250-114*.

## RACEWAY WIRING SYSTEMS

It takes some practice to develop techniques and skills for the installation of a conduit electrical system. The raceway is usually installed as a complete system, and then the wires are installed. This is to prevent wire insulation from becoming damaged during installation. A raceway system must also be installed so that there are no obstructions or sharp corners that will prevent the wires from being installed easily. When a sharp corner is necessary, an ell fitting with a removable cover is used.

Stranded wires are more flexible than solid wires. For this reason, only stranded wires should be installed in a raceway system. Stranded wires are easier to install. For a raceway system, conductors size No. 8 AWG and larger is required to be stranded, *NEC Section 310-3*.

Installing a raceway system involves the following steps and skills.

- Select the appropriate type of raceway for the application.
- Cut the exact length of conduit to fit between boxes, panels, and fittings.
- Prepare the cut end for installing in a box or fitting. Threading may be required when using rigid conduit and IMC.
- Bend the conduit where necessary.
- Install the conduit so that it fits closely to the surface upon which it is installed. Exposed conduit must have a neat appearance after it is installed.
- Provide access fittings at appropriate intervals to aid the installing of wire.
- Choose an appropriate bracket, strap or fastener to se-

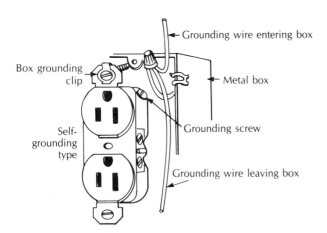

**Figure 9-61** A grounding wire is not required to the receptacle grounding screw if the receptacle is of the self-grounding type.

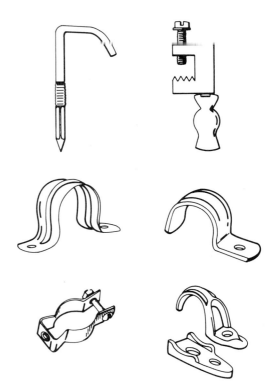

**Figure 9-62** Brackets, straps, and clips for attaching conduit to the building

cure the conduit to the surface of the building, Figure 9-62.
- Install the wire so as not to damage the insulation.

## Rigid Metal Conduit and IMC

Raceway systems using rigid metal conduit and IMC can be used for all applications except in those areas where the conditions are corrosive. Particular applications include areas exposed to physical abuse, and where the conduit riser must support an overhead service drop or lighting fixtures.

Rigid metal conduit and IMC are best cut to length with a hacksaw (18 teeth-per-inch blade), or a metal cutting band saw. A pipe cutter may also be used. After cutting, any burrs or sharp edges must be removed with a round file, reamer, or another acceptable tool, *NEC Section 346-7*. Threadless connectors are available for rigid metal conduit and IMC but, usually, the electrician threads the conduit on the location. Conduit can be threaded with a hand threader or a power threader. Conduit threads are tapered ¾ in (19 mm) per ft (305 mm).

## CONDUIT BENDING

An electrician must learn to make the common types of conduit bends while in the field. Most electricians can bend ½-in and ¾ in trade size rigid metal conduit and IMC by using a hand bender, Figure 9-63. Larger sizes are usually bent by using a mechanical, electrical, or hydraulic bender, Figure 9-64. All bender manufacturers provide tips and instructions for making the various types of bends. The common types of bends are the kick, stub, back-to-back, offset, and saddle. See Figure 9-65.

The more bends in a run of conduit, the harder it will

**Figure 9-64** Large sizes of rigid metal conduit, IMC, and EMT are bent using mechanical, hydraulic, or electric benders.

be to install the wires. For this reason, the *National Electrical Code* limits the number or degrees of bend between access openings to 360° maximum, *Sections 345-11 and 346-11*. Table 9-12 lists the common bends and the angle of bend required.

## Problem 9-5

Determine the total number of degrees of bend in the run of conduit shown in Figure 9-66A.

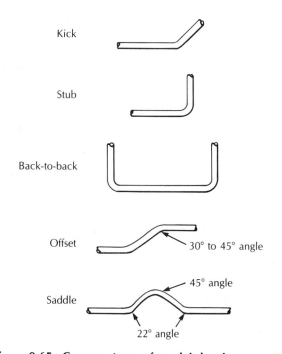

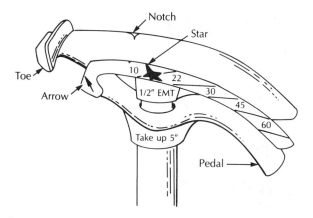

**Figure 9-63** Hand bender is used for smaller sizes of EMT, rigid metal conduit, and IMC.

**Figure 9-65** Common types of conduit bends

Unit 9   FARM WIRING METHODS AND MATERIALS   197

Table 9-12 Angle required for common conduit bends

| Bend | Angle |
|---|---|
| Kick | Up to 90° |
| Stub or 90° | 90° |
| Offset 30 × 30 | 60° |
| 45 × 45 | 90° |
| 60 × 60 | 120° |
| Saddle | 90° |

Figure 9-67 Conduit union or three-piece connector

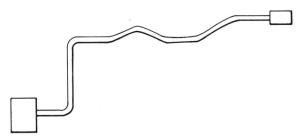

Figure 9-66A Determine the total number of degrees of bend in this run of conduit (Problem 9-5).

performing this task. Once in a while, it becomes impossible to turn the next piece onto the previous one. The problem is solved by joining the conduit with a conduit union often called a *three-piece coupling,* Figure 9-67.

Several popular devices are available for fastening rigid metal conduit and IMC to the structure. Beam clamps are used to attach wiring materials to I-beams, steel channel, and other structural steel, Figure 9-68. These devices eliminate the need to drill holes into the steel.

IMC and rigid metal conduit must be supported within 3 ft (914 mm) of each outlet box, junction box, cabinet, or fitting. The conduit must be supported at least every 10 ft (3.05 m), *NEC Section 345-12.* An exception in the case of straight runs of rigid metal conduit is found in *NEC Table 346-12,* represented in this text as Table 9-13. These increased support spacings are permitted provided the long span does not apply undue stress on the supports.

## Solution

The electrician is required to estimate the angle used for an offset, Figure 9-66B.

An electrician installing agricultural wiring seldom must bend large sizes of conduit. A hand bender called a *hickey* can be used for bending rigid metal conduit and IMC in sizes larger than ¾ in.

Manufactured bends with threads are available in the larger trade sizes. Most distributors carry 30°, 45°, and 90° bends. An offset can be made using two 30° or 45° bends and a nipple of the desired length.

A threaded conduit system is installed by starting at one end and assembling one piece of conduit or fitting after another. Each part is turned onto the previous part. Pipe wrenches and all-purpose pump pliers are best for

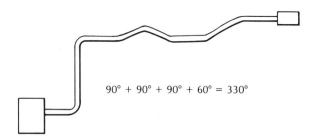

90° + 90° + 90° + 60° = 330°

Figure 9-66B The solution to Problem 9-5

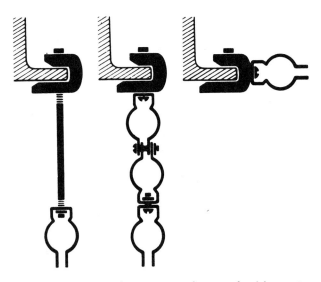

Figure 9-68 Beam clamps are used to attach wiring materials to structural steel.

**Table 9-13** (*NEC Table 346-12*) **Supports for rigid metal conduit**

| Conduit Size (inches) | Maximum Distance Between Rigid Metal Conduit Supports (feet) |
|---|---|
| ½–¾ | 10 |
| 1 | 12 |
| 1¼–1½ | 14 |
| 2–2½ | 16 |
| 3 and larger | 20 |

For SI units: (Supports) one foot = 0.3048 meter.

Reprinted with permission from NFPA 70-1984, *National Electrical Code®*, Copyright © 1983, National Fire Protection Association, Quincy, MA. This reprinted material is not the complete and official position of the NFPA on the referenced subject which is represented only by the standard in its entirety.

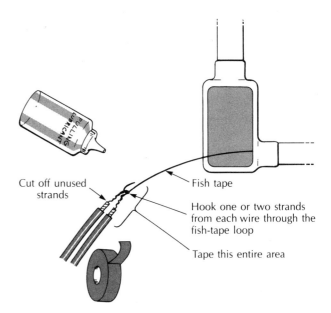

Figure 9-69 Using a fish tape to pull wires into a run of conduit

## INSTALLING WIRE IN CONDUIT

All of the required wires in a run of conduit are usually installed at the same time. This may involve the wires for several circuits. When large size wires are involved, it may be easier to install them one at a time.

Before installing the wires, tape them together to prevent snagging as the wires go through fittings. If the run of conduit is short and has few bends, the wires can usually be pushed into the conduit. But, if the run is long, and there are several bends, the wires will have to be pulled.

A fish tape is generally used for pulling wire into conduit. The fish tape is pushed into the conduit. For a relatively easy pull, the wire is fastened to the tape and pulled through the conduit. Pulling lubricant can be applied to the wire as it enters the conduit to make it slide more easily and to help prevent damage to the wire insulation, Figure 9-69.

Occasionally, a pull is too hard to be accomplished by hand. Mechanical, electrical, and other types of pulling devices can be used to pull the wires into the conduit. First, the fish tape is pushed through the conduit. The cable or rope of the mechanical or electrical puller is then pulled into the conduit, Figure 9-70. The wires are fastened to the cable or rope and then pulled into the conduit. Pulling is best accomplished by two people; one does the pulling while the other feeds the wire into the conduit. When wires are pulled mechanically, care must be taken not to damage the wires or the conduit.

## INSTALLING ELECTRICAL METALLIC TUBING (EMT)

The wall of electrical metallic tubing (EMT) is too thin to support threads; therefore, it is joined using threadless connectors. EMT is cut to length with a hacksaw (24 teeth-per-inch blade is recommended) or a power band saw. The burr remaining after cutting must be removed, *NEC Section 348-11*. This is usually accomplished with a round file, although most electricians simply use a pair of pliers with sharp edges, Figure 9-71. EMT should *not* be cut with a pipe cutter because it is likely to deform the end of the conduit.

Conduit bodies for rigid metal conduit and IMC are used with EMT. An EMT connector is turned into the threaded hub and the EMT is slipped into the connector, Figure 9-72.

Figure 9-70 Mechanical puller for pulling large wires into conduit

**Figure 9-71** To ream the burr from EMT, push a pair of pliers into the end of the EMT and rotate until the burr has been removed.

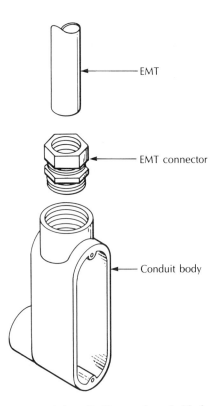

**Figure 9-72** EMT is installed into a threaded hub using an EMT connector.

EMT is not suitable for areas in which it would be subject to abuse or corrosive conditions. It is not recommended for use in areas where animals are housed unless out of reach of the animals.

Support for EMT must be provided at least every 10 ft (3.05 m), *NEC Section 348-12*. EMT in small trade sizes should be supported at more frequent intervals. EMT must also be supported within 3 ft (914 mm) of each box, cabinet, or similar device. Straps, hangers, and pipe supports for EMT are similar to those used for rigid conduit and IMC. Sizes of supports are different for the same trade size. For example, ½-inch rigid metal conduit has an outside diameter nearly the same as the outside diameter of ¾-in EMT.

Bends of not more than a total of 360° are permitted between access openings. The same types of benders are used for EMT as for rigid metal conduit and IMC, but an EMT bending shoe is required. An electrician can usually bend EMT in trade sizes up to 1¼ in. For the larger sizes, factory-made bends are available from an electrical distributor. If an electrician must frequently bend EMT, rigid metal conduit, and IMC up through 2-in trade size, mechanical benders are available with shoes for each type.

An EMT hand bender can also be used for rigid metal conduit and IMC. Table 9-14 lists the bender size required for all three types of raceway.

Good bends are the result of practice. EMT has a thin wall and, therefore, poor bending techniques can result in ripples or other deformation of the conduit cross section at the bend. If the cross section is reduced because of a poor bend, the conduit is not permitted to be installed, *NEC Section 348-9* and *346-10*. There is also a minimum bend radius in inches, Table 9-15. All manufactured conduit benders, except hickeys, automatically produce a bend with a minimum radius or larger than minimum radius.

An EMT hand bender has the amount of "take-up" marked on the bender. This take-up is used in bending

**Table 9-14** EMT bender size required for bending several types of raceway

| EMT Bender Size (inches) | Raceway Trade Size (inches) | | |
|---|---|---|---|
| | EMT | Rigid | IMC |
| ½ | ½ | — | — |
| ¾ | ¾ | ½ | ½ |
| 1 | 1 | ¾ | ¾ |
| 1 ¼ | 1 ¼ | 1 | 1 |

Table 9-15 Minimum bend radius for conduit (in inches) to the inside edge of the finished bend

| Conduit Size (inches) | Minimum Radius (inches) | |
|---|---|---|
| | One-shot Bender | Other |
| ½ | 4 | 4 |
| ¾ | 4 ½ | 5 |
| 1 | 5 ¾ | 6 |
| 1 ¼ | 7 ¼ | 8 |
| 1 ½ | 8 ¼ | 10 |
| 2 | 9 ½ | 12 |
| 2 ½ | 10 ½ | 15 |
| 3 | 13 | 18 |
| 3 ½ | 15 | 21 |
| 4 | 16 | 24 |

stubs. It is the radius of the bend to the outside of the conduit, Figure 9-73. The *NEC* minimum radius of bends is to the inside of the bend.

## INSTALLING RIGID NONMETALLIC CONDUIT (PVC)

Rigid nonmetallic conduit (PVC) is used extensively in agricultural wiring because it provides a dusttight, watertight, and noncorrosive conduit system. It is inexpensive, lightweight, and, with practice and the correct tools, it is easy to install.

PVC is generally cut to the proper length with a hacksaw, a power band saw, or a PVC cutter. A cutter is a good investment for an electrician who installs a considerable amount of PVC conduit. Burrs must be removed from the cut edge of the conduit. Usually, a PVC conduit cutting tool does not leave a burr. A fine-toothed handsaw is an easy way to make a good square cut.

Nonmetallic conduit is not threaded. It is joined by applying cement to the conduit and sliding it into the fitting. The sequence of steps for joining PVC conduit is shown in Figure 9-74. Nonmetallic boxes of all types are available with threadless hubs. An FSC box for one device is shown in Figure 9-75. This type of installation is

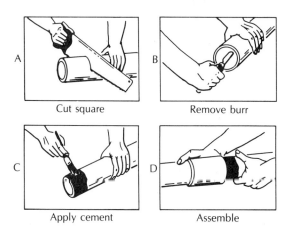

Figure 9-74 Sequence of steps for joining nonmetallic conduit

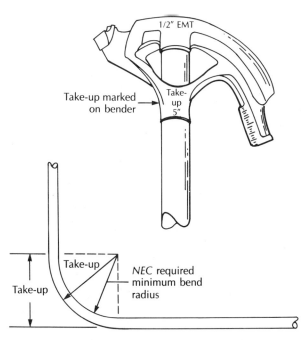

Figure 9-73 The take-up marked on the bender is the radius of the conduit to the outside of the bend, and is used in bending stubs.

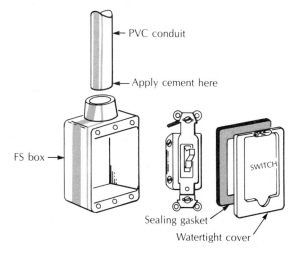

Figure 9-75 PVC conduit and FS box with sealing gasket and cover make a switch box waterproof.

Unit 9 FARM WIRING METHODS AND MATERIALS

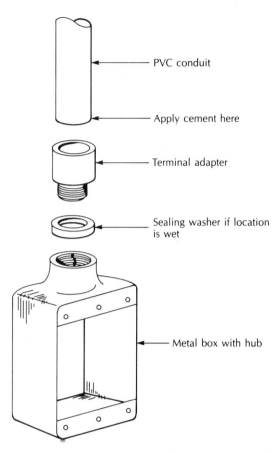

**Figure 9-76** PVC conduit and metal boxes can be used together by using a threaded PVC terminal adapter.

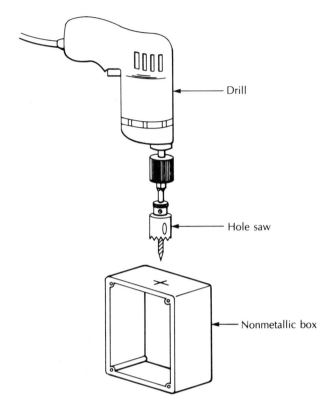

**Figure 9-77** A hole saw is used to make a hole in a nonmetallic box.

ideal for a surface-mounted switch or a receptacle outlet in a milkhouse or milking parlor where the walls are occasionally washed down. Metal boxes may also be used with PVC conduit by installing a threaded PVC terminal adapter, Figure 9-76.

Nonmetallic boxes are available without holes or hubs. These are convenient because holes are drilled only where needed. Unused holes or hubs must be plugged, *NEC Section 110-12*. PVC plugs are available to simply cement in place. Holes are drilled in nonmetallic boxes using conduit trade size hole saws, Figure 9-77. A ⅜-in variable speed drill works well for drilling holes in PVC boxes. Figure 9-78 shows two methods of connecting PVC conduit to a nonmetallic box with a drilled hole. Metal conduit and PVC can be joined as shown in Figure 9-79.

Nonmetallic conduit must be supported at more frequent intervals than metal conduit. *NEC Table 347-8*, represented in this text as Table 9-16, gives the maximum spacing of supports for the various trade sizes of nonmetallic conduit. The same types of supports can be used for nonmetallic conduit as for metal conduit. Nonmetallic straps are available for use in areas where corrosion is a particular problem. The conduit must be supported within 3 ft (914 mm) of boxes, cabinets, conduit bodies, and so forth.

Thermal expansion can be a problem with nonmetallic conduit and aluminum conduit. The electrician must be aware of the problem and take the necessary steps to avoid damage to the wiring system. The thermal expansion in a short run of conduit is not enough to cause problems. However, in long runs approaching 50 ft (15.24 m) or more the conduit must be made expandable. Expansion fittings are available for this purpose, and one should be installed for each 50-ft (15.24-m) run of straight conduit, Figure 9-80. Proper layout of the conduit system can help to avoid thermal expansion problems, Figure 9-81. The temperature in farm buildings can vary greatly from season to season and, therefore, thermal expansion is a problem to be considered.

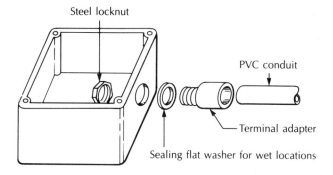

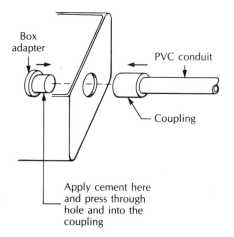

**Figure 9-78** Two methods of connecting PVC conduit to a nonmetallic box with a drilled hole

Table 9-16 (*NEC Table 347-8*) **Supports for rigid nonmetallic conduit**

| Conduit Size (inches) | Maximum Spacing Between Supports (feet) |
|---|---|
| ½–1 | 3 |
| 1 ¼–2 | 5 |
| 2 ½–3 | 6 |
| 3 ½–5 | 7 |
| 6 | 8 |

For SI units: (Supports) one foot = 0.3048 meter.

Reprinted with permission from NFPA 70-1984, *National Electrical Code®*, Copyright © 1983, National Fire Protection Association, Quincy, MA. This reprinted material is not the complete and official position of the NFPA on the referenced subject which is represented only by the standard in its entirety.

**Figure 9-80** Expansion fitting for nonmetallic conduit

Nonmetallic conduit is not permitted to support the weight of lighting fixtures and other electrical equipment, *NEC Section 347-3(b)*. Fixtures and equipment should be supported independently of the conduit system or by threaded rigid metal conduit.

Nonmetallic conduit must be heated in order to bend it. Even though it is flame retardant, it will burn if overheated. PVC conduit should *never* be heated with an open flame. Electric heaters are available for heating PVC conduit, Figure 9-82. Once the conduit has softened, it can be bent into a kick, stub, offset, or saddle. Nonmetallic conduit manufacturers have bending equip-

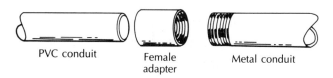

**Figure 9-79** Joining metal conduit to PVC conduit

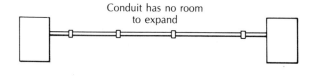

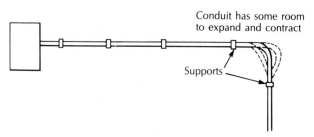

**Figure 9-81** Install nonmetallic conduit so that the conduit has room to expand and contract.

Unit 9   **FARM WIRING METHODS AND MATERIALS**   203

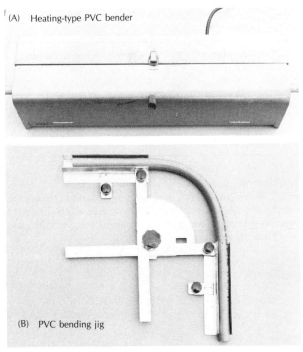

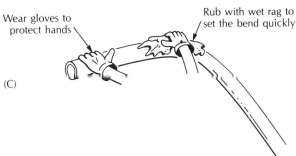

**Figure 9-82** Electric heater is used to soften conduit, and then it is formed into the correct bend by hand or by using a bending jig.

ment and instructions for successful bending. At first, PVC bending is awkward but, with practice, most electricians find PVC fast and easy to bend.

Preformed bends are available from electrical distributors. The common preformed bends are 30°, 45°, and 90°. Almost every type of bend can be formed using these combinations of angles.

## PULL BOXES

Pull boxes are used in conduit systems for making splices and taps, and for providing access into the conduit system. Pull boxes are particularly useful when large size wires must be pulled into a conduit system.

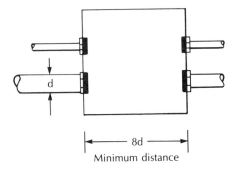

**Figure 9-83** Pull box with a straight pull

The pull box must be large enough so that the wires can be pulled in without damaging the insulation. The rules for sizing pull boxes are found in *NEC Section 370-18*. The following are the rules for sizing a pullbox for a straight pull or an angle (U) pull.

- For straight pulls, the box must be at least eight times the diameter of the largest raceway when the raceways are ¾ in or larger in diameter, and contain No. 4 AWG or larger conductors, Figure 9-83.

- For angle pulls, Figure 9-84, the distance between the raceways and the opposite wall of the box must be at least six times the diameter of the largest raceway. To this value, it is necessary to add the diameters of all other raceways entering the same wall of the box. This requirement applies when the raceways are ¾ in or larger in diameter, and contain No. 4 AWG or larger conductors.

## Problem 9-6

Determine the minimum size for the pull box shown in Figure 9-85.

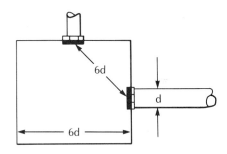

**Figure 9-84** Pull box with an angle pull

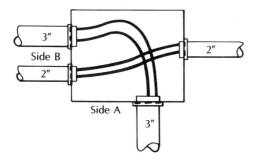

**Figure 9-85** Combination straight and angle pull

## WATERTIGHT WIRING MATERIALS AND INSTALLATION

A watertight and dusttight wiring system is required in animal confinement buildings, milkhouses, milking rooms, and other agricultural buildings where there is excessive dust or moisture with dust, or where periodic sprayings and cleaning take place, *NEC Section 547-3*. Nonmetallic conduit and UF cable with nonmetallic boxes are the wiring systems most widely used for this type of wiring. Nonmetallic material is used as often as possible because of the corrosive nature of water mixtures in agricultural buildings.

In some areas, there is a limited variety of nonmetallic boxes available which can be made dusttight and watertight. The wiring system may have to be a combination of metallic and nonmetallic boxes. A nonmetallic box for use with UF cable should have threaded openings or hubs. A box with no openings at all is also acceptable. FS boxes with nonthreaded hubs are awkward to use with UF cable. Typical boxes are shown in Figure 9-86. Holes can be drilled in nonmetallic boxes with a drill and hole saw (as shown in Figure 9-77).

## Solution

This exercise is only necessary if the wire size is No. 4 AWG or larger, and the conduit trade size is ¾ in or larger. Assume that the wire in both the 3-in and 2-in conduit is No. 4 or larger.

The length of side A is determined by the conduit entering side B.

```
6 × largest conduit diameter
plus diameter of other conduit on same side:
6 × 3 in       = 18 in
plus              2
Total length     20
```

This is a tricky problem. The 2-in conduit is a straight pull. Therefore, it is necessary to calculate the minimum dimension required, and take whichever dimension from this or the previous calculation is larger.

```
8 × conduit diameter
8 × 2 in = 16 in
```

Therefore, side A must be at least 20 in in length. The length of side B is determined by the conduit entering side A.

```
6 × 3 in = 18 in
```

The pull box must have minimum dimensions of 18 in by 20 in.

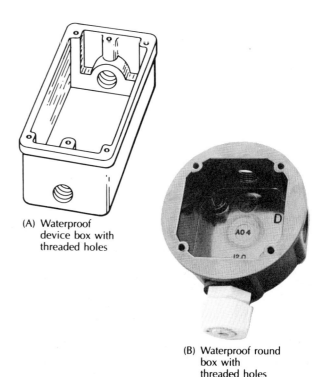

(A) Waterproof device box with threaded holes

(B) Waterproof round box with threaded holes

**Figure 9-86** Waterproof and dusttight nonmetallic boxes with threaded holes

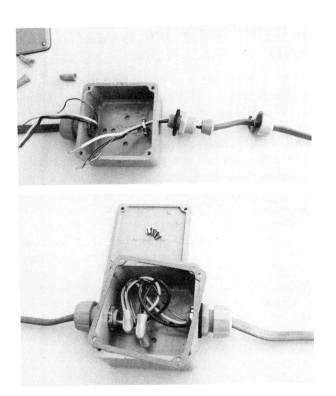

**Figure 9-87** Installing a waterproof UF cable connector in a waterproof and dusttight junction box

**Figure 9-88** Waterproof and dusttight incandescent lighting receptacle

A watertight and dusttight connector must be used where the UF cable enters the box. A completely nonmetallic, corrosion-resistant, watertight UF cable connector is shown in Figure 9-87.

Watertight switch and receptacle covers must be installed where moisture is a particular problem. An assembly similar to that shown in Figure 9-75 is used for UF cable. Watertight and dusttight lamp receptacles are available with lamp ratings up to 150 W, Figure 9-88. The wiring compartment of these fixtures gets hot when incandescent lamps larger than 60 W are installed. It is not recommended to install lamps larger than 60 W when UF cable is used as the wiring method. For cases where higher wattage lamps will be used, the circuit wire entering the box of the fixture should have a 110°C (230°F) rating, and conduit rather than cable should be used.

## REVIEW

Refer to the *National Electrical Code* when necessary to complete the following review material. Write your answers on a separate sheet of paper.

1. An electrician plans to use cable to run a circuit in a barn, but the area where the cable is to be installed is often moist. Of the following cable types, which is the most appropriate for this application?
    a. NM
    b. RU
    c. TW
    d. UF

2. Of the following, which is the minimum size copper AWG THHN conductor required for a 50-A circuit in a dry location?

a. 3
   b. 4
   c. 5
   d. 8

3. Type SE cable can be used as the branch-circuit conductor for interior wiring, according to which *NEC* section?

4. A wire, Type THWN-THHN, is installed in a moist environment. If the wire size is No. 6 AWG copper, the maximum ampacity under this condition is:
   a. 55 A
   b. 65 A
   c. 70 A
   d. 75 A

5. A run of conduit contains eight 2-wire lighting circuits. If the wires are No. 12 AWG THW copper, then the maximum permitted ampacity of the wire is:
   a. 15 A
   b. 17.5 A
   c. 20 A
   d. 25 A

6. The wires supplying a motor on a grain dryer are contained within conduit, but they are exposed to an ambient temperature of 130°F (54°C). The ampacity of No. 8 AWG THHN wire in a dry environment is:
   a. 55 A
   b. 44 A
   c. 39 A
   d. 31 A

7. The type of cable most generally used for residential wiring where the environment is dry is which of the following types?
   a. UF
   b. NMC
   c. MC
   d. NM

8. The primary problem with aluminum cable for wiring branch circuits is which of the following?
   a. Its cost
   b. Its light weight
   c. It has a lower resistance than copper wire
   d. Its rapid rate of thermal expansion and slightly nonelastic behavior

9. Aluminum wire can be directly connected to a switch for a general-use lighting circuit if the switch is marked CO/ALR, according to which *NEC* section?

10. The flexible cord which is appropriate for use on an automatic milker washer in a milkhouse is which of the following?
    a. SO
    b. HPD
    c. SP-1
    d. SW

11. Three-wire electrical cable used in residential wiring has in addition to a bare grounding wire, insulation with which of these colors?
    a. Black, red, and white
    b. Black, white, and green
    c. Red, green, and white
    d. White, black, and blue

12. Rigid steel conduit and rigid aluminum fittings and equipment may be used in the same wiring system, according to *NEC* section?

13. The copper wire with TW insulation which is permitted to supply a maximum load current of 15 A is size No. ____?____ AWG.
    a. 10
    b. 18
    c. 12
    d. 14

14. Which of the following is the standard length of rigid metal conduit?
    a. 8 ft (2.44 m)
    b. 10 ft (3.05 m)
    c. 12 ft (3.66 m)
    d. 20 ft (6.1 m)

15. What is the minimum trade size for EMT?
    a. ¼ in
    b. ⅜ in
    c. ½ in
    d. ¾ in

16. An electrician makes the following bends between two access fittings: one 45° × 45° offset, two stubs, and a 60° kick. Are there too many bends in this run of conduit, according to the *NEC*? Justify your answer.

17. A ventilation fan installed in a poultry house is to be wired in conduit. To prevent the transmission of vibration of the fan into the wiring system, a short piece of which of the following should be used to make the final connection to the motor?
    a. Rigid nonmetallic conduit
    b. Flexible metal conduit
    c. EMT
    d. Liquidtight flexible conduit

18. The minimum size rigid nonmetallic conduit permitted to contain sixteen No. 14 AWG THW wires is which of the following?
    a. ¾ in
    b. 1 in
    c. 1¼ in
    d. 1½ in

19. A conduit contains three wires No. 4 AWG XHHW, and six wires No. 10 AWG THW. The minimum size rigid metal conduit permitted for these wires is which of the following?
    a. 1 in
    b. 1¼ in
    c. 1½ in
    d. 2 in

20. Assuming that a circuit has been wired with NM cable, according to the *National Electrical Code,* if you open an electrical outlet box on the wall and find a white insulated wire connected to a single-pole switch, the white wire
    a. is a neutral wire.
    b. is to be considered hot.
    c. is a grounding wire.
    d. connected to a switch is a violation.

21. Where in the *NEC* does it state that making a tight bend in NM cable around a staple is a violation?

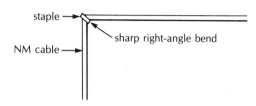

22. The free conductor ends connected to the switch shown in the following diagram must be not less than 6 in (152 mm) long, according to which *NEC* section?

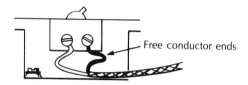

23. When using nonmetallic boxes, clamping of the cable to the box is not required where the cable is supported within a distance from the box of which of the following?
    a. 8 in (203 mm)
    b. 4 in (102 mm)
    c. 12 in (305 mm)
    d. 18 in (457 mm)

24. When NM cable is being run on interior building surfaces, it must be stapled or secured at intervals not exceeding which of the following?
    a. 3 ft (914 mm)
    b. 3½ ft (1.07 m)
    c. 4 ft (1.22 m)
    d. 4½ ft (1.37 m)

25. A neutral wire is a wire which has been intentionally grounded. According to the *National Electrical Code,* this neutral wire color must be which of the following?
    a. White or gray
    b. Green
    c. Bare
    d. Not black or red, but any other color

26. The minimum size device box permitted for the box in the diagram on the top of page 210 is which of the following? (The grounding wire is not shown in order to simplify the diagram, but must be included.)
    a. 4 × 1½ × 2      c. 3 × 2 × 2¾
    b. 3 × 2 × 2½      d. 3 × 2 × 3½

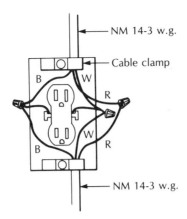

27. In this review problem and the next, choose the circuit which is wired to comply with the provisions of the *National Electrical Code*. The circuits are wired with 14-2 and 14-3 cable.

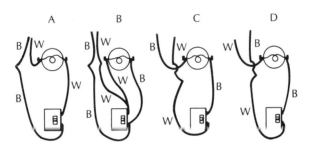

28.

Common in this location for all switches

210   Unit 9   *Farm Wiring Methods and Materials*

29. For the following diagram, what is the minimum acceptable length dimension? What is the minimum acceptable width dimension?

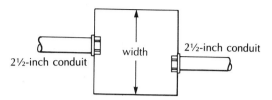

30. For the following diagram, determine the following dimensions.
    a. Dimension "a" must be at least ___?___ inches.
    b. Dimension "b" must be at least ___?___ inches.
    c. Dimension "c" must be at least ___?___ inches.

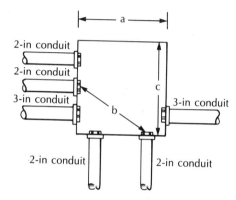

# UNIT 10 EQUIPMENT GROUNDING AND BONDING

## OBJECTIVES

After studying this unit, the student will be able to

- explain how an effective grounding system protects against electrical shock.
- define the terms commonly used when discussing grounding.
- list acceptable grounding conductor materials.
- explain what equipment and materials require grounding.
- explain the methods of identifying grounding wires.
- size an equipment grounding wire for a circuit.
- list the factors involved in electrical shock.
- explain the operation of a ground-fault circuit interrupter.
- list the receptacles or equipment which must be protected by a ground-fault circuit interrupter.

## IMPORTANCE OF GROUNDING

A good grounding system is extremely important in farm wiring. A grounding system acts to protect people and animals from electrical shock, and helps to reduce the possibility of an electrical fire. When an electrical system is installed correctly and is functioning properly, no measurable amount of current will flow in the grounding system.

Two types of grounding are essential for a farm electrical system:

1. *System grounding,* where one of the wires is intentionally connected to the earth

2. *Equipment grounding,* where all metal parts, materials, and equipment, which are not intended to carry current, are connected to the system ground at the main service entrance

This unit deals with electrical equipment grounding. Grounding the electrical system is discussed in Unit 4. Diagrams of all the types of electrical systems used on farms, showing the grounded wire, are included in Unit 3.

Grounding seems to be one of the most confusing aspects of electrical wiring. But, it really is not as complicated as it may appear. The principal importance of grounding is that there is a reliable electrical path from every piece of metal equipment all the way back to the electrical system grounded conductor at the main service, Figure 10-1.

## Terminology

Part of the confusion about grounding results from a lack of understanding of the terms used. The following are definitions of some basic terms.

*Ground.* An electrical connection to the earth. This connection is usually best if the earth is damp.

*Ground fault.* An accidental or unintentional connection of a circuit conductor to the earth, or to the grounding conductor which is connected to the earth.

*Equipment grounding conductor.* A conductor connecting all metal which is likely to become energized to the system grounding electrode of the main service. The grounding conductor is intended to carry current only when a ground fault occurs.

*System grounded conductor.* A current-carrying wire which has been intentionally grounded to the earth. This is the neutral wire for most farm electrical systems. This wire is permitted to be connected to the earth only at the main service panel.

*Bonding.* The permanent joining of metal parts to form an adequate electrical conducting path.

## HOW THE GROUNDING SYSTEM FUNCTIONS

The grounding conductor acts as a protective shield between electrical wires and people or animals near the wires. For example, consider the situation where an electric motor is operating at 120 V. The motor is supplied by a hot wire and a neutral wire. If a ground fault occurs which involves the hot wire, the metal frame of the motor becomes live. If a person or animal should touch the motor frame, the circuit will be completed through the earth, Figure 10-2 (A). The person or animal will suffer a shock. An equipment grounding conductor attached to the motor frame prevents the occurrence of shock. The equipment grounding conductor provides a low-resistance path for the current to follow back to the power panel, Figure 10-2 (B). Usually, the circuit breaker trips or the fuse blows the instant the ground fault occurs.

Single- and 3-phase equipment operating at 240 V or more can create a serious electrical shock hazard if not properly grounded. This problem is illustrated in Figure 10-3 (A), which shows two electrically powered bunk feeders in a beef feedlot. The motors are powered by single-phase current at 240 V requiring two hot wires. In time, a ground fault develops in each motor, causing different hot wires to become faulted at each motor

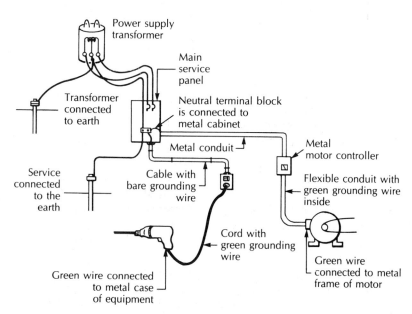

**Figure 10-1** All exposed metal likely to become energized is connected to the neutral terminal block of the main service panel by the grounding conductor.

Unit 10   Equipment Grounding and Bonding   213

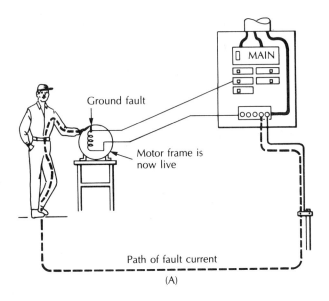

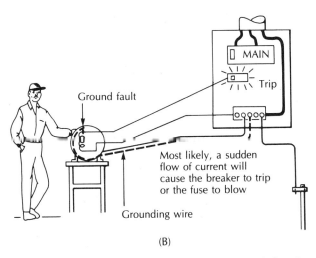

**Figure 10-2** (A) This ground-faulted nongrounded electric motor has a live frame; the person suffers a shock. (B) Because the motor is grounded, the breaker trips when the fault occurs. The person does not suffer a shock.

frame. Wire "A" is faulted at motor No. 1, and wire "B" is faulted at motor No. 2. Now both bunk feeders are live with 240 V between them. A human or an animal touching or perhaps even standing near either feeder could suffer a *harmful or fatal shock*. This problem is a source of stray voltages detected on equipment and in the ground on many farms.

A good grounding system prevents a hazardous voltage from developing between faulted electrical equipment. When a ground fault occurs, sufficient current usually flows to trip a circuit breaker or blow a fuse,

Figure 10-3 (B). The first fine-print note to *NEC Section 250-1* states that one purpose of grounding is to cause an overcurrent device to open if a ground fault occurs. The second fine-print note states that grounding is intended to prevent a voltage from developing between exposed metal parts of an electrical system or equipment and the earth.

## WHAT MUST BE GROUNDED?

Any exposed noncurrent-carrying metal parts of fixed equipment, and metal parts of the electrical system likely to become energized, must be grounded if any of the following conditions exist, *NEC Section 250-42*.

- If within 8 ft (2.44 m) vertically or 5 ft (1.52 m) horizontally of the ground or metal object which could be contacted by a person
- If the location is damp or wet
- Where in electrical contact with metal such as a metal grain bin or metal building
- If the wiring is metal raceway
- If any wire operates at more than 150 V to ground

For most wiring systems, particularly in farm environments, at least one of these conditions exists, thus requiring grounding.

Interior metal water piping systems, and other interior metal piping systems which are likely to become energized, are required to be grounded, *NEC Section 250-80*. This requirement includes the metal vacuum line used to operate milking machines in a dairy. It may also include metal stalls and feeders in animal barns.

A metal interior water piping system must be connected to the service grounding electrode, the service grounding electrode wire, or the service grounding terminal block. The bonding jumper used must be the same size as the service grounding electrode wire. If the service is grounded to the metal water pipe and there are no nonmetallic sections in the water pipe, then a bonding jumper is not necessary. Bonding the metal water piping system is shown in Figure 10-4.

Other metal piping systems are considered as grounded by the equipment grounding conductor supplying the equipment connected to the pipe. For example, a metal vacuum line is considered to be grounded by the grounding wire for the vacuum pump motor, provided there are no insulating sections in the pipe. This procedure is permitted by *NEC Section 250-80(b)*.

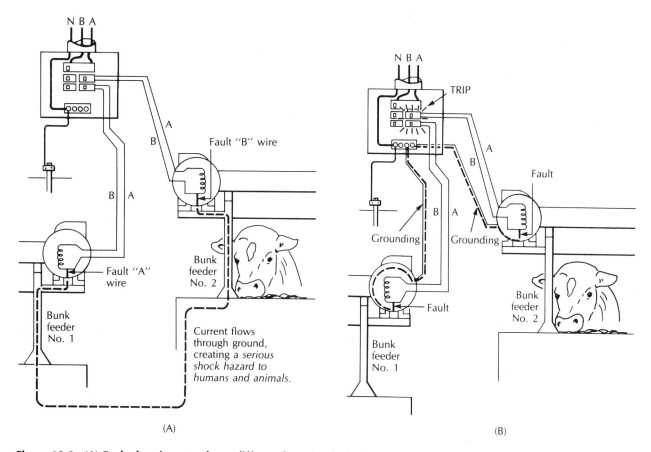

**Figure 10-3** (A) Each electric motor has a different hot wire faulted to the frame. There are 240 V between the two bunk feeders. (B) Because the motor frames are grounded, the breakers trip when each fault occurs.

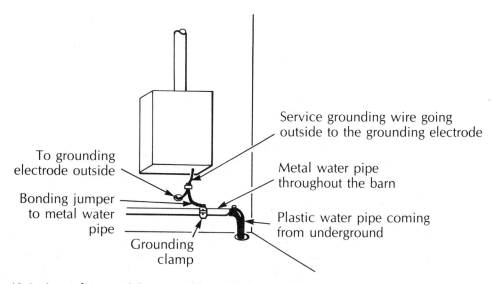

**Figure 10-4** A metal water piping system in a building must be bonded to the electrical system ground.

Unit 10   Equipment Grounding and Bonding   215

## SUPPLEMENTAL GROUNDING ELECTRODES

Extra grounding electrodes are not uncommon on farms. Grounding for a lightning protection system is a common example. The electrical system is required to have a grounding electrode which is separate from the lightning system ground, *NEC Section 250-86*. But, it is recommended that these electrodes be bonded together to form a single grounding system. If an electrical system grounding electrode and the lightning system grounding electrode are within 6 ft (1.83 m) of each other, they are required to be bonded together, *NEC Section 250-46*.

Electrically operated equipment on a farm is permitted to have its own grounding electrode, *NEC Section 250-91(c)*. But, this grounding electrode does not take the place of a grounding conductor back to the service ground. The earth is not permitted to serve as the grounding electrode.

A typical example of farm equipment which may have a separate grounding electrode is an electrically heated stock waterer. If the waterer is supplied with a plastic water pipe, then a separate grounding electrode at the waterer is recommended. This installation minimizes the chance of a voltage developing between the metal frame of the waterer and the adjacent earth, Figure 10-5. Defective stock waterers are another source of stray voltages on farms. The problem is particularly serious because animals will refuse to drink if they experience a shock from a waterer. Adequate water consumption for animals is essential for high productivity.

## EQUIPMENT GROUNDING

The grounding conductor can be a metal raceway or a metal box, cabinet, or frame of equipment. The equipment grounding conductor may be copper or other corrosion-resistant conductor, *NEC Section 250-91(b)*. However, there is a restriction on the use of aluminum wire for equipment grounding. Aluminum wire cannot be used where it is in direct contact with masonry or the earth, or where subject to corrosive conditions. When aluminum wire is exposed and used for grounding outside, it is not permitted within 18 in (457 mm) of the earth, *NEC Section 250-92(b)(2)*.

Equipment grounding wires can be solid or stranded, insulated or bare. If the wire is insulated, it is required to have insulation which is green or green with yellow stripes. There is an exception, however. If the wire is larger than No. 6 AWG, it can be identified as a ground-

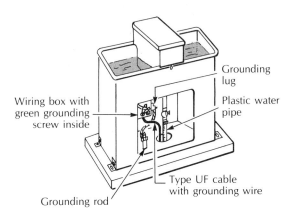

**Figure 10-5** Grounding an electrically heated stock waterer with a supplementary grounding electrode is recommended if the waterer is not supplied with a metal water pipe.

ing wire at the time of installation at every location where it is accessible. According to *NEC Section 250-57(b)*, acceptable means of identification are

- by stripping the insulation from the entire exposed wire, leaving the wire bare; or
- by covering the entire exposed wire with green tape or paint, Figure 10-6.

All fittings, conduit splices, and any other connections in the grounding conductor must be made tight using proper tools, *NEC Section 250-92(b)(1)*. Flexible conduit is not as good a grounding conductor as rigid metal conduit, IMC, or EMT. Therefore, limitations are placed on its use as a grounding conductor.

According to *NEC Sections 250-91(b)* and *350-5*, flexible metal conduit can be used as an equipment grounding conductor, provided the following conditions are met.

- The conduit is not more than 6 ft (1.83) long.
- The wires inside the conduit are protected at not more than 20 A.
- The conduit is terminated in fittings approved for grounding.

If these limitations are not satisfied, then a grounding wire must be provided to bond around the flexible metal conduit, *NEC Section 350-5*. Manufacturers of flexible conduit connectors state in their literature if the fitting is UL listed as suitable for grounding.

Limitations as to where liquidtight flexible metal conduit can be used as an equipment grounding conductor are stated in *NEC Section 250-91(b), Exception No.*

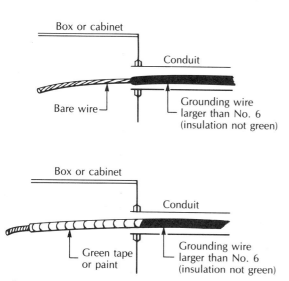

Figure 10-6 Insulated wires larger than No. 6 and not having green insulation can be identified as grounding wires at all points where they are visible.

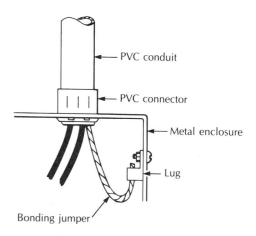

Figure 10-7 A bonding jumper can be terminated at a lug installed inside an enclosure.

2, and *Section 351-9*. Liquidtight flexible metal conduit is permitted as a grounding conductor if all of the following conditions are met

- Trade sizes 1 ¼ in and smaller.
- Length is not greater than 6 ft (1.83 m).
- Conduit and fittings are approved for grounding.
- Circuit conductors contained within ⅜ in to ½ in trade size are not protected at more than 20 A.
- Circuit conductors contained within ¾ in to 1 ¼ in trade size are not protected at more than 60 A.

Several methods are used for installing and terminating a grounding jumper wire around flexible conduit. The equipment bonding or grounding jumper can be run inside the conduit or it can be run on the outside. If run on the outside, the jumper wire is not permitted to be more than 6 ft (1.83 m) long. The bonding jumper must be run with the flexible conduit; it is not permitted to take a different route. Special fittings are manufactured for terminating the bonding wire on the outside of the conduit.

A number of methods are acceptable for terminating an equipment bonding jumper or grounding wire at an enclosure. The methods apply to equipment grounding wires run with nonmetallic conduit, as well as flexible conduit. The grounding wire can be terminated at a grounding screw, if available, or by installing a lug in the enclosure, as shown in Figure 10-7.

The improper mounting of a grounding lug in an enclosure is often a point of failure in the grounding system. The lug must be held in place with a screw or bolt which is used for no other purpose, *NEC Section 250-114(a)*. Before the grounding lug is installed, any paint or nonconductive coatings must be removed from the box so there will be good electrical contact between the lug and the metal box, *NEC Section 250-118*.

Pressure-type connectors and clamps are required for connections and terminations of grounding conductors. Connections or fittings that depend upon solder are not permitted, *NEC Section 250-113*.

Grounding wires passing through a box or enclosure are permitted to be connected to the box or enclosure, but continuity must not be interrupted by removing a device from the box, *NEC Section 250-114*. Several methods of connecting grounding wires in an enclosure are shown in Figure 10-8.

Bonding is required around nonmetallic enclosures, loosely joined raceway, or any other wiring which could cause an interruption in the grounding path. Bonding is permitted inside the wiring system or on the outside as long as the wire is not subject to physical abuse. If the wiring installer applies good judgment, an effective grounding system will most likely be established.

## SIZING EQUIPMENT GROUNDING CONDUCTORS

The sizing of an equipment grounding wire is based upon the size of the overcurrent device protecting the circuit. *NEC Table 250-95*, represented in this text as Table 10-1, lists the minimum wire size permitted for various overcurrent device ratings. In manufactured electrical cable, the manufacturer has already installed

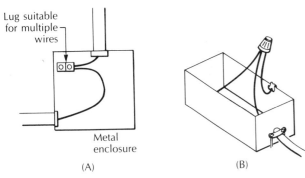

Figure 10-8 Two methods are shown to ground a metal box and maintain grounding continuity to the remainder of the circuit.

the correct size grounding wire, based upon the maximum size overcurrent device rating which can be used for that cable. When raceway is permitted as a grounding conductor, the cross-sectional area of metal is adequate for the largest size wires permitted in the raceway. Usually, the only time a grounding wire must be sized is when it is used in nonmetallic conduit or for bonding jumpers.

## Problem 10-1

Determine the minimum size copper grounding wire for a 30-A circuit consisting of THW No. 10 copper wires in PVC conduit.

Table 10-1 (*NEC Table 250-95*) Minimum size equipment grounding conductors for grounding raceway and equipment

| Rating or Setting of Automatic Overcurrent Device in Circuit Ahead of Equipment, Conduit, etc., Not Exceeding (Amperes) | Size | |
|---|---|---|
| | Copper Wire No. | Aluminum or Copper-Clad Aluminum Wire No.* |
| 15 | 14 | 12 |
| 20 | 12 | 10 |
| 30 | 10 | 8 |
| 40 | 10 | 8 |
| 60 | 10 | 8 |
| 100 | 8 | 6 |
| 200 | 6 | 4 |
| 300 | 4 | 2 |
| 400 | 3 | 1 |
| 500 | 2 | 1/0 |
| 600 | 1 | 2/0 |
| 800 | 0 | 3/0 |
| 1 000 | 2/0 | 4/0 |
| 1 200 | 3/0 | 250 MCM |
| 1 600 | 4/0 | 350 " |
| 2 000 | 250 MCM | 400 " |
| 2 500 | 350 " | 600 " |
| 3 000 | 400 " | 600 " |
| 4 000 | 500 " | 800 " |
| 5 000 | 700 " | 1 200 " |
| 6 000 | 800 " | 1 200 " |

*See installation restrictions in Section 250-92(a).

Reprinted with permission from NFPA 70-1984, *National Electrical Code*®, Copyright © 1983, National Fire Protection Association, Quincy, MA. This reprinted material is not the complete and official position of the NFPA on the referenced subject which is represented only by the standard in its entirety.

## Solution

From *NEC Table 250-95*, (Table 10-1 in this text), the minimum size copper grounding wire is No. 10 AWG.

## Problem 10-2

A 7-ft (2.13-m) length of liquidtight flexible metal conduit extends from a fusible disconnect switch to an electric furnace. The wires inside the conduit are THHN No. 4 copper, protected by a 70-A fuse. Determine the minimum size copper equipment grounding conductor to bond around the flexible metal conduit.

## Solution

The liquidtight flexible metal conduit is more than 6 ft (1.83 m) long; therefore, a bonding jumper is required. There will be a grounding lug in the furnace; thus, the bonding jumper should be run inside the flexible conduit.

A 70-A overcurrent device is not listed in *NEC Table 250-95;* therefore, the next larger rating must be used. For this example, use the grounding wire size required for a 100-A overcurrent device. The minimum size copper grounding wire for this 70-A circuit is No. 8 AWG.

## ELECTRICAL SHOCK

Electrical shock is caused by the flow of electricity through the body or a part of the body. Current passing through the body can cause burns, muscular reaction, and injury to bodily organs. Electricity usually takes the shortest route through the body to complete the circuit. If the current travels through the head or central part of the body, *severe injury* or *even death can occur*.

Human skin which is dry has a high resistance to the flow of electricity. With a voltage of less than 50 V, the current flow through the body is generally not enough to be harmful to a human. This is not necessarily true for an animal standing in a moist environment. The moist hoof or foot of a farm animal offers little resistance to the flow of current.

The main factors involved in electrical shock to a person or an animal are:

- Voltage
- Condition of skin or body surface
- Path the current takes
- Amount of surface area contact with source of voltage
- Duration of the electrical shock

For humans, voltage and condition of the skin usually determine the amount of current that flows. An average person usually will just begin to feel the sensation of shock if 1 mA (0.001 A) flows through the body. Shock usually becomes painful with a flow of 8 mA (0.008 A). Muscular contraction (the person is unable to let go) almost always occurs with a current flow of 15 mA (0.015 A). Greater flows of current through the body are usually much more serious. An effective grounding system on a farm is important for the protection of people and animals because of the existing damp and wet conditions.

## GROUND-FAULT CIRCUIT INTERRUPTER (GFCI)

A circuit breaker, called a *ground-fault circuit interrupter* (GFCI), can prevent serious shock to a person or animals under certain conditions. The GFCI works on the principle that both wires supplying a single-phase load must carry the same number of amperes if the circuit is operating properly. If a ground fault occurs either to the grounding wire, or through a person, some of the current will take an alternate route back to the system grounding electrode, Figure 10-9. One wire now carries less current than the other. If the difference in current between the two wires is more than 5 mA (0.005 A), the GFCI will trip the circuit.

The GFCI functions in the same manner as a clamp-on ammeter. Both the hot wire and the neutral wire pass through the current transformer (CT) sensing coil inside the GFCI, Figure 10-10. As long as the current is the

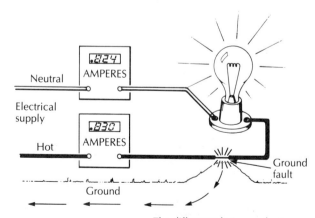

The difference between the two meters is 6 mA (0.006 A)

**Figure 10-9** The neutral wire carries less current than the hot wire when a ground fault occurs.

same in both wires, the current transformer output is zero. But, if a ground fault occurs, the CT sends an output current to trip the breaker, Figure 10-10.

Ground-fault circuit interrupters are available for 120-V circuits involving one hot wire and a neutral, and for 240-V circuits involving two hot wires. A 120-V, single-pole GFCI fits into the same size space as a standard single-pole breaker. The neutral wire from the circuit is connected to the *load neutral* terminal on the GFCI. The long white wire marked *panel neutral,* attached to the GFCI, is run to the service *panel neutral* block. The hot wire is run to the *load power* terminal, Figure 10-11.

Ground-fault circuit interrupters are available in the form of a receptacle outlet, Figure 10-12. Any equipment plugged into the GFCI receptacle has ground-fault protection. This type of GFCI also protects other outlets on the same circuit.

A portable GFCI is recommended for persons using power tools outside in damp or wet locations. The portable GFCI is plugged into an outlet, and the power tool is plugged into the GFCI, Figure 10-13.

A ground-fault interrupter only detects current flowing through the grounding wire or through the earth. It detects current not flowing in the neutral wire. The neutral wire and the grounding wire are not permitted to be electrically connected, except at the main service, *NEC Sections 250-23(a)* and *250-61(b).* A connection between the neutral wire and the grounding wire on a GFCI circuit will cause the GFCI to trip. Tripping occurs because some current flows back to the panel on the neutral wire, and some on the grounding conductor.

## *Where GFCI Protection is Required*

The *National Electrical Code* requires ground-fault circuit interrupter protection for specific receptacle out-

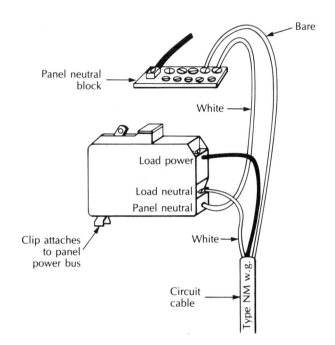

**Figure 10-11  Wiring a 120-V circuit to a GFCI.**

lets in a dwelling, for pools, and for construction sites, *NEC Sections 210-8, 305-4,* and *680-6.*

GFCI protection is required for new construction and new circuits added in the designated areas of a dwelling. GFCI protection is required for all 120-V, 15-A and 20-A, single-phase dwelling receptacle outlets installed

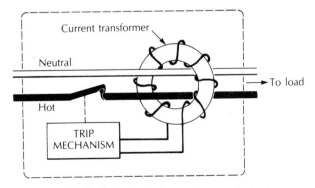

**Figure 10-10  Both the hot wire and the neutral wire pass through the GFCI.**

**Figure 10-12  This receptacle outlet is also a ground-fault circuit interrupter.**

## Nuisance Tripping of a GFCI

It takes only 5 mA (0.005 A) of current leakage from the hot wire to the ground to cause a ground-fault circuit interrupter to trip. A small amount of leakage current may be difficult to avoid in some normal circuits. Hand-held power tools do not cause a tripping problem if the tool is maintained in good condition. Some stationary motors, such as a bathroom vent fan or fluorescent lighting fixtures, may produce enough leakage to cause nuisance tripping. Another problem may be a long circuit with many splices. If possible, keep GFCI circuits less than 100 ft (30.5 m) long. To avoid nuisance tripping, a GFCI should not supply

- circuits longer than 100 ft (30.5 m).
- fluorescent or other types of electric-discharge lighting fixtures.
- permanently installed electric motors.

Installing a GFCI to prevent electrical shock from farm equipment seems like a good idea, but nuisance tripping may become a serious problem. Loss of critical ventilation can be fatal to animals. Stock waterers may freeze in northern climates if the GFCI trips. The installer must carefully consider the effects of loss of power to an agricultural circuit before installing GFCI protection. The most effective shock prevention system for agricultural equipment and circuits is a good equipment grounding conductor run with the circuit wires and connected to all metal agricultural equipment. Experience has shown that stray voltage problems on farms are usually eliminated or reduced to harmless levels if the wiring is installed properly and there is a good equipment grounding system.

**Figure 10-13   A portable GFCI provides protection at any receptacle outlet.**

- in garages, but not including an outlet in the ceiling for a garage door opener, *NEC Section 210-8(a)(2)*.
- in bathrooms, *NEC Section 210-8(a)(1)*.
- on the outside of the dwelling where they can be reached from the ground, *NEC Section 210-8(a)(3)*.
- where located within 20 ft (6.10 m) of the inside edge of a pool, *NEC Section 680-6(a)(3)*.
- for lighting fixtures within a certain distance of the pool, *NEC Section 680-6(b)*.
- for the water recirculating pump motor, *NEC Section 680-6(a)(1)*.

## REVIEW

Refer to the *National Electrical Code* when necessary to complete the following review material. Write your answers on a separate sheet of paper.

1. An electric motor powering a metal feeder for animals develops a ground fault between a hot wire and the metal frame of the motor. Which of the following situations will occur?
   a. The motor and feeder will become a live shock hazard if not grounded.
   b. The overcurrent device will most likely open if the motor is properly grounded.
   c. The motor would not be a shock hazard if properly grounded.
   d. All of these.

2. The shock hazard to humans and animals is practically eliminated by the grounding

conductor which produces a path of \_\_\_?\_\_\_ resistance to ground rather than through the body.

3. Name the type of conductor used to connect the noncurrent-carrying metal parts of equipment, raceway, and other enclosures to the grounding electrode conductor at the service.

4. Fine-print notes at the end of which *NEC* section explain the purpose of grounding the electrical system and the grounding conductor?

5. In which part of the electrical system in the following diagram must an equipment grounding wire be installed?

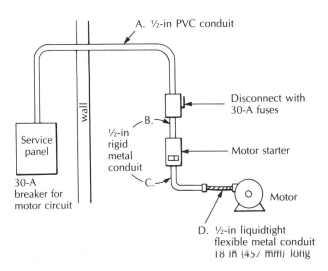

6. Assume that the motor in the previous diagram is in a dry, dust-free area, and ¾-in liquidtight flexible metal conduit is used instead of ½ in. How does this change the answer to Problem No. 5, and why?

7. Plastic water pipe enters a barn; therefore, the building electrical system is grounded to a ground rod driven in a damp location outside the building. State the *NEC* section which requires that the metal interior water piping system be bonded to the grounding electrode wire.

8. A grounding wire for a circuit must be bare, or with insulation which is \_\_\_?\_\_\_ in color, or \_\_\_?\_\_\_ with yellow stripes.

9. A grounding lug added to an enclosure is not permitted to be held in place by a screw which is used to mount the enclosure, according to which *NEC* section?

10. When grounding to an enclosure, any paint or other nonconductive coating must be removed to assure good electrical contact between the equipment grounding wire and the enclosure, according to which *NEC* section?

11. A 150-A feeder to a subpanel is run in PVC conduit. What is the minimum size AWG copper equipment grounding wire for this feeder?

12. Electrical current flowing through the body causes electrical shock. Name the two major factors which determine the amount of current that will flow through the body.

13. A ground-fault circuit interrupter for a dwelling circuit is designed to trip when a ground fault exceeds how many milliamperes (amperes)?

14. In the process of removing a 120-V plug from the wall, a person experiences a shock by accidentally making contact with the hot and the neutral blade simultaneously. A current of 6 mA flows from one blade to the other through the person's hand (refer to the diagram). If the circuit is protected by GFCI, will it trip? Explain.

15. Name the term used to describe the permanent joining of metallic parts to form an electrically conductive path which will assure electrical continuity.

# UNIT 11
# FARM LIGHTING

> ## *OBJECTIVES*
>
> After studying this unit, the student will be able to
>
> - give an example of how light affects a plant or animal.
> - explain light, wavelength, and color.
> - define the meaning and use of the terms *footcandle* and *lumen*.
> - measure the level of illumination with a light meter.
> - list several factors which affect the quality of light.
> - name the light sources used on farms.
> - compare lamp life, efficacy, and color output of light sources.
> - identify common types of lighting fixtures.
> - describe the installation of a common lighting fixture.
> - draw a simple, 1-lamp electric-discharge circuit, and label the parts.
> - determine the number of lighting fixtures that can be installed on a circuit.
> - give an example of the use of a dimmer, a time clock, and a lighting contactor.

Farmers use light to control behavior and growth of plants and animals. High levels of light are needed for plant growth and flowering of some varieties. Flowering can often be controlled by the length of the lighted day. The application of relative lengths of alternating periods of lightness and darkness as they affect the growth and maturity of an organism is called *photoperiodism*. Some animals exhibit photoperiodic behavior. The length of the lighted day affects the production of hormones which control growth in cattle. Lengthening the day with artificial light in winter increases the rate of growth of cattle. The start of laying for hens is controlled artificially with light. High levels of light are needed in dairies for cleaning animals and equipment, but low levels of light are essential in a poultry egg-laying house.

Because of the importance of lighting in agriculture, automatic controls may be required. Some lighting circuits may be several hundred feet in length. Corrosion-resistant fixtures may be required, particularly in controlled environment animal housing. Many fixtures and light sources with different characteristics are available. An understanding of the principles of light and lighting design are essential for effective and reliable agricultural lighting installations.

## WHAT IS LIGHT?

*Light* is a form of energy called *electromagnetic radiation*. Light travels in waves similar to electric sine waves. It travels at a speed of about 186 300 miles per second (300 million meters per second). Light which is visible to the human eye is actually only a small part of the total electromagnetic spectrum, Figure 11-1. X rays, ultraviolet rays, heat waves, microwaves, radar (UHF waves), and radio and television waves are also part of the electromagnetic spectrum. These waves and rays differ only in frequency or rate of vibration.

Everyone is familiar with the tuner on a radio. Different stations are assigned to different frequencies. Your favorite station, for example, may be assigned to operate at a frequency of 940 kilohertz (kHz), which is 940 000 cycles per second. Electromagnetic radiation is also emitted by a wire carrying normal 60-hertz (60-Hz) electrical current. The frequency of this radiation is only 60 cycles (complete sine waves) per second. Green light, for example, has a frequency of 600 terahertz (THz) (numeral 6 followed by 14 zeros). These numbers are too large to talk about comfortably. Therefore, scientists use the term *wavelength* instead of frequency to describe electromagnetic radiation. The wavelength is the length or distance the radiation travels through space in the time it takes to complete one cycle, Figure 11-2.

All electromagnetic radiation travels at the same speed through space; therefore, the faster the radiation vibrates, the shorter is the wavelength. Equation 11.1 relates speed, frequency, and wavelength. In the case of electromagnetic radiation, the speed is constant at 300 million meters per second.

$$\text{Wavelength} = \frac{\text{Speed}}{\text{Frequency}} \qquad \text{Eq. 11-1}$$

Electric, radio, and light waves are compared in Table 11-1.

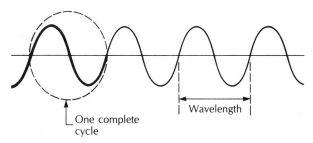

1 Hz = number of cycles per second

**Figure 11-2** Light travels through objects and space as waves.

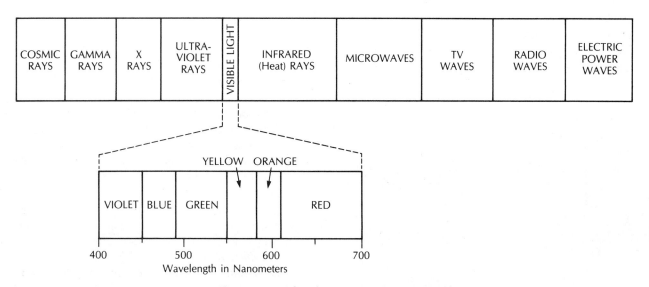

**Figure 11-1** The electromagnetic spectrum

Table 11-1  Comparison of common types of electromagnetic radiation

| Type of Radiation | Frequency | Wavelength |
|---|---|---|
| Electric wave | 60 hertz (Hz) | 5 million meters (Mm) |
| Radio wave | 940 kilohertz (kHz) | 319 meters (m) |
| Green light | 600 terahertz (THz) | 500 nanometers (nm) |
|  |  | (500 × 0.000 000 001) |

## COLOR

The *nanometer* (nm) is the unit of measure of the wavelengths of visible light. Light occupies the portion of the electromagnetic spectrum ranging from about 400 nm to 700 nm. The nerve endings in the retina of the normal human eye are affected differently by the different wavelengths of light. The eye sees 400 nm as violet, 500 nm as green, and 700 nm as red (Figure 11-1).

When the eye sees all the colors at the same time, they appear only as white light. Light from the sun or an incandescent lamp contains all of the colors. Passing this light through a triangular piece of glass, called a *prism*, breaks up, or separates, white light into its colors, Figure 11-3. Raindrops act as a prism when the sun shines on them, producing a rainbow.

The normal human eye does not see all colors equally well. The eye is most sensitive to green light and yellow light, as shown in Figure 11-4. The eye is not sensitive enough to see wavelengths shorter than 400 nm (ultraviolet radiation) or longer than 700 nm (infrared radiation or heat). A chicken is sensitive to red light but not very sensitive to blue light. Many insects see violet and ultraviolet, but not yellow, orange, or red. This is why outdoor yellow bug lamps attract fewer insects than normal lamps.

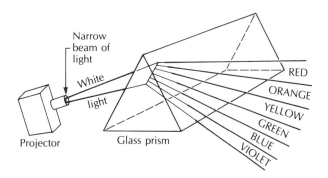

**Figure 11-3  A prism breaks up white light into colors.**

## Why Objects Have Color

Light can be either reflected or absorbed by an object, or it can be transmitted through the object. The color of an object appears as the color of light reflected. A ripe tomato appears red because it absorbs most of the violet, blue, and green light, and reflects mostly red light.

Every color can be produced by an object reflecting just the right amount of various colors. The eye sees all the colors at the same time. It cannot see each individual color, so the eye interprets the combination as a new color. For example, if green light and red light are mixed by shining both colors on a projection screen, one on top of the other, the eye will see yellow. In a color television, any color can be produced by mixing lights of various intensities of red, green, and blue (the *primary* colors), Figure 11-5. If the three primary colors are projected onto a screen at the same time, a spot of white light will be produced.

## MEASURING LIGHT

Many terms are used by scientists and engineers when working with light and lighting design. From a practical application standpoint, however, two terms must be understood: lumens and footcandles.

The *lumen* (lm) is the unit of measure of the rate of production or flow of light energy. If, for example, a room is to be made light enough for reading a book, the lamps in the room must produce a sufficient rate of flow of light energy to make the room bright. A car engine must produce a rate of energy flow or power to maintain a speed of 55 miles (88 km) per hour.

The candlepower or intensity of a light source must first be understood to gain a clear understanding of lumens and footcandles. *Candlepower*, measured in the unit *candela* (cd), is approximately the intensity of an ordinary burning wax candle. The lumen is the rate at which light energy falls upon a 1-ft$^2$ surface, placed 1 ft

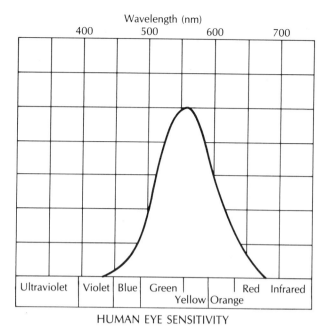

**Figure 11-4** The human eye is most sensitive to yellow light and green light.

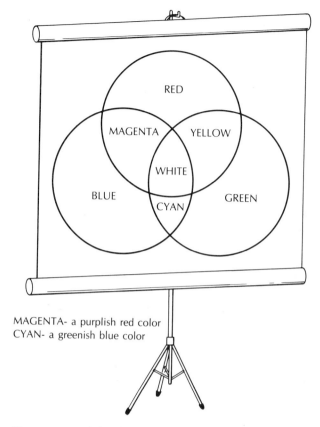

MAGENTA- a purplish red color
CYAN- a greenish blue color

**Figure 11-5** Mixing the three primary colors in the proper amounts will produce any other color.

from a 1-cd light source, Figure 11-6. A scientifically acceptable and calibrated light source is used by lighting manufacturers, but an ordinary candle serves to demonstrate the lumen.

It is more useful in lighting design to know the rate of production of light energy of a lamp than the candlepower of the lamp. Therefore, manufacturers publish in their lamp catalogs the lumen output of their lamps. Generally, lumen output decreases the longer a lamp is used. Manufacturers publish the initial or beginning lumen output of lamps and, often, the mean or average lumen output. This mean or average lumen output is a more realistic value for the purpose of lighting design.

The *footcandle* (fc) is a measure of the level of illumination of a surface. Adequate illumination is what we strive to achieve in order to have sufficient light to perform a task. Footcandle levels to perform various agricultural tasks are published by the American Society of Agricultural Engineers, and the Illuminating Engineering Society. A footcandle is the amount of illumination resulting when 1 lm of light falls upon a 1-ft² area. The surface shown in Figure 11-6 has a level of illumination of 1 fc.

A *footcandle meter* is used to measure the level of illumination, Figure 11-7. The meter is placed on the desired surface with the light-sensing element facing upward. If the illumination level is desired at the floor, then set the meter on the floor. Be careful not to let shadows affect the meter. Many illumination levels are specified at table or desk height. The meter must be placed at the appropriate level for the measurement. This level is often called the *work plane*.

The metric system of measurements uses the unit candela for light intensity, and the unit lumen for rate of

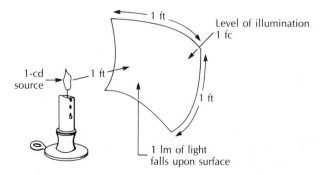

**Figure 11-6** One fc of illumination is produced when 1 lm of light falls upon a 1-ft² surface.

Unit 11  Farm Lighting  227

Figure 11-7 A footcandle (or lux) meter is used to measure the level of illumination of a surface.

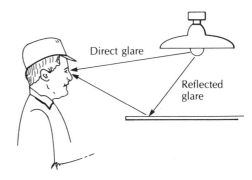

Figure 11-8 A lighting system should not cause glare.

flow of light energy. The *lux* (lx) is used as the unit of illumination. The lux is the level of illumination produced by 1 lm falling on a 1-m² surface. One lux is equal to 0.093 fc. One footcandle is equal to 10.8 lx.

## QUALITY OF LIGHT

Quality of light is often as important as quantity. Light quality is affected by the type of lamp and lighting fixture, the placement of the fixture, and the area to be lighted. First, consider reflectance and glare.

The reflectance of the surfaces in an area can have a significant effect upon the overall energy required to light the area. In general, it is desirable to have surfaces that are high in reflectance. Reflectance means that more of the light from the lamps will be reflected to the area to be illuminated rather than being absorbed and wasted. Dark colors absorb light, while pale colors reflect much of the light received.

*Glare* is bright light which hits the eyes directly. Glare can come directly from the lamp or it can be reflected from a shiny surface, Figure 11-8. Glare is not a problem in most agricultural buildings, unless people will be working for long periods of time in an area where glare occurs. Glare at a desk can cause such problems as eye fatigue and, sometimes, headaches.

*Contrast* is how well something stands out from the background. Black printing on white, pale green, or pale yellow paper is an example of good contrast. The words are easily visible. When the contrast is low between the object to be seen and its background, high levels of illumination are required. A high level of illumination (50 fc) is required in a dairy milking area so that the farm operator can detect certain milk disorders in a sampling taken before the milking machine is placed on the cow.

Extreme contrast created by a poorly designed lighting system can cause eye fatigue. An example is a desk lamp which aims light only at the desk surface. When all other lights in the room are turned off, the field of view that the eye sees now is the brightly lighted desk surface, and the darkness of the surrounding room. The eye fatigues rapidly under these conditions. The problem can be eliminated by providing some other lighting in the room. Sharp contrast of light and dark areas in a room can be a problem where people must work in the area for long periods of time. A small amount of general background lighting is important to reduce sharp contrast.

Shadows must be eliminated in primary seeing areas if a good lighting level is to be achieved. Lighting fixtures must be located so that light can reach the desired area without the operator creating shadows. If lighting fixtures are installed in the wrong locations, lamp wattage sometimes is increased to create more light in the desired area. This leads to wasted energy. With the proper placement of fixtures, the correct level of illumination can be achieved with the minimum energy input.

Color is another factor which affects the quality of light. The color of an object is determined by the colors of light reflected. An object can appear its natural color when illuminated by a white light source, such as the sun. A white light source contains all colors. Artificial light sources do not necessarily produce all colors. The true color of an object will be distorted if the light source does not produce a balance of colors.

A mercury-vapor lamp with a clear glass bulb produces very little red light. Red apples placed under this light will look brown. Obviously, clear glass mercury-vapor lamps should not be used to light a fruit stand at night. The wrong type of fluorescent lamps near a supermarket meat case can have a similar undesirable effect.

## LAMP LIFE AND EFFICACY

Lamp life varies widely among different types of light sources. Some lamps will last for years without burning out, while others only last a few weeks. Lamp life should be compared with lamp cost, light output, and cost or convenience of replacement when making lamp selection. Manufacturers test a large number of lamps to determine average lamp life. *Lamp life* is defined as the hours of operation until 45% of the lamps have burned out. An ordinary incandescent light bulb rated for 1 000 hours' life could burn out in less time or it could last longer. Manufacturers' lamp catalogs list the length of life for each lamp.

*Efficacy* of a lamp is light output divided by electrical power input. Lamp output is given in lumens. The lumen, being a measure of the rate of production of lighting energy is, therefore, a unit of measure of power (light output power). The *input power* is the power in watts drawn by the lighting fixture. Lamp efficacy is, therefore, measured in lumens per watt. The term *efficiency* is sometimes used. Efficiency is given in decimals, or percents, while *efficacy* is the proper term when using units such as lumens per watts. It is desirable for the lumens per watt to be as high as possible for a lamp and fixture.

Efficacy varies for different types of lamps, and for different sizes of the same type of lamp. The reason for this is because all of the electrical energy is not converted to visible electromagnetic radiation. In the case of the incandescent lamp, most of the electrical energy is converted to heat or infrared radiation, Figure 11-9. This is a phenomenon of physics that cannot be avoided. The efficacy of an incandescent lamp can be increased by designing the filament to operate at a higher temperature. This will produce more visible light and a smaller proportion of heat, but the filament will burn out sooner at the higher temperature. Lamp manufacturers, therefore, design incandescent lamps for the maximum efficacy, while maintaining a length of life of 750 hours to 1 500 hours.

## LIGHTING DESIGN

An adequate amount of light must be provided in an area to achieve an adequate level of illumination. If all the light produced by a lamp reached the surface to be illuminated, it would take 1 lm to produce 1 fc for each square foot ($ft^2$) of surface.

Consider, for example, an area of 50 $ft^2$. How many lumens are required to produce 1 fc of illumination?

$$50 \text{ ft}^2 \times 1 \text{ lm/ft}^2 = 50 \text{ lm}$$

But, what if 60 fc are needed? One lumen produces 1 fc; therefore, 60 lm produce 60 fc.

$$50 \text{ ft}^2 \times 1 \text{ lm/ft}^2 \cdot \text{fc} \times 60 \text{ fc} = 3\,000 \text{ lm}$$

All of the light from the lamp does not find its way to the surface to be illuminated; some light is trapped in the fixture and never gets out. Light is absorbed by the ceiling, walls, floor, and furnishings. Light is lost through windows. The lamp and fixture get dirty in time, and

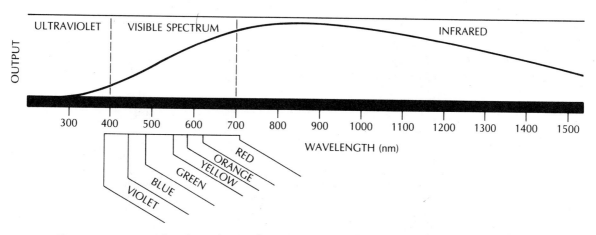

**Figure 11-9** Most of an incandescent lamp output is in the infrared portion of the spectrum.

light is absorbed by the dirt. The output of most lamps decreases with time.

Light-loss factors must be considered when determining the number of lumens that are required in an area to achieve the desired footcandles. As a general rule, only one-third to one-half of the light from the lamps will reach the desired surface. If walls and ceiling are highly reflective, and the ceiling is only about 10 ft (3.05 m) high, about one-half of the light will be available to illuminate the floor. A fixture with a good reflector which directs the light downward will also get half of the light to the floor. If walls and ceiling are not highly reflective, only about one-third of the light will reach the floor. See Equation 11.2.

> Lamp lumens required = Area × footcandles × LLF
> **Eq 11.2**

(LLF is light-loss factor.)
Reflective walls and ceiling, and clean area:

> LLF = 2

Not very reflective walls and ceiling, and not very clean area:

> LLF = 3 or more

Area is in square feet

## Problem 11-1

A milkhouse has an area of 300 ft² (27.9 m²), an 8-ft (2.44-m) ceiling, and highly reflective walls and ceiling. The area is considered clean. Determine the number of 2-lamp fluorescent fixtures required to obtain a level of illumination of 50-fc (540 lx) if each lamp has an output of 3 200 lm. Choose a light-loss factor of 2.

## Solution

> Lamp lumens required = 300 ft² × 50 fc × 2
> = 30 000 lm

or

> 27.9 m² × 540 lx × 2 = 30 000 lm

To find the number of lamps, divide the total lumens by the lumens per lamp.

> $$\frac{30\ 000\ \text{lm}}{3\ 200\ \text{lm/lamp}} = 9.4 \text{ lamps}$$
> Round this to 10 lamps

There are two lamps per fixtures. Therefore, five fixtures are required to achieve a level of illumination of 50 fc (540 lx).

This is only an approximate method for determining the number of fixtures. An illuminating engineer should be consulted when the approximate method is not considered adequate.

## INCANDESCENT LAMP

The incandescent lamp has a filament made of tungsten wire, Figure 11-10. The filament is heated white hot by passing electricity through it. The filament would burn if it were in air, but there is either a vacuum inside the bulb or it is filled with an inert gas. Nitrogen and argon are the most common filling gases.

Ordinary lamps have soft glass bulbs, and they must be protected from splashing water, rain, or snow. If liquid contacts the hot bulb, the bulb will shatter. Hard glass bulbs and shatterproof bulbs are available.

The filament must be provided with extra support

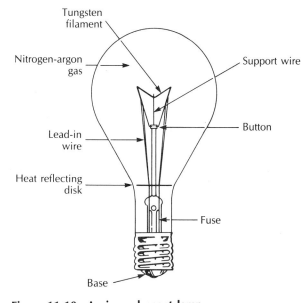

Figure 11-10 An incandescent lamp

when installed in an area where there is unusual vibration. An ordinary filament will fail if jolted with a mild blow or vibration. These types of lamps are designated "rough service" lamps.

An incandescent filament actually evaporates because it is usually heated to more than 4 600°F (2 538°C). The evaporated tungsten collects on the inside of the glass bulb, causing it to darken. This material absorbs light as the lamp gets older. Finally, the filament wears so thin that it cannot carry the current and it burns out. A light bulb which has severe inside blackening should be replaced because it is getting close to failure, and much of the light it produces is being absorbed.

The efficacy of an incandescent lamp ranges from less than 9 lumens per watt (9 lm/W) to more than 30 lm/W. Larger lamps have higher efficacy because their filament can be operated at a higher temperature without decreasing the life of the lamp. For a given filament, an increase in temperature increases efficacy and light output, but decreases the life of the bulb. Increasing the circuit voltage increases current flow, operating temperature, and light output. Table 11-2 shows the effect a change of voltage has on incandescent lamp light output, power use, and life. Table 11-3 shows the lumens, life, and efficacy for incandescent lamps, and other types of lamps commonly used in agriculture, as described in this unit.

Incandescent lamps are manufactured with voltage ratings of 120 V, 125 V, and 130 V. Table 11-2 shows that a 5% voltage reduction from 120 V to 115 V reduces light output by about 15%. But, the life of the lamp is almost doubled. When extended lamp life is desired, a 125-V or 130-V lamp can be used on a 115-V or 120-V circuit. Light output will be reduced. Therefore, this practice should be used only where there is plenty of light.

Several types of incandescent lamps used for different agricultural applications are shown in Figure 11-11. While common incandescent bulbs will shatter if water contacts them when lighted, spot and flood lamps with a

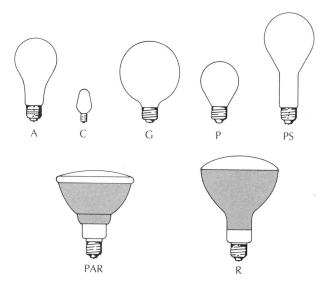

Figure 11-11 Incandescent lamp types used in agriculture

parabolic aluminized reflector (PAR) are made of hard glass suitable for unprotected outside use.

Incandescent lamps for common use have medium or mogul bases. The medium base is not permitted for lamps with a wattage above 300, *NEC Section 410-53*. Incandescent lampholders are not permitted to be connected to a circuit operating above 150 V between conductors except in industrial establishments, *NEC Section 210-6(a)*. Therefore, farm incandescent lamps are operated at 120 V.

## QUARTZ INCANDESCENT LAMP

The quartz incandescent tube is most frequently used outside when it is used for agricultural lighting. It has high efficacy for an incandescent lamp (see Table 11-3), and it makes objects appear their natural color. An important advantage is that its light output remains nearly constant throughout the life of the lamp. The filament is contained within a quartz glass tube, Figure 11-12. These lamps must be handled with care and must be kept

Table 11-2 An ordinary 120-V, 100-W incandescent lamp operating at overvoltage and undervoltage

| Voltage | Watts | Life (hours) | Lumens | Lumens/Watt |
|---------|-------|--------------|--------|-------------|
| 110 | 88 | 2 325 | 1 250 | 14.2 |
| 115 | 94 | 1 312 | 1 450 | 15.2 |
| 120 | 100 | 750 | 1 690 | 16.9 |
| 125 | 106 | 525 | 1 960 | 18.5 |
| 130 | 112 | 375 | 2 210 | 19.7 |

Table 11-3 Lumens, life, and efficacy for the lamps commonly used in agriculture for general illumination

| Watts | Color | Length (feet)† | Initial Lumens | Life (hours) | Efficacy (lumens/watt) |
|---|---|---|---|---|---|
| Incandescent, standard inside frosted | | | | | |
| 25 | | | 235 | 1 000 | 9 |
| 40 | | | 480 | 1 500 | 12 |
| 60 | | | 840 | 1 000 | 14 |
| 75 | | | 1 210 | 850 | 16 |
| 100 | | | 1 670 | 750 | 17 |
| 150 | | | 2 850 | 750 | 19 |
| 200 | | | 3 900 | 750 | 19 |
| 300 | | | 6 300 | 1 000 | 21 |
| 500 | | | 10 750 | 1 000 | 21 |
| 1 000 | | | 23 100 | 1 000 | 23 |
| Incandescent PAR-38 | | | | | |
| 150 Spot | | | 1 100 | 2 000 | 7 |
| 150 Flood | | | 1 350 | 2 000 | 9 |
| Quartz incandescent | | | | | |
| 500 | | | 10 550 | 2 000 | 21 |
| 1 000 | | | 21 400 | 2 000 | 21 |
| 1 500 | | | 35 800 | 2 000 | 24 |
| Fluorescent, rapid start | | | | | |
| 40 | CW & WW | 4 | 3 150 | 20 000 | 78 |
| 40 | CWX | 4 | 2 200 | 20 000 | 55 |
| 60 | CW & WW | 4 | 4 300 | 12 000 | 71 |
| 60 | CWX | 4 | 3 050 | 12 000 | 50 |
| 35 | CW & WW | 4 | 2 850 | 20 000 | 81 |
| 35 | CWX | 4 | 2 000 | 20 000 | 57 |
| Fluorescent, U-line | | | | | |
| 40 | CW & WW | 2 | 2 850 | 12 000 | 71 |
| 40 | CWX | 2 | 2 020 | 12 000 | 50 |
| Fluorescent, instant start | | | | | |
| 60 | CW & WW | 8 | 5 600 | 12 000 | 93 |
| 60 | CWX | 8 | 4 000 | 12 000 | 66 |
| 75 | CW & WW | 8 | 6 300 | 12 000 | 84 |
| 75 | CWX | 8 | 4 500 | 12 000 | 60 |
| 95 | CW & WW | 8 | 8 500 | 12 000 | 89 |
| 95 | CWX | 8 | 6 100 | 12 000 | 64 |
| 110 | CW & WW | 8 | 9 200 | 12 000 | 83 |
| 110 | CWX | 8 | 6 550 | 12 000 | 59 |
| 180 | CW | 8 | 12 300 | 10 000 | 68 |
| 215 | CW | 8 | 14 500 | 10 000 | 67 |
| 215 | WW | 8 | 13 600 | 10 000 | 63 |
| Mercury-vapor | | | | | |
| 40 | Deluxe | | 1 140 | 24 000 | 28 |
| 50 | Deluxe | | 1 575 | 24 000 | 31 |
| 75 | Deluxe | | 2 800 | 24 000 | 37 |
| 100 | Deluxe | | 4 300 | 24 000 | 43 |
| 175 | Deluxe | | 8 500 | 24 000 | 48 |
| 175 | Clear | | 7 900 | 24 000 | 45 |
| 250 | Deluxe | | 13 000 | 24 000 | 52 |
| 250 | Clear | | 12 100 | 24 000 | 48 |
| 400 | Deluxe | | 23 000 | 24 000 | 57 |
| 400 | Clear | | 21 000 | 24 000 | 52 |
| 700 | Deluxe | | 43 000 | 24 000 | 61 |
| 1 000 | Deluxe | | 63 000 | 24 000 | 63 |

(Table 11-3 Continued)

| Watts | Color | Length (feet)† | Initial Lumens | Life (hours) | Efficacy (lumens/watt) |
|---|---|---|---|---|---|
| Metal-halide | | | | | |
| 175 | | | 14 000 | 7 500 | 80 |
| 250 | | | 20 500 | 7 500 | 82 |
| 400 | | | 34 000 | 15 000 | 85 |
| 1 000 | | | 105 000 | 10 000 | 105 |
| High-pressure sodium | | | | | |
| 50 | | | 3 300 | 20 000 | 66 |
| 70 | | | 5 800 | 20 000 | 82 |
| 100 | | | 9 500 | 20 000 | 95 |
| 150 | | | 16 000 | 24 000 | 106 |
| 150* | | | 12 000 | 12 000 | 80 |
| 200 | | | 22 000 | 24 000 | 110 |
| 215* | | | 19 000 | 12 000 | 88 |
| 250 | | | 27 500 | 24 000 | 110 |
| 400 | | | 50 000 | 24 000 | 125 |
| 1 000 | | | 140 000 | 24 000 | 140 |
| Low-pressure sodium | | | | | |
| 35 | | | 4 800 | 18 000 | 137 |
| 55 | | | 8 000 | 18 000 | 145 |
| 90 | | | 13 500 | 18 000 | 150 |
| 135 | | | 22 500 | 18 000 | 166 |
| 180 | | | 33 000 | 18 000 | 183 |

*suitable for use in a mercury-vapor fixture
CW—cool white
WW—warm white
CWX—deluxe cool white

Deluxe—produces extra red light
Clear—produces very little red light

†1 ft = 0.3048 mm

clean. A quartz tube must never be handled with bare hands, because the oils on the skin will be absorbed by the quartz glass. Clean cotton or canvas gloves should be worn.

## ELECTRIC-DISCHARGE LAMP

The *National Electrical Code* often refers to electric-discharge lamps. These include fluorescent lamps, mercury-vapor lamps, and other similar lamps. Light is produced by passing electrical current from one electrode to another through a gas. Lightning is an electrical discharge with current flowing through the air. Electric-discharge lamps operate with the aid of auxiliary equipment, such as a ballast or a transformer. The ballast is a coil of electric wire which controls the flow of current through the lamp.

## FLUORESCENT LAMP

The fluorescent lamp is one of the most efficient types of lighting available. The fluorescent lamp is from

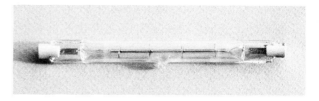

Figure 11-12 A quartz incandescent lamp has a tungsten filament supported inside a quartz tube.

three to four times more efficient than the incandescent lamp. That is, the fluorescent lamp produces from three to four times as much light for the same electrical energy consumed. The efficacy ranges from about 40 lm to 95 lm per watt (see Table 11-3).

A fluorescent lamp consists of a glass tube with a cathode at each end. The glass tube is filled with an inert gas. The glass tube also contains a few droplets of mercury that vaporize when the tube warms up. The inside surface of the tube is coated with a phosphor powder.

A fluorescent tube produces light when an electric current passes from one cathode to the other through the tube. As the electric current travels through the tube, it hits the mercury that has been vaporized, causing it to give off ultraviolet rays. When these ultraviolet rays strike the phosphor powder on the inside surface of the tube, the phosphor emits light, Figure 11-13.

All fluorescent lamps operate in this manner once they are started, but there are different ways to start a fluorescent lamp. There are three types of fluorescent lighting units, based upon their method of starting!

1. Preheat type—the lamp lights a few seconds after being turned on
2. Rapid-start type—the lamp lights almost instantly
3. Instant-start type—the lamp lights instantly

The preheat-type fluorescent lamp will not start until the cathodes are heated enough so that they can emit electrons. When the cathodes become heated, a current flows through the tube, Figure 11-14.

A starter is needed in order to preheat the cathodes. It can be a manual starter, such as the one on a desk fluorescent lamp, or it can be automatic. There are several types of preheat starter switches, but the one most widely used is called the "sealed-glow switch," Figure 11-15.

The following describes how the glow switch operates. Inside of the can-shaped enclosure of the starter is the glow-switch starter. It consists of an electrical capacitor and a bimetallic switch. The bimetallic switch is enclosed in a gas-filled glass bulb. The gas is usually neon or argon, the same as in the fluorescent tube.

When the lamp is turned on, the bimetallic strip begins to glow. When it glows, it heats up. As it heats up, the free end begins to move towards the stationary contact pole. Once the bimetallic strip makes contact with the stationary pole, electricity travels through the cathodes at each end of the tube. The current traveling through the cathodes causes them to heat up.

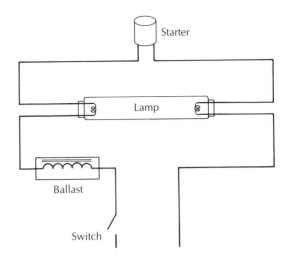

**Figure 11-14** Preheat-type of fluorescent circuit

When the bimetallic strip makes contact with the stationary pole, the bimetallic strip quits glowing. When this happens, the strip begins to cool. As it cools, the free end moves away from the stationary pole and the circuit is broken.

It takes only a few seconds for the glow switch to complete this sequence, but it is long enough to allow the cathodes in the fluorescent tube to heat enough so that an electric current will flow through the tube.

The ballast is another important part of each fluorescent lighting unit. Its basic function is to limit the amount of current flowing through a fluorescent lamp. When the current begins to flow through the lamp, electrical resistance through the tube decreases. The more the resistance decreases, the more the current flow increases, and the lower the electrical resistance becomes. The flow of current will keep on increasing until a fuse blows or the tube destroys itself, unless something is done to prevent the current from increasing out of control. The ballast is the part that keeps the electrical current under control.

The rapid-start fluorescent lamp does not require a starter. Instead, the cathodes are designed to heat quickly. The switch is turned on. Because the cathodes heat quickly, current begins to flow through the tube almost instantly. A special ballast is required on a rapid-start fluorescent lamp which will heat the cathodes quickly, and keep them heated while they are turned on, Figure 11-16.

The instant-start fluorescent lamp provides light instantly when the switch is turned on. Instead of depending upon preheating the cathode as with the other two

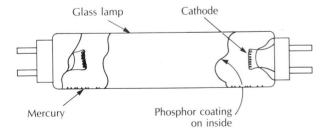

**Figure 11-13** Electricity flows from one cathode to the other, producing ultraviolet light which makes the phosphor glow to give visible light.

234  Unit 11  Farm Lighting

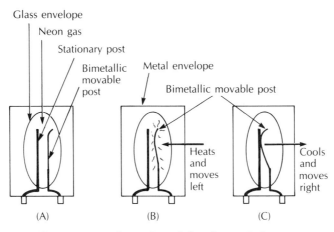

Figure 11-15 Operation of the glow-switch starter

types, it depends upon high voltage. A special transformer-type ballast provides this high voltage, Figure 11-17.

Fluorescent lamps come in several diameters. Generally, there is a code letter and number on the lamp, such as T12. The T means that the lamp is tubular, and the 12 is the diameter of the lamp in eighths of an inch (in). The T12 lamp is 12/8 in, or 1½ in diameter.

Fluorescent lamps are affected by cold weather. The fluorescent lamp performs best when the surface of the glass tube is at a temperature of 100°F (38°C). This occurs when the air temperature is from 70°F to 80°F (21°C to 27°C), Figure 11-18. Light output drops slightly as the temperature increases. But, when the temperature drops below 60°F (16°C), light output decreases rapidly. In a strong draft, light output of a fluorescent lamp can fall to 50%, at 50°F (10°C). Low-temperature starters and ballasts are usually necessary to start a fluorescent lamp when the temperature is below 50°F (10°C).

Protecting a fluorescent lamp in a tight fixture holds heat around the lamp. Lamps in these fixtures usually perform quite well at temperatures near or below 0°F (−18°C). Jacketed lamps are made for applications where the bare lamp must be exposed to the weather. The lamp is contained within a glass jacket which holds the heat next to the lamp.

The lamp life of an electric-discharge lamp is affected by the number of times the lamp is started. Each time the lamp starts, some of the special and necessary coating on the cathode is blown away. This material appears as a darkened area at each end of a fluorescent lamp. When this material is gone, the lamp cannot start; it just blinks. If this blinking continues for a long period of time, the ballast may become overheated. All ballasts should be thermally protected so they will shut off before they become dangerously overheated. Fluorescent fix-

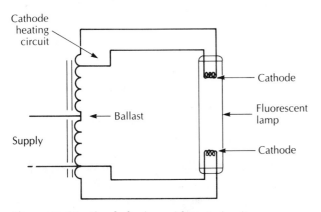

Figure 11-16 Simple basic rapid-start circuit

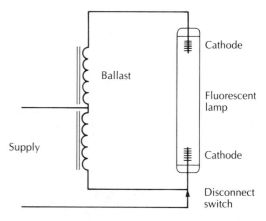

Figure 11-17 Simple basic instant-start (cold cathode) circuit

Unit 11  Farm Lighting  235

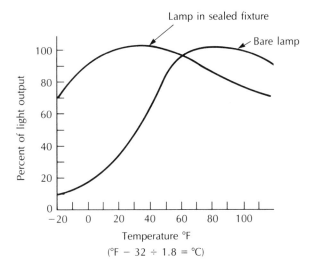

**Figure 11-18** The light output of a fluorescent lamp decreases as the air around the lamp gets colder.

tures installed inside are required to be thermally protected, *NEC Section 410-73(e)*.

Fluorescent lamps should not be turned on and off frequently. Incandescent lamps are best suited for that type of use. Fluorescent lamps should be installed in areas where the lights are generally left on for more than an hour at a time. A severely darkened area at the end of a fluorescent lamp is an indication that the lamp is near the end of its life. A slight darkening may appear early in lamp life, and should not cause concern.

A fluorescent lamp will produce great quantities of certain colors, depending upon the type of phosphors used to coat the inside of the tube. Some colors are emitted at a specific wavelength, as indicated by the tall spikes on the graph in Figure 11-19. If fluorescent light were separated into its colors, a complete rainbow would be visible as in the incandescent lamp. However, the red would be very bright for the incandescent lamp, and quite dim for the fluorescent. The fluorescent lamp would have some peculiar bright color lines on the rainbow. In the example in Figure 11-19, the color lines are violet, blue, green, and yellow.

Fluorescent lamps are available in varying shades of white, depending upon the type of phosphor in the lamp. They are also available in most colors. It is important to choose a fluorescent lamp which is compatible with the colors of the room to be lighted. A cool white lamp can make red look rather displeasing, because this lamp produces more blue light than red.

Colors can be grouped as warm colors (red, orange, yellow), and as cool colors (violet, blue, green). The words *cool* and *warm* are used to describe the various shades of fluorescent lamps. A warm lamp should be chosen for a room with warm colors.

Some lamps are marked "Deluxe," such as Deluxe Cool White. The word *deluxe* means that phosphors have been added which produce red light. At first, it might seem that the deluxe cool white lamp is good only for cool colors but, because of the added red-producing phosphors, this is a good all-purpose lamp for most colors.

If the greatest amount of light is the objective, and color is not important, the cool white or the warm white fluorescent lamps are the best choice. These lamps generally produce the most light for the electricity used.

Lamp color is printed on one end of the tube. A system of abbreviations has been developed to describe the colors. The following are some common colors (the X designates "deluxe").

CW—Cool White
CWX—Deluxe Cool White
WW—Warm White
WWX—Deluxe Warm White
W—White
D—Daylight

## Special Circuits

Most fluorescent lamps are operated by one of the circuits previously discussed: the preheat, rapid-start, or instant-start circuits. Variations of these circuits, however, are available for special applications.

**Dimming circuits.** The light output of a fluorescent lamp can be adjusted by maintaining a constant voltage on the cathodes, and controlling the current passing through the lamp. Devices such as thyratrons, silicon-controlled rectifiers, and autotransformers can provide this type of control. The manufacturer of the ballast should be consulted about the installation instructions for dimming circuits. Dimming control for fluorescent lamps requires a special dimming ballast.

**Flashing circuit.** The burning life of a fluorescent lamp is greatly reduced if the lamp is turned on and off frequently. Special ballasts are available which maintain a constant voltage on the cathodes, and interrupt the arc current to provide flashing.

**High-frequency circuit.** Fluorescent lamps operate with a higher efficacy at frequencies above 60 Hz. The gain in efficacy varies according to the lamp size and type. Solid-state ballasts provide a high-frequency output to the lamp; therefore, solid-state ballasts operate

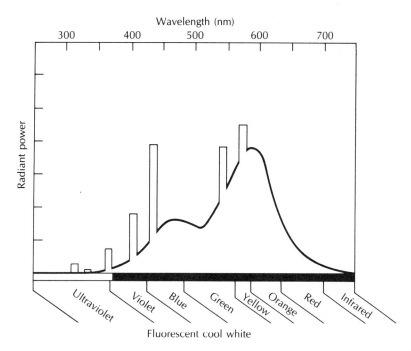

**Figure 11-19** The fluorescent lamp produces a rainbow of colors; especially blue, green, and yellow. The spikes indicate that extra amounts of violet, blue, green, and yellow are produced.

with a higher efficacy than conventional autotransformer ballasts.

*Direct-current circuit.* Fluorescent lamps can be operated on a dc power system if the proper ballasts are used. A ballast for this type of system contains a current-limiting resistor which provides an inductive kick to start the lamp.

## Class P Ballast

*NEC Section 410-73(e)* discusses thermal protection for fluorescent fixtures installed indoors. The National Fire Protection Association had reported that a frequent cause of electrical fires was the overheating of fluorescent ballasts. To lessen this hazard, Underwriters' Laboratories, Inc. established a standard for a thermally protected ballast, which is designated as a Class P ballast.

The Class P ballast has an internal protective device which is sensitive to the ballast temperature. This device opens the circuit to the ballast if the average ballast case temperature exceeds 90°C (194°F) when operated in a 25°C (77°F) ambient temperature. After the ballast cools, the protective device is automatically reset. As a result, a fluorescent lamp with a Class P ballast is subject to intermittent off-on cycling when the ballast overheats.

External fuses can be added for each ballast so that a faulty ballast can be isolated to prevent the shutdown of the entire circuit because of a single failure.

## Sound Rating

All ballasts emit a hum which is caused by magnetic vibrations in the ballast core. Ballasts are given a sound rating (from A to F) to indicate the intensity of the hum. The most quiet ballast has a rating of A. The need for a quiet ballast is determined by the ambient noise level of the location where the ballast is to be installed. For example, the additional cost of the A-rated ballast is justified when it is to be installed in a doctor's waiting room. In a bakery work area, however, a ballast with a C rating is acceptable. In a factory, the noise of an F ballast probably will not be noticed.

## Power Factor

The ballast limits the current through the lamp by providing a coil with a high reactance in series with the lamp. An inductive circuit of this type has an uncorrected power factor of from 40% to 60%; however, the power factor can be corrected to within 5% of unity by the addition of a capacitor. In any installation where there are to be a large number of ballasts, it is advisable to install ballasts with a high power factor.

## Stroboscopic Effect

Fluorescent lamps produce light only when electrical current is flowing. Sixty-Hz alternating current stops flowing for an instant 120 times every second. A fluorescent lamp, therefore, turns on and shuts off 120 times each second. This blinking is so fast that most people cannot detect it.

The blinking can sometimes be a problem when fluorescent lighting is used as the only source of light in an area where a machine with rotating parts is operating. The blinking fluorescent lamps may act like a timing light used to set the timing of the spark for a tractor or automotive engine. A rotating pulley or wheel may appear to be standing still even though it is turning very fast. This phenomenon is called the *stroboscopic effect*.

The stroboscopic effect of fluorescent lamps can be eliminated by installing 2-, 3-, 4-, or 6-lamp fluorescent fixtures. In these fixtures, two lamps are operated from a single ballast. The current flowing through each lamp experiences a slight phase shift so that it will not be zero in each lamp at the same time. Therefore, when one lamp is momentarily off, the outer is still lighted.

The stroboscopic effect can be minimized with a 3-phase system. If an area requires several circuits, then the A phase is used for some circuits, the B phase for others; and the C phase for still others. With this wiring arrangement, the blinking cannot cause a problem around rotating machinery.

Multilamp fluorescent fixtures are usually constructed with two lamps for each ballast. Two types of ballast circuits are in common use: lead-lag, and series-sequence. The lead-lag type is generally used for preheat and rapid-start lamps. The series-sequence circuit is common with instant-start fluorescent fixtures.

## Ballast Losses

A typical fluorescent ballast for two 40-W lamps draws a current of 0.8 A at 120 V for a total of 96 VA. The difference between the lamp rating of 80 W and the power input of 96 VA represents the ballast losses. These losses are both reactive and resistive, and will vary with the ballast type and manufacturer. To assure a sufficient load capacity and to simplify the calculations, an allowance of 100 VA is usually made in load calculations for a luminaire using two 40-W fluorescent lamps. Losses are lower with a solid-state ballast.

## Troubleshooting Fluorescent Lamps

Most fluorescent lamp problems can be found quite easily by following a few simple procedures.

1. Identify the type of fluorescent lamp (preheat, rapid-start or instart-start).
2. Make sure that the circuit is live and the switch is on.
3. Check to make sure that the lamp terminals are making good contact at both ends of the fixture.
4. If the lamp blinks or does not light at all, and the ends of the lamp are black, then the lamp needs replacing.
5. If the lamp has a starter and both ends of the lamp glow but the lamp does not light, then the starter is most likely defective.
6. If all of the previous steps are followed and the lamp still does not light, the problem can be the ballast. Sometimes the ballast will show signs of overheating; other times it will not. A voltmeter is necessary to make sure the ballast is receiving power, and that the proper voltage is delivered to opposite ends of the fixture. A noisy ballast is often defective and should be replaced.

## HIGH-INTENSITY DISCHARGE LAMP

The high-intensity discharge lamp (HID) is an important lighting source for inside and outside agricultural locations. There are three types of HID lamps: 1) mercury-vapor, 2) metal-halide, and 3) high-pressure sodium. Major advantages of these lamps are long life, high lighting efficacies, and little effect from ambient temperature.

## MERCURY-VAPOR LAMP

The mercury-vapor lamp is most widely used for outside lighting. However, it is ideal for inside lighting, particularly where the ceiling is 12 ft (3.66 m) or more in height. The mercury-vapor lamp is generally the least expensive HID type. Because it is an electric-discharge lamp, a ballast is required.

The mercury-vapor lamp consists of a quartz arc tube inside a glass bulb, Figure 11-20. A small amount of liquid mercury inside the arc tube is vaporized by the heat of the electric arc flowing from one end of the tube to the other. This electric arc causes the mercury to give off visible light and ultraviolet radiation. The ultraviolet radiation cannot escape, because it does not pass through the outer glass bulb. When the lamp is turned on, a small

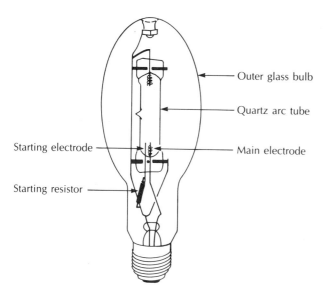

**Figure 11-20** Mercury-vapor lamp

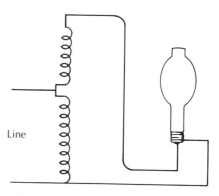

**Figure 11-21** A mercury-vapor, single-lamp circuit

arc is struck between the starting electrode and the main electrode next to it. The small arc ionizes* the argon gas to support an arc between the main electrodes. The heat from the main arc vaporizes the mercury and the lamp becomes brighter. It takes several minutes for the lamp to reach full brightness.

A major disadvantage of the mercury vapor lamp and other HID lamps is the problem of relighting the lamp once it has been turned off. If an HID lamp is turned off and then immediately turned on again, it will not relight immediately. The electric arc just cannot leap from one end of the arc tube to the other when the tube is hot; it must cool off first. This generally takes up to five minutes. The lamp will cool off faster and restrike sooner in cold weather. Because of delayed restrike time and slow warm-up, these lamps are best suited for applications where they are left on for long periods of time.

A typical mercury-vapor, single-lamp circuit is shown in Figure 11-21. The ballast acts as a transformer to provide the proper voltage to get the lamp started. Then the ballast acts to control the flow of current through the lamp. Occasionally, a ballast will fail. Signs of overheating are often visible when the ballast fails.

Mercury-vapor lamps have efficacies generally in the range of 25 lm to 65 lm per watt. Some energy is lost in the ballast, and a 175-W lamp and fixture draws about 200 W. As a general rule, the total fixture draws 10% to 15% more than the lamp wattage.

Color must be considered when selecting mercury-vapor lamps. The clear glass mercury bulb produces practically no red light. Red objects simply do not appear their natural color under these lamps. Some lamps have the inside of the glass bulb coated with a phosphor that produces red light. Much of the ultraviolet radiation produced by the arc tube is converted to red light by this phosphor coating. Clear and color-corrected mercury-vapor lamp outputs are compared in Figure 11-22.

Troubleshooting mercury fixtures can be confusing, but a clear understanding of the operation of the fixture makes the procedure easier. The problem is usually a burned-out lamp, a loose or broken wire, a bad photoelectric switch, or a burned-out ballast. As with all electric-discharge lamps, lamp life is shortened by frequent starts. Most mercury-vapor lamps have a rated life of 24 000 hours. This is about seven years for a yard light controlled with a photoelectric switch. The photoelectric switch must be positioned so that light will not be reflected into it at night. This will cause the light to turn off. The photoelectric switch has three wires: a neutral, which is white or gray; a hot input wire; and an output wire to the ballast. Shorting together the input and output wires (these wires are usually red or black) will bypass the photoelectric switch and the lamp should start. If it does not, then there is a problem other than with the photoelectric switch.

Mercury-vapor lamps produce short wavelength ultraviolet radiation (shorter than 300 nm) which is *dangerous to health*. This radiation is trapped inside the lamp because it cannot pass through the glass bulb. However, if the bulb is broken, it is possible for the lamp to remain lighted. A broken bulb allows the radiation to

---

*A gas, such as air, argon, and others, must be ionized before electricity will flow through it. The molecules of the gas are ionized when they become electrically charged. Air becomes ionized just before lightning strikes.

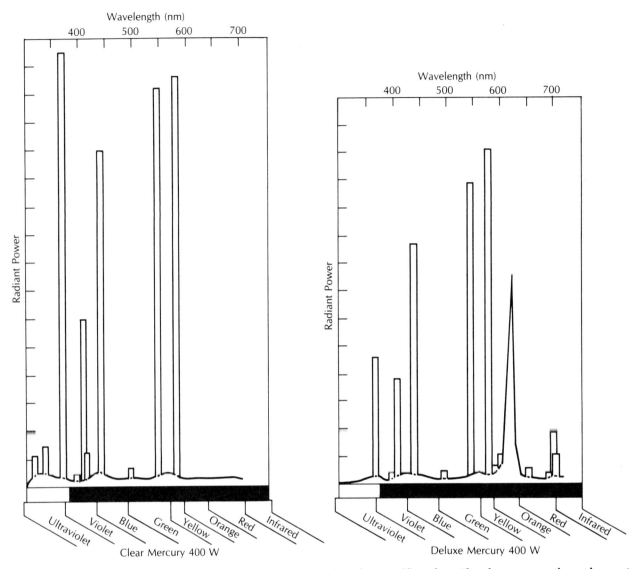

**Figure 11-22** Most of the light of a mercury-vapor lamp is produced as specific colors. The clear mercury lamp does not produce red.

escape, and *extended exposure to the rays can cause burning of the eyes and skin*. A mercury-vapor lamp must be turned off *immediately* if the glass bulb is broken. Safety lamps are available which burn out immediately if the glass bulb is broken.

Mercury-vapor lamps are available which can be screwed into a 120-V lampholder. The ballast is contained inside the lamp. The advantage is that incandescent lamps can be replaced with mercury-vapor lamps without rewiring. The disadvantages are that the self-ballasted mercury-vapor lamps are more expensive than standard mercury lamps. In addition, rated life is not as great, and efficacy ranges only between 10 lm and 25 lm per watt. Light output is about the same as the incandescent lamps they replace.

## *METAL-HALIDE LAMP*

The metal-halide lamp operates much like a mercury-vapor lamp. In fact, both are sometimes installed in the same fixture. This should not be done, however, without the recommendation of the fixture manufacturer. The major advantages of the metal-halide lamp are good color and high efficacy. Metal-halide lamps produce a balance of colors which makes colored objects appear nearly their true color. A major application of these lamps is sports field lighting. They would also be ideal for lighting a produce stand at night. Efficacy ranges from 80 lm to 105 lm per watt (see Table 11-3).

Metal-halide lamps have a shorter rated life than mercury-vapor lamps, and light output generally decreases rapidly as the lamps age. Metal-halide lamps are more costly than mercury-vapor lamps.

The metal-halide lamp derives its name from the materials used inside the quartz arc tube to produce the light. Three metals are used, in addition to mercury: sodium, thallium, and indium. These metals are combined with iodine, which is a member of the halide chemical family. Thus, the name, metal-halide lamp. Each manufacturer, however, uses a different trade name for these lamps.

The materials inside the arc tube tend to corrode the electrodes. This corrosive effect is minimized by disconnecting power to the starting electrodes inside the arc tube. This is done with a bimetallic switch, Figure 11-23. The bimetallic switch is activated by the heat of the lamp. The bimetallic switch must be in the upper end of the lamp when it is in the fixture. When ordering metal-halide lamps, it must be known whether the lamp will be installed with the base up or the base down. If the bimetallic switch does not open, the electrodes will corrode quickly and the lamp will burn out.

## HIGH-PRESSURE SODIUM LAMP

The high-pressure sodium lamp has an arc tube constructed of a translucent ceramic material. This arc tube contains xenon gas as well as the metals mercury and sodium, which produce light when vaporized with an electric arc. A typical high-pressure sodium lamp is shown in Figure 11-24. Lamp efficacy ranges from 60 lm to 140 lm per watt, and lamp life is rated at 24 000 hours (see Table 11-3). The lamp produces several colors, but an abundance of orange makes the lamp appear somewhat orange in color. Objects lighted by these lamps appear nearly their natural color. They are ideal for farm applications where high levels of illumination are required.

A special ballast is required for a high-pressure sodium lamp which will produce high voltage for starting. These lamps warm up more quickly than the other types. They restrike when turned off and on again in less than two minutes.

High-pressure sodium lamps are made for use in conventional farm mercury-vapor fixtures. They cost about three times as much as the mercury-vapor lamps they replace, and they produce about twice as much light. Efficacy of these special high-pressure sodium lamps is a little less than conventional high-pressure sodium lamps, and the rated life is only about one-half.

## LOW-PRESSURE SODIUM LAMP

Low-pressure sodium lamp efficacy ranges from about 130 lm to 180 lm per watt (see Table 11-3). The major drawback is color. Low-pressure sodium lamps

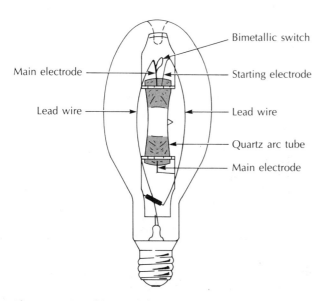

**Figure 11-23** This type of metal-halide lamp must be operated in the base-down position.

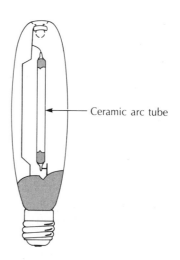

**Figure 11-24** High-pressure sodium lamp

produce only orange light and, therefore, objects do not appear their natural color. Rated lamp life is about 18 000 hours. Low-pressure sodium lamps are of the electric-discharge type requiring a special ballast and fixture. These lamps are suited for areas where high levels of illumination are required, and color is not important. They are frequently used for security lighting.

## *LAMP DATA*

The types of lamps discussed are not compared in Table 11-3. These are only the common lamps which are used for agricultural applications. A manufacturer's lamp catalog contains data on additional lamps as well as current prices. Selection of a lamp is often based upon:

- Lamp characteristics such as restrike time, warm-up time, suitability for cold weather, and so forth
- Light output and efficacy
- Rated life
- Colors produced
- Lamp replacement cost

Lamp manufacturers have illuminating engineers available to assist with the design of large or unique lighting systems.

The values for efficacy of the lamps listed in Table 11-3 are based upon initial lumens, and ballast losses are not considered for the electric-discharge lamps. Overall efficacy based upon average light output and including ballast losses for the electric-discharge lamps could be as much as 20% lower than the figures listed in Table 11-3.

## *LIGHTING FIXTURES*

The purposes of a lighting fixture are to hold the lamp, provide the correct power to the lamp, protect the lamp, and direct the light in the proper direction. Electric lamps and auxiliary equipment, such as transformers, must be considered a source of heat. An appropriate fixture must be selected so that a fire hazard is not created near combustible materials. Fixtures must be constructed, installed, or guarded so that combustible material will not be heated to a temperature above 194°F (90°C), *NEC Section 410-5*.

Enclosed fixtures are required in many agricultural locations to keep out dust and moisture. *NEC Section 410-4(a)* calls attention to some barns as areas where moisture is a problem and fixtures must be marked "suitable for wet location." Sealed fixtures are available for all types of lamps suitable for damp, wet, and even corrosive locations. Noncorrosive sealed fixtures must be used in farm buildings where areas are sprayed with water for cleaning, such as dairy milking areas, and some areas of hay barns.

## *AGRICULTURAL BUILDING ILLUMINATION*

The level of illumination or footcandles for agricultural buildings is dependent upon the use of the area. Animal feeding areas should be lighted to help increase feed intake, and reduce the chance of injury to animals. Sufficient light must be provided around machinery to help prevent injury. Areas where farm produce is cleaned and graded must be provided with adequate light. A particularly critical area is any location where equipment must be sanitized, such as a dairy. An area where too much light is undesirable is a poultry egg-laying house. Lighting requirements for specific agricultural locations are discussed in later units.

## *LIGHTING CIRCUIT LOADS*

Lighting fixtures and lampholders installed inside agricultural buildings usually are not of the heavy-duty type (660 W or more). Therefore, circuits for lighting fixtures are required to be rated at 15 A or 20 A, *NEC Section 210-21(a)*. The number of circuits required depends upon the number of lighting outlets which can be installed on a circuit. First, the current drawn by each lighting outlet or fixture must be determined. The current rating of an electric-discharge fixture is printed on the ballast. The fixture ampere rating must be used, not the wattage rating of lamps, *NEC Section 210-22(b)*. A 4-lamp fluorescent fixture has two ballasts. Add together the current rating of each to get the fixture total current rating. A ballast with a high power factor draws less current than one which does not have a power factor correcting capacitor. The less current drawn by the fixture, the more fixtures that can be installed on a circuit.

The load of an incandescent lampholder depends upon the maximum rated lamp which is permitted in the fixture. The load on open lampholders depends upon the maximum lamp wattage likely to be installed. For a medium-base, screw-shell lampholder the maximum wattage is 300 W. At 120 V, this is a load of 2.5 A. A smaller load can be used, provided there is justification that a larger load is not likely to be installed. A case in point is a poultry egg-laying house where the maximum lamp wattage used is 60 W, and the typical lamp wattage used is 25 W or 40 W.

Assume that 2-lamp, 40-W fluorescent fixtures are rated at 0.84 A. How many of these fixtures are permitted to be installed on a 15-A circuit? Most lighting loads are considered to be continuous loads. That is, they are often on for three hours or more at a time. Under these conditions, the circuit can be loaded only to 80% of its rating, *NEC Section 210-22(c)*. This is 12 A maximum load for a 15-A circuit, and 16 A for a 20-A circuit.

$$0.80 \times 15 \text{ A} = 12 \text{ A}$$
$$0.80 \times 20 \text{ A} = 16 \text{ A}$$

The fluorescent fixtures which draw 0.84 A each are usually to be operated for more than three hours; therefore, this must be considered a continuous load. The number of fixtures per circuit is determined by dividing 12 A by the ampere rating of the fixture.

$$\frac{12 \text{ A per circuit}}{0.84 \text{ A per fixture}} = 14.3 \text{ fixtures per circuit}$$

This number is rounded down to the whole number 14. Therefore, 14 of these fixtures may be installed on a 15-A circuit. For a 20-A circuit, the maximum permitted number of these fixtures would be 19.

$$\frac{16 \text{ A per circuit}}{0.84 \text{ A per fixture}} = 19 \text{ fixtures per circuit}$$

## Problem 11-2

A beef farmer provides light inside his feeder barn to stimulate animal growth. The fixtures are 400 W mercury-vapor type, each drawing 1.9 A at 240 V. The number of fixtures is 24. Determine the minimum number of 20-A circuits required.

## Solution

First, determine the total connected lighting load.

$$24 \text{ fixtures} \times 1.9 \text{ A per fixture} = 45.6 \text{ A}$$

This is considered a continuous load; therefore, the 20-A circuit is permitted to be loaded only to 16-A. Divide 16 A into the total connected load to find the minimum number of circuits.

$$\frac{45.6 \text{ A}}{16 \text{ A per circuit}} = 2.85 \text{ circuits}$$

The number 2.85 is rounded up to the next whole number. Therefore, three circuits are required to supply this lighting load.

## WIRING INCANDESCENT FIXTURES

Incandescent lighting fixtures are mounted directly over a junction box. Some fixtures can be mounted directly to the junction box with screws, while others require a fixture strap or a fixture stud, Figure 11-25.

All lighting fixtures must be securely fastened in place, *NEC Section 410-15*. Therefore, the fixture must be fastened to a suitable surface or to structural supports of the building. It is important for the box to the securely fastened in place, particularly if it will be required to support a fixture, *NEC Section 370-13*. A fixture weighing up to 50 lb (22.7 kg) may be supported by a box, provided the box is adequately supported, *NEC Section 410-16(a)*. Fixtures heavier than 50 lb (22.7 kg) are required to be secured independent of the box. A fixture stud may be secured to the hanger, not the box, and is capable of supporting weights up to the limit stated by

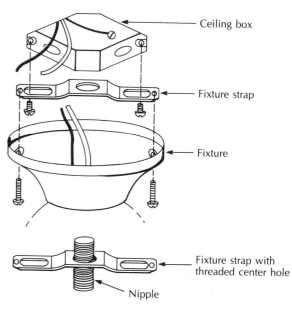

**Figure 11-25 A fixture strap is often used to mount lighting fixtures.**

Unit 11  Farm Lighting  243

the manufacturer. Unusually heavy fixtures require construction of suitable support brackets or frames attached directly to the structural frame of the building.

Pendant-type incandescent fixtures are permitted to be supported by a stranded, rubber-covered cord not smaller than No. 18 AWG, Figure 11-26. The cord must be connected to the outlet box in a manner which will not put strain on the wire splices, *NEC Section 410-27*. A strain-relief connector or a knot in the wire can provide proper support. The wire splice at the junction box in the ceiling is required to be soldered.

Recessed incandescent fixtures can pose a *fire hazard* if they are not of a special design or if not installed properly. The main problems are to prevent the fixture from making contact with flammable material, and to prevent thermal insulation from covering the fixture. Thermal insulation traps heat in the fixture. Insulation is not permitted within 3 in (76 mm) of the fixture, *NEC Section 410-66*. There is an exception in the case of fixtures designed, tested, and identified for direct contact with thermal insulation. Fixtures must be kept at least ½ in (12.7 mm) from combustible materials, such as ceiling tile and wood joists. Proper installation of a recessed incandescent fixture is shown in Figure 11-27.

Recessed incandescent fixtures are required to be thermally protected, *NEC Section 410-65(c)*. The thermal protector disconnects power to the lamps in the event of potentially dangerous overheating. Fixtures are required to indicate if they are thermally protected. Fixtures which have been designed, tested, and identified

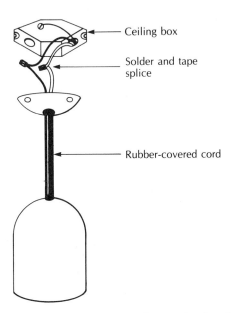

**Figure 11-26** A pendant fixture is permitted to be supported by a rubber-covered flexible cord not smaller than No. 18 AWG.

for direct contact with thermal insulation are not required to be thermally protected.

The circuit wires entering a recessed lighting fixture must not become overheated. Some fixtures are designed to remain cool enough that wire with insulation rated at 60°C (140°F) is suitable. The lighting fixture manufacturer is required to give the temperature rating of the fixture wiring box if the minimum acceptable tempera-

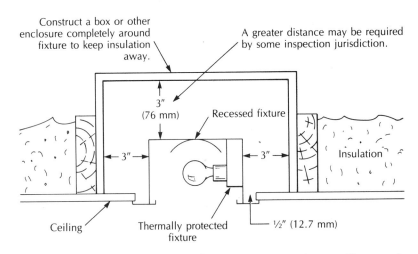

**Figure 11-27** A recessed incandescent lighting fixture must be installed so that it will not make contact with thermal insulation.

ture is more than 60°C (140°F). Branch-circuit wires are not permitted to run through junction boxes provided with the recessed fixture, unless specifically permitted by the manufacturer, *NEC Section 410-11*. Only the wires required to supply the fixture should enter the box.

## WIRING FLUORESCENT FIXTURES

Fluorescent fixtures supplied by the manufacturer are prewired and ready to mount in place and connect to power. Many fixtures are not corrosion resistant and, therefore, are not suitable for installation in damp agricultural locations. Suitable fixtures for the environment are required by *NEC Section 410-4*.

A ballast is essential auxiliary equipment for all electric-discharge lamps. The ballast need not be in the fixture. If the ballast is not contained within the fixture, it must be accessible and installed in a noncombustible enclosure, *NEC Section 410-54*. The ballast produces heat and, therefore, it must be located where the heat will escape.

Fluorescent fixtures attached directly to the ceiling surface may have circuit wires supplied directly to the fixture or to a junction box above the fixture. Figure 11-28 shows a fluorescent fixture mounted over a junction box in the ceiling. A hole in the back of the fixture is required in order to provide adequate access to the interior of the junction box without removing the fixture, *NEC Section 410-14(b)*. It may be necessary to make an opening of adequate size with a hole saw or knockout punches. The fixture must be installed so that it completely covers the box, *NEC Section 410-12*. Fluorescent fixtures can be installed end to end to form a continuous row, Figure 11-29. Only one junction box is required to supply all the fixtures in the row.

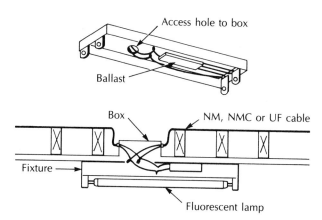

**Figure 11-28** Fluorescent fixture mounted directly over a junction box

The area within 3 in (76 mm) of the ballast requires wire with a 90°C (194°F) rating. When supplying fixtures in a continuous row, it is not possible to supply other fixtures without the wire passing within 3 in (76 mm) of the ballast. Examples of types of wire permitted for this application are listed in *NEC Section 410-31*. Type THW insulation is considered to have a 90°C (194°F) rating for this application only, as stated in *NEC Table 310-13*.

The circuit wires running through fixture enclosures are only permitted to supply the fixtures in the row and no other outlets, *NEC Section 410-31*. If other outlets are to be served, a junction box must be provided. A cable or raceway may be attached directly to a fluorescent fixture. Examples of typical installations are shown in Figures 11-30 and 11-31. It must be remembered that if Types NM, NMC, and UF cable contain wire with a 60°C (140°F) rating, the cables are not permitted within 3 in (76 mm) of the ballast.

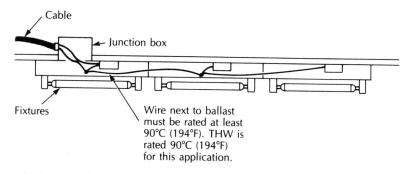

**Figure 11-29** A continuous row of fluorescent fixtures supplied from one junction box

Unit 11 Farm Lighting

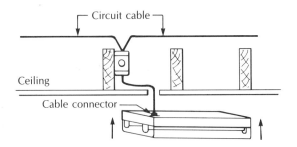

**Figure 11-30** A surface-mount fluorescent fixture can be supplied with cable or raceway.

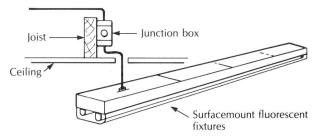

**Figure 11-31** One supply wire may serve several fixtures in a continuous row. However, the circuit wire passing by the ballast must be rated at 90°C (194°F).

Recessed fluorescent fixtures are available to fit into 2 ft × 4 ft (0.61 m × 1.22 m) and 2 ft × 2 ft (0.61 m × 0.61 m) suspended ceilings. These fixtures are required to be fastened directly to the ceiling framing members (T bars), *NEC Section 410-16(c)*. Screws, bolts, rivets, or approved clips are used to fasten the fixture to the T bars. The fixtures are usually wired by means of flexible metal conduit, as shown in Figure 11-32.

Industrial-type florescent fixtures may be wired by means of raceway or they can be cord connected to an overhead outlet, Figure 11-33. If the fixture is cord connected, the outlet must be directly above the fixture. The cord must be completely visible for its entire length, *NEC Section 410-30(c)*. The overhead receptacle is required to be of the grounding type.

High-intensity discharge fixtures installed in agricultural areas are usually supported by a rigid metal conduit or an IMC nipple with threaded hubs between the fixture and the box, Figure 11-34. The box may be fastened directly to the ceiling or structural support. The box is permitted to be secured by conduit entering opposite

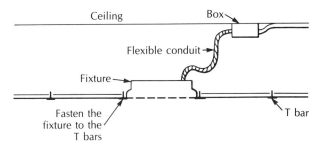

**Figure 11-32** Installation of a fluorescent fixture in a suspended ceiling

sides where the conduit is fastened within 18 in (457 mm) of the box, *NEC Section 370-13*.

## *LIGHTING CONTROLS*

Proper control of a lighting system is very important for many agricultural applications. Typical control devices are snap switches, photoelectric switches, time clocks, and lighting contactors.

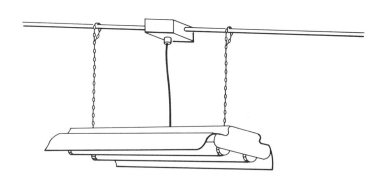

**Figure 11-33** An industrial fluorescent fixture, supported by chains, with cord to receptacle above

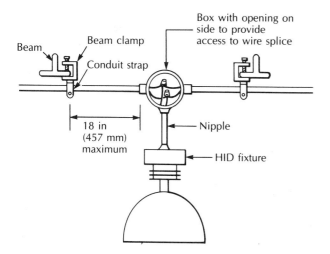

Figure 11-34 Installation of a high-intensity discharge fixture

## Snap Switches

Ordinary snap switches have rating of 15 A and 20 A. Double-pole switches are available to control two circuits with one switch. Snap switches provide only manual control. Automatic control is required in many situations.

## Photoelectric Switches

Lights which will be operated all night are simply operated with a photoelectric switch. These switches come as an integral part of a single fixture or they can be installed separately to control several fixtures. A photoelectric control must be selected with an ample wattage or ampere rating to handle the load.

## Time Clocks

Time clocks come in many different types. The most popular type are 24-hour timers, and 7-day timers, Figure 11-35. The 24-hour timer can turn lights on and off several times during a single day, if desired. Some 24-hour timers have a skip-a-day feature. Lights can be controlled Monday through Friday, for example, and set to skip Saturday and Sunday. Time clocks have self-contained switch contacts rated at 20 A or more per contact. A double-pole time clock has two contacts and can control two circuits. A 4-pole unit can control four circuits at

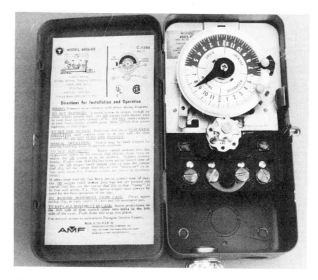

Figure 11-35 A 24-hour time clock control can be used to operate lighting circuits in an agricultural building.

the same time. A time clock has a manual switch inside so that the farm worker can turn lights on and off, if desired.

## Lighting Contactors

Lighting contactors are useful when many circuits must be controlled at the same time. A lighting contactor is a magnetically operated switch, Figure 11-36. The

Figure 11-36 A magnetically operated lighting contactor can control several circuits at the same time.

Unit 11 Farm Lighting 247

magnetism created by passing current through a coil closes the switch contacts. Contactors which control up to twelve circuits rated at 20 A for each contact are common. The contactor can be controlled with a snap switch, time clock, photoelectric switch, or other control device. A typical circuit diagram for a poultry house is shown in Figure 11-37. Four circuits are controlled with a 4-pole lighting contactor operated by a time clock.

## Dimmer Controls

Dimmer controls for incandescent lamps have few practical applications for agriculture except for decorative lighting. Dimmer controls are suggested on occasion for poultry lighting. Dimmer controls must have an ampere or a wattage rating sufficient to handle the intended load. The dimmer produces heat when it is used to control an entire circuit. An aluminum faceplate with fins is often needed to keep the dimmer from overheating.

Dimmer controls are variable resistors, variable transformers, or fast-acting electronic switches. Most common dimmers are of the electronic type. They turn the electricity on and off 120 times each second, Figure 11-38. Turning the dial controls the length of time the circuit is on.

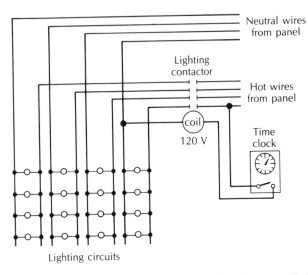

Figure 11-37 Diagram of four 20 A circuits controlled with a lighting contactor operated by a time clock

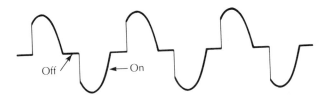

Figure 11-38 An electronic dimmer switch turns the electrical supply on and off 120 times each second. The result is that part of the ac sine wave is missing.

## REVIEW

Refer to the *National Electrical Code* when necessary to complete the following review material. Write your answers on a separate sheet of paper.

1. The length of the lighted day affects the
    a. flowering of some plants.
    b. growth rate of some breeds of cattle.
    c. start of egg laying of chickens.
    d. All of these

2. Light is only a small part of the total
    a. sonic spectrum.
    b. electromagnetic spectrum.
    c. lumatomic spectrum.
    d. magnetic spectrum.

3. Visible light varies in wavelength from 400 to 700
    a. millimeters.
    b. nanometers.
    c. angstroms.
    d. meters.

4. An object will appear the color of light
   a. reflected.
   b. absorbed.
   c. transformed.
   d. adsorbed.

5. A unit of measure of illumination is the
   a. candela.
   b. footcandle.
   c. lumen.
   d. watt.

6. One projector shines red light on a screen, and another shines green light on a screen so the two colors overlap, producing the new color:
   a. Yellow
   b. Blue
   c. Magenta
   d. Cyan

7. The human eye is most sensitive to which of the following colors?
   a. Green and blue
   b. Violet and red
   c. Yellow and green
   d. Orange and blue

8. Lamp efficacy is measured in
   a. watts per lumen.
   b. lumens per hour.
   c. lumens per watt.
   d. watts per hour.

9. Name two major factors of light quality that can cause eye fatigue.

10. An animal resting barn has a floor area of 7 500 ft$^2$ and will be lighted to 5 fc using 150-W incandescent lamps in fixtures that direct the light downward. If 2 lm per square foot are required to get 1 fc (light loss factor = 2), and each lamp has an output of 2 850 lm, the total number of lamps required is:
    a. 14
    b. 20
    c. 26
    d. 32

11. Name two examples of electric-discharge lamps.

12. Operating voltage is stamped on the top of a filament lamp. Which of the following lamps of the same wattage and rated life should last the longest if operated at 120 V?
    a. 115-V lamp
    b. 120-V lamp
    c. 125-V lamp
    d. 130-V lamp

13. The type of lamp for an application is often chosen on the basis of high efficacy because
    a. the operating cost will be lower.
    b. the energy consumed by the lamp is lower.

c. fewer fixtures may be required.
d. All of these

14. A reflector floodlight suitable for use outside exposed to the weather is a lamp with which of the following types?
   a. G
   b. R
   c. PAR
   d. PS

15. A person installing a quartz incandescent tube in a fixture should wear
   a. sun glasses.
   b. shockproof gloves.
   c. clean cotton or canvas gloves.
   d. a hat.

16. A fluorescent lamp on an electric range is started by pressing the start switch and holding it in for a few seconds until the ends of the lamp begin to glow. When the hand is removed from the switch the lamp will light. This fluorescent lamp is which of the following types?
   a. Rapid-start type
   b. Automatic-start type
   c. Instant-start type
   d. Preheat-start type

17. A T10 fluorescent lamp has a tube diameter of:
   a. 1 in
   b. 1¼ in
   c. 1½ in
   d. 1¾ in

18. Which part (A, B, C, or D) of the following diagram represents a rapid-start ballast?

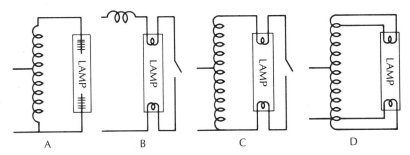

19. Does the light output of a fluorescent lamp increase or decrease as the surrounding temperature drops below 60°F (16°C)?

20. A deluxe fluorescent lamp or deluxe mercury-vapor lamp is one which
   a. always produces more light than a nondeluxe lamp.
   b. produces extra amounts of red light.
   c. has a longer life than a nondeluxe lamp.
   d. is manufactured from a more durable glass.

21. A fluorescent lamp ballast installed inside is required by the *NEC* to be:
   a. Thermally protected

    b. A class P ballast
    c. Both A and B
    d. None of these

22. Florescent fixtures with the same lamp wattage rating are available with high power factor ballasts, and low power factor ballasts. Which of the following statements is correct?
    a. More fixtures with high power factor ballasts can be installed on a circuit than with low power factor ballasts.
    b. More fixtures with low power factor ballasts can be installed on a circuit than with high power factor ballasts.
    c. The fixture with the high power factor ballast draws more current.
    d. The fixtures both draw the same current.

23. The major advantage of the high-pressure sodium lamp is its
    a. best color balance of any lamp.
    b. longer lamp life as compared to a mercury-vapor lamp.
    c. bright light the instant the switch is turned on.
    d. very high lighting efficacy.

24. A building contains twelve fluorescent fixtures which draw 1.65 A at 120 V. Under normal conditions the lights are operated six hours per day. The minimum number of 20-A lighting circuits is:
    a. 1
    b. 2
    c. 3
    d. 4

25. Unless designed and tested for direct contact with thermal insulation, the insulation is required to be kept at least 3 in (76 mm) away from the fixture, according to which *NEC* section?

# UNIT 12
# STANDBY ELECTRICAL POWER SYSTEM

## OBJECTIVES

After studying this unit, the student will be able to

- list the essential parts of a standby electrical power system.
- choose the proper size of engine for an alternator.
- explain the purpose and operation of a transfer switch.
- select a standby alternator for a farm or home.
- size equipment and install a standby electrical power system.

Electricity is essential to farm operation and family living. Ventilating fans, water pumps, household furnaces and refrigerators, and other vital equipment require uninterrupted electrical service. Storms, accidents, or equipment failure at times can cause power outages. A power outage that lasts overly long can cause suffocation of animals in controlled environment shelters, food spoilage, frozen pipes, and loss of production. A properly sized, installed, and maintained standby electric power system can eliminate most of the financial loss and inconvenience resulting from a power outage.

## EQUIPMENT

A *standby electrical power* system consists of an alternator, an engine or tractor to power the alternator, and a transfer switch. The transfer switch provides a safe way to connect the alternator to the farm or home wiring. Electrical power suppliers and the *National Electrical Code* require a method of electrical load transfer which will prevent the accidental interconnection of the alternator to the power lines. Generally, this is accomplished with a transfer switch. In addition to the necessary equipment, a power failure alarm may be desirable where an outage cannot be tolerated for more than a short time. A tractor-operated standby power system for a farm is shown in Figure 12-1.

## ALTERNATOR

An *alternator* is an electric generator which produces alternating current (ac). Alternators for use on farms are available with an engine or they can be powered with a tractor. The investment is lower for tractor-powered units, but they must be connected before every use.

*Alternator size* refers to the electrical power output. Ratings are expressed in kilowatts (kW). Some alternators for farm use have two kW ratings; for example, 45/25. The higher rating (45) is the short-time overload capacity. The lower rating (25) is the continuous rated

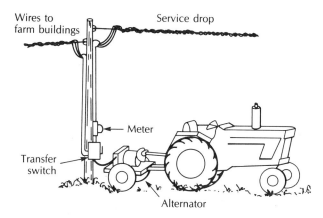

Figure 12-1 Tractor-powered alternator and transfer switch at a farm central distribution pole

output. Electric motors draw more current when they start than when running. An alternator must have sufficient overload capacity for starting motors.

The alternator must be matched to the type of electrical service and voltage. Most farm and home electrical systems are single phase, 120/240 V. Small, portable engine-alternator units generally have outputs rated at 120 V. Some farms and businesses are served with a combination of single and 3 phase power. Generally, one alternator can be selected to provide both types of service.

Rotation speed of the armature of the alternator depends upon design. Small alternators usually operate at 3 600 revolutions per minute (r/min). These are 2-pole alternators. Large alternators have four poles and, therefore, turn only half as fast (1 800 r/min). The alternator must operate at the proper speed, or the cycles per second (60 Hz) and voltage will not be correct.

Alternators must be protected against overload. It is easy to turn on more load than the alternator is designed to handle. Properly sized, factory-installed circuit breakers or fuses on the alternator will protect against alternator damage due to overloads.

Some alternators have a commutator and brushes, and others are solid state. Direct current is required to create a magnetic field necessary for alternator operation. A commutator and brushes are required on many alternators to obtain the direct current. With some models, however, diodes are placed on the rotating armature and brushes are not necessary. A commutator and brushes require maintenance, but diodes are sensitive and subject to electrical damage.

Proper voltage is essential for safe operation of electrical equipment, particularly motors. The alternator should be equipped with a voltmeter to make sure the output voltage is within acceptable limits.

## ENGINES AND TRACTORS

The engine used to power the alternator must be of the proper size. The engine or tractor should develop 2½ hp for each kilowatt of electric power produced by the alternator. An alternator with a continuous 30-kW rating would require a 75-hp engine. For alternators larger than 75 kW, 2 hp per kilowatt is usually sufficient.

The alternator can be operated from the power-take-off (PTO) of a tractor. This has the advantage of lower cost at the time of the initial investment, and it eliminates the care and maintenance of an additional engine. However, during a power outage, the tractor powering the alternator will not be available for other uses. Trailer mounting of the alternator provides portability. Used with an arc welder and power tools, farm equipment can be repaired in the field.

Alternators are designed to operate at a constant fixed speed; therefore, the tractor must have a tachometer to indicate when the correct speed has been obtained.

The direct-connected, engine-driven unit is ready to operate at a moment's notice. Because it is a complete unit, it can be designed with automatic starting and automatic transfer switch if the need warrants the extra expense. It is important to test these units periodically to make certain the engine will start in an emergency.

The choice of engine is important for greatest efficiency and least care. For alternators up to 15 kW, an air-cooled engine is recommended because of the lower maintenance requirements. For alternators larger than 15 kW, a liquid-cooled engine is usually preferred. Gasoline, LP gas, and diesel engines are available for powering standby alternators. LP gas generally burns cleaner than gasoline; this tends to promote longer engine life and reduced frequency of maintenance. Gasoline forms a varnish in the system when in storage for long periods of time, which may keep the engine from starting in an emergency. This problem can be prevented with proper maintenance. Diesel units are costly initially, except in the very large engines, but operating and maintenance costs are low. Generally, it is best to select an engine which uses the fuel type most commonly available on the farm.

ing feedback of alternator voltage on power lines, which would endanger the life of a person working on the lines. It also eliminates the possibility of the alternator being damaged when normal power is restored, thus protecting the alternator investment.

The transfer switch, when located at the central distribution pole or at the main service, Figure 12-2, will serve any electrical load on the farm. Seldom is the standby alternator sized large enough to operate all of the usual loads; therefore, all but essential equipment should be turned off. The transfer switch must be sized according to the rating of the service equipment and wires where it is installed. Common sizes are 100 A, 200 A, and 400 A. Sometimes, the lower transfer switch contacts for the alternator are adequately sized to handle alternator current, but smaller than the service contacts. The transfer switch can be located at the service entrance of a particular building, but only that building will be served by the alternator. It is also possible to wire one alternator to several circuits within a building, to be connected to the transfer switch if only those circuits need to be powered during an outage. The inside of a transfer switch is shown in Figure 12-3.

A cord must be provided at the transfer switch to connect to the portable alternator. A cabinet can be provided at the transfer switch to store the alternator cord.

Figure 12-2　A transfer switch located at a central distribution pole, with a cabinet for storing the alternator wires

## TRANSFER SWITCH

A standby alternator must be connected to the wiring in a way that will prevent the accidental interconnection of the alternator and the power lines, *NEC Sections 702-6* and *230-83*. A double-throw transfer switch is generally the most practical way to connect the alternator to the wiring. The transfer switch will keep the alternator isolated from the power lines at all times, thus prevent-

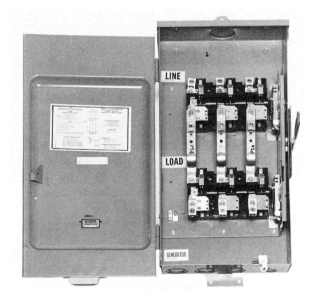

Figure 12-3　Terminals are clearly marked in a UL listed transfer switch. A neutral terminal usually must be ordered separately if a neutral is required. (*Courtesy of Midwest Electric Products, Inc.*)

**Figure 12-4** The transfer switch is installed at the top of the central distribution pole, and the alternator is permanently connected. (*Courtesy of Ronk Electric Industries*)

When pole-top current transformers are used for metering, the wires do not extend down the pole. Pole-top transfer switches are available for these installations, Figure 12-4. A handle at the bottom of the pole is used to operate the switch. The wires from the alternator must extend up the pole to the transfer switch.

## SIZING THE ALTERNATOR

The alternator may be sized to power the entire electrical load on the farm but, generally, only the most essential equipment and lights are considered. Even this essential equipment, in most cases, need not be operated at the same time. But, for a growing farm or business, it is not wise to size the alternator at the minimum requirement or, in the future, it may be inadequate.

Many methods are used for determining the proper size alternator for a particular application. Sizing alternators would be easier if it were not for the fact that electric motors draw up to six times as much current when starting as when they reach running speed. The alternator must be capable of supplying these short-time, high-power demands. Table 12-1 can be used as a general guide for alternator selection but, remember, this table is based upon averages, and the size may not be adequate for farms or homes with high electrical demands.

To size the alternator more accurately, list the essential lights and equipment which must operate during a outage. Not all the essential equipment will need to be operated at the same time. For example, the feeding operation on a dairy farm can be delayed until after the milking is finished but, in this case, the water pump and milk tank compressor will probably continue to operate for some time after milking is completed. For a typical farm, there probably will be two or three lists of equipment. For each equipment list determine the wattage requirement. Size the alternator to the largest wattage requirement.

The nameplate on most appliances and equipment, except motors, gives the wattage requirement. If the wattage of equipment and appliances cannot be determined, then Table 12-2 can be used to estimate the wattage. Remember, motors have high power requirements when starting. Table 12-3 lists approximate starting and running wattages for common motors.

Selecting the proper alternator size for a farm first requires grouping the necessary farm tasks which can be performed independently. A typical dairy farm electrical load might involve milking, feeding, and dwelling oper-

**Table 12-1 Guide for selecting typical farm and home alternator. The residence is included as a part of each farm operation listed.**

| | |
|---|---|
| Dairy farm | |
|     40–50 cows | 15–25 kW |
|     60–100 cows | 20–30 kW |
|     120–200 cows | 50–75 kW |
| Poultry farm | |
|     8–10 000 birds | 15–20 kW |
|     15–18 000 birds | 20–30 kW |
|     25–30 000 birds | 30–50 kW |
| Hog farm | 15–25 kW |
| Beef farm (mechanical feeding) | 25–50 kW |
| Greenhouse (no refrigeration) | 25–50 kW |
| Residence (nonfarm) | |
|     Without electric range | 4–8 kW |
|     With electric range and/or air conditioning | 10–15 kW |

Unit 12    Standby Electrical Power System    255

Table 12-2  Electrical power requirements of household appliances and equipment

| Appliances | Watts | Horsepower |
|---|---|---|
| Refrigerator | | ¼ |
| Freezer | | ¼ to ½ |
| Water pump | | ½ to 2 |
| Furnace oil burner | | ⅙ |
| Furnace blower | | ¼ to ½ |
| Electric clothes dryer | 1 500–5 000 | |
| Clothes washer | | ¼ to ½ |
| Water heater | 1 000–6 000 | |
| Electric range | 3 000–12 000 | |
|     Small surface unit | 1 000–1 400 | |
|     Large surface unit | 1 500–2 400 | |
|     Oven | 2 500–4 000 | |
| Microwave oven | 1 200–1 900 | |
| Television | 200–600 | |
| Dishwasher (may have a booster heater) | | ⅙ |
| Electric fan | 75–300 | |
| Electric heaters | 500 and up | |
| Electric iron | 500–1 200 | |
| Coffee maker | 1 000 | |
| Mixer | 100–175 | |
| Electric skillet | 1 200 | |
| Toaster | 1 100 | |
| Incandescent lamps | (W on lamp) | |
| Fluorescent lamps* | (W on lamp x 1.2) | |
| Mercury yard lamps* | 200, 300, or 450 | |

*The fixture draws more power than the lamp wattage because of the ballast.

ations. The following procedure for determining alternator size provides for a modest amount of extra capacity for future equipment.

- Starting wattage of largest motor. (If there are two or more motors of the same size, list the starting wattage of only one. Include the running wattage of the other motors in the next step.)

Table 12-3  Power requirements of single-phase 120-V and 240-V electric motors

| Motor (horsepower) | Starting (watts) | Running (watts) |
|---|---|---|
| ⅙ | 860 | 215 |
| ¼ | 1 500 | 300 |
| ⅓ | 2 000 | 400 |
| ½ | 2 300 | 575 |
| ¾ | 3 350 | 835 |
| 1 | 4 000 | 1 000 |
| 1 ½ | 5 000 | 1 500 |
| 2 | 7 500 | 2 000 |
| 3 | 11 000 | 3 000 |
| 5 | 15 000 | 4 500 |
| 7 ½ | 21 000 | 7 000 |

- Running wattage of other motors
- Nameplate wattage of appliances and equipment
- Wattage of lights

The following example illustrates how this procedure works. The dairy farm load has been broken down into three essential operations.

**Milking Operation**

| | | |
|---|---|---|
| Bulk tank compressor | 5 hp | 15 000 W |
| Bulk tank paddle | ¼ hp | 300 |
| Ventilation fan | ½ hp | 575 |
| Vacuum pump | 2 hp | 2 000 |
| Water pump | 1 hp | 1 000 |
| Water heater | 5 000 W | 5 000 |
| Lights | 500 W | 500 |
| Total demand | | 24 375 W |

**Feeding Operation**

| | | |
|---|---|---|
| Silo unloader | 5 hp | 15 000 W |
| Bunk feeder | 5 hp | 4 500 |
| Elevator | ½ hp | 575 |
| Bulk tank compressor | 5 hp | 4 500 |
| Bulk tank paddle | ¼ hp | 300 |

| | | |
|---|---|---|
| Water pump | 1 hp | 1 000 |
| Lights | 200 W | 200 |
| Total demand | | 26 075 W |

**Dwelling**

| | | |
|---|---|---|
| Water pump | 1 hp | 4 000 W |
| Freezer | ⅓ hp | 400 |
| Refrigerator | ¼ hp | 300 |
| Furnace | ¼ hp | 300 |
| Range (one surface unit) | 1 400 W | 1 400 |
| Lights | 750 W | 750 |
| Total demand | | 7 150 W |

Note that the largest total power demand occurred during the feeding operation; therefore, the alternator is sized to this demand. An alternator with a continuous rating of 25 kW would be the minimum size because alternators are generally designed to withstand short-time overloads, such as the starting of a motor. A 30-kW alternator would most likely be adequate for emergency duty on this dairy farm with some extra capacity.

## AUTOMATIC STARTING

A standby electrical power system can be made completely automatic to serve critical loads which cannot be without power for more than a very short time. A poultry farmer may want to operate the water pump and ventilation fans with an automatic standby system. Loss of ventilation and water can be tolerated for only a few minutes when the weather is very hot.

An automatic standby electrical system consists of an engine-driven alternator sized to the entire load to be supplied. An engine starting control and an automatic load transfer control are also required. A typical installation is shown in Figure 12-5. Standby electrical power equipment manufacturers will provide assistance in sizing the alternator and selecting control equipment.

An automatic standby electrical power system should be installed to handle only the critical farm loads. These loads can be separated from the main loads by using a separate small panelboard next to the main panelboard. The automatic load transfer control can be installed between these two panels, Figure 12-6. The critical circuit panel is energized from the main panel, unless a power outage occurs. Thus, the alternator starts and supplies only the critical circuit panel.

If the critical loads consist mainly of motors, the alternator must have sufficient capacity to handle the starting current of all motors at the same time. This problem can be eliminated by controlling some motors with time-delay relays. Time-delay relays can be set to turn on the motors in sequence to minimize the initial load starting requirements. The cost of installing this sequencing equipment should be compared to the savings resulting from the reduction of alternator size.

## INSTALLATION

Standby electrical power systems for agriculture are considered to be optional standby systems, *NEC Article 702*. Actually, *Article 702* applies only to permanently installed engine-driven alternator sets. This article does not necessarily apply to alternators which are power-takeoff (PTO) driven from a tractor, *NEC Section 702-1*.

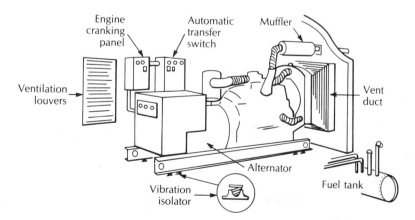

**Figure 12-5** Indoor engine-alternator installation with automatic controls. Manual start-up unit would be the same except for the automatic controls.

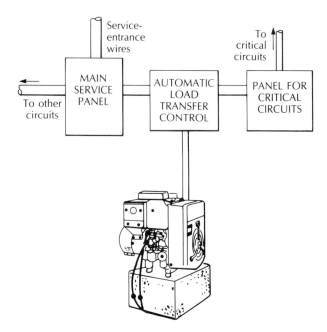

Figure 12-6 The critical circuits originate in a separate panel which is controlled by an automatic transfer switch.

Several requirements are present in the *National Electrical Code* which apply to all standby electrical power installations. These requirements are discussed later in this unit.

Electrical power suppliers are concerned about any generation equipment connected to their power system. They are concerned about the safety of personnel working on their system, and the reliability of service to their customers. Therefore, they have installation procedures which must be followed.

The American Society of Agricultural Engineers, Engineering Practice 364, "Installation and Maintenance of Farm Standby Electric Power," is covered in this unit, insofar as electrical installation is concerned. EP 364 provides additional information concerning the mounting, maintenance, and operation of standby electrical generation equipment.

## SIZING THE TRANSFER SWITCH

The transfer switch serves as a means of disconnecting the farm feeders and circuits from the source of electrical supply. The handle is moved from the supply or line position to the off position. The transfer switch must, therefore, have an ampere rating not less than the rating of the service it controls. If the transfer switch is installed on a central distribution pole rated at 200-A, then the transfer switch supply and load terminals must be rated at 200-A. A 400-A service requires a transfer switch with 400-A supply and load terminals.

The generator terminals of the transfer switch are required to have an ampere rating not less than the ampere rating of the generator. A 20-kW alternator has a continuous output current rating of 87 A. Therefore, a transfer switch with 100-A generator terminals is sufficient. If this transfer switch is used on a 200-A service, the transfer switch would be rated at 200 A on the supply and load terminals, and 100 A on the generator terminals.

The transfer switch terminals are labeled "Line," "Load," and "Generator." *NEC Section 702-6* and *230-83* require that a transfer switch be installed in such a manner as to prevent an accidental connection between the power supplier's lines and the generator. Electric utilities will provide their particular requirements concerning the connection of standby alternators. Requirements of both the *National Electrical Code* and the electric utility must be followed.

## WIRING TO THE ALTERNATOR

The standby alternator must be protected from overcurrent, *NEC Section 445-4(a)*. Usually, the manufacturer provides the overcurrent protection as an integral part of the alternator. In this case, the size of overcurrent protection is used to size the wires leading from the alternator to the transfer switch. For example, if an alternator is protected by a 50-A circuit breaker, then the alternator wire must be rated for 50-A. This procedure is permitted by *Exception No. 2* to *NEC Section 445-5*.

Overcurrent protection for the alternator must be installed if it is not provided as a part of the alternator. In this situation, the wire leading from the alternator to the transfer switch must be sized at 115% of the nameplate current rating of the alternator. The nameplate current rating can be determined from the output voltage and the continuous kilowatt (kW) or kilovolt-ampere (kVA) rating by using Equation 12.1.

$$\text{Full-load current} = \frac{\text{Continuous kVA} \times 1\,000}{\text{Output Voltage}} \qquad \text{Eq. 12.1}$$

Permanently installed, engine-driven electric plants should be mounted on a concrete base. Vibration-damping pads should be placed between the skid and the concrete base to minimize the transfer of vibration to other

**Table 12-4** Ampere rating of flexible SO cord with three copper wires supplying a 120/240-V single-phase, 3-wire system

| AWG No. | Amperes |
|---------|---------|
| 14 | 18 |
| 12 | 25 |
| 10 | 30 |
| 8 | 40 |
| 6 | 55 |
| 4 | 70 |
| 2 | 95 |

**Table 12-5** Minimum size copper grounding electrode and bonding conductor for standby alternators

| Generator Wire Size Copper AWG No. | Copper Grounding Electrode and Bonding Wire Size |
|---|---|
| 2 or smaller | 8 |
| 1 or 0 | 6 |
| 2/0 or 3/0 | 4 |
| 3/0 through 350 MCM | 2 |

equipment. The exhaust must be vented to the outside in a manner which will not cause a *fire hazard* where the pipe extends through the wall.

Tractor-driven alternators are generally stationary or mounted on a trailer or a three-point hitch. Stationary units should be mounted on a concrete base. For mobile units, the mounting should be of sufficient size and stability to withstand pulling over rough terrain and to withstand torque or turning stresses experienced when full loads are applied.

A permanently mounted alternator is connected to a transfer switch by means of single wires inside conduit. Flexible copper wire should be used where the alternator is portable. Type SO cord is available in sizes up to No. 2 AWG. According to *NEC Table 400-5*, Type SO No. 2 copper cord has a rating of 95 A. The wire used to connect to a portable alternator should be highly flexible. Table 12-4 gives ampere ratings for common sizes of Type SO copper flexible cord when supplying a 120/240-V, single-phase, 3-wire electrical system.

## GROUNDING

A standby electric alternator providing power for a farm wiring system is required to be grounded if the electrical system it supplies is required to be grounded, *NEC Section 250-5(d)*. A standby alternator is considered to be a separately derived system, and specific grounding instructions are given in *NEC Section 250-26*.

The frame of the alternator is required to be bonded to the grounded wire, *NEC Section 250-26(a)*. The grounded wire for most systems is the neutral. The size of the bonding wire is determined from *NEC Table 250-94*, based upon the size of wires connecting the alternator to the transfer switch. The bonding of the frame of the alternator is generally done by the manufacturer, but this should be verified at the time of installation. Table 12-5 lists the minimum size grounding electrode and bonding conductor for typical standby alternators installed on farms.

The alternator must be grounded, and it can use the same grounding electrode as the electrical system supplied, *NEC Section 250-26(c)*. The grounding electrode conductor for the standby electrical system extends from the grounding terminal in the transfer switch to the grounding electrode. The grounding electrode conductor size is determined from text Table 12-5 or *NEC Table 250-94*.

## MAINTENANCE

The standby alternator must be kept clean and in good running order at all times so that it will be ready for immediate use. An accumulation of dust and dirt on the alternator can cause overheating during operation.

Tractor-driven alternators should be operated at least every three months to blow out dust, to dry out moisture which may have entered the alternator, to remove oxidation from slip-rings, and to familiarize the operator with start-up procedures. The alternator should be operated under load.

Manually operated engine-powered units should be operated under load at least once every month. A unit should be operated long enough to enable the engine to attain normal operating temperature. An automatic engine-alternator set can be test operated with automatic controls. These units should be operated once each week.

## POWER FAILURE ALARMS

Power failure alarms should be installed where manual start-up alternators are used and power outages cannot be tolerated for more than a few minutes, such as in animal confinement housing. The alarm can be installed

Unit 12 Standby Electrical Power System

on critical electrical circuits to indicate directly the loss of power, or it can sense some adverse effect of the power outage, such as a rise in temperature. The alarm system can be operated by a battery or a pressurized gas cylinder.

Sometimes an outage can occur in one building and not on the entire farmstead. In the event of this type of outage, an alarm in the residence would not sound. To be safe, an alarm should be installed in each building where it is critical that electric service not be interrupted.

## REVIEW

Refer to the *National Electrical Code* when necessary to complete the following review material. Write your answers on a separate sheet of paper.

1. A standby power system consists of an alternator, an engine to power the alternator, and what additional component?

2. The engine used to power a 30-kW standby alternator should have a minimum rating of what horsepower?
    a. 30 hp
    b. 45 hp
    c. 60 hp
    d. 75 hp

3. A 4-pole alternator will have a rotational speed of:
    a. 1 200 r/min
    b. 1 800 r/min
    c. 2 400 r/min
    d. 3 600 r/min

4. The purpose of the transfer switch is to provide a means of connecting a standby alternator to a wiring system, and to prevent the interconnection of the alternator to which electrical system?

5. A beef farmer has a silo with a 5-hp electric motor, a mechanical feeder with a 3-hp motor, a 1-hp water pump, and 500 W of lighting. The minimum size standby alternator required to supply this load is which of the following?
    a. 15 kW
    b. 20 kW
    c. 30 kW
    d. 50 kW

6. An alternator with a continuous rating of 30 kW, 120/240 V, singlephase has a continuous full-load current rating of which of the following?
    a. 80 A
    b. 100 A
    c. 125 A
    d. 200 A

7. A standby alternator is permanently mounted and connected to the transfer switch with single wires in conduit. The minimum size copper THW wire for an alternator with a 100 A rating is No. ____?____ AWG.
    a. 1
    b. 2
    c. 3
    d. 4

8. The wire connecting an alternator to the transfer switch is No.0 AWG copper. The minimum size grounding electrode conductor for the alternator is No. ___?___ AWG.
   a. 8
   b. 6
   c. 4
   d. 2

9. A standby alternator is considered to be a separately derived system, and grounding for the alternator is covered in which *NEC* section?

10. A farm standby electric engine-alternator set permanently mounted must meet the requirements of which *NEC* section?

# UNIT 13
# PHASE CONVERTERS

## OBJECTIVES

After studying this unit, the student will be able to

- describe the purpose of a phase converter.
- give at least one major limitation of a phase converter.
- discuss several conditions where it is advisable to install a phase converter.
- name the two major types of phase converters.
- select a phase converter and motor to power a load.
- wire a phase converter circuit.

*Phase converters* are used on farms to operate 3-phase electric motors from a single-phase electrical supply, Figure 13-1. Three-phase electric motors, particularly in larger sizes, are generally less expensive and more maintenance free than single-phase motors of the same horsepower rating. Besides, large-horsepower, single-phase electric motors are not readily available. A farmer installing a grain dryer requiring a 20-hp motor probably will have to install a 3-phase motor. Ideally, this farm should be supplied with 3-phase power, but the electric utility primary lines at the farm may only be able to supply single-phase power. Three-phase can usually be provided, if the farmer is willing to share the cost of extending the additional primary wire. If this cost is too great, a phase converter may be an acceptable alternative.

## PHASE CONVERTER LIMITATIONS

Phase converters are not the solution to all farm electrical power problems. Some types of loads are difficult to power with a 3-phase motor operating from a phase converter. These are loads which require high starting torque, such as silo unloaders, and gutter cleaners.

The phase converter limits the starting current draw of the motor. When starting current is limited, motor starting torque is greatly reduced. However, many phase converter applications can be used for easy-starting farm loads, such as fans or pumps. A grain dryer is a typical example. Some equipment can be started empty, with the load applied after the motor has obtained full operating speed.

Another limitation to the use of phase converters is

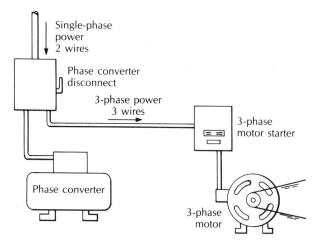

Figure 13-1 A phase converter provides a means of powering a 3-phase motor from a single-phase electrical supply.

that voltages between phase wires to the 3-phase electric motor are generally not exactly balanced. Voltage unbalance generally results in slightly unbalanced currents within the different motor windings. Slight unbalances need not be of concern if properly sized motor running overcurrent is provided.

## WHEN TO CHOOSE A PHASE CONVERTER

There are various reasons for choosing to install a phase converter.

- The cost of primary line extension to provide 3-phase power to the farm is much higher than it is for a phase converter installation.
- The 3-phase electrical energy utility rate, plus the cost of rebuilding the farm electrical service, is enough higher than it is for single phase that the phase converter is the less costly solution.
- A single-phase motor is available, but it has such a high starting current that the electric utility will not permit its operation because it causes objectionable primary line voltage drop. This voltage drop causes the lights to flicker at neighboring farms and residences every time the motor starts. A 3-phase motor operated from a phase converter draws less current, so that line flicker may not be a problem.
- A single-phase electric motor for the job is not available, and a 3-phase motor is the only alternative.
- Temporary 3-phase service is required until permanent, utility-supplied 3-phase service is available.
- New equipment is provided with a 3-phase motor as an integral part of the unit, and replacement with a single-phase motor is either too difficult or more costly than a phase converter installation.
- Several motors are required, and the cost of 3-phase motors plus a phase converter is less than it is for single-phase motors.

## TYPES OF PHASE CONVERTERS

Two types of phase converters are available: static, and rotary. Each type offers advantages for specific applications.

### Static Phase Converter

The *static phase converter,* Figure 13-2, has no moving parts. It consists of a cabinet containing capacitors and, sometimes, an autotransformer. Static phase converters are selected for a specific size of 3-phase motor. It is possible to operate more than one motor from a single static phase converter, but extra capacitors are required, as well as careful balancing of the line currents. More than one motor for a static phase converter should not be installed unless recommended by the phase converter manufacturer.

A simple static phase converter uses a bank of capac-

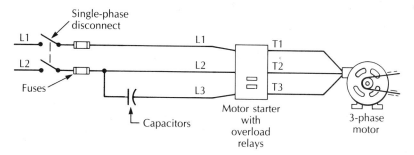

Figure 13-2 A simple static phase converter uses a bank of capacitors to develop the third phase.

itors to supply one of the 3-phase motor lead wires (a bank of capacitors consists of several capacitors connected in parallel). The other two motor lead wires are supplied directly from the single-phase line.

Excessive line current to the motor can occur with this type of phase converter. To minimize the risk of damage to the motor due to excessive current in one or more motor windings, two precautions are recommended.

- Provide properly sized fuses or thermal overload relays in each of the three wires to prevent motor damage due to excessive winding current.
- Load the motor to only 75% of its nameplate horsepower rating.

The proper-horsepower 3-phase motor, operated from a capacitor-type static phase converter, can be determined using Equation 13.1. First, the horsepower rating of the farm equipment must be determined.

$$\frac{\text{Static phase converter}}{\text{motor horsepower}} = \frac{\text{Equipment horsepower required}}{0.75} \quad \text{Eq. 13.1}$$

## Problem 13-1

A large poultry house ventilation fan requires a 7 ½ hp electric motor. A 3-phase motor will be operated from single-phase power, using a capacitor-type phase converter. Determine the minimum recommended horsepower rating for the 3-phase motor for this application.

## Solution

A motor operated from a capacitor-type static phase converter should be loaded to not more than 75% of its horsepower rating. Use Equation 13.1 to size the motor.

$$\frac{\text{Static phase converter}}{\text{motor horsepower}} = \frac{7.5 \text{ hp}}{0.75} = 10 \text{ hp}$$

A 10-hp, 3-phase electrical motor should be selected to power this load.

The second type of static phase converter uses an autotransformer in addition to capacitors. The main advantage of this type is that the amount of capacitance and the autotransformer voltage can be adjusted to balance motor line currents so the motor can develop its full horsepower rating. As an example, a 10-hp submersible pump may be operated at full nameplate rating, provided proper balancing procedures are followed. Significant variation in motor load will require rebalancing. The manufacturer should provide adjustment instructions. A clamp-on type ammeter is required to make this adjustment. An autotransformer-type static phase converter is shown in Figure 13-3.

Three-phase motors operated from static phase converters generally have reduced starting torque. The starting torque can be improved by adding extra capacitors while the motor is starting. Once the motor reaches full

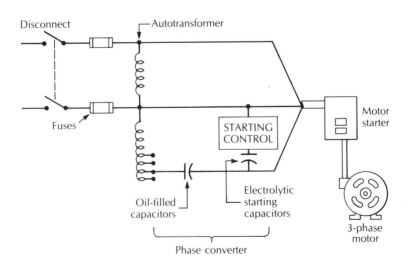

**Figure 13-3** An autotransformer-type static phase converter with starting capacitors

operating speed, these capacitors are disconnected. Manufacturers provide some starting capacitors and the control circuit as an integral part of the phase converter. However, with extra starting capacitors, a 3-phase motor can develop about as much starting torque as it would if operating from a utility-supplied 3-phase supply.

Two types of capacitors are used in phase converters: oil filled, and electrolytic. The *oil-filled capacitors* remain in the circuit continuously. They are large and can dissipate the heat produced by the alternating current flowing in the circuit. *Electrolytic capacitors* can only be energized for a short time, or they will become overheated and *fail*. This type provides a large amount of capacitance in a small size and, therefore, it is used as a starting capacitor. The electrolytic type must *not* be energized continuously. Static phase converters should be limited to applications where the load power requirement is fairly constant.

## Rotary Phase Converter

The *rotary phase converter* consists of a rotating unit and a bank of oil-filled capacitors, Figure 13-4. The rotating unit is usually a 3-phase motor without an external shaft. The single-phase lines are connected to two leads of the rotating unit, as shown in Figure 13-5. The capacitors are connected between one of the single-phase leads and the third lead. The rotating unit acts as a generator and, with the aid of the capacitors, produces a third phase in the third lead. The 3-phase converter leads can then be connected by a 3-phase motor through a motor starter.

**Figure 13-4** Rotary phase converter (*Courtesy of Ronk Electrical Industries*)

An advantage of the rotary phase converter is that more than one 3-phase motors can be operated from a single phase converter. The motors can also drive loads which have a fluctuating power requirement. This is the major advantage of rotary converters over the static type.

The size of the rotary phase converter is determined by the horsepower rating of the largest motor to be started and the total of all motors to be operated. For example, assume that a farm materials handling system requires a single 10-hp motor and two 5-hp motors. This load can be supplied with a single rotary phase converter with a single motor rating of 10 hp. This phase converter must also have a total connected load rating of at least 20 hp. When starting the motors, the largest is started first, and then each of the others individually. Manufacturers may rate their phase converters in kilovolt-amperes (kVA) rather than in horsepower.

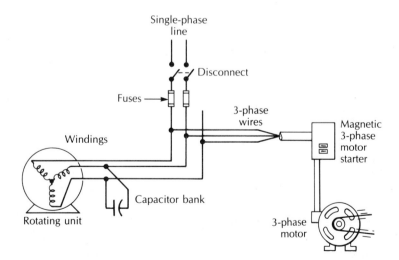

**Figure 13-5** A rotary phase converter consists of a rotating unit and a bank of capacitors.

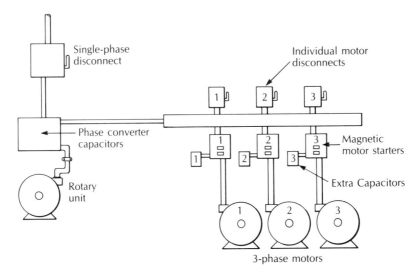

Figure 13-6 Extra oil-filled capacitors may be required at each motor when a single rotary phase converter supplies several 3-phase motors.

The phase converter must first be turned on with the rotating unit reaching full speed before any motor is turned on. Then, each motor is started individually. Another motor is not turned on until the previous motor has reached full operating speed. Typically, a rotary phase converter can supply two times its individual motor rating. Extra oil-filled capacitors are usually required when several motors are operated from a single rotary phase converter, Figure 13-6. The manufacturer provides instructions for multi-motor installations.

Three-phase motors operated from a rotary phase converter develop only about half as much starting torque as when operated directly from a utility-supplied 3-phase supply. However, starting torque can be improved by adding a starting capacitor unit to any motor requiring additional torque. These units generally consist of electrolytic capacitors which are disconnected automatically when the motor achieves full operating speed. Starting torque can also be improved by installing an oversized phase converter. A 5-hp motor will develop higher starting torque when operated from a 10-hp phase converter than from a 5-hp phase converter.

The rotary phase converter has a minimum horsepower rating for a single motor installation. Generally, the smallest motor which can be operated alone is one-fourth the horsepower rating of the rotary phase converter. The manufacturer should be consulted if smaller motors are to be operated alone from a rotary phase converter.

## Problem 13-2

A grain dryer has a fan requiring a 25-hp, 3-phase electric motor. The grain dryer has three additional motors rated at 5 hp, 3 hp, and 1 hp. Determine the minimum size rotary phase converter required to supply all four motors.

## Solution

The minimum size rotary phase converter is determined by the size of the largest motor which, in this case, is 25 hp. The individual motor horsepowers must be added together to make sure that they do not exceed two times the phase converter rating, which is 50 hp.

$$25 + 5 + 3 + 1 = 34 \text{ hp}$$

The manufacturer must be consulted to determine if extra capacitors are required for the additional motors.

## 480-volt Phase Converters

The phase converters discussed thus far connect to a 240-V, single-phase system and produce power suitable for operation of 240-V, 3-phase motors. Phase converters are available which produce power suitable to operate 480-V, 3-phase motors from a 240-V, single-phase supply. Some require a transformer to step up the voltage from 240 V to 480 V.

## PHASE CONVERTER WIRING

Information pertaining directly to the wiring of phase converters is not contained in the *National Electrical Code*. All of the previous diagrams in this unit show the minimum necessary components of a phase converter installation:

- A single-phase disconnect must be installed which will disconnect power from the entire installation.
- The phase converter.
- If more than one motor is supplied by the phase converter, a separate disconnect and overcurrent protection is required for each motor.
- A controller for each motor.

The wiring of a phase converter installation requires good judgment. *NEC Article 430* provides most of the information necessary for a safe and efficient installation. The requirements of *NEC article 460* must be met because capacitors are used in the phase converters. *NEC Article 450* requirements must be met if a transformer is included as a part of the phase converter.

### Sizing the Wires

Two sets of wires must be sized in a 1-motor phase converter installation: 1) the 3-phase wires to the motor, and 2) the single-phase wires supplying the phase converter. The first step in the procedure involves the determination of the motor full-load current. This is found in *NEC Table 430-150*. For example, the full-load current of a 10-hp, 230-V, 3-phase motor is 28 A. The full-load current of the motor must be multiplied by 1.25 (125%) to determine the minimum ampere rating of the motor circuit wire, *NEC Section 430-22(a)*. See Equation 13.2

| Minimum ampere rating of motor circuit wire $= 1.25 \times$ motor full-load current | Eq. 13.2 |
|---|---|

For the 10-hp motor in the example, the minimum rating of the wire is 35 A.

| Minimum ampere rating of motor circuit wire $= 1.25 \times 28 = 35$ A |
|---|

Using copper wire with TW insulation, look up the minimum wire size in *NEC Table 310-16*. The minimum wire size is AWG No. 8, which is rated at 40 A.

The question now is, how much current will flow on the single-phase wires? To get a clue to the answer, look up the full-load current of a 10-hp, 230-V, single-phase motor in *NEC Table 430-148*. The single-phase, 10-hp motor would draw 50 A; nearly twice as much as the 3-phase, 10-hp motor. If the wire size were based upon the single-phase motor full-load current, then the minimum single-phase wire rating would be 62 A.

| $50 \times 1.25 = 62$ A |
|---|

The same result would be obtained by multiplying the 10-hp, 3-phase, 230-V motor full-load current by 2.2 (220%). The single-phase input wires to the phase converter supplying one motor are determined by Equation 13.3.

| Minimum ampere rating of phase converter input wires, one motor $= 2.2 \times$ motor full-load current | Eq. 13.3 |
|---|---|

### Problem 13-3

A phase converter supplies a 25-hp, 230-V, 3-phase motor. The circuit wires are copper with THWN insulation. Determine the minimum size of 1) the 3-phase wire from the phase converter to the motor, and 2) the single-phase input wire to the phase converter.

### Solution

The full-load current of the motor from *NEC Table 430-150* is 68 A. Determine the minimum ampere rating of the motor circuit wire from Equation 13.2.

| $1.25 \times 68 = 85$ A |
|---|

The minimum size copper THWN wire from *NEC Table 310-16* is No. 4 AWG, which is rated at 85 A. Determine the minimum ampere rating of the phase converter input wires from Equation 13.3

| $2.2 \times 68 = 150$ A |
|---|

The minimum size copper THWN wire from *NEC Table 310-16* is No. 0 AWG.

A phase converter that supplies more than one motor is only slightly more involved than the case of a single motor. Consider an installation similar to that shown in Figure 13-6. The wires are shown in schematic form in

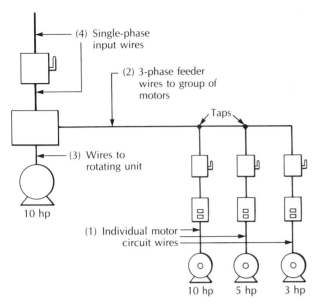

**Figure 13-7** Several different sizes of wire may be required for a phase converter installation involving several motors.

Figure 13-7. The wires to the individual motors, as shown by (1) in Figure 13-7, are determined by using Equation 13.2, as discussed previously.

The minimum ampere rating of a feeder wire, as shown by (2) in Figure 13-7, supplying two or more motors is required to be the full-load currents of the motors added together, plus one-fourth of the largest motor full-load current, *NEC Section 430-24*. Equation 13.4 is stated in a slightly different form.

| Minimum motor feeder ampere rating    **Eq. 13.4** |
|---|
| = 1.25 × Full-load current of largest motor |
| + Full-load current of all other motors |

If there are two or more motors of the largest horsepower, then multiply the current of only one of them by 1.25.

## Problem 13-4

The 230-V, 3-phase motors of Figure 13-7 are rated 10 hp, 5 hp, and 3 hp. Determine the minimum size copper THWN three-phase feeder conductor.

## Solution

The full-load current of each motor from *NEC Table 430-150* is:

| 10 hp, 230 V, 28 A |
|---|
| 5 hp, 230 V, 15.2 A |
| 3 hp, 230 V, 9.6 A |

The minimum feeder current rating is determined by using Equation 13.4

| Minimum motor feeder ampere rating |
|---|
| = (1.25 × 28) + 15.2 + 9.6 |
| = 35 + 15.2 + 9.6 = 59.8 or 60 A |

The minimum size motor THWN copper feeder wire from *NEC Table 310-16* is No. 6 AWG.

The wire from the phase converter capacitor panel to the rotating unit, as shown by (3) in Figure 13-7, will not carry more current than the motor feeder. Therefore, the minimum size for this wire is the same as it is for the motor feeder wire.

The single-phase input wire to a phase converter supplying a group of motors carries almost twice the current as the 3-phase feeder supplying the motors. The single-phase input wire is sized at 2.0 (200%) times the sum of the full-load currents of all the motors, Equation 13.5

| Phase converter input wire ampere     **Eq. 13.5** |
|---|
| rating for group of motors |
| = 2 × Full-load current of all motors |

## Problem 13-5

Determine the minimum size copper THWN single-phase input wire size for the installation described in Problem 13-4.

## Solution

Add the full-load currents of all motors and multiply by two, as in equation 13.5

| Phase converter input wire ampere rating for |
|---|
| group of motors |
| = 2 × (28 + 15.2 + 9.6) |
| = 2 × 52.8 = 105.6 or 106 A |

The minimum size phase converter THWN copper input wire from *NEC Table 310-16* is No. 2 AWG, which is rated at 115 A.

Most phase converter installations involve 230-V 3-phase motors. Thus, the previous calculations apply to the sizing of the circuit wires. The procedure varies only slightly when 460-V, 3-phase motors are supplied from a 230-V, single-phase electrical supply.

- For 460-V motors, find the motor full-load current in the 460-V column of *NEC Table 430-150*.
- Use Equations 13.6 through 13.9 for determining the following.

Single 460 V motor supplied by phase converter from a 230 V supply:

> Minimum ampere rating of motor circuit wires     **Eq. 13.6**
> = 1.25 × Motor full-load current

> Minimum ampere rating of phase converter input wires, one motor     **Eq. 13.7**
> = 4.4 × Motor full-load current

Two or more 460 V motors supplied by one phase converter from a 230 V supply:

> Individual motor circuit wires; use Equation 13.6

> Minimum motor feeder ampere rating     **Eq. 13.8**
> = 1.25 × Full-load current of largest motor
> + Full-load current of all other motors

> Phase converter input wire ampere rating for group of motors     **Eq. 13.9**
> = 4 × Full-load current of all motors

## PHASE CONVERTER DISCONNECT

A means must be provided to disconnect all ungrounded single-phase input wires for the phase converter, *NEC Section 460-8(c)*. This disconnect can serve as the motor circuit disconnect provided only one motor is supplied by the phase converter. This disconnect must also be within 50 ft (15.24 m) and within sight from the motor controller, *NEC Section 430-102*. Individual motor circuit disconnects are required when one phase converter supplies several motors, *NEC Section 430-112*. These individual motor circuit disconnects are in addition to the phase converter main single-phase disconnect (as shown in Figure 13-6).

The main phase converter disconnect should be a single-phase fusible safety switch or a suitable circuit breaker. This disconnect switch must be rated in horsepower, *NEC Section 430-109*. A circuit breaker rated in amperes is acceptable, but is less desirable as a disconnect. The disconnect switch must have a single-phase horsepower rating not smaller than the sum of the horsepower ratings of the motors to be supplied, *NEC Section 430-112*.

## Problem 13-6

Determine the horsepower rating required for the phase converter disconnect supplying motors with ratings of 25, 5, 3, and 1 horsepower.

## Solution

> Disconnect switch rating
> = 25 + 5 + 3 + 1 = 34 hp

Choose a disconnect switch which has a *single-phase* horsepower rating not smaller than 34 hp. Single-phase horsepower is used because the switch is installed in the single-phase input wire.

Single-phase and 3-phase horsepower ratings are not the same. A 3-phase disconnect switch with a 10-hp rating cannot be used as the disconnect for a 10-hp, single-phase motor. The single-phase motor draws 1.73 times as much current as the 3-phase motor. Therefore, a single-phase, 10-hp disconnect switch must have a higher current rating than a 10-hp, 3-phase disconnect switch. Table 13-1 gives approximate equivalent single- and 3-

Table 13-1 Approximate equivalent 3-phase and single-phase horsepower ratings for 240-V safety switches

| Horsepower | |
|---|---|
| Three phase | Single phase |
| 3 | 1 ½ |
| 7 ½ | 3 |
| 10 | 5 |
| 15 | 7 ½ |
| 20 | 10 |
| 30 | 15 |
| 40 | 20 |
| 50 | 30 |
| 60 | 35 |
| 75 | 40 |
| 100 | 60 |
| 125 | 70 |
| 150 | 85 |
| 200 | 115 |

phase horsepower ratings. For example, a disconnect switch rated at 100 hp, three phase can usually be used as the disconnect for a 60-hp, single-phase motor load.

## OVERCURRENT PROTECTION

Overcurrent protection must be provided for each ungrounded single-phase input wire to the phase converter. Time-delay fuses are usually installed in the phase converter disconnect switch. The time-delay fuse size is determined by the same procedure used to size the single-phase input wires, Equation 13.3, 13.5, 13.7, or 13.9. The next standard fuse size larger than the value determined by one of the input wire ampere rating equations may be chosen to protect the input wires. In Problem 13-5, the proper size time-delay fuse for the No. 2 AWG copper single-phase input wires is 110 A.

Overcurrent protection for individual motor circuit wires is determined by the procedures required in *NEC Article 430*, as discussed in Unit 17.

## MOTOR CONTROLLERS

The controller for a motor supplied by a phase converter is permitted to be any appropriate type described in *NEC Article 430*. Some types of motor controllers do not provide adequate overload protection for the motor. It is extremely important in a phase converter installation that some type of overload protection sensing motor current be provided. Three types of motor controllers that provide this protection are:

- Magnetic motor starter with properly sized thermal overload relays
- Manual motor starter with properly sized thermal overload relays
- Safety switch with horsepower rating equal to or greater than motor horsepower, and time-delay fuses sized not greater than 1.15 (115%) times the motor full-load current

A rotary phase converter must always be turned on and reach full operating speed before any motor supplied is turned on. Should there be a temporary power interruption, it is important that the motors be turned off, or the phase converter may become *damaged* when power is restored. A manually operated controller will not turn the motors off during a temporary power outage. When power is restored, the motors will try to start. *Magnetic motor starters must always be used to control motors supplied by a rotary phase converter.* Also, do not use 2-wire control, because this will cause immediate restarting of the motor. Time-delay relays may be used with 2-wire control to prevent immediate restarting of the motor. When power is lost, the magnetic motor starters will shut off the motors so they cannot restart when power is restored. The motors must be restarted manually; or an automatic restart control can be added to start each motor, in sequence, once the rotary phase converter has been restarted.

## INSTALLING EXTRA CAPACITORS

Extra capacitors are often required when two or more motors are operated from one rotary phase converter. These capacitors are most easily installed at the magnetic motor starter. A disconnect is required for the capacitor bank unless the capacitors are on the load side of the motor running overcurrent protective device, *NEC Section 460-8(b)(1), Exception*.

The capacitors may change the amount of current drawn by the motors. The amount of current from the supply up to the capacitors will be affected, but not the current flow from the capacitors to the motor. The capacitors are to be connected to the load side of the thermal overload heaters (Figure 13-8). Therefore, **the current**

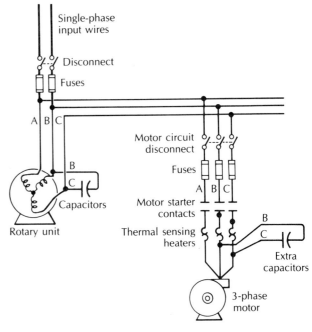

**Figure 13-8** Extra capacitors, if required, should be connected on the load side of the thermal sensing heaters in the motor starter.

flow through the thermal overload heaters may be affected. The overload heaters' sensing current setting may need to be adjusted if the addition of these capacitors significantly changes one or more of the motor line current values, *NEC Section 460-9, Exception.*

## GROUNDING

A grounding path must be provided for all exposed metal parts of the phase converter installation, including phase converter, capacitors, disconnects, motor controllers, and motors. Procedures for grounding are covered in *NEC Section 250,* as discussed in detail in Unit 10.

## REVIEW

Refer to the *National Electrical Code* when necessary to complete the following review material. Write your answers on a separate sheet of paper.

1. Is the purpose of a phase converter to provide a means of operating a 3-phase electrical motor from a single-phase electrical supply?

2. Name the two types of phase converters.

3. Capacitors are used in phase converters. Which type of capacitor can carry alternating current continuously?

4. Is it true that a 3-phase electric motor supplied by a phase converter generally does not develop as much starting torque as it would if operated from utility-supplied 3-phase power?

5. Which of the following is the major reason for choosing a phase converter rather than utility-supplied 3-phase power?
   a. Its overall cost is lower.
   b. The power supplier does not produce 3-phase power.
   c. A single-phase motor of the required size is not available.
   d. The farm is not wired for 3-phase power.

6. A grain dryer fan requires a 15-hp electric motor. Determine the minimum size 3-phase electric motor and capacitor-type static phase converter required to power this load.

7. Identify the static phase converter which, through proper adjustment of the number of capacitors, allows a motor to develop full nameplate horsepower.

8. If a farmer wants to operate several motors from one phase converter, then which type of phase converter should be chosen?

9. A phase converter is used to supply, 7 ½-, 3-, and 2-horsepower, 3-phase, 230-V electric motors. Determine the minimum size copper THW single-phase input wires.

10. A farmer can buy a used 30-hp rotary phase converter from a neighbor. If the phase converter is in good operating condition, should this phase converter be used to supply a 5-hp, 3-phase motor?

11. Would a 10-hp rotary phase converter be adequate to supply the motors in Problem No. 9 of this review?

12. Which of the following types of controllers should be used for a motor supplied by a rotary phase converter?
    a. Circuit breaker

b. Fusible switch
   c. Manual motor starter
   d. Magnetic motor starter

13. A 3-phase, 3-pole disconnect switch can be used on the input wires to a phase converter if a single-phase, 2-pole disconnect is not available. Determine the minimum 3-phase horsepower rating required for a 3-phase disconnect used on the input wires to a rotary phase converter supplying motors rated at 15, 10, 3, and 2 horsepower.

14. A phase converter supplies one 15-hp, 230-V, 3-phase electric motor. Determine the recommended size time-delay fuses used to protect the single-phase converter input wires.

15. Extra oil-filled capacitors are required for a phase converter supplying several motors. Refer to the diagram and indicate the most appropriate location (A, B, C, or D) to connect the capacitors to the motor circuit.

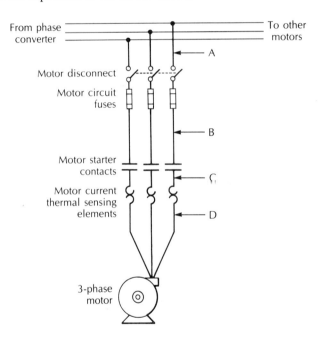

# Unit 14
# TRANSFORMERS

## OBJECTIVES

After studying this unit, the student will be able to

- state the purpose of a transformer.

- explain the principle of mutual induction.

- determine the output voltage of a transformer if the input voltage and turns ratio are known.

- determine the full-load current of a transformer given the kVA and voltages of the primary and secondary windings.

- identify the common types of transformers from their schematic diagrams.

- read transformer winding diagrams and connect a transformer for the desired primary and secondary voltage.

- choose the proper transformer taps to obtain the desired output voltage.

- connect buck and boost transformers to obtain desired voltage for a single-phase application.

- choose the correct transformer kVA for the application, given the voltage, current, and phase requirement of a load.

- size overcurrent protection for dry-type transformers operating at 600 V or less.

- size the feeder conductor for the transformer and wires from the transformer to loads.

- properly ground a transformer, and the secondary electrical system produced by the transformer.

The purpose of a transformer is to change electrical voltage to a different value. For example, a farmer has a large, 480-V, 3-phase motor powering a well. The motor is in a building, and the farmer wants one 120-V circuit for a few lights and a receptacle outlet. A transformer is used to lower the voltage from 480 V to 120 V for the lighting circuit, Figure 14-1. The controls for furnaces and air-conditioning units are often operated at 24 V,

Figure 14-1 Dry-type transformers

Figure 14-2. A small transformer inside the equipment lowers the line voltage to 24 V for the control circuit. Transformers are frequently used inside electronic equipment.

## HOW THE TRANSFORMER WORKS

A clear understanding of how transformers work is necessary in order to wire them properly in an electrical system. Understanding input and output current and grounding are particularly troublesome. A dual-voltage transformer can be ruined when power is applied, if the connections are made improperly.

An important property of electricity is that a magnetic field is produced around a wire in which electrical current is flowing, Figure 14-3. The more current that flows, the stronger is the magnetic field. An even stronger magnetic field can be produced by winding the wire into a coil. Now the magnetic fields of adjacent wires add together to form one strong magnetic field.

The electrical current flowing in a transformer is alternating current. The current flows first in one direction, stops, then reverses and flows in the other direction. The magnetic field around the winding is constantly in motion. Figure 14-4 shows the magnetic field during one cycle. Notice that the north and south poles of the magnetic field reverse when the flow of current reverses.

Another property of electricity is important to the operation of a transformer. When a magnetic field moves across a wire, a voltage is induced into the wire, Figure 14-5. If the wire forms a complete circuit, current will flow in the wire. If a second coil of wire is placed in a moving magnetic field, then a voltage will be induced in this second coil, Figure 14-6. This phenomenon is called *mutual induction*. Alternating current in one winding produces a moving magnetic field that induces a voltage in a second winding. Electrical energy is converted into a magnetic field and then converted back into electrical energy in a second winding. The trick is to do this with little or no loss of energy.

The magnetic field loses strength quickly in air; therefore, a special steel core is used. The core is composed of thin sheets of a silicon-steel alloy. The magnetic field is concentrated in the core, and energy losses are reduced to a minimum. Figure 14-6 shows the two windings separated. Most transformers have one winding placed directly over the other to further reduce the loss of energy, as shown in Figure 14-7.

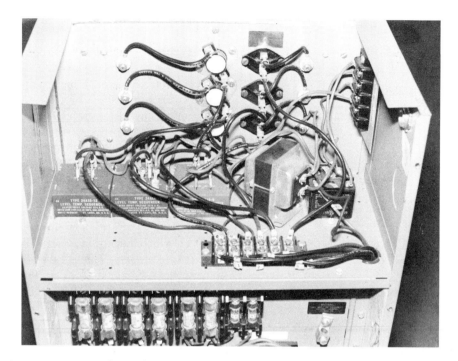

**Figure 14-2** Control transformers are common in appliances and electrical equipment.

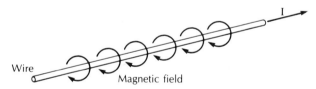

**Figure 14-3** When electrical current flows through a wire, a magnetic field is built up around the wire.

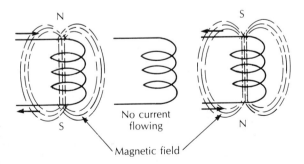

**Figure 14-4** When alternating current is flowing in the coil, the magnetic field is constantly moving. (The arrows indicate electron flow.)

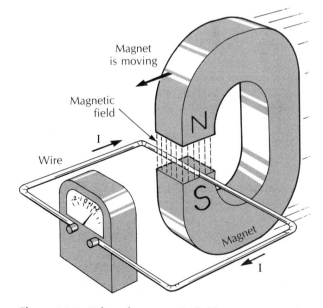

**Figure 14-5** When the magnetic field moves across this wire, an electrical current flows in the wire.

Unit 14  Transformers  275

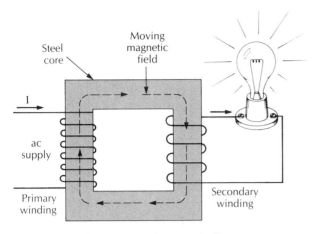

Figure 14-6 The two transformer windings are on separate parts of the silicon-steel core.

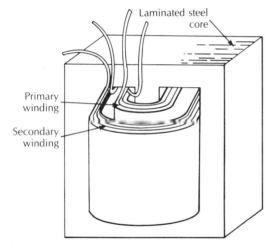

Figure 14-7 In most transformers, the two windings are placed one over the other to reduce energy losses.

## VOLTAGE AND TURNS RATIO

The input winding to a transformer is called the *primary winding*. The output winding is called the *secondary winding*. If there are more turns of wire on the primary than on the secondary, the output voltage will be lower than the input voltage. This is illustrated in Figure 14-8 for a step-down and a step-up transformer. Notice that the winding with the greater number of turns has the higher voltage. In Figure 14-8, one winding has twice as many turns as the other. In one case the voltage is stepped down to half, while in the other the voltage is stepped up to double.

It is important to know the ratio of the number of turns of wire on the primary winding as compared to the secondary winding. This is called the *turns ratio* of the transformer, Equation 14.1. The actual number of turns is not important, just the turns ratio.

$$\text{Turns ratio} = \frac{\text{Number of turns on the primary}}{\text{Number of turns on the secondary}} \quad \text{Eq. 14.1}$$

The step-down transformer of Figure 14-8 has 14 turns on the primary, and 7 turns on the secondary; therefore, the turns ratio is 2 to 1, or just 2. The step-up transformer has 7 turns on the primary and 14 on the secondary; therefore, the turns ratio is 1 to 2, or 0.5. If one voltage and the turns ratio are known, the other voltage can be determined with Equations 14.2 or 14.3.

$$\text{Primary voltage} = \text{Secondary voltage} \times \text{turns ratio} \quad \text{Eq. 14.2}$$

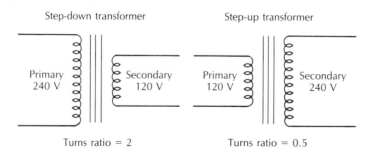

Figure 14-8 Schematic diagrams of step-down and step-up transformers.

$$\text{Secondary voltage} = \frac{\text{Primary voltage}}{\text{Turns ratio}} \quad \text{Eq. 14.3}$$

Use these equations to verify the voltages in Figure 14-8.

## *Problem 14-1*

A step-down transformer has a turns ratio of 4 to 1 or 4. If the transformer secondary voltage is 120 V, determine the primary voltage.

## *Solution*

Use Equation 14.2 to solve for the primary voltage.

$$\text{Primary voltage} = 120 \text{ V} \times 4 = 480 \text{ V}$$

The turns ratio tells us that the primary voltage is four times as great as the secondary voltage.

## *TRANSFORMER RATINGS*

Transformers are rated in volt-amperes (VA) or kilovolt-amperes (kVA). This means that the primary and the secondary winding are designed to withstand the VA or kVA rating stamped on the transformer nameplate. The primary and secondary full-load currents usually are not given. The installer must be able to calculate the primary and secondary currents from the nameplate information.

When the volt-ampere (or kilovolt-ampere) rating is given, along with the primary voltage, then the primary full-load current can be determined, using Equation 14.4 (for a single-phase transformer) or Equation 14.5 (for a 3-phase transformer).

Single phase:

$$\text{Full-load current} = \frac{\text{VA rating}}{\text{Voltage}} \quad \text{Eq. 14.4}$$

or

$$\text{Full-load current} = \frac{\text{kVA} \times 1\,000}{\text{Voltage}}$$

Three phase:

$$\text{Full-load current} = \frac{\text{VA rating}}{1.73 \times \text{Voltage}} \quad \text{Eq. 14.5}$$

or

$$\text{Full-load current} = \frac{\text{kVA} \times 1\,000}{1.73 \times \text{Voltage}}$$

## *Problem 14-2*

A single-phase transformer with a 2-kVA rating has a 480-V primary, and a 120-V secondary. Determine the primary and secondary full-load currents of the transformer.

## *Solution*

Use Equation 14.5 to solve for both primary and secondary currents.

$$\text{Primary full-load current} = \frac{2 \text{ kVA} \times 1\,000}{480 \text{ V}} = 4.17 \text{ A}$$

$$\text{Secondary full-load current} = \frac{2 \text{ kVA} \times 1\,000}{120 \text{ V}} = 16.67 \text{ A}$$

It may seem strange at first, but the transformer current will be higher in the winding which produces the lower voltage. This concept is important to understand in order to avoid transformer or conductor overloading. The primary and secondary transformer full-load currents are also related by the turns ratio, as shown in Equation 14.6.

$$\text{Primary full-load current} = \frac{\text{Secondary full-load current}}{\text{Turns ratio}} \quad \text{Eq. 14.6}$$

## *TYPES OF TRANSFORMERS*

Transformers are of the *dry type* or *oil filled*. From 2% to 5% of the electrical energy is lost in a transformer, mostly due to the resistance of the windings. Large transformers circulate oil through the windings to remove the heat. Dry transformers use air for cooling. Heat is moved from the windings to the case by conduction in smaller sizes of the dry type. Large dry-type transformers actually allow air to circulate through the windings, Figure 14-9. Oil-filled transformers are used by the electric utility, and for industrial or large commercial applications.

Figure 14-9 Air can circulate around and through the windings of a large, dry-type transformer.

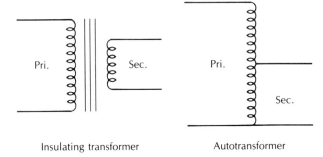

Figure 14-10 Two basic types of transformers are the insulating transformer and the autotransformer.

Transformers installed for applications on a farm electrical system are almost always dry-type transformers. Only dry-type transformers operating at less than 600 V are discussed in this unit.

Common two-winding transformers are often called *insulating transformers*. The primary winding and the secondary winding are separate and not connected. An *autotransformer* has its windings interconnected so that the primary and the secondary share the same windings. Autotransformers, therefore, have an electrically connected primary and secondary. These two basic types of transformers are shown in Figure 14-10. A major advantage of autotransformers over the insulating types is their lighter weight and compact size. Autotransformers are used for electric-discharge lighting ballasts.

A special type of autotransformer, called a *grounding autotransformer* or *zigzag transformer*, is occasionally used to create a neutral wire or a ground for an ungrounded 480-V, 3-phase system. These transformers are found occasionally in industrial wiring. Standard insulating transformers can be used to make a zigzag transformer. The wiring of these transformers is covered in *NEC Section 450-5*.

*Control transformers* are special insulating transformers commonly used to supply power for control of appliances, equipment, and motor starters, Figure 14-11. A control transformer is required when the control circuit voltage is different from the line voltage supplied. Common control circuit voltages are 24 V and 120 V. A 120-V control circuit to a start-stop push-button station may be desirable for a 125-hp, 480-V, 3-phase motor powering an irrigation pump. The control transformer would step down the 480 V to 120 V for the control circuit.

Figure 14-11 Control transformers are commonly used in appliances, equipment, and motor starters.

Control transformers are designed to withstand short-duration overloads with minimal output voltage drop. Motor starter solenoid coils draw six to eight times as much current when they are closing as is required to hold them closed.

*Constant output voltage* transformers or voltage regulating transformers produce a nearly constant output voltage, even though the input voltage may not be constant. The voltage supplied by the utility typically will fluctuate up and down a few percent during the day. This voltage fluctuation is of little concern except for certain equipment, such as electronic computers. Installing a constant output voltage transformer to supply sensitive equipment will eliminate undesirable voltage fluctuations. Special filters can also be added to these transformers to eliminate voltage spikes and electrical noise caused by other equipment operating on the electrical system, Figure 14-12.

## CONNECTING TRANSFORMER WINDINGS

Transformer wiring diagrams are printed on the transformer nameplate which may be affixed to the outside of the transformer or printed inside the cover to the wiring compartments. The lead wires or terminals are marked with the letters H and X. Those lettered H are the primary (high-voltage) leads, and those lettered X are the secondary (low-voltage) leads.

Some transformers have two primary and two secondary windings (as shown in Figure 14-13) so they can be used for several applications. These are called *dual-voltage transformers*. Connections must be made correctly with dual-voltage transformers. If connected improperly, it is possible to create a dead short that will usually ruin the transformer when it is energized.

Consider a dual-voltage transformer rated at 240/480 V on the primary, and 120/240 V on the secondary. Each of the two primary windings is, therefore, rated at 240 V. Each secondary winding is rated at 120 V. The transformer must be connected so that each primary winding receives the proper voltage. In Figure 14-13, the transformer is shown with the primary windings connected in series, with H1 and H4 connected to a 480-V supply. The voltage across H1 and H2 is 240 V and the voltage across H3 and H4 is 240 V. Each winding is receiving the proper voltage. With each primary winding receiving the proper 240 V, each secondary winding will have an output of 120 V. Connecting the secondary windings in series produces 240 V across X1 and X4.

Now consider a case where the primary voltage available is 480 V, but the desired output is 120 V. In this case, the primary windings are connected in series, as in Figure 14-13. The secondary windings are, however, connected in parallel, Figure 14-14. This is accomplished by connecting X1 to X3, and X2 to X4. If this is not done properly, a 240-V dead short will occur. A voltmeter can be used to make sure the connection is correct. Connect X1 to X3, and then connect a voltmeter between X2 and X4. Energize the primary and read the

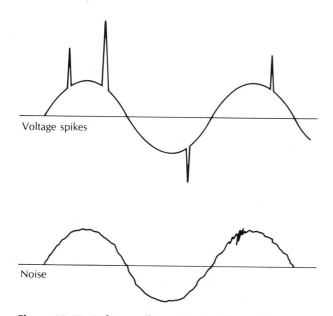

Figure 14-12  Voltage spikes and noise distort the normal alternating-voltage sine wave.

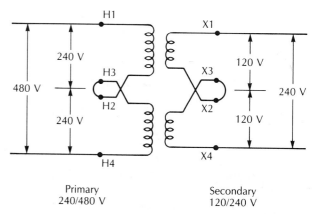

Figure 14-13  The windings are connected in series to obtain the higher of the rated transformer voltages.

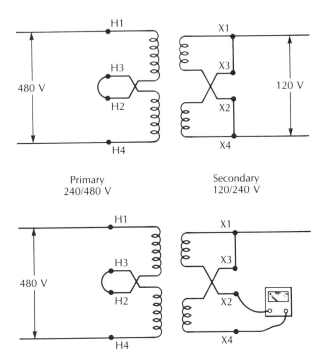

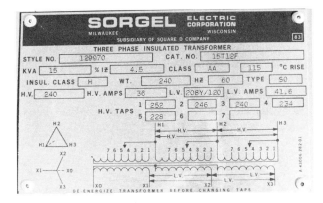

Figure 14-15 Nameplate of a 3-phase transformer

Figure 14-14 The secondary windings are connected in parallel for an output of 120 V. A voltmeter can be used to make sure the transformer is connected properly.

voltmeter. If the connection is correct, the voltmeter will read zero. If the voltmeter reads something other than zero, check all primary and secondary connections to make sure they are connected exactly as indicated by the manufacturer.

The primary on the example transformer has two windings; therefore, it can also be connected for a 240-V supply. The primary windings must be placed in parallel by connecting H1 to H3 and H2 to H4. If this is not done as indicated on the transformer nameplate, the magnetic fields created by each winding will oppose each other. The magnetic fields work together when the windings are properly placed in parallel.

## THREE-PHASE TRANSFORMERS

Changing the voltage of a 3-phase system can be done with a 3-phase transformer or with single-phase transformers. Three-phase transformers are generally designed and constructed for specific voltages. For example, a transformer may have a 480-V delta primary and a 120/208-V wye secondary. A typical nameplate for this type of transformer is shown in Figure 14-15.

The 3-phase transformer has one core with three sets of windings. A primary and a secondary winding are placed one on top of the other on each of the three legs of the core, Figure 14-16. The secondary windings are connected in either wye or delta, as required by the load to be supplied. The primary is connected in wye or delta, depending upon the type of electrical system available. Common 3-phase transformer connections, listing primary windings first, are: delta-delta, wye-delta, and delta-wye. A wye-wye connection is usually not recommended. In a wye-wye connection, a third harmonic current may occur, causing possible current overloading and damage to the primary neutral wire. A delta-wye transformer can usually be substituted. Always be sure to consult the transformer manufacturer before installing a wye-wye connection.

## Problem 14-3

A building is supplied with a 480-V, 3-phase electrical system. Many 120-V circuits are needed; therefore, it is decided to use a 3-phase transformer to step down the voltage to supply a 100-A, 120/208-V panelboard. Which of the following transformers is suitable for this application: a delta-delta, a wye-delta, or a delta-wye?

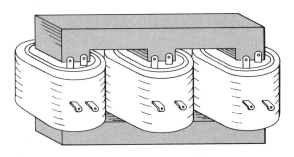

Figure 14-16 Three-phase transformer construction

## Solution

A 120/208-V system is a wye system. Therefore, the secondary must be wye connected. Usually, it does not matter if the supply system is wye or delta connected. A wye-wye connection is not recommended; therefore, a delta-wye is a likely choice. It is always best, however, to consult a transformer manufacturer's representative to make sure the most appropriate transformer is selected for an application.

Single-phase transformers can be used to form a 3-phase transformer bank. It is important that single-phase transformers are identical when connecting them to form a 3-phase system. They should be identical in voltage, kVA, impedance, manufacturer, and model number. Transformer *impedance* is the combined effect of resistance and inductance, and is given in percent.

Connecting single-phase transformers to form a 3-phase bank must be done with *extreme caution*. It is essential that the windings be connected in a certain way only. Reversing a winding can damage the transformer. Figure 14-17 shows three individual single-phase transformers connected to step down 480-V delta to 240-V delta. It is best to obtain a 3-phase transformer rather than to try to connect single-phase transformers. To illustrate the complexity, standard dual-voltage, single-phase transformers are used to change a 480-V, 3-phase delta to a 120/208-V 3-phase wye in Figure 14-18.

## WINDING TAPS

Transformers, except for small sizes, are often supplied with winding taps to compensate for abnormally low or high primary voltage. Assume, for example, that a transformer is rated 480 V primary and 240 V secondary. This means that 240 V will be the output if the input is 480 V. But, what if the input is only 444 V? The turns ratio for this transformer is 2 to 1; therefore, using Equation 14.3, the output will be 222 V.

$$\text{Secondary voltage} = \frac{444 \text{ V}}{2} = 222 \text{ V}$$

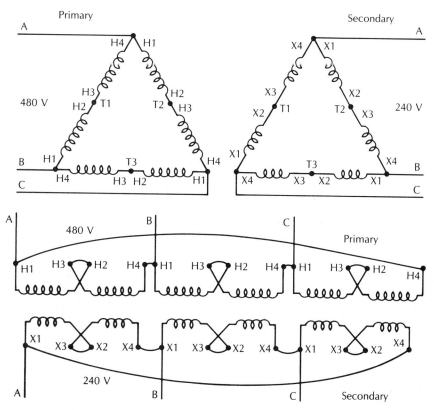

**Figure 14-17** Dual-voltage, single-phase transformers with 240/480-V primary windings, and 120/240-V secondary windings are shown connected to form a 480-V delta to 240-V delta, 3-phase, step-down transformer bank.

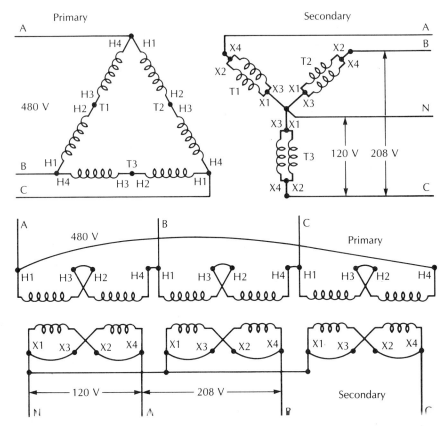

Figure 14-18 Dual-voltage, single-phase transformers with 240/480-V primary windings, and 120/240-V secondary windings are shown connected to form a 480-V delta to 120/208-V wye, 3-phase, step-down transformer bank.

In order to get an output of 240 V with an input of only 444 V, the turns ratio will have to be changed to 1.85 to 1. The purpose of the tap connections, usually on the primary, is to change easily the transformer turns ratio. A typical single-phase transformer nameplate with primary taps is shown in Figure 14-19.

Winding taps each make a 2½% change in the voltage. A transformer will often have two taps above normal voltage and four taps below normal voltage. A transformer usually comes preconnected for normal voltage. If an abnormal voltage is present, it is up to the installer to change the tap connections.

## Problem 14-4

A 25-kVA, single-phase transformer is used to supply 120/240 V, 3 wires, to a 100-A panelboard. The primary voltage is only 450 V instead of the normal 480 V. Show how to connect the transformer of Figure 14-19 to compensate for the low input voltage.

## Solution

The primary windings must be connected in series because the normal voltage should be 480 V. Therefore, check the transformer nameplate for series tap connections. Move down the voltage column until the actual input voltage fits between two numbers. Usually, the higher voltage on the chart is used. If the lower number is used, the output will be greater than 240 V. The proper connection is shown in Figure 14-20 with a jumper wire placed between taps 5 and 6. If the input had been 480 V, the jumper would have been between taps 3 and 4.

Assume that the input is 225 V and the desired output is 120 V. This requires the primary windings to be placed in parallel. This time, the parallel high-voltage tap connection chart on the nameplate is used. Mark the location on the chart where the actual input voltage fits. Then take the next higher tap connection on the chart. Figure 14-21 shows the proper connection.

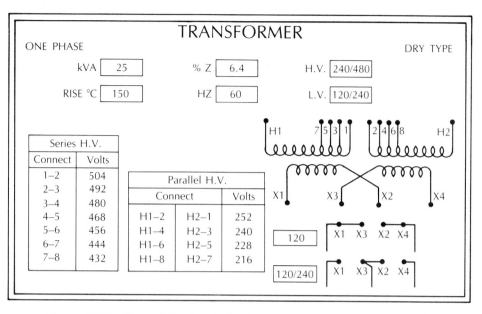

Figure 14-19 Nameplate of a single-phase transformer with primary taps.

## BUCK AND BOOST TRANSFORMERS

A *buck and boost transformer* is an insulating transformer which can be connected as an autotransformer. The buck and boost transformer is used to make small adjustments in voltage either up or down. For example, a machine has an electric motor which requires 208 V, but the electrical supply is 240 V. If ordering the machine with a 240-V motor is expensive, a less costly solution to the problem may be to buck the voltage from 240 V down to 208 V with a buck and boost transformer.

Low voltage resulting from voltage drop can be corrected with a buck and boost transformer, although this is not a good practice except in unusual circumstances. Voltage drop on wires is wasted energy and should be avoided.

Buck and boost transformers for single-phase applications have a dual-voltage primary rated at 120/240 V. A choice of two sets of secondary voltages is available, depending upon the amount of boosting or bucking required: 12/24 V and 16/32 V. Three-phase applications from 380 V to 500 V requires the use of a buck and boost transformer with a 240/480-V primary and a 24/48-V secondary. A typical buck and boost transformer is shown in Figure 14-22.

A buck and boost transformer, when used as an autotransformer for bucking or boosting, can supply a load which requires several times the kVA rating of the transformer. The maximum kVA rating of the load supplied depends upon the full-load current rating of the transformer secondary and the operating voltage of the load. Each manufacturer supplies load current and kVA data for buck and boost transformers for all combinations of input and output voltages.

The manufacturer also supplies complete wiring diagrams for both single- and 3-phase applications. The

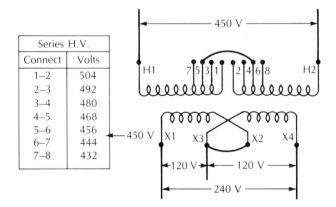

Figure 14-20 Single-phase transformer showing proper tap connection for an input of 450 V, and a 3-wire 120/240-V output.

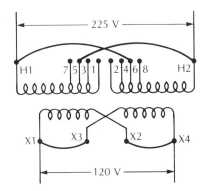

| Parallel H.V. | |
|---|---|
| Connect | Volts |
| H1–2  H2–1 | 252 |
| H1–4  H2–3 | 240 |
| H1–6  H2–5 | 228 |
| H1–8  H2–7 | 216 |

Figure 14-21  Single-phase transformer showing proper tap connection for an input of 225 V, and an output of 120 V.

transformer of Figure 14-22 is shown boosting, and the transformer of Figure 14-23 is shown bucking.

## CHOOSING kVA RATING FOR THE LOAD

The kVA rating of a transformer for a particular application depends upon the phase, voltage, and current requirement of the load. With these three quantities known, the kVA can be calculated by using either Equation 14.7 or Equation 14.8.

$$\text{Single-phase kVA} = \frac{\text{Volts} \times \text{Amperes}}{1\,000} \quad \text{Eq. 14.7}$$

$$\text{3-phase kVA} = \frac{1.73 \times \text{Volts} \times \text{Amperes}}{1\,000} \quad \text{Eq. 14.8}$$

Assume that a 120/240-V, 3-wire, single-phase, 100-A panelboard will be supplied from a 480-V service through a step-down transformer. Determine the minimum kVA rating of the transformer.

The transformer is single-phase, and the higher of the two output voltages is used in the calculations. The kVA is determined by using Equation 14.7.

$$\text{Single-phase kVA} = \frac{240\text{ V} \times 100\text{ A}}{1\,000} = 24\text{ kVA}$$

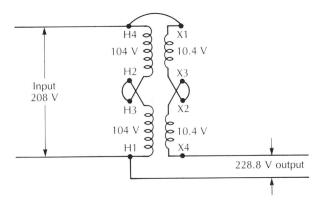

Figure 14-22  A buck and boost transformer connected as an insulating transformer to produce a 24-V output, and as an autotransformer to boost 208 V up to 228.8 V.

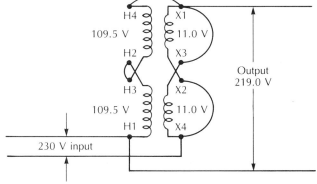

Figure 14-23  A buck and boost transformer with a 10 to 1 turns ratio connected to buck 230 V down to 219 V. Note that the secondary winding is parallel connected.

The next standard transformer size larger than this minimum kVA requirement is chosen. In this example, a 25-kVA, single-phase transformer will be used. This could be the same as the transformer shown in Figure 14-19. Common transformer kVA ratings are shown in Table 14-1.

## Problem 14-5

A farm grain-drying and storage center is supplied with a 277/480-V, 3-phase system. Two 20-A, 120-V circuits are required for lights and receptacle outlets. Determine the minimum kVA rating of the single-phase, 480-V-to-120-V step-down transformer.

## Solution

Two 20-A circuits are required for a total load requirement of 40 A at 120 V. The minimum kVA requirement is determined by using Equation 14.7.

$$\text{Single-phase kVA} = \frac{120 \text{ V} \times 40 \text{ A}}{1\,000} = 4.8 \text{ kVA}$$

From Table 14-1, choose a 5-kVA, single-phase transformer.

## Problem 14-6

A machine has a 480-V, 3-phase electrical motor as an integral part of the machine. The total machine load requirement is 10 A at 480 V. If the building has a 240-V, 3-phase electrical system, determine the minimum-kVA 3-phase transformer required.

## Solution

The load requirement is 10 A, 480 V, 3 phase; therefore, use Equation 14.8.

$$\text{3-phase kVA} = \frac{1.73 \times 480 \text{ V} \times 10 \text{ A}}{1\,000} = 8.3 \text{ kVA}$$

From Table 14-1, choose a 9-kVA, 3-phase, 240-V-to-480-V step-up transformer.

Three single-phase transformers can be used to supply a 3-phase load. This is frequently the case when low kVA rating are required, such as in Problem 14-6. When single-phase transformers are used, the rating of each transformer must be not smaller than one-third the 3-phase kVA required. In Problem 14-6, the 3-phase load requirement is 8.3 kVA. Therefore, the single-phase transformer rating must be at least 2.8 kVA.

$$\frac{8.3 \text{ kVA}}{3} = 2.8 \text{ kVA}$$

The next larger common size single-phase transformer is 3 kVA. Therefore, three single-phase, 3-kVA, 240-V-to-480-V step-up transformers can be connected to form a 3-phase transformer bank. The transformer bank will have a 3-phase rating of 9 kVA.

## OVERCURRENT PROTECTION

Wiring a transformer circuit is one of the most difficult of wiring tasks, unless the installer understands transformer fundamentals. This unit deals with dry-type transformers operating at 600 V and less. Rules for siz-

**Table 14-1 Common transformer kVA ratings**

| Transformer kVA | |
|---|---|
| Single phase | Three phase |
| 0.10 | — |
| 0.15 | — |
| 0.25 | — |
| 0.50 | — |
| 0.75 | — |
| 1 | — |
| 1.5 | — |
| 2 | — |
| 3 | 3 |
| 5 | — |
| — | 6 |
| 7.5 | — |
| — | 9 |
| 10 | — |
| 15 | 15 |
| 25 | 25 |
| — | 30 |
| 37.5 | 37.5 |
| — | 45 |
| 50 | 50 |
| 75 | 75 |
| 100 | — |
| — | 112.5 |
| — | 150 |
| 167 | — |
| — | 225 |
| 250 | — |
| — | 300 |
| — | 500 |
| — | 750 |
| — | 1 000 |

ing overcurrent protection for this type of transformer are covered in *NEC Section 450-3(b)*. It must be noted that these rules apply only to the transformer itself, and not necessarily to the input and output circuit wires. Sizing and protecting transformer input and output wires is covered in the next section.

Three methods of providing overcurrent protection for transformers is covered by the *National Electrical Code*. Both the primary and the secondary windings must be protected. The procedure begins by calculating the primary and the secondary full-load current, using Equation 14.4 for single-phase transformers, and Equation 14.5 for 3-phase transformers.

A transformer can be protected by one overcurrent device on the primary side rated at not more than 1.25 (125%) times the primary full-load current, Figure 14-24. This overcurrent device can be a set of fuses in a panelboard, a fusible switch, or a circuit breaker.

Consider that a 25-kVA, single-phase transformer has a 480-V primary and a 120/240-V, 3-wire secondary. The primary full-load current is 52 A.

$$\text{Primary full-load current} = \frac{25 \text{ kVA} \times 1\,000}{480 \text{ V}} = 52 \text{ A}$$

$$\text{Maximum size overcurrent device} = 52 \text{ A} \times 1.25 = 65 \text{ A}$$

The maximum size overcurrent device is 65 A. Checking a fuse catalog or *NEC Section 240-6*, 65 A is not a standard size. However, *Exception No. 1* to *NEC Section 450-3(b)(1)* permits the next higher standard size overcurrent device to be chosen. Therefore, the maximum size overcurrent device permitted for this situation is 70 A. A smaller size overcurrent device could have been used; for example, 60 A. In fact, the overcurrent device can be as small as desired as long as it is large enough to satisfy the load requirements.

If the primary full-load current is less than 9 A, the primary overcurrent device is not permitted to exceed 1.67 (167%) times the primary full-load current, *NEC Section 450-3(b)(1), Exception No. 1*. If the primary current is less than 2 A, the overcurrent device is not permitted to exceed 3.0 (300%) times the primary full-load current, *NEC Section 450-3(b)(1)*.

Consider the case of a 3-kVA transformer stepping down 480 V to 120 V to supply one 20-A single-phase circuit. The primary full-load current is 6.25 A.

$$\text{Primary current} = \frac{3 \text{ kVA} \times 1\,000}{480 \text{ V}} = 6.25 \text{ A}$$

The overcurrent device is sized at 125% of the primary full-load current.

$$\text{Overcurrent size} = 6.25 \text{ A} \times 1.25 = 7.8 \text{ A}$$

The next standard size overcurrent device larger than 7.8 A may be used as long as it does not exceed 167% of the primary current.

$$6.25 \text{ A} \times 1.67 = 10.4 \text{ A}$$

A 10-A time-delay fuse is the maximum size permitted to protect the 3-kVA transformer. Time-delay fuses are used when the overcurrent device is sized at less than 15 A.

The overcurrent device protecting the primary of a transformer is permitted to be sized as large as 2.50 (250%) times the primary full-load current, provided the transformer secondary is also protected, Figure 14-25. The transformer secondary overcurrent device is sized at 1.25 (125%) times the secondary full-load current. The next size larger overcurrent device is permitted except where the secondary current is less than 9 A. In this case, the overcurrent device is not permitted to be larger than 1.67 (167%) times the transformer secondary circuit.

Consider again the 3-kVA, single-phase, 480-V-to-

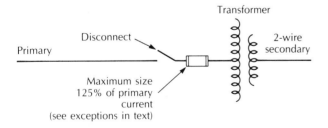

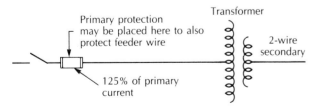

**Figure 14-24** A transformer may be protected with one overcurrent device on the primary, and sized at not more than 125% of the primary full-load current

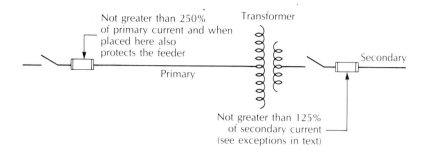

**Figure 14-25** If the transformer is protected on the secondary with an overcurrent device rated at not more than 125% of the secondary current, the primary overcurrent device may be sized as large as 250% of the primary current.

120-V step-down transformer of the previous example. Assume that the input wire is No. 14 AWG copper with THW insulation, and that it is protected by a 15-A circuit breaker. The circuit breaker clearly is too large to serve as the sole protection for the transformer. But, the 15-A circuit breaker is less than 250% of the transformer full-load current.

$$6.25 \text{ A} \times 2.50 = 15.6 \text{ A}$$

An overcurrent device installed to protect the transformer secondary at 125% of the secondary full-load current will satisfy the Code requirements for transformer protection, *NEC Section 450-3(b)(2)*.

$$\text{Secondary full-load current} = \frac{3 \text{ kVA} \times 1\,000}{120 \text{ V}}$$
$$= 25 \text{ A}$$
$$\text{Secondary overcurrent device} = 25 \text{ A} \times 1.25$$
$$= 31 \text{ A}$$

A 30-A fuse or circuit breaker at the transformer secondary will provide adequate transformer protection. Recall that the purpose of installing the transformer was to provide one 20-A circuit. It would then make sense to provide a 20-A fuse or circuit breaker at the transformer secondary, Figure 14-26.

Some transformers may have a thermal protector installed by the manufacturer to sense transformer overload and open the input circuit. The transformer nameplate will indicate if it is thermally protected. A thermally protected transformer requires secondary overcurrent protection sized at 1.25 (125%) times the secondary full-load current, the same as in the previous example.

The thermally protected transformer, with an impedance of 6% or less, is permitted to have the primary protected at not greater than six times the primary full-load current. If the transformer impedance is more than 6% but less than 10%, the primary protection must not exceed four times the primary full-load current, Figure 14-27.

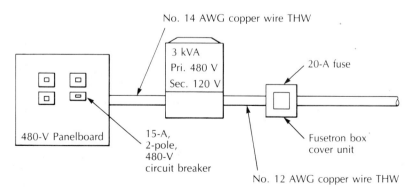

**Figure 14-26** Transformer supplying one circuit is protected on both the primary and secondary

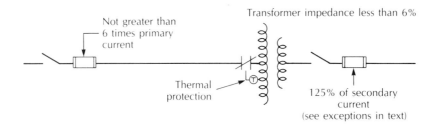
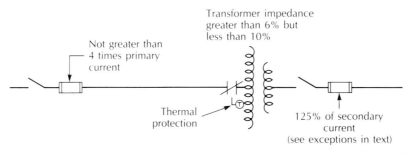

Figure 14-27 Overcurrent protection for a transformer with factory-installed thermal protection on the primary

## WIRE SIZE AND PROTECTION

Electrical wires must be adequate to supply the load and protected from overcurrent, according to the wire ampere rating as given in *NEC Table 310-16*. Therefore, wire ampere rating must also be considered when sizing transformer overcurrent devices. An example will help to illustrate the procedure of sizing feeder wires and overcurrent protection.

A 100-A, 120/208-V, 3-phase panelboard is supplied from a 480-V, 3-phase system through a 37.5-kVA step-down transformer, Figure 14-28.

The first procedure is to determine the primary and the secondary full-load currents.

$$\text{Primary full-load current} = \frac{37.5 \text{ kVA} \times 1\,000}{1.73 \times 480 \text{ V}} = 45 \text{ A}$$

$$\text{Secondary full-load current} = \frac{37.5 \text{ kVA} \times 1\,000}{1.73 \times 208 \text{ V}} = 104 \text{ A}$$

The 120/208-V panelboard in the example is rated at 100 A; therefore, it must be protected at not more than 100 A, *NEC Section 384-16(a)*. An overcurrent device must be provided on the secondary side of the transformer because the transformer is 3 phase and the feeder consists of more than two wires, *NEC Sections 384-16(d)* and *240-3, Exception No. 5*. A fusible disconnect switch is installed adjacent to the transformer. If 100-A fuses are installed in the disconnect switch, a main breaker will not be needed in the 100-A panelboard. The secondary feeder wire must have an ampere rating of at least 100 A. No. 3 AWG copper with THWN insulation is adequate.

The primary overcurrent device is not permitted to exceed 250% of the primary full-load current.

$$45 \text{ A} \times 2.5 = 112 \text{ A}$$

But, there is no reason to protect the primary at 100 A, which is the largest size fuse that does not exceed 112 A. If the fuses are sized at 110 A, then the wire must be rated for 110 A. In this example, find a copper THWN wire from *NEC Table 310-16* which is rated for at least 45 A. No. 8 AWG copper THWN wire is rated at 50 A. Therefore, install 50-A fuses in the 480-V panelboard.

The disconnect switch in Figure 14-28 must be located so that the wire from the transformer is not more than 25 ft (7.62 m) in length, *NEC Section 240-21, Exception No. 8*. This tap conductor must have an ampere rating sufficient to supply the load; in this case, at least 100 A. It must end at a circuit breaker or set of fuses with a rating not greater than the ampere rating of the tap conductor. The tap conductor must be enclosed in raceway.

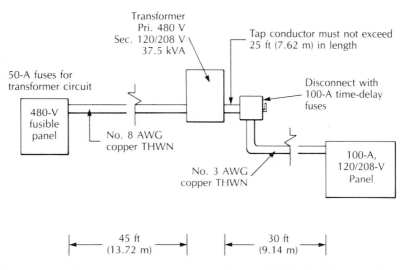

**Figure 14-28** A 3-phase transformer supplying a 120/208-V, 100-A panelboard

If the 100-A panelboard of Figure 14-28 is within 25 ft (7.62 m) of the transformer, the disconnect switch can be eliminated, and a set of 100-A main fuses or a 100-A main circuit breaker can be installed in the panelboard, as shown in Figure 14-29. The minimum size secondary wire would be No. 3 AWG copper THWN.

A 2-wire secondary from a single-phase transformer is permitted to be protected by the transformer primary overcurrent device, *NEC Section 240-3, Exception No. 5*. However, there is a *caution*. The primary overcurrent device must be sized so that the secondary wire cannot become overloaded. Equation 14.9 can be used to determine the maximum size primary overcurrent device rating.

$$\text{Maximum primary fuse rating} = \frac{\text{Secondary voltage}}{\text{Primary voltage}} \times \text{Secondary wire ampacity}$$

Eq. 14.9

An example will help to illustrate how a primary fuse can protect the secondary circuit wire. Consider a 2-kVA, 480-V-to-120-V, single-phase step-down transformer supplying one circuit for lights and receptacle outlets.

First, determine the primary and the secondary full-load current for the 2-kVA transformer.

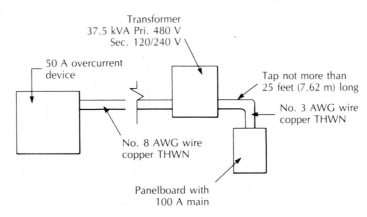

**Figure 14-29** The panelboard is kept within 25 ft (7.62 m) of the transformer and the fusible disconnect is eliminated. A 100-A main breaker is installed in the panelboard.

Unit 14  Transformers  289

Primary full-load current
$$= \frac{2\text{ kVA} \times 1\,000}{480\text{ V}} = 4.2\text{ A}$$

Secondary full-load current
$$= \frac{2\text{ kVA} \times 1\,000}{120\text{ V}} = 16.7\text{ A}$$

If the secondary wire is No. 14 AWG copper, it has a maximum load ampere rating of 15 A. Equation 14.9 is used to determine the maximum size primary fuse permitted.

Maximum primary fuse rating
$$= \frac{120\text{ V}}{480\text{ V}} \times 15\text{ A} = 3.75\text{ A}$$

If the secondary wire is No. 12 AWG copper, then it has a maximum load ampere rating of 20 A.

Maximum primary fuse rating
$$= \frac{120\text{ V}}{480\text{ V}} \times 20\text{ A} = 5\text{ A}$$

But, if the primary fuse is sized at 5 A, the transformer could become overloaded. The best choice would be to size the primary fuse at 4 A, and install a No. 12 AWG secondary wire. The maximum secondary current will be 16 A, which is slightly below the secondary full-load current. This transformer circuit is shown in Figure 14-30.

Sizing transformer circuit conductors and overcurrent protection requires careful thought and good judgment.

## MOUNTING TRANSFORMERS

Transformers must be installed in an area that will minimize the possibility of physical damage, *NEC Section 450-8(a)*. It must be remembered that transformers are not 100% efficient and, therefore, they produce heat. They must be installed in an area where there is free circulation of air around the transformer, *NEC Section 450-9*. Dry-type transformers rated at 600 V and less, and not more than 112.5 kVA, should be mounted on fire-retardant material. Transformers rated at more than 112.5 kVA shall be installed in a transformer room of fire-retardant construction, *NEC Section 450-21*. Transformers mounted outdoors shall have a waterproof enclosure and, if rated at more than 112.5 kVA, shall be spaced at least 12 in (305 mm) from combustible materials of buildings.

## GROUNDING

The alternating-current system derived from the transformer is required to be grounded the same as the building electrical service entrance, *NEC Section 250-26*. *NEC Section 250-5* specifies the wire which must be grounded. The grounded wire, usually the neutral, must be connected to a grounding electrode. This grounding connection can be made at the transformer, or at a service panel or disconnect switch supplied from the transformer, Figure 14-31.

The grounding electrode shall be as close as possible to the point where the grounding electrode conductor connects to the grounded circuit wire. Acceptable grounding electrodes listed in *NEC Section 250-26(c)* are:

- The nearest available effectively grounded structural metal member of the structure
- The nearest available effectively grounded metal water pipe

If neither of these grounding electrodes is available, then one of the grounding electrodes described in *NEC Sections 250-81* and *250-83* shall be used.

The size of the grounding electrode conductor is based upon the size of the output conductor from the transformer, using *NEC Table 250-94*.

A bonding jumper is required between the grounding electrode conductor of the derived system and the metal enclosure of the transformer, *NEC Section 250-26(a)*. This connection is permitted to be made at a panelboard or disconnect supplied by the transformer.

The transformer is required to be grounded and bonded at the transformer when the secondary has no disconnect or overcurrent protection, but is protected only by the primary overcurrent device.

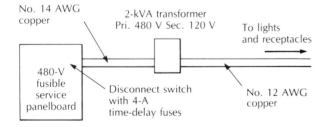

Figure 14-30 The primary fuse is sized to protect the transformer and the secondary wire.

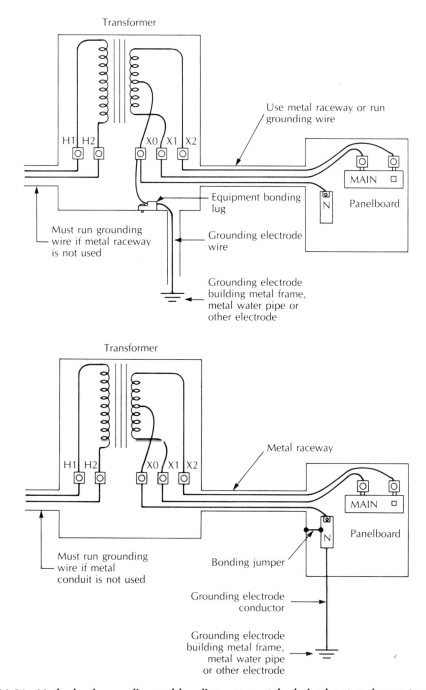

Figure 14-31 Methods of grounding and bonding a separately derived system from a transformer

## CONTROL TRANSFORMERS

Transformers used only for the purpose of supplying power to control motors and equipment come under the category of a Class 1 circuit, *NEC Section 725-11*. The installation of these transformers is covered by *NEC Article 450*, as discussed in this unit. Control transformers are available with a fuseholder on the primary side to provide transformer protection, as required in *NEC Sections 725-11, 725-12,* and *430-72*.

A grounding electrode and grounding electrode conductor are not required for these transformers as long as the transformer is rated at not more than 1 000 VA (1 kVA), *Exceptions* to *NEC Section 250-26*. Grounding can be accomplished by bonding the wire to be grounded to the metal frame of the transformer or enclosure.

## REVIEW

Refer to the *National Electrical Code* when necessary to complete the following review material. Write your answers on a separate sheet of paper.

1. Explain the purpose of a transformer.

2. As electrical current flows through a wire, is a force field built up around the wire?

3. Describe what will happen if a magnetic field is moved across a wire.

4. A transformer operates on the principal of what kind of induction?

5. A transformer has a primary to secondary turns ratio of 15 to 1. If the primary is rated at 240 V, then the secondary voltage is:
   a. 120 V
   b. 24 V
   c. 16 V
   d. 12 V

6. A transformer has a primary rated at 240 V and a secondary rated at 120 V. The primary to secondary turns ratio is:
   a. 4
   b. 2
   c. 0.5
   d. 0.25

7. The primary full-load current of a 3-phase, 480-V-to-120/208-V step-down transformer with a 30-kVA rating is:
   a. 36 A
   b. 45 A
   c. 62 A
   d. 83 A

8. A control transformer has a secondary rated 10.4 A at 24 V. The primary is rated 480 V, and the primary to secondary turns ratio is 20 to 1. Of the following, which is the primary full-load current?
   a. 2.08 A
   b. 1.22 A
   c. 0.80 A
   d. 0.52 A

9. Of the following diagrams A, B, C, and D, which represents an autotransformer?

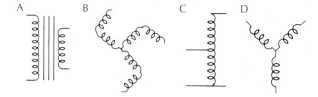

10. Of the diagrams A, B, C, and D for Problem No. 9, which represents an insulating transformer?

11. In the following diagram, show in pencil how to connect the 120/240-V primary and the 12/24-V secondary to obtain a 24-V output when the input to the transformer is 240 V.

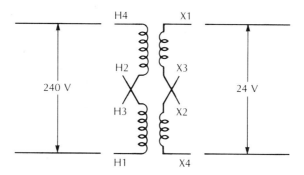

12. The 3-phase transformer in the following diagram is connected:
    a. Wye-wye
    b. Wye-delta
    c. Delta-wye
    d. Delta-delta

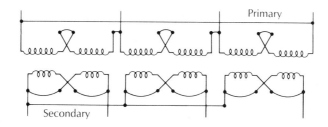

13. A single-phase step-down transformer is wired for a 120/240-V, 3-wire output. If the input voltage is only 440 V, show how to connect the primary so that the output will be close to but will not exceed 240 V.

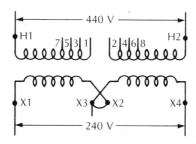

14. What is the name of the insulating transformer that can be connected as an autotransformer to raise 208 V up to 230 V?

15. A building is supplied with a 480-V, 3-phase electrical system. The kVA rating of a single-phase transformer required to supply a 60-A, 3-wire, 120/240-V panelboard is:
    a. 10 kVA        c. 20 kVA
    b. 15 kVA        d. 37.5 kVA

16. A 6-kVA, 3-phase transformer is used to step up 240 V to 480 V to supply a machine with a 3-phase motor. The secondary of the transformer is protected with 8-A time-delay fuses. What is the maximum size circuit breaker permitted to be installed to protect the primary side of the transformer?
   a. 15 A
   b. 20 A
   c. 30 A
   d. 40 A

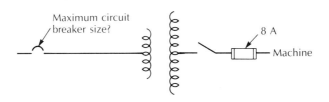

17. A 2-kVA, single-phase transformer is used to supply one 120-V circuit for lights and receptacles. The building is served with 240 V, 3 phase. The secondary circuit is protected by 8-A fuses on the primary side of the transformer. The minimum size secondary copper TW wire is No. ___?___ AWG.
   a. 10
   b. 12
   c. 14
   d. 16

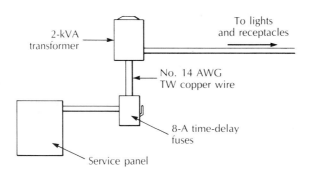

18. The buck and boost transformer has a primary rated at 120/240 V, and a secondary rated at 12/24 V. (See the diagram at the top of page 295.) Of the following, which is the output voltage of the transformer?
   a. 71 V
   b. 107 V
   c. 114 V
   d. 119 V

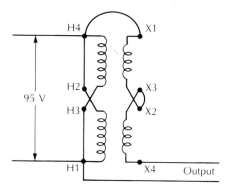

19. On a separate sheet of paper, make a sketch of the transformer shown below, and draw in the grounding and bonding wires required for this single-phase, 480-V primary, 120/240-V secondary, 37.5-kVA transformer.

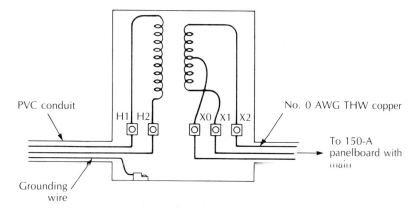

20. The minimum size copper grounding electrode conductor and equipment bonding jumper for the transformer in Problem No. 19 is:
    a. 0
    b. 2
    c. 4
    d. 6

Unit 14 Transformers 295

# UNIT 15
# HEATING AND COOLING

## OBJECTIVES

After studying this unit, the student will be able to

- list the units of measurement used in heating and cooling.
- list the three methods of heating spaces, animals, and humans.
- name at least four types of electric heating.
- describe the types of thermostats used for heating and cooling equipment control.
- list the components of an oil or gas heater.
- describe the installation of electric heating equipment, including the sizing of conductors and overcurrent protection.
- list at least three sources of heat other than heaters commonly found on farms.
- describe the installation of an electric forced-air heating system which utilizes available heat in the building.
- explain at least two farm uses of heating cable installed in concrete or asphalt floors and slabs.
- describe the installation of electric heating cable in floors and slabs.
- size conductors and electrical components for an air conditioner.
- explain the function of a heat pump.

## HEATING UNITS OF MEASUREMENT

Heat moves from a warm object to one that is cooler. The greater the difference of temperature between two objects, the greater is the rate of flow of heat. Temperature difference is to the flow of heat as voltage difference between two wires is to the flow of electricity. Two temperature scales are used in the United States: Fahrenheit (F), and Celsius (C); but only the Celsius scale is used in most other countries. The two temperature scales are compared in Figure 15-1.

Materials conduct heat at different rates or, in reverse, materials offer different resistances to the flow of

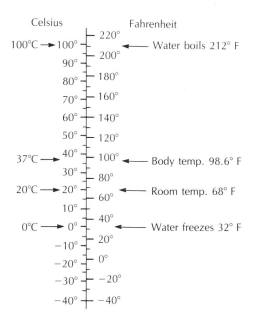

Figure 15-1 Celsius and Fahrenheit temperature scales

Table 15-1 Conversion factors for heating and cooling units

| Multiply | By | To Find |
|---|---|---|
| Btu | 1.055 | kJ |
| kCal | 4.187 | kJ |
| kWh | 3 600 | kJ |
| kWh | 3.6 | MJ |
| kW | 3 413 | Btu/hr |
| Tons | 12 000 | Btu/hr |
| Tons | 3.516 | kW |
| kW | 3 600 | kJ/hr |
| Btu/hr | 1.055 | kJ/hr |
| kCal/hr | 4.187 | kJ/hr |

heat. The resistance of a material to the flow of heat is called the *R value*. Thermal insulation is rated according to its R value.

Heat, which is energy, is measured in several units. Common units for measuring heat are: British thermal units (Btu), kilocalories (kCal), kilowatthours (kWh), and kilojoules (kJ). The kilojoule (kJ) is the SI metric unit of heat.

Electric heaters are rated in kilowatts (kW), while fuel-fired furnaces are rated in Btu per hour (Btu/hr). Heating and cooling load calculations are generally made in Btu/hr. Air-conditioning units are rated in Btu/hr for the smaller sizes, and in tons for larger sizes. A *ton* of refrigeration is the amount of heat absorbed by a 1-ton block of ice in a 24-hour period. Heating and cooling equipment is, therefore, rated in heat per unit time.

Two units will most likely be used someday for rating heating and cooling equipment, and making load calculations. These units are the *kilowatt* (kW) and *kilojoule* (kJ). The *joule* (J) is a small unit of measure of heat; therefore, the megajoule (MJ) may be used at times instead of kilojoule. One megajoule is equal to 1 000 kJ. Table 15-1 provides conversion factors for the various units used in heating and cooling.

## PRINCIPLES OF TRANSFERRING HEAT

Moving heat from one object to another involves three methods, all of which are used in agriculture. These methods are convection, conduction, and radiation.

### Convection

*Convection* is the moving of heat through a fluid such as air. Convection is the main method of moving heat for most heating and cooling systems. Forced-air or forced-convection heating systems use a fan to move air through ducts or blow the air throughout the space to be heated. Warm air rises and, therefore, air can be moved throughout an area without the aid of a fan. This is called *natural convection*.

### Conduction

*Conduction* is the moving of heat from one object to another object it touches. Heat moves by conduction through walls, floors, windows and ceilings. Animals cool themselves by conduction in summer by lying on the cool ground. Electric heating cable in the floor can be used to heat baby piglets by conduction.

### Radiation

*Radiation* includes the processes of emission, transmission, and absorption of radiant energy. Heat is emitted from an object and travels in the form of electromagnetic radiation. Therefore, heat can be moved by radiation directly from the source to a person, an animal, or an object. Heat from the sun reaches earth by radiation through the emptiness of space. Radiation is important, because an object can be heated without heating the entire area.

## HEATING WITH ELECTRICITY

Heat is produced electrically by passing current through a wire containing a desired amount of resistance.

Tungsten filaments are used in infrared heat lamps and quartz infrared tubes. Nichrome wire is used as the resistance element in most other electric resistance heaters, Figure 15-2.

A 2-kW electric resistance heater designed to operate at 235 V has a resistance of 27.7 Ω and produces 2 kWh of heat in 1 hour. This heat is equivalent to 6 826 Btu/hr, or 7.2 MJ. The amount of heat produced in watthours is determined by using Equation 2.22. The current draw of the heater is 8.5 A, as determined by using Ohm's Law, Equation 2.7.

$$\text{Heat} = I^2 \times R \times t$$
$$\text{Heat} = (8.5 \text{ A})^2 \times 27.7 \text{ Ω} \times 1 \text{ hr} = 2\ 000 \text{ Wh}$$
$$\text{Heat} = \frac{2\ 000 \text{ Wh}}{1\ 000} = 2 \text{ kWh}$$

## TYPES OF ELECTRIC HEAT

### Baseboard Heater

A popular type of electric heat is the resistance baseboard unit. This type is attached to the outside wall near the floor. Baseboard heaters are best suited for dwellings and clean areas, such as an office in a farm building, Figure 15-3.

### Self-contained Heater

Self-contained electric unit heaters are available for ceiling or wall mounting. These units are popular for agricultural applications. They come in a variety of sizes. A fan forces air over the heating elements and out into the room, Figure 15-4.

### Electric Furnace

An electric furnace is a self-contained unit containing resistance heating elements, and a blower capable of developing sufficient pressure to force the warm air through a duct system. These units are popular for heating milking facilities. Heat is forced through ducts into the milking parlor pit where the operator is working. Operating cost is generally low when an electric furnace heating system is designed to utilize heat removed from the milk by the milk-cooling system.

### Infrared Heater

Infrared heaters are used extensively around farms. They are used mostly for temporary heat in such work areas as a shop, Figure 15-5, and for heating newborn animals. Infrared heaters can consist of simply a heat

Metal fins

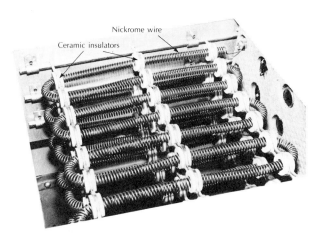

Figure 15-2 Electric resistance heating elements

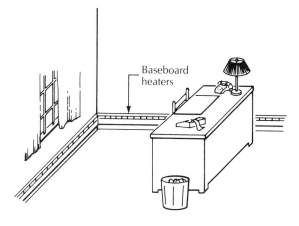

Figure 15-3 Electric baseboard heaters in a farm office

**Figure 15-4** An electric unit heater with a fan is frequently used for agricultural applications.

**Figure 15-6** A proper fixture for a heat lamp has a guard to prevent the lamp from making contact with combustible materials.

## Ceiling Cable

Ceiling cable is another form of radiant electric heat. Resistance cable is embedded in the ceiling plaster. The ceiling is heated to about 40°C (104°F), and radiant energy heats objects in the room. The cable is stapled to the base plaster and is then covered with a ½-in (13-mm) finish plaster coat, Figure 15-8. This type of heat is most suitable for dwellings and similar areas.

## Floor Cable

Electric resistance floor-heating cable can provide comfort in unheated areas where workers may be per-

**Figure 15-5** An infrared heater can provide temporary heat in a normally unheated area, such as a farm shop.

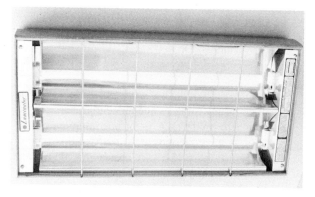

**Figure 15-7** Infrared heater with quartz tubes

lamp in a fixture with a guard to prevent the lamp from making contact with combustible materials, Figure 15-6. Infrared fixtures are available which contain quartz infrared lamps or metal sheathed nichrome heating elements, Figure 15-7. These figures are permanently mounted, and have ratings up to several kW. An important aspect of the infrared heater is that heat is directed at a particular area, and the entire room is generally not heated.

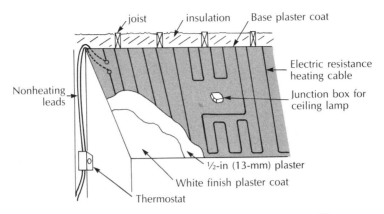

Figure 15-8 Electric resistance cable in a plaster ceiling

forming a task at a fixed location, Figure 15-9. An example of this application would be in a building where fruit and vegetable grading and packaging take place. Ice and snow removal is another important application of these cables, Figure 15-10. Ice buildup in animal passage areas can be eliminated with heating cable in concrete. Swine producers often install electric cable in the floor under baby piglets. Thermostats can be installed in the floor to regulate the proper temperature, Figure 15-11. A conduit is placed in the concrete with the thermostat remote-sensing element inserted inside.

## THERMOSTATS

A *thermostat* is a switch operated by a change of temperature. Thermostats are of the heating type and the cooling type. A heating thermostat closes a switch that turns on a heater when the temperature falls. The thermostat opens when the temperature rises. A cooling thermostat closes when the temperature rises, and turns on an air conditioner, a refrigeration unit, or a ventilation fan. The thermostat switch opens when the temperature falls. The symbols for both types of thermostats are shown in Figure 15-12. Some thermostats are both heating and cooling type.

Thermostats are of the low-voltage type and the line-voltage type. Many furnaces and air conditioners contain a transformer to produce 24 V for operating control relays. A low-voltage thermostat is used for this type of control system, Figure 15-13. A line-voltage thermostat operates at the voltage supplied to the heater or air conditioner. Line-voltage thermostats are often rated up to

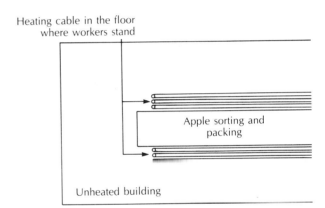

Figure 15-9 Electric heating cable in the floor used to keep workers warm in an unheated building

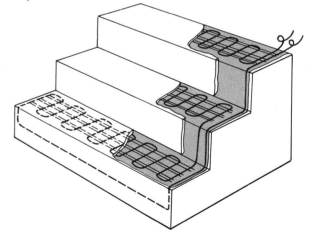

Figure 15-10 Electric heating cable in concrete steps prevents slippery ice buildup.

300    Unit 15    Heating and Cooling

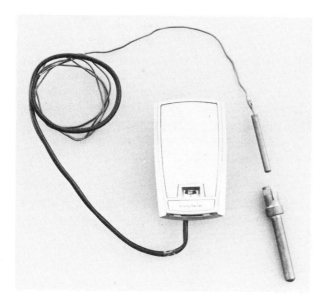

Figure 15-11  A thermostat placed in the concrete near the heating cable controls floor temperature.

22 A at 240 V, Figure 15-14. Line-voltage thermostats also have a maximum horsepower rating which must not be exceeded.

## SIZING CIRCUIT WIRES AND OVERCURRENT PROTECTION

Electric heating may consist of a central furnace that provides heat for the entire building or heated area of a building. Electric heating may consist of separate heat for each room or area within a building. In this case, a separate thermostat is provided in each room. Individual room electric heating may consist of heating cable, unit heaters, infrared heaters, or baseboard heaters. There may be one heater in a room, or several heaters. This information must be known, along with the voltage and

Figure 15-13  Low-voltage thermostat

ampere ratings of the electric heating units, before planning the circuits and wiring.

The branch-circuit wire supplying an individual electric heating device must have a rating not less than 1.25 (125%) times the ampere rating of the heater, plus any

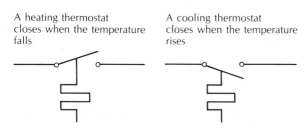

Figure 15-12  Symbols for a heating thermostat and a cooling thermostat. Think of the folded portion below the switch contact as expanding when the temperature rises.

Figure 15-14  Line-voltage thermostat

motors on the unit, *NEC Section 424-3*, Figure 15-15. As an example, a single-phase electric unit heater is rated at 7.5 kW, 240 V, and has a fan motor which draws 1 A at 240 V. The actual full-load current is usually provided on the nameplate. In this case, the heater draws 31.3 A plus 1 A for the fan, which gives a total load of 32.3 A. The supply conductor must have an ampere rating of at least 125% of the total heater current, which for this example is 40 A.

> Heater wire rating
> = 1.25 × Total heater ampere rating
> = 1.25 × 32.3 A = 40 A

The minimum wire size is found in *NEC Table 310-16*. For copper wire, the minimum size is No. 8 AWG THW.

The circuit breaker or fuses protecting the circuit must have an ampere rating of at least 125% of the heater full-load current, *NEC Section 424-3(b)*. There is an exception in the case where the overcurrent device is rated to carry 100% of its ampere rating continuously. The heater circuit for the example could be protected with 40-A fuses or a 40-A, 2-pole circuit breaker.

Several electric heaters may be supplied by a single branch circuit; however, the circuit is not permitted to exceed a rating of 30 A, *NEC Section 424-3(a)*. The same 125% rating applies as in the previous example, except that the total load is the total load of all heaters on the circuit. An example will illustrate how the minimum size circuit conductor is determined.

A farm office is heated with two baseboard electric heaters installed on the same circuit and controlled by a thermostat, Figure 15-16. One heater is rated at 8.3 A,

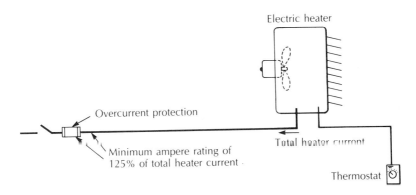

Figure 15-15 Selecting branch-circuit wire size and overcurrent protection for an electric heater

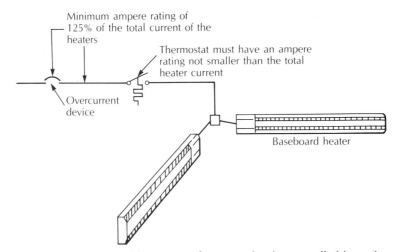

Figure 15-16 Two baseboard heaters on the same circuit, controlled by a thermostat

2 000 W, and the other is rated at 6.3 A, 1 500 W. Both are rated at 240 V.

The total load of the heaters is 14.6 A. The wire size is based upon 125% of this value, or 18.3 A. The minimum size copper wire permitted is No. 12 AWG protected with a 20-A circuit breaker or fuses.

Infrared heaters are frequently used for agricultural applications. If more than one infrared heater is installed on a single circuit, the circuit may be rated as high as 50 A, *NEC Section 424-3(a)*. This does not apply to a dwelling. The circuit conductors and overcurrent protection are sized at not less than 125% of the total heater load on the circuit.

It must be remembered that most line-voltage thermostats are usually rated at not more than 22 A, and many have a much lower rating. If the heater load exceeds the rating of the thermostat, then a heating contactor will be required. A *heating contactor* is an electrically operated switch. High-wattage electric heaters are usually supplied with a contactor, Figure 15-17. A line-voltage thermostat operating a heating contactor is shown in the diagram of Figure 15-18.

The wires to the line-voltage thermostat in Figure 15-18 are tapped directly from the heater circuit supply wires just ahead of the contactor. This thermostat wire carries line voltage; therefore, it must meet the installation requirements of the *NEC*. However, the only current that will flow on this thermostat wire during normal operation is the current drawn by the coil of the contactor. This current is seldom more than 1 A. This is considered a Class 1 remote-control wire, *NEC Section 725-11(b)*.

If shorted, the Class 1, remote-control, line-voltage thermostat wire can cause the heater to remain on. A possible *fire hazard* could be created. Therefore, the thermostat wire is required to be enclosed in conduit, *NEC Section 725-18*. The conduit may be rigid metal, IMC, EMT, or rigid nonmetallic. Flexible conduit is also permitted; however, liquidtight flexible metal conduit is required in areas which are damp or subject to corrosion.

The line-voltage thermostat wire No. 14 AWG and larger operating a heating contactor is permitted to be tapped directly from the heater circuit wires. The thermostat wire is not required to have its own overcurrent protection, provided the overcurrent device protecting the heater circuit is not more than three times (300%) the ampere rating of the thermostat wire, *NEC Section 725-12, Exception No. 3*.

## Problem 15-1

A single-phase electric unit heater is rated 7.5 kW at 235 V, and has a full-load current rating of 32 A. The

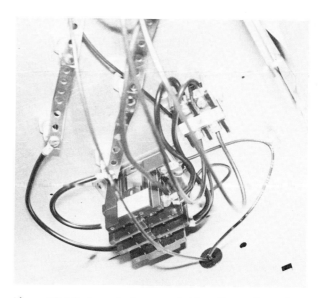

**Figure 15-17** A contactor operated by a line-voltage thermostat operates this unit heater.

heater is controlled with a contactor operated with a line-voltage thermostat. The heater circuit wire is No. 8 AWG copper THW, and the overcurrent device is rated at 40 A. Determine the minimum size TW copper wire permitted for the thermostat if there are no fuses in the heater to protect the thermostat wire.

## Solution

The heater circuit overcurrent device is not permitted to exceed three times the thermostat wire ampere rating. From *NEC Table 310-16*, a No. 14 AWG TW copper wire has a rating of 20 A.

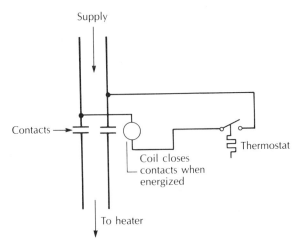

**Figure 15-18** Schematic diagram of a line-voltage thermostat controlling a heating contactor

```
Maximum rating of heater circuit
overcurrent protection
                         = 3 × 20 = 60 A
```

No. 14 AWG copper TW wire is, therefore, permitted for the thermostat.

The line-voltage thermostat wire must be installed according to the wiring methods of *Chapter 3* of the *NEC, Section 725-14*. Therefore, any appropriate conduit wiring method discussed earlier in this text is permitted.

The Class 1, remote-control thermostat wire will require overcurrent protection if the 300% requirement of *NEC Section 725-12, Exception No. 3* cannot be satisfied. A fuse rated for the voltage of the circuit is installed in a fuseholder mounted next to the contactor. The requirements for sizing the remote-control thermostat wire are summarized in Figure 15-19.

## Problem 15-2

Two 3 200-W, 240-V quartz infrared heaters are installed in a barn and supplied by one circuit. The heaters are controlled with a contactor operated by a line-voltage thermostat. Determine the following:

- Minimum heater circuit wire size
- Minimum size circuit breaker for the heater circuit
- Minimum size remote-control thermostat wire size not requiring separate overcurrent protection

All of the wire is THWN copper in conduit, Figure 15-20.

## Solution

The ampere rating of each infrared heater was not given in the problem, but it can be calculated easily.

$$\text{Current} = \frac{3\ 200\ \text{W}}{240\ \text{V}} = 13.3\ \text{A}$$

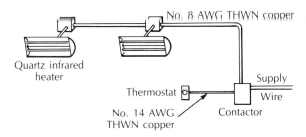

**Figure 15-20** Two quartz infrared heaters operated with a contactor controlled by a line-voltage thermostat

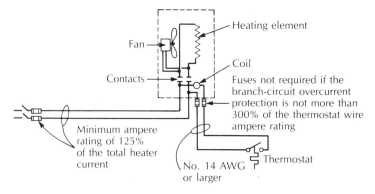

**Figure 15-19** Sizing the wires and overcurrent protection for an electric heating circuit

There are two heaters on the circuit; therefore, the total current is 26.6 A.

First, the minimum size branch-circuit wire is determined. The minimum ampere rating of the THWN copper wire must be 33 A.

$$\text{Branch-circuit wire} = 1.25 \times 26.6 \text{ A} = 33.3 \text{ A}$$

From *NEC Table 310-16*, the minimum size wire is No. 8 AWG copper THWN.

Next, the minimum size circuit breaker must be selected. The circuit breaker must have a rating of at least 33 A in this example. Make sure, also, that the branch-circuit wire has a rating not smaller than the circuit breaker selected. For this problem, a 40-A circuit breaker will be selected.

Finally, the minimum size line-voltage thermostat wire must be selected. The circuit breaker is permitted to have a rating three times that of the thermostat wire ampere rating. Therefore, divide the circuit breaker ampere rating by three, and look up the minimum wire size in *NEC Table 310-16*.

$$\text{Thermostat wire ampere rating} = \frac{40 \text{ A}}{3} = 13.3 \text{ A}$$

Find a copper THWN wire with an ampere rating of at least 13.3 A. Obviously, for this circuit, a No. 14 AWG wire is adequate.

## SUPPLEMENTARY OVERCURRENT PROTECTION

The resistance elements in resistance-type electric space heating equipment are not permitted to be protected by an overcurrent device rated at more than 60 A, *NEC Section 424-22(b)*. Considering that 60 A is 125% of the heater current, the maximum heater current is 48 A. Therefore, the heating elements in high-wattage electric heating equipment are subdivided and individually protected within the heating equipment. Some resistance-type electric boilers are permitted to have heating elements that draw up to 120 A, *NEC Section 424-72(a)*.

## DISCONNECT

An electric heater is required to have a disconnect that will open any ungrounded (hot) wire to the heater. This disconnect is required to be within sight of the electric heating equipment, or it must be capable of being locked in the open (off) position. This is required of electric heating equipment containing a motor rated at more than 1/8 hp. This is also required for high-wattage heating equipment which has the heating elements subdivided and individually protected, Figure 15-21(A).

A circuit breaker or a fusible switch not in sight from the electric heating equipment is permitted to serve as the disconnect provided the equipment does not have a motor larger than 1/8 hp, and the heating elements are not individually protected. Such equipment includes small unit heaters, infrared heaters, baseboard heaters, and ceiling heating electric cable, Figure 15-21(B).

A switch on the electric heating equipment with a marked off position that opens all ungrounded (hot) wires may serve as the disconnect. The circuit breaker or fusible switch protecting the heater circuit does not need to be in sight from the heater, but it must be accessible, Figure 15-21(C).

A line-voltage thermostat may be temperature operated only, or it may also have a manual off position. When the thermostat is turned to the off position, it cannot be turned on by a change in temperature. A thermostat with a manual off position is permitted to serve as a disconnect for electric heating equipment, provided it opens all ungrounded (hot) wires.

Line-voltage thermostats are commonly available as single-pole and 2-pole devices. A single-pole thermostat with an off position is permitted to serve as the disconnect only for a 120-V electric heating unit. Most electric heating equipment is rated at 240 V; therefore, a 2-pole thermostat with an off position is required if the thermostat is to serve as the disconnect. A single-pole thermostat without a marked off position is permitted to control 240-V electric heating equipment, but it does not serve as a disconnect.

## HIGH-TEMPERATURE LIMIT SWITCH

Most space heating equipment contains a high-temperature limit switch to shut off power to the heater if the temperature becomes excessive. Failure of a furnace blower, for example, will result in a sudden increase in the temperature within the furnace. The high-temperature limit switch opens the circuit and shuts off the heat before the danger of fire can develop. The high-temperature limit switch is wired in series with the electric heating element. The manufacturer installs the switch adjacent to the heating element or in the upper portion of the heater where heat is most likely to become excessive.

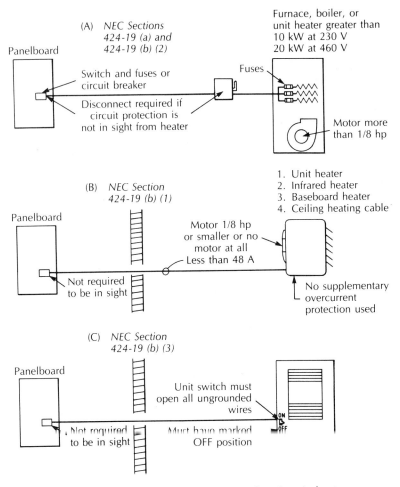

Figure 15-21 Disconnect requirements for electric heaters

Figure 15-22 shows a high-temperature limit switch in a baseboard electric heater.

Sometimes the high-temperature limit switch fails and becomes permanently open. When this happens, the heater will not operate. The high-temperature limit switch should be tested with a continuity tester to make sure that it is closed when troubleshooting a heater that fails to operate.

## INSTALLING ELECTRIC HEATING EQUIPMENT

Electric heating equipment installed on farms must be protected from physical abuse, *NEC Section 424-12(a)*. This also applies to circuit conductors. Electric heating equipment must be resistant to corrosion and moisture when installed in damp areas containing livestock, *NEC Section 424-12(b)*. Moisture may accumulate in moving parts, such as contactors, causing rust and eventual failure. Contactors installed in damp areas should be periodically treated with a suitable lubricant to prevent rust and to keep them operating freely.

Electric heating equipment, as well as all types of heating equipment, must be kept a safe distance from combustible materials, *NEC Section 424-13*.

Grounding is extremely important for all equipment installed in agricultural locations. All exposed noncurrent-carrying metal parts of electric heating equipment and the wiring must be grounded, *NEC Section 424-14*.

Figure 15-22 A high-temperature limit switch in series with the heating element prevents overheating.

## FUEL-FIRED HEATERS

Space heaters fired with gas or oil are frequently used for agricultural applications. Common types are unit heaters, infrared heaters, and furnaces. Unit heaters and infrared heaters are usually gas fired, while central furnaces may be oil or gas fired. Central furnaces may be of the air circulation type, requiring a heating duct system, or they may be of the hot-water type.

The hot-air central furnace heating system requires heating ducts to deliver air to the desired areas. A means must be provided to return cooler air to the furnace room. Electricity requires a complete circuit for proper operation, and a hot-air heating system also requires a complete circuit. An inadequate air return can cause inadequate heating of an area.

Radiators are installed in each area to be heated with a hot-water system. These radiators are often in the form of baseboard units that, from the outside, appear similar to electric baseboard heaters. A circulating pump forces the hot water throughout the piping and radiator system. When heat is required in an area, the circulating pump turns on. This allows cooler water to enter the furnace. A thermostat senses this cooler water, and turns on the furnace burner motor. Individual rooms or areas of a building may have separate thermostats. The hot-water piping system is subdivided into zones, each of which is controlled with a thermostat. Hot water to each zone is controlled by a zone valve.

The heating contractor usually furnishes and installs oil and gas heating equipment and controls. A 120-V circuit is usually required for a furnace. Low-voltage thermostats are generally used. Electric motors are a part of these systems; therefore, a disconnect is required within sight from the furnace, *NEC Sections 430-102 and 422-26*. If the motor is not thermally protected, then running overcurrent protection for the motor is required, *NEC Section 430-32(c)*. A fusible disconnect can be installed near the furnace. Time-delay fuses sized at 1.15 (115%) times the motor nameplate current will provide satisfactory overload protection.

A separate circuit for the furnace, while not required by the *NEC*, is highly recommended. Some local codes may indeed require a separate circuit for the furnace.

## ELECTRIC HEATING CABLE IN FLOORS AND SLABS

Electric heating cable for floors and slabs has numerous practical agricultural applications. Slippery entrances through which animals pass can be eliminated in northern climates with heating cable embedded in a concrete slab. Swine producers use heating cable embedded in the concrete floor to provide heat for baby piglets. Workers in a cool area can be kept comfortable by providing a heated floor. This is particularly practical in areas where fruits and vegetables are cleaned, sorted, and packaged.

Electric heating cable for floors is available in the form of a ready-to-install mat to provide either 10 W or 20 W per square foot, Figure 15-23. The mats are generally 16 in (406 mm) in width, and up to 150 ft (45.7 m) long. Ten watts per square foot (107 W/m$^2$) are adequate for supplemental heating in northern climates, while 20 W per square foot (215 W/m$^2$) are usually adequate to provide total heating for most inside areas.

Heat mats for swine farrowing pens usually provide about 40 W per square foot (430 W/m$^2$). A *farrowing pen* is a stall containing a sow and an 18-in (457-mm) area on both sides of the sow for baby piglets. The piglets are attracted to the sides by the heat, thereby reducing the chance of the sow lying on the piglets. This is one reason why heat is extremely important for a swine farrowing area. Heat mats are embedded in the concrete beneath the areas for the piglets, Figure 15-24.

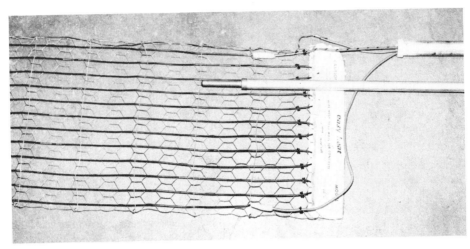

Figure 15-23 Electric cable mat ready to install in a concrete floor

Heat mats for outside use for snow and ice melting are available in 18-in (457-mm) or 36-in (914-mm) widths, and lengths up to 60 ft (183 m). They are available in 40 W or 60 W per square foot (430 W/m$^2$ or 645 W/m$^2$). These heat mats can be embedded in concrete or laid under asphalt.

Electric heating cable installation in the floor is covered in *NEC Section 424-44*. Electric heating cable manufacturers generally provide detailed installation instructions.

A typical electric heating cable floor installation is shown in Figure 15-25. There must be a firm base beneath the cable. This can be established with a 6-in (152 mm) layer of sand or fine gravel. A polyethylene vapor barrier should be placed over the sand or gravel base. Place a 2-in (50.8-mm) thick layer of rigid polystyrene or polyurethane insulation in the area where the heating cable is to be installed. The insulation board should be anchored so that it will not float to the top when concrete is poured. A base layer of about 2 in (50.8 mm) of concrete is poured over the insulation. The heating cable is anchored to the base layer, and a finish layer of at least 2 in (50.8 mm) is poured over the heating cable. The total thickness of concrete should be 4 in (102 mm).

The nonheating cable leads must be protected where they emerge from the concrete, *NEC Section 424-44(e)*. This protection may consist of rigid nonmetallic conduit. However, if the conduit is subject to abuse, rigid metal conduit, IMC, or Schedule 80 nonmetallic conduit should be used.

Floor heating cables may be controlled by a thermostat with a sensing element in the concrete floor. Figure 15-26 shows a typical installation with the lead wires protected, and a conduit extending into the floor containing the thermostat sensing equipment. Thermostat kits are available for sensing floor temperature (see Figure 15-11).

Heating cable installed in slabs for snow and ice removal is permitted to have a higher watts-per-square-foot value than is permitted for installations within buildings, *NEC Section 426-20(a)*. A base layer of at least 2 in (50.8 mm) of asphalt or concrete, and 1 ½-in (38-mm) finish coat is the minimum requirement, *NEC Section 426-20(c)*. Heating cables must not bridge expansion joints in the concrete unless adequate provision is made for normal expansion and contraction.

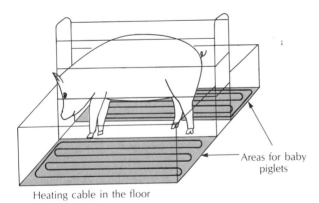

Figure 15-24 Floor heating electric cable is used to keep baby piglets warm and away from the sow when the piglets are not nursing

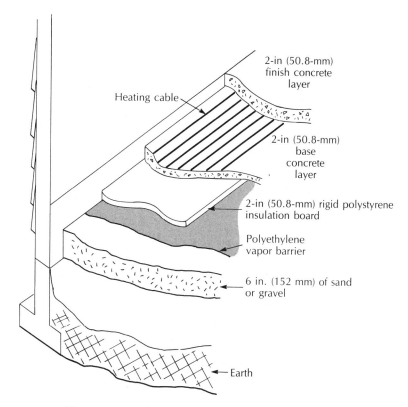

**Figure 15-25  Electric heating cable floor installation**

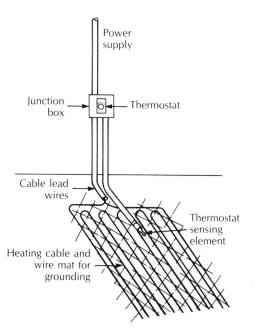

**Figure 15-26  Thermostat sensing element and lead wires must be protected with conduit**

Branch-circuit wires to the electric heating cable should be in conduit. This conduit may be embedded in the concrete or in the earth beneath the slab. Watertight junction boxes are available to install in the conduit with access from the surface. For most installations on the farm, a junction box in the slab is not necessary. The lead wires supplied with the cable are usually long enough to reach a junction box inside an adjacent building. Only 1 in to 6 in (25.4 mm to 152 mm) of the nonheating lead are permitted to be embedded in the concrete, unless the lead wire has a grounding sheath, Figure 15-27. See *NEC Section 426-22(b)*.

The wattage and proper operating voltage are marked on electric heating cable for easy identification. The lead wires, however, are color coded for easy identification, *NEC Section 424-35*. The following color coding applies to all types of heating cable.

```
120 V—yellow
208 V—blue
240 V—red
277 V—brown
```

Unit 15  Heating and Cooling

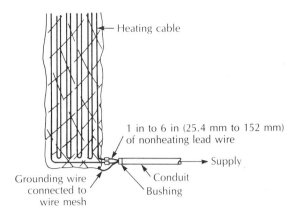

**Figure 15-27** Only the nonheating leads are permitted in the conduit.

## HEAT SOURCES IN AGRICULTURAL BUILDINGS

Animal, product, and natural heat may be available in significant quantities to augment a space heating system. A mature dairy cow can produce up to 3 000 Btu of heat per hours. People at normal work produce 1 000 Btu per hour. Fruit and vegetable products, such as apples or potatoes, produce heat while in storage. This *heat of respiration*, as it is called, must be removed from the storage by a refrigeration system. The heat from the refrigeration system may be used for space heating of an area, or it may be used for heating water. Water preheaters attached to a refrigeration system are commonly used on dairy farms.

Natural heat, particularly from the sun, can be put to good use; but, this requires careful planning prior to construction of farm buildings. Heating cable in a slab may not be necessary if the animal entrance to a building is on the south or west side of the building. Windows can be arranged to catch sunlight in the winter. Sunlight should be kept out in the summer, because it can add a significant heat load which must be removed by an air-conditioning system.

Heat from lights and machinery can also be used for heating if properly planned into the heating system. Grouping the machinery into a single room used as a heating plenum introduces this heat into the heating system. A *plenum* for heating on a farm is an enclosed space in which various sources of natural and mechanical heat are utilized together to augment the heating system.

Animals and humans give off moisture into a room as well as heat. An area cannot be closed up tight to conserve heat, because some ventilation is required to eliminate excess moisture. Unvented, portable fuel-fired heaters produce water and carbon dioxide as products of combustion. These heaters are suitable only in well-ventilated spaces. Ventilation for animals is discussed more thoroughly in a later unit. Cooperative Extension Service personnel in virtually every predominantly agricultural county in the United States and Canada are available to provide assistance in planning livestock housing.

## PIPE-HEATING CABLES

Electric heating cables for pipelines on farms are generally of the plug-in type. This type is not covered in the *National Electrical Code*. If the cable is permanently connected to the wiring system, it then comes under *NEC Section 427*.

Pipeline heating tape should be wrapped around the pipe in a spiral pattern, and secured at intervals with friction tape. The cable must never cross over itself; if it does, it will generally overheat and become damaged. The heat tape should be covered with insulation to conserve heat. Heat tapes for pipelines are available with a thermostat so they will operate only when the temperature nears freezing. An electrical outlet must be located conveniently to supply power to the heating cable. It is important that a metal pipe be grounded in the event a hot wire accidentally contacts the pipe. Grounding will cause a fuse to blow or a circuit breaker to trip, thus preventing a voltage on the pipe which could endanger animals.

## ENERGY GUIDE

Air conditioners and household refrigeration appliances contain an energy guide to aid the consumer in selecting energy-efficient models. Refrigeration appliances contain a guide which gives the estimated annual operating cost. This is shown on a scale with the range of operating costs for other manufacturers' models with the same capacity.

Air conditioners contain an energy efficiency rating (EER) which compares the cooling capacity with the input power required to operate the unit. It is desirable for this number to be as large as possible. This EER number is listed on a scale showing the range of other models.

## REFRIGERATION PRINCIPLES

*Refrigeration* is the removal of heat from a room or an object. A block of ice placed in a space will cool the air because the ice absorbs about 144 Btu of heat per hour for every pound of ice that melts. The heat absorbed to simply melt the ice is called *latent heat*.

A modern refrigeration system uses a special refrigerant gas. A refrigeration compressor simply pumps the refrigerant throughout the system. The refrigerant repeatedly changes back and forth from a liquid to a gas. When a liquid evaporates, it absorbs heat. When it condenses, it gives off heat.

The compressor pumps refrigerant vapor into the condenser, where it changes to a liquid and gives off heat. The liquid then travels to the expansion valve. The pressure suddenly drops at the expansion valve, and the liquid begins to evaporate. The liquid refrigerant is completely evaporated by the time it leaves the evaporator. Heat is absorbed by the refrigerant in the evaporator. The refrigerant vapor returns to the compressor where it begins another cycle. The refrigerant, therefore, absorbs heat at the evaporator and gets rid of heat at the condenser. Common refrigerants used for agricultural applications are R-12, R-22, and ammonia.

Common applications of refrigeration on the farm are air conditioning and farm product cooling and storage. Cooling for eggs, milk, apples, and potatoes are typical farm refrigeration applications.

## HERMETIC COMPRESSOR

Refrigeration systems must be airtight to prevent air from leaking into the system and refrigerant from leaking out. The compressor can be a real problem inasmuch as it is difficult to prevent leakage at the shaft bearings to the compressor. The problem is prevented in most air conditioners and farm refrigeration units by sealing the motor and compressor into one container. The only openings are a refrigerant inlet and outlet. This is called a *hermetically sealed* motor and compressor, Figure 15-28. Special requirements for wiring hermetic motor compressors are contained in *NEC Article 440*.

## REFRIGERATION CONTROLS

Controls for air conditioners and refrigeration units are installed by the refrigeration specialist or the manufacturer. The electrician supplies power to the equipment. Thermostats and pressure switches are generally used to control the refrigeration system. A thermostat is often used to turn the system on and off. Pressure switches, sensing either low pressure on the suction side of the compressor or high pressure on the discharge side, are often used to detect abnormal conditions, Figure 15-29.

Figure 15-28 A hermetically sealed motor and compressor

Unit 15 Heating and Cooling 311

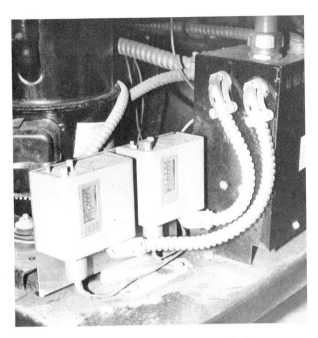

Figure 15-29 A refrigeration pressure switch

The pressure at various locations inside the refrigeration system is dependent upon the temperature. A pressure switch is often used to control the operation and, therefore, the evaporator temperature. The low-pressure switch is used instead of a thermostat. A dual-pressure switch, sensing both high and low pressure, is commonly used. The high-pressure switch is usually used to detect abnormal operation.

## NAMEPLATE WIRE SIZE AND PROTECTION INFORMATION

The nameplate of a refrigeration system contains information which is different from that given on a typical motor nameplate. A typical nameplate for a refrigeration unit is shown in Figure 15-30. A refrigeration compressor-condenser unit usually contains a hermetically sealed compressor, and a small fan motor to circulate air through the condenser coil. The full-load current as well as phase and voltage are given on the nameplate. The minimum circuit current rating is given on the nameplate. This value of current is used to select the wire from *NEC Table 310-16*. The maximum size branch-circuit overcurrent device rating is also given on the nameplate. The overcurrent device may be smaller than the value given, but not larger. The locked-rotor current (LRC) is used in sizing the disconnect.

## DISCONNECT

A disconnect switch must be located within sight from the motor compressor and where it is readily accessible, *NEC Section 440-14*. The disconnect must be rated in horsepower. The horsepower must be determined because the refrigeration unit usually consists of more than one motor, and the horsepower rating of the compressor motor often is not given.

The first step in determining the equivalent horsepower of the refrigeration unit is to determine the total full-load current of the unit. For the nameplate of Figure 15-30, this is 27.0 plus 2.2, which is 29.2 A. Multiply this total rated load current by 1.15, *NEC Section 440-12(a)*.

$$1.15 \times 29.2 \text{ A} = 34 \text{ A}$$

Now, compare the value of 34 A with the branch-circuit selection current from the motor nameplate, which is 37 A. The larger of these values is used to determine the equivalent horsepower from *NEC Table 430-148* or

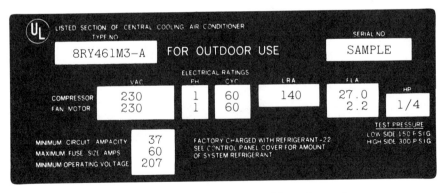

Figure 15-30 Nameplate for a refrigeration unit

*Table 430-150* (these tables are found in Unit 17 of this text). This example is a single-phase, 230-V unit; therefore, use *NEC Table 430-148*. Move down the 230-V column until a value as large or larger is found. This corresponds to 7 ½ hp.

Next, the equivalent horsepower must be determined using the locked-rotor current of 140 A from Figure 15-30. Find the 230-V column in *NEC Table 430-151* (Table 15-2 in this text) and move down the column until a value as large or larger is reached. For this example, 140 A would have a 10-hp rating.

Choose whichever horsepower rating is larger: 7 ½ or 10. The disconnect switch must be rated at least 10 hp and at least 230 V. A disconnect sized for a refrigeration unit in combination with another load, such as resistance heaters, is determined in a similar manner according to *NEC Section 440-12(b)*.

## MOTOR OVERLOAD PROTECTION

The compressor motor must be protected from overload, *NEC Section 440-52(a)*. A protection system is often installed at the factory. Alternate protection is a thermally protected compressor motor, or a time-delay fuse sized at not more than 125% of the motor-compressor full-load current.

## ROOM AIR CONDITIONERS

A room air conditioner is usually of the plug-in type. It may be supplied with an individual circuit, or it may be supplied by an existing general circuit. There are, however, minimum requirements for these circuits. The disconnect for a room air conditioner is permitted to be the attachment cord and plug, *NEC Section 440-13*. Air

Table 15-2 (*NEC Table 430-151*) Conversion table of locked-rotor currents for selecting disconnecting means and controllers as determined from horsepower and voltage ratings

| Motor Locked-Rotor Current Amperes* | | | | | | | Max. HP Rating |
|---|---|---|---|---|---|---|---|
| Single Phase | | Two or Three Phase | | | | | |
| 115 V | 230 V | 115 V | 200 V | 230 V | 460 V | 575 V | |
| 58.8 | 29.4 | 24 | 18.8 | 12 | 6 | 4.8 | ½ |
| 82.8 | 41.4 | 33.6 | 19.3 | 16.8 | 8.4 | 6.6 | ¾ |
| 96 | 48 | 43.2 | 24.8 | 21.6 | 10.8 | 8.4 | 1 |
| 120 | 60 | 62 | 35.9 | 31.2 | 15.6 | 12.6 | 1 ½ |
| 144 | 72 | 81 | 46.9 | 40.8 | 20.4 | 16.2 | 2 |
| 204 | 102 | — | 66 | 58 | 26.8 | 23.4 | 3 |
| 336 | 168 | — | 105 | 91 | 45.6 | 36.6 | 5 |
| 480 | 240 | — | 152 | 132 | 66 | 54 | 7 ½ |
| 600 | 300 | — | 193 | 168 | 84 | 66 | 10 |
| — | — | — | 290 | 252 | 126 | 102 | 15 |
| — | — | — | 373 | 324 | 162 | 132 | 20 |
| — | — | — | 469 | 408 | 204 | 162 | 25 |
| — | — | — | 552 | 480 | 240 | 192 | 30 |
| — | — | — | 718 | 624 | 312 | 246 | 40 |
| — | — | — | 897 | 780 | 390 | 312 | 50 |
| — | — | — | 1 063 | 924 | 462 | 372 | 60 |
| — | — | — | 1 325 | 1 152 | 576 | 462 | 75 |
| — | — | — | 1 711 | 1 488 | 744 | 594 | 100 |
| — | — | — | 2 153 | 1 872 | 936 | 750 | 125 |
| — | — | — | 2 484 | 2 160 | 1 080 | 864 | 150 |
| — | — | — | 3 312 | 2 880 | 1 440 | 1 152 | 200 |

*These values of motor locked-rotor current are approximately six times the full-load current values given in Tables 430-148 and 430-150.

Reproduced by permission from NFPA 70-1984, *National Electrical Code®*, Copyright © 1983, National Fire Protection Association, Quincy, MA. This reprinted material is not the complete and official position of the NFPA on the referenced subject which is represented only by the standard in its entirety.

conditioners rated to operate below 250 V single phase are permitted to be attachment plug connected, *NEC Section 440-60*. These units are required to be supplied by a grounding-type attachment plug and cord from a grounded electrical circuit, *NEC Section 440-61*.

The nameplate of the air conditioner gives the full-load current of the unit and the branch-circuit selection current. The current is not permitted to exceed 40 A at 250 V single phase if it is cord and plug connected, *NEC Section 440-62(a)*.

Room air conditioners must be supplied with a branch circuit having an overcurrent device not larger than the rating of the wire. For example, a No. 10 AWG copper wire must be protected at not more than 30 A.

Consider that a single air conditioner is on a branch circuit which supplies only the air conditioner. The full-load current of the air conditioner is not permitted to exceed 80% of the circuit rating, *NEC Section 440-62(b)*. For example, a 20-A circuit is permitted to supply an air conditioner with a full-load current up to 16 A.

An air conditioner may be connected to an existing general-purpose circuit. This circuit must be one which is lightly loaded. The full-load current of the air conditioner is not permitted to exceed 50% of the current rating, *NEC Section 440-62(c)*. For a 20-A circuit, the air conditioner full-load current would not be permitted to exceed 10 A.

The length of the attachment cord for an air conditioner is restricted to not more than 10 ft (3.05 m) for a 120-V unit, and 6 ft (1.83 m) for a 208-V or 240-V unit, *NEC Section 440-64*. The unit off switch for the air conditioner must be located not more than 6 ft (1.83 m) above the floor if the attachment plug is to serve as the disconnect, *NEC Section 440-63*. If the air conditioner is mounted more than 6 ft (1.83 m) above the floor, a switch rated in horsepower sufficient for the unit may be installed in the circuit within sight from the air conditioner.

## HEAT PUMP

A *heat pump* is a space heating system and a central air conditioner in one unit. The heat pump is actually a refrigeration system. Heat pumps are of two types: air to air, and water to air. The air-to-air system has an outside coil which takes heat from the air and, by means of the refrigerant, discharges the heat into an inside air duct. Thus the name air to air.

The water-to-air heat pump supplies water to one refrigeration coil. Sources of water are wells, rivers, and lakes. In the winter, the heat pump removes heat from the water and delivers it to the inside heating ducts. In summer, heat is removed from the air flowing through the ducts and delivered to the water flowing through the other coil.

In northern climates, a heat pump is augmented with resistance heating elements in the ductwork. If the heat pump is unable to supply adequate heat, the resistance heaters turn on to supply the additional heat necessary.

The installation of the heat pump and electric resistance heaters follows the same requirements as a refrigeration unit. Information for sizing the branch-circuit wires is found in *NEC Sections 440-34* and *440-35*. The disconnect is sized according to *NEC Section 440-12(b)*.

## REVIEW

Refer to the *National Electrical Code* when necessary to complete the following review material. Write your answers on a separate sheet of paper.

1. A room used as an office is located adjacent to the milkhouse with a masonry wall in between. The office is maintained at a temperature of 61°F (16°C) and the milkhouse is maintained at 46°F (8°C). Heat flow through the wall will be
   a. from the milkhouse to the office.
   b. from the office to the milkhouse.
   c. stopped if the door between the office and milkhouse is kept closed.
   d. increased if the wall contains thermal insulation.

2. Four space heaters have heat output ratings in different units of value. The space heater with the greatest output is
   a. 6 kW             c. 20 000 kJ/hr
   b. 20 000 Btu/hr    d. 4 500 kCal/hr

3. Name four types of electric heat.

4. Heat is moved from an object or area to another by which three heat transfer methods?

5. A nichrome heating wire has a resistance of 11 Ω, and draws 21.3 A. The kW rating of the heater is which of the following?
    a. 2 kW
    b. 3 kW
    c. 4 kW
    d. 5 kW

6. A 2 000-W infrared heater is used to keep newborn lambs warm in a sheep barn. If the heater operates at 120 V, the minimum size copper TW wire required to supply the infrared heater is No. ___?___ AWG.
    a. 16
    b. 14
    c. 12
    d. 10

7. The minimum size branch-circuit fuse permitted for the previous problem is which of the following?
    a. 20 A
    b. 25 A
    c. 30 A
    d. 35 A

8. An electric heater draws 44 A at 230 V, and is operated with a magnetic contactor controlled by a line-voltage thermostat. The heater is supplied by No. 6 AWG THWN copper wires, and protected by 60-A fuses. The thermostat wire is No. 12 AWG copper tapped directly from the branch-circuit wires. This is permitted by which *NEC* section?

9. An electric heater mounted near the ceiling of a room has a rating of 12 kW, single phase, and draws 50 A at 240 V. Does this heater contain fuses inside? Explain.

10. A 2.5-kW, 240-V wall electric heater installed in an office meets the *NEC* disconnect requirement if the heater:
    a. Has a 2-pole unit switch with an off position
    b. Has a 2-pole line voltage thermostat with an off position
    c. Circuit is protected with a circuit breaker in another room
    d. Has any of these

11. An electric heater does not operate, and the high-temperature limit switch is suspected of being defective. The limit switch is tested with a continuity tester. Describe the results.

12. Two 250-W, 240-V floor heating electric mats are required for each swine farrowing crate. The maximum number of farrowing crates that can be supplied by a 20-A circuit is which of the following?
    a. 4         c. 7
    b. 5         d. 9

13. A typical heating cable mat for outside ice and snow melting has a watts-per-square-foot rating of:
    a. 10
    b. 20
    c. 30
    d. 40

14. What is the minimum concrete finish-coat thickness when poured over electric cable?

15. The lead wires for a 240-V electric heating cable are color coded:
    a. Yellow
    b. Brown
    c. Red
    d. Blue

16. List three sources of heat on a farm from other than a space heater.

17. A herd of dairy cows produces 1 500 pounds of milk every time the herd is milked. The amount of heat removed from the milk by the milk tank cooling system is:
    a. 30 000 Btu
    b. 40 000 Btu
    c. 70 000 Btu
    d. 90 000 Btu

18. A refrigeration unit nameplate is rated at 230 V, single phase, and has a full-load current rating of 16 A. The locked-rotor current is 65 A. The minimum rating of the disconnect switch is required to be:
    a. 3 hp
    b. 5 hp
    c. 7.5 hp
    d. 10 hp

19. A 120-V room air conditioner draws 9.5 A. Is it permissible to plug this unit into an existing general-purpose 20-A circuit? Which *NEC* section supports your answer?

20. Describe a heat pump.

# UNIT 16
# ELECTRIC MOTORS

## OBJECTIVES

After studying this unit, the student will be able to

- state the purpose of an electric motor.
- identify the parts of an electric motor.
- name the basic types of electric motors.
- explain how an induction motor works.
- list the common types of single-phase induction motors.
- connect dual-voltage motors for operation at either voltage.
- reverse the direction of rotation of an electric motor.
- explain the meaning of nameplate information.
- list several factors that affect motor efficiency.
- choose the correct motor enclosure for an application.
- explain how motor speed can be varied.

## PURPOSE OF AN ELECTRIC MOTOR

An *electric motor* is a device for converting electrical power into mechanical power. An electric motor will try to deliver the required power even at the risk of self-destruction. Therefore, an electric motor must be protected from self-destruction. Motors may be ruined by physical damage to the windings but, usually, the enemy of a motor is excessive heat in the windings. Overheating breaks down the thin varnishlike insulation on the windings. When the insulation fails, the motor fails. Overheating is the result of excessive current flow or inadequate ventilation. Accumulation of dust and dirt on and in the motor can reduce ventilation and heat removal.

## MOTOR PARTS

Many types of electric motors are used for agricultural applications, and they are not all alike. However, all motors have some basic essential parts, Figure 16-1. The *frame* of the motor holds all the parts in place, and

Figure 16-1 Parts of an electric motor

provides a means of mounting the motor to machinery. The frame also conducts heat produced within the motor to the surrounding air. Some motors have *fins* to help get rid of heat even faster. The *stator* is a laminated magnetic core which holds the electrical windings. Electricity flowing through these windings produces a magnetic field.

The *rotor* is the part that turns and, in most farm motors, has no windings, Figure 16-2. The rotor shaft extends from one end of the motor for connection to the driven device or machine. Some motors, such as a grinder, have a shaft extending from both ends. Direct-current (dc) motors and some alternating-current (ac) motors have windings on the rotor. These are called *armatures*. One end of the armature has a commutator and brushes, Figure 16-3.

The *bearings* hold the rotating shaft to the motor frame. Different types are sleeve, roller, and ball bearings. The choice of bearings depends upon the application and mounting position of the motor.

A *fan* inside the motor moves air over the windings to remove heat (part 4 in Figure 16-1). Heat is produced by the resistance of the windings: Heat = $I^2 \times R \times t$; where t = time (see Eq. 2.22). In an open motor, air is drawn in at one end, moved over the windings, and blown out the other end. For a totally enclosed motor, the air picks up heat from the windings and takes it to the frame, where it gets to the outside by conduction.

A *motor terminal housing* is provided to connect the branch-circuit wires to the wires from the motor windings. Minimum terminal housing dimensions are specified in *NEC Section 430-12*. A means of terminating an equipment grounding wire is also required at the terminal housing for new motors, *NEC Section 430-12(e)*.

Figure 16-2 Rotor of an electric motor

318 Unit 16 *Electric Motors*

Figure 16-3 An electric motor with an armature

A *nameplate* provides the necessary information for motor application and circuit wiring. The nameplate may also contain diagrams for connecting and reversing the motor. This information may also be provided in the terminal housing.

## MOTOR TYPES

Three types of motors are used in agriculture: induction, synchronous, and direct current.

### Induction Motor

The *induction motor* is the motor most commonly used in agriculture. The speed of rotation of an induction motor is fairly constant, but it does vary somewhat with loading. As the motor is loaded, it slows down slightly. The induction motor is discussed in detail in this unit.

### Synchronous Motor

The *synchronous motor* runs at a constant speed regardless of the load on the motor. Synchronous motors usually have an armature. The most common farm application of a synchronous motor is in an electric clock or timer.

### Direct-current Motor

*Direct-current motors* are used in electric vehicles, and for applications where variable speed is required. An application would be a variable-speed motor operating an auger to meter high-protein supplement into a cattle feeding system. Although the motor requires direct current, the supply is usually alternating current. A solid-state rectifier in the motor controller changes ac to dc.

One type of direct-current motor can also operate on alternating current. This type is called a *universal motor*. Its most common application is in such power tools as an electric drill. It is also referred to as a *series-wound* dc motor. The stator or field winding is wired in series with the armature winding. The speed of these motors is not constant; the more they are loaded, the slower they turn. A big advantage, however, is that they develop very high torque at low speeds (*torque* is twisting force). Torque is observed every time an electric drill is forced through tough material.

## THREE-PHASE INDUCTION MOTOR

The most basic 3-phase induction motor has three sets of windings, with each phase connected to a different set of windings, Figure 16-4. The current in each winding is 120 electrical degrees out of phase with the current in the other windings. The current flowing through the windings creates an electromagnet with a north pole and a south pole. Since this motor has one north pole and one south pole, it is a 2-pole motor.

Unit 16  Electric Motors  319

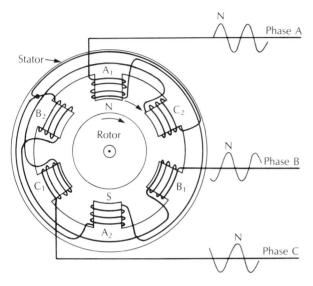

Figure 16-4 Simplified view of a 2-pole, 3-phase induction motor. It is a 2-pole motor because at any one time there is one north and one south magnetic pole. The stator magnetic field will rotate from A1 to B1 to C1 and then back to A1, 60 times each second.

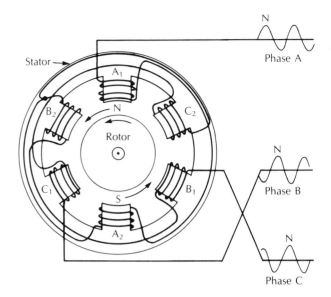

Figure 16-5 The direction of rotation of the magnetic field is reversed by reversing any two phase wires of a 3-phase induction motor. Compare the rotation of this motor with that in Figure 16-4.

Assume that when the alternating current reaches a maximum, a strong north pole is created in the winding next to the rotor, as in Figure 16-4. Notice that there is a north pole at stator pole A1, but not at pole B1 or pole C1. As time progresses, a strong north pole will appear at pole B1 and then at pole C1. This sequence repeats itself 60 times each second, or 3 600 times per minute. The magnetic field in the stator rotates around the motor at a speed of 3 600 r/min.

A 4-pole motor has twice as many windings, and it actually creates two sets of north and south poles at the same time. In one alternating-current cycle, the pole moves halfway around the stator. It takes a second ac cycle for the magnetic field to complete a full revolution of the stator. The magnetic field rotates at 30 revolutions per second, or 1 800 revolutions per minute.

## Reversing the Rotation

Now, let us see how the motor can be reversed. Figure 16-5 is a 2-pole motor with phase wires B and C reversed. The north pole will move from stator pole A1 to pole C1, and then to pole B1. The magnetic field is now rotating counterclockwise. In Figure 16-4, the magnetic field is rotating clockwise. Therefore, the direction of rotation of the magnetic field for a 3-phase induction motor is reversed by changing any two input phase wires.

## Why the Rotor Turns

How does the rotor turn? Electrical power is not applied directly to the rotor. Electrical current is produced in the rotor by mutual induction, similar to a primary winding of a transformer producing current in the secondary winding. The rotor actually has a path in which current can flow. This shape of this path is similar to a squirrel cage; thus, the name squirrel-cage motor, Figure 16-6. The squirrel cage is not obvious because it is hidden by the laminated steel in the rotor. This laminated steel is needed to keep the magnetic field strong, and to give the rotor weight like a flywheel.

The rotating magnetic field of the stator cuts across the squirrel-cage conductors, and current is induced into the squirrel cage. This current simply flows from one end of the squirrel cage to the other, forming a complete loop, Figure 16-7. The current flowing in the squirrel cage produces its own magnetic field. This new magnetic field tries to follow the rotating stator field. Thus, the rotor turns in the same direction as the rotating stator magnetic field. Reversing the field rotation reverses the rotor direction.

**Figure 16-6** Squirrel cage inside a rotor with the shaft and steel core removed

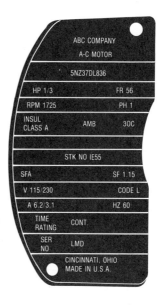

**Figure 16-8** Nameplate of an electric induction motor

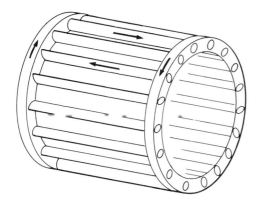

**Figure 16-7** Current flow is induced into the squirrel cage of the rotor by the rotating magnetic field created by the stator windings.

The rotor of an induction motor never rotates as fast as the stator magnetic field. This difference in speed is called *slip*. If the rotor were rotating at the same speed as the field, then the field would not be cutting across the rotor conductors. Recall from Unit 14 that a wire must be cutting across a magnetic field for current to be induced into the wire. Therefore, it is impossible for the rotor of an induction motor to be in synchronization (to rotate at the same speed) with the stator field. There must be slip. Notice the motor nameplate in Figure 16-8. The rotor turns at 1 725 r/min when operating at full load. The stator magnetic field of this 4-pole motor is actually rotating at 1 800 r/min.

A motor running idle with no load will almost, but not quite, come up to synchronous speed. The motor represented by the nameplate in Figure 16-8 would probably rotate idle at about 1 790 r/min. As the motor is loaded, more torque is required. The rotor will slip more and more, thus slowing down. The stator field cuts the squirrel cage faster, thus producing greater rotor current. This produces more torque. An induction motor will rotate slower and slower the more it is loaded.

## *Motor Current*

The instant a motor is turned on, current flows through the stator windings. The amount of current depends upon the impedance of the windings. The wire in the winding has resistance, and that limits the flow of current. But self-induction also limits the current flow. A magnetic field builds up around each wire because of the current flowing in the wire. This magnetic field cuts across adjacent wires, inducing a voltage in the adjacent wires. This is called *self-induction,* and it produces a voltage in the wire opposite to the applied voltage. This self-induced voltage resists the flow of current. The combined effect of the resistance of the wire and the self-induction is the impedance of the winding that resists the flow of current.

The current draw of the motor would be constant, as shown in Figure 16-9, if the rotor did not turn. This is called the *locked-rotor current* of the motor. The rotor

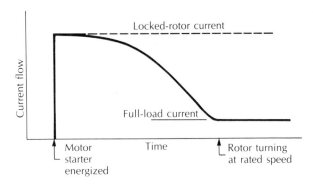

Figure 16-9 The starting current of a motor decreases as the rotor speeds up

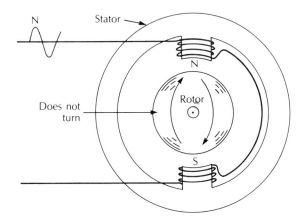

Figure 16-10 The poles alternate in a single-phase induction motor, but they do not rotate around the stator. The rotor does not turn in either direction.

has a magnetic field induced by the stator magnetic field. As the rotor begins to turn, its magnetic field cuts across the stationary stator winding. The rotor magnetic field induces a voltage in the stator winding that also opposes the applied voltage. The faster the rotor turns, the greater is the reverse voltage in the stator winding. The current flow in the stator winding decreases until the rotor reaches full operating speed. After that, the motor current remains constant. The solid line in Figure 16-9 shows the actual motor current during starting. This start-up may occur in a fraction of a second or it may take several seconds, depending upon the load. The locked-rotor current is about five to six times the full-load current for most induction motors.

## SINGLE-PHASE INDUCTION MOTOR

The single-phase induction motor is most commonly used for agricultural applications. A 2-pole, single-phase motor is shown in Figure 16-10. As the 60-Hz current reverses direction, the north pole and the south pole of the stator windings alternate. The stator magnetic field appears to be alternating back and forth rather than rotating, as in the case of the 3-phase motor. The rotor does not turn in either direction, it remains still. The single-phase motor will not start. The motor will run, however, once the rotor has started turning. The rotor may be started either clockwise or counterclockwise, and it will turn in that direction. Therefore, a single-phase motor needs a method of starting, and then it can run by itself on single-phase power.

A starting winding is added to the single-phase motor to get it started, Figure 16-11. The impedance of the starting winding is made different from the main running winding. This causes the current in the starting winding to be out of phase with the current in the running winding. A single-phase motor is temporarily turned into a 2-phase motor to get it started. Now the stator magnetic field will rotate. A capacitor is commonly used to shift the current in the starting winding out of phase with the current in the running winding. (Refer to the section on capacitance in Unit 2). This type of motor is called a *capacitor start, induction run motor.*

## REVERSING A SINGLE-PHASE MOTOR

Reversing the lead wires to the starting winding reverses the direction of rotation of the rotor, Figure 16-12. Directions are contained on the nameplate for clock-

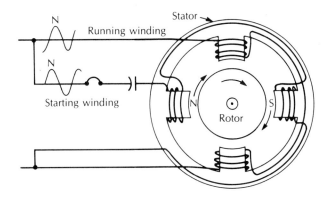

Figure 16-11 The capacitor in the starting winding shifts the current out of phase with the running winding so the magnetic field will rotate around the stator.

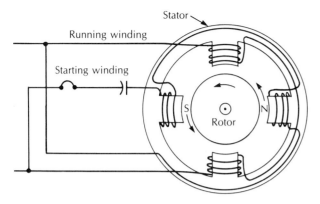

Figure 16-12 Reversing the starting winding lead wires causes the magnetic field to rotate around the stator in the opposite direction.

wise and counterclockwise rotation. However, some single-phase electric motors cannot be reversed in this manner.

## MOTOR TORQUE

A motor must develop enough turning force to start a load and to keep it operating under normal conditions. The manufacturer designs an electric motor to produce adequate torque for different types of loads. A graph can be drawn of the torque developed by the motor at various rotor r/min, Figure 16-13. The *locked-rotor torque* is the torque available to get a load or machine started. This is one of the most important considerations when choosing a motor for a farm application. Single-phase motors are discussed later in this unit, from lowest to highest starting torque. The breakdown torque is not a consideration when selecting a motor. However, it is used by manufacturers in determining the rated horsepower of a motor. If the load torque requirement exceeds the breakdown torque, the motor will stall.

A motor is designed to operate at the full-load torque. A continuous-duty motor will operate indefinitely at full-load torque without overheating. If the motor is oversized for the load, it will produce less than the full-load torque. If the motor is overloaded, it will develop more than the full-load torque. Look closely at Figure 16-13 and notice that the induction motor slows down when overloaded, and speeds up when underloaded.

Many single-phase motors have a starting winding that is disconnected when the motor achieves about three-quarters of operating r/min. A centrifugal switch attached to the rotor shaft is often used to disconnect the starting winding. This switching point is easily noticeable on a single-phase induction motor torque-speed graph, Figure 16-14.

## TYPES OF SINGLE-PHASE MOTORS AND THEIR APPLICATIONS

Single-phase motors are of several different types to meet different load requirements. The motor must have sufficient starting torque for the load. The motor should be as free as possible from maintenance problems, as well as having the lowest practical cost.

The split-phase induction motor is commonly used for easy-starting loads. A typical farm application for motors with low starting torque is the powering of ventilation fans. A schematic diagram of a split-phase induction motor is shown in Figure 16-15. The starting wind-

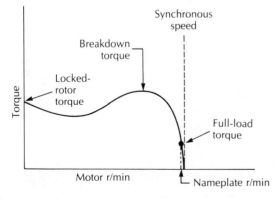

Figure 16-13 Torque of a 3-phase motor, and some single-phase motors, as the rotor accelerates from zero to full speed.

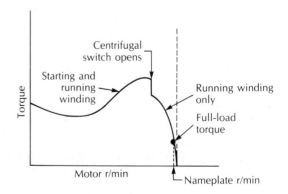

Figure 16-14 Torque of a single-phase motor as the rotor accelerates from zero to full speed

Unit 16 Electric Motors 323

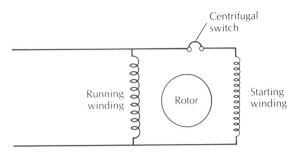

**Figure 16-15** Schematic diagram of the windings of a single-phase, split-phase induction motor

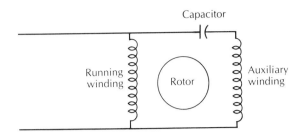

**Figure 16-17** Schematic diagram of the windings of a single-phase, permanent split-capacitor motor

ing has many turns of small size wire. The starting winding is not designed for continuous operation; therefore, a centrifugal switch disconnects the winding before the motor reaches full operating speed.

The permanent split-capacitor induction motor has a starting winding which also acts as a running winding in addition to the main running winding. This motor has a slightly higher starting torque than the split-phase type. An oil-filled capacitor is either attached to the outside of the motor or sometimes placed inside the motor frame, Figure 16-16. This type of motor does not have a centrifugal switch, thus reducing maintenance problems. It is commonly used to power variable-speed ventilation fans, and heating system blowers. A schematic diagram of a permanent split-capacitor motor is shown in Figure 16-17.

The capacitor start, induction run motor, discussed earlier in this unit, is probably the most commonly used single-phase motor for farm applications. It has good starting torque. An electrolytic capacitor is used in the starting winding, which is disconnected before the motor reaches full operating speed. The capacitor is often attached to the surface of the motor, Figure 16-18, but sometimes it is placed in the motor terminal housing or inside the motor frame. A schematic diagram of a capacitor start, induction run motor is shown in Figure 16-19.

Silo unloaders, refrigeration units, gutter cleaners and similar hard starting loads require a motor which develops high starting torque. The two-value capacitor induction run motor is frequently used for this type of load. An auxiliary winding is used for starting, but it is also operated continuously. These motors are available

**Figure 16-16** A permanent split-capacitor motor has an oil-filled capacitor connected in series with the auxiliary winding.

**Figure 16-18** A single-phase capacitor start, induction run motor

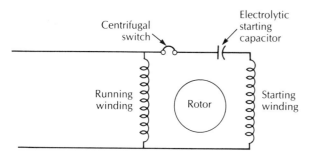

**Figure 16-19** Schematic diagram of the windows of a single-phase capacitor start, induction run motor

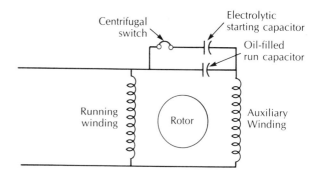

**Figure 16-21** Schematic diagram of the windings of a single-phase, two-value capacitor induction motor

in sizes larger than 1 hp, and they use both electrolytic and oil-filled capacitors. Both types of capacitors are used during starting, but a centrifugal switch disconnects the electrolytic capacitors before the motor reaches full operating speed. The capacitors may be attached to the outside of the motor, Figure 16-20, or they may be placed in the terminal housing. A schematic diagram of a two-value capacitor induction run motor is shown in Figure 16-21.

## SHADED POLE MOTOR

The shaded pole motor is available only in very small sizes, usually not larger than 1/20 hp. Its most frequent application is in ventilation fans for range hoods and bathrooms, electrical equipment, and similar applications. Shaded pole motors have a special wire loop attached to the poles of the stator, Figure 16-22. These motors are inexpensive to build, but they develop very low starting torque. Sometimes when they fail to start, lubrication of the bearings will solve the problem. Normally, these motors are not reversible.

## REPULSION START, INDUCTION RUN MOTOR

The repulsion start, induction run motor develops very high starting torque. These motors are expensive to build because they have an armature, a commutator, and brushes (see Figure 16-3). The current that flows through the armature windings is induced by the stator field. The motor starts due to a strong repelling effect between the stator magnetic field and the armature magnetic field. Once the rotor achieves nearly full operating speed, a centrifugal switch shorts all the armature windings together. The motor then operates like an induction motor. Repulsion start motors are reversed by changing the location of the brushes on the commutator. The bracket

**Figure 16-20** A single-phase, two-value capacitor induction motor

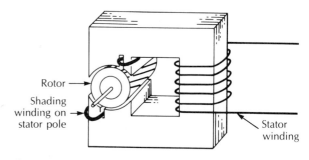

**Figure 16-22** Shaded pole single-phase motor

Unit 16 *Electric Motors* 325

holding the brushes is moved to the reverse location. The schematic diagram for a repulsion start, induction run motor is shown in Figure 16-23.

## DUAL-VOLTAGE MOTORS

Many single- and 3-phase motors are designed to be operated at two different voltages. A typical dual-voltage motor may be rated at 115/230 V. This motor may be connected for operation from a 115-V supply or a 230-V supply. Dual-voltage motors have two sets of running windings. The running windings are connected in parallel for the lower voltage operation, and in series for the higher voltage operation, Figure 16-24. Each winding of a 115/230-V motor is designed to be operated at 115 V. Examination of Figure 16-24 reveals that each winding receives 115 V even when the motor is connected to a 230-V supply. These motors can be reversed by simply reversing the leads to the starting winding.

Three-phase motors are also commonly available as dual-voltage motors. The most common voltage combination is 230/460. Three-phase motor windings may be connected to form a wye or a delta. The wye-connected dual-voltage motor is most common for farm applications. This type of motor usually has nine leads available in the terminal housing, Figure 16-25. A single-voltage wye-connected motor has only three leads available.

A connection diagram is usually shown on the motor nameplate for both the higher and lower voltages. At the higher voltage the windings are connected in series, and at the lower voltage they are connected in parallel, Figure 16-26.

Delta-connected 3-phase motors are available as single voltage or dual voltage. They are also three-lead and nine-lead motors, as shown in Figure 16-27. The dual-voltage delta motor windings are connected in parallel for the lower voltage, and in series for the higher voltage, Figure 16-28. A six-lead, single-voltage, delta-connected motor is available for a special type of starting

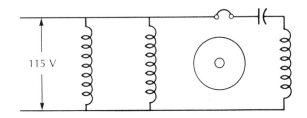

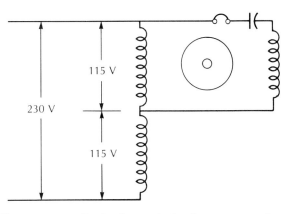

**Figure 16-24** Single-phase, dual-voltage motor shown connected for operation at 115 V and at 230 V

called wye-delta starting. Dual-voltage motors started in this manner must have twelve leads available at the terminal housing.

Dual-voltage motors should be operated at the higher voltage whenever possible. The motor power will be the same, but only half as much current will be drawn. This means that the motor circuit wire can be sized smaller. Looking up the current draw of a single-phase, 5-hp motor in *NEC Table 430-148* shows that the motor draws 56 A at 115 V. When the motor is operated at 230 V it draws only 28 A or half as much. Recall that power is proportional to current times voltage. If the voltage is doubled, the current must be cut in half to maintain the same power.

## NAMEPLATE INFORMATION

The National Electrical Manufacturers Association (NEMA) has developed design and rating standards, so that motors supplied by different manufacturers can be compared for basic minimum performance and used interchangeably. A typical motor nameplate is shown in

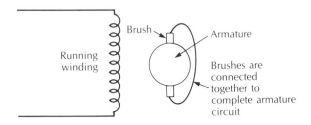

**Figure 16-23** Schematic diagram of the windings of a single-phase repulsion start, induction run motor

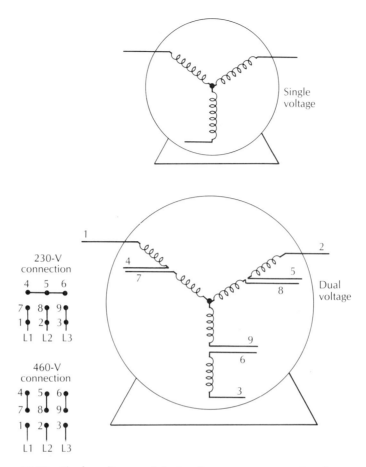

Figure 16-25 Single-voltage and dual-voltage wye-connected 3-phase motor

Figure 16-29. An electric motor nameplate may not contain all of the detailed information discussed next, but the following is a general description of motor nameplate data.

**VOLTS:** The proper operating voltage. This may be a single value, such as 115 V, or a dual voltage, such as 115/230 V.

**AMPS:** The full-load current in amperes (A) of the motor operating with the proper supply voltage. A single voltage motor will have one current value listed, and a dual-voltage motor will have two values, such as 11.4/5.7. This means that the motor will draw 11.4 A when connected at 115 V, and 5.7 A when connected at 230 V.

**HP:** The design full-load horsepower (hp) of the motor. It is determined from the value of the breakdown torque from the motor torque-speed curve. The SI metric unit for electric motor power is the kilowatt (kW). The conversion of 1 hp is equal to 0.746 kW (rounded).

Motors are available in fractional horsepower sizes and integral horsepower sizes. *Fractional* simply means fractions of a horsepower, such as ¼ or ½. *Integral* comes from the word integer or whole number, such as 1, 5, 20, or 75.

Common horsepower sizes are given in *NEC Tables 430-148* and *430-150*. The list in Table 16-1 contains the common motor sizes used for agricultural applications. The equivalent SI metric kW value is also given.

**PHASE (PH):** Specifies whether the motor is single phase or 3 phase.

**RPM:** Full-load speed of the rotor in revolutions per minute (r/min). This is the speed of the motor corresponding to the full-load torque from the torque-speed curve.

**HZ:** The design operating frequency of the electrical supply in cycles per second; usually 60 hertz (Hz). A motor must not be operated at a frequency other than the nameplate value unless permitted by the manufacturer.

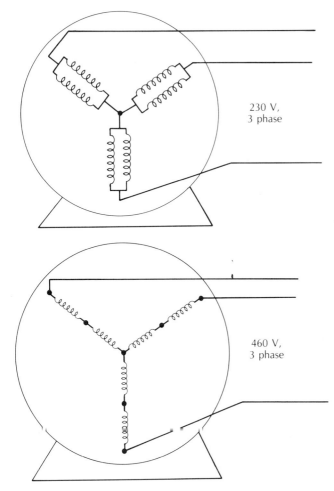

**Figure 16-26** Dual-voltage, wye-connected 3-phase motor connected for operation at 230 V and 460 V

**FRAME (FR):** These are standard frame numbers used by motor manufacturers to make sure motors are interchangeable. Common frame numbers for fractional horsepower sizes are 42, 48, and 56. Letters may be added before or after the number to specify motor dimensions and type of mounting. The frame number divided by 16 is the height in inches from the center line of the shaft to the bottom of the mounting. Common induction motors in use today are generally designated as T-frame motors, such as 213T for a 5-hp single-phase motor. Other motors may have what is called a U frame. The U frame usually has a heavier construction than the T frame. It is important when replacing a motor to obtain a replacement with the same frame number so that the new motor will bolt into the same place as the old motor without alterations.

**HOURS OR DUTY:** Most motors for farm applications should be designed for continuous duty. There are applications where the motor will be operated for only a few minutes. Motor size and cost can be reduced if it is designed for intermittent duty. The actual hours or minutes of operation will be stated. If the motor is operated longer than the stated period of time, the maximum safe operating temperature of the motor may be exceeded.

**TEMPERATURE RISE (°C):** Sometimes the nameplate will state temperature rise or degree C rise. This indicates the rise in temperature of the motor above a 40°C (104°F) ambient temperature when operating at full-load conditions. If the surrounding temperature is 40°C (104°F) and the motor temperature rises an additional 40°C (104°F), then the frame of the motor will be 80°C (176°F). This is too hot to touch. Generally, if a motor is not too hot to touch it is actually operating quite coolly. A motor rated 40°C (104°F) can often be overloaded 10% to 15% without damaging the motor. A motor rated 50°C or 55°C (122°F or 131°F) rise is actually operating near the breakdown temperature of the insulation on the windings. Motors with a high-temperature rise rating should never be overloaded.

Ambient temperature is sometimes printed on a motor nameplate. This has a different meaning than rise above ambient. Essentially, *ambient* means that the motor is suitable for application where the surrounding temperature does not exceed the temperature marked on the nameplate. For example, a nameplate may be marked 40°C ambient (104°F). The motor must not be placed in an environment with a temperature above 40°C (104°F) without consulting the manufacturer.

**SERVICE FACTOR (SF):** This is a better indication of safe overload rating of an electric motor than temperature rise. The service factor is a multiplier which, when applied to the rated horsepower, indicates a permissible horsepower loading. A service factor of 1.0 means that a motor does not have any continuous overload capacity. A service factor of 1.15 means a motor could be operated with a 15% overload without danger of overheating. The service factor is considered when sizing the motor overload protection. A replacement motor for a farm machine should have a service factor the same as that of the old motor. If not, consult the machine manufacturer for a recommended replacement. A farm-duty motor could have a service factor of 1.35 or more.

**INSULATION CLASS:** The insulation on motor wires is rated in several classes, depending upon the maximum safe operating temperature which will not cause premature breakdown. Common insulation classes are A, B, F,

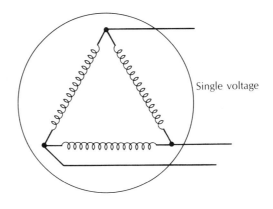
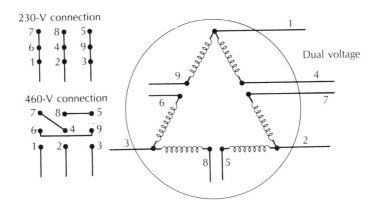

**Figure 16-27** Single-voltage and dual-voltage, delta-connected 3-phase motor

and H. Class A insulation has the lowest temperature rating. Most farm-duty motors have class A or B insulation. The insulation class is considered when assigning a motor service factor. The life of a motor is dependent upon the life of the insulation. Frequent overheating of the motor will shorten the motor's normal operating life.

**CODE LETTER:** The code letter is based upon the locked-rotor current drawn by the motor. The code letter is used to determine the maximum rating of the motor branch-circuit protection. The code letter is determined by multiplying the supply voltage by the locked-rotor current and then dividing by the motor horsepower. This value must also be divided by 1 000 to give kilovolt-amperes per horsepower (kVA/hp). The approximate locked-rotor current can be found in *NEC Table 430-151* (included in Unit 15 of this text). The code letter is found in *NEC Table 430-7(b)*.

The code letters may be used to calculate the approximate motor instantaneous starting current (locked-rotor current). Consider an example of a 5-hp, 3-phase, 230-V motor with code letter J on the nameplate.

Here is how to calculate motor starting current from the code letter. Equation 16.1 gives motor locked-rotor current.

$$I = \frac{kVA \times hp \times 1\,000}{E \times 1.73}$$ **Eq. 16.1**

(Omit 1.73 for a single-phase motor.)

The kilovolt-amperes per horsepower corresponding to code letter J is looked up in *NEC Table 430-7(b)*. For letter J, the range is 7.1 to 7.99.

$$I = \frac{7.1 \times 5 \times 1\,000}{230 \times 1.73} = 89.2 \text{ A minimum}$$

$$I = \frac{7.99 \times 5 \times 1\,000}{230 \times 1.73} = 100.4 \text{ A maximum}$$

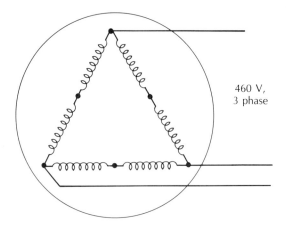

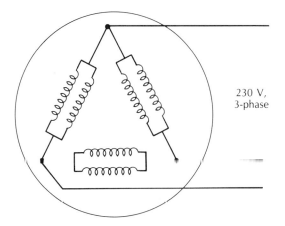

Figure 16-28 Dual-voltage, delta-connected 3-phase motor connected for operation at 230 V and 460 V

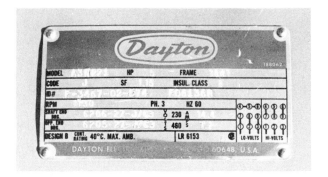

Figure 16-29 Nameplate from a 3-phase electric motor

The instantaneous starting current of the motor in the example ranges from 89.2 A to 100.4 A.

**DESIGN:** A design letter may be given on the nameplate or in the manufacturer's literature. This letter is assigned according to NEMA design standards. The design letter is an indication of the shape of the motor torque-speed graph, the level of starting torque, and the level of starting current. Most squirrel-cage induction motors are design A or B. The design B motor generally has a lower starting current than the design A, with approximately the same starting torque.

**THERMAL PROTECTION:** Many motors have built-in thermal protection to disconnect power to the motor in the event the motor windings become too hot. Some thermal protectors sense motor current, while another type is a temperature-sensitive switch placed in the motor windings. Some thermal protectors are manual resetting, while others are automatic. With the automatic type, the motor will restart when it cools. Some farm motors have a thermal protector in the windings which must be wired in as a part of the control circuit. These wires will be in the terminal housing with the lead wires, and are marked thermal protection or thermostat. The wiring of these thermal protectors is discussed in Unit 17 on the subject of motor control.

## EFFICIENCY

More than half of electrical energy is used to power electric motors; therefore, motor efficiency is of vital concern. An electric motor is usually designed for maximum efficiency when it is operated at 80% to 100% of its rated load. Fractional-horsepower motors often reach maximum efficiency at 100% to 120% of full load. Efficiency decreases rapidly when a motor is operated signif-

Table 16-1 Common motor sizes used for agricultural applications

| Horsepower (hp) | Kilowatts (kW) |
|---|---|
| 1/6 | 0.124 |
| 1/4 | 0.187 |
| 1/3 | 0.249 |
| 1/2 | 0.373 |
| 3/4 | 0.56 |
| 1 | 0.75 |
| 1 1/2 | 1.12 |
| 2 | 1.50 |
| 3 | 2.25 |
| 5 | 3.75 |
| 7 1/2 | 5.6 |
| 10 | 7.5 |

icantly under its rated load. Therefore, matching a motor to the load is important to achieve maximum efficiency.

Three major sources of motor losses are: 1) winding losses, 2) magnetic core losses, and 3) mechanical losses. The losses in the windings occur from the resistance of the wire. As current flows through the windings, heat is produced (Heat = $I^2 \times R \times t$, Eq. 2.22). The manufacturer selects wire size and materials to minimize this heating effect. The user of electric motors can keep them clean so that they will run cooler. If operating temperature is minimized, the winding resistance will be minimized.

Losses in the stator and rotor core are controlled by the manufacturer. A high-efficiency motor will be more expensive because the manufacturer has used materials in the motor with better magnetic qualities. Heat is produced in the core by eddy currents that are induced by the rotating magnetic fields. The core is laminated with insulation between each layer to reduce these heat-producing eddy currents.

Mechanical losses result from friction in bearings, gears, and belt and chain drives. Bearings should be lubricated according to the manufacturer's recommendations in order to reduce losses. This lubrication practice applies to the entire machine, not just the motor. Chain drives should be adjusted to proper tension. Belt drives should not be loose, but neither should they be bowstring tight. Proper adjustment will reduce friction and losses.

## ENCLOSURES FOR MOTORS

The type of motor enclosure for an agricultural application is extremely important. Dust, excessive moisture, and rodents must not be permitted to enter a motor. Rodents are a particular problem. Openings in motors should be not larger than ¼ in (6.35 mm). Basically, if a pencil can be inserted into the motor, the opening is too large. Common motor enclosures used for agricultural applications are the open, dripproof, splashproof, totally enclosed, or explosionproof types.

An open motor is shown in Figure 16-30. Air is drawn through the motor to cool the windings. These motors must only be used inside where dust is not present. They are often used for furnace blowers. These motors must not be used where water will be sprayed for cleaning. In general, open motors are not recommended for use in agricultural buildings.

A dripproof motor is shown in Figure 16-31. Air is drawn through the motor to cool the windings, but the openings are located in such a way that rain will not enter the motor. These motors are suitable for outside use, but

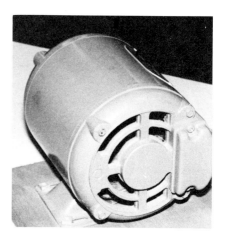

Figure 16-30  A motor with an open enclosure

only in dust-free areas. The openings must be small enough so that rodents cannot enter. It is possible for water to be splashed into the motor. A similar type is the splashproof motor which has the openings located in such a way that water cannot be easily splashed inside. Many dripproof motors are actually splashproof motors.

Totally enclosed motors are permitted in areas where dust and settlings may be present around the motor. Air is not drawn through the motor for cooling. Instead, air inside the motor is circulated through the windings carrying heat to the case where it moves to the outside by conduction. Many totally enclosed motors have a fan on one end which blows air over the outside of the case, Figure 16-32. This is called a totally enclosed fan-cooled (TEFC) motor. This type of motor enclosure is recommended for dusty or wet areas in or around farm buildings.

Figure 16-31  A motor with a dripproof enclosure

Figure 16-32 A motor with a totally enclosed, fan-cooled (TEFC) enclosure

Explosionproof motors must be used in areas where hazardous vapors could cause an explosion or fire. A typical example is a motor on a fuel-dispensing pump. Wiring in areas where there are hazardous vapors present is called Class I wiring. Gasoline is further identified as a Group D vapor, *NEC Section 500-2*. An explosionproof motor will state "suitable for Class I areas" on the nameplate. If it is to be used in an area where gasoline vapors are present, the nameplate must also state that the motor is suitable for Group D vapors.

## SOFT-START MOTORS

The high starting current of a large motor can be up to six times the full-load current at full-load speed. This high starting current may cause excessive voltage drop on the electric power supplier's primary lines. The result is voltage flicker. The most noticeable effect is a dimming of incandescent lights while the motor is starting, or the shrinking of a television image on the screen. The problem is that this dimming also occurs for neighboring electric customers. If the voltage flicker becomes annoying to customers, the power supplier will impost a limit upon the maximum size across-the-line motor which a customer may install. An industrial customer may be limited to the number of times in one day and the hours when a large motor may be started.

An *across-the-line motor starter* is one which receives full operating voltage. A common motor starting technique often acceptable to the power supplier is to start the motor at reduced voltage. The motor draws much less starting current when started at reduced voltage, Figure 16-33. The starting torque is also reduced; therefore, this type of starting will not work on some farm loads.

A soft-start single-phase motor has special controls which limit the current draw of the motor during starting. These motors are available in sizes up to 50 hp. The electric power supplier must be contacted prior to the purchase of a large size motor to make sure the primary wires and transformer are adequate. A 3-phase motor operating from a phase converter also draws less current during starting than the same motor operating directly from utility-supplied 3-phase power.

Special starters are available which start a motor in such a way that the inrush current is limited. Methods of starting motors with reduced inrush current are: 1) part winding, 2) wye-delta, 3) primary resistor, and 4) autotransformer starting. The wye-delta technique works with 3-phase motors only, while the others work with both single- and 3-phase motors.

## VARIABLE-SPEED MOTORS

Variable speed may be desirable for some farm applications. An electric drill is a universal motor, and its speed can be varied by lowering the input voltage. Usually, speed in an electric drill is controlled with a solid-state electronic device called a *silicon controlled rectifier* (SCR controller). The input current of the motor would look much like that of the light dimmer shown in Unit 11. Direct-current motors may be used for special applications such as feed-supplement metering where variable speed is desirable.

Speed may be continuously varied with a variable belt drive. A handle is turned which changes the diame-

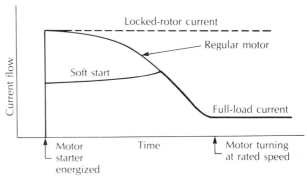

Figure 16-33 Starting current of a regular motor and a soft-start motor

ter of a pulley in the drive unit. This type of drive is most desirable where high power requirements are required. The motor is always operating at full rated speed even though the output shaft speed is variable. This ensures adequate available torque for the load.

At one time, alternating-current induction motors were generally not capable of significant speed variability, but electronic devices have overcome that barrier. A permanent split-capacitor motor is capable of some speed variation. A common application is a variable-speed fan. The input power to the motor is controlled in a manner similar to the light dimmer shown in Unit 11. A thermistor, rather than a variable resistor, is used to control the fan speed. A *thermistor* is a resistor whose resistance changes as the temperature changes. A variable resistor can be used instead of a thermistor in the electronic circuit to select the desired speed.

To understand how the variable-speed, permanent split-capacitor motor works, it is necessary to understand the torque required to turn a fan at different speeds. A fan requires very little torque to get started, but as it turns faster it pushes more air and requires more torque. Figure 16-34 is a torque-speed curve for a fan and an electric motor. The operating point is where the fan load curve and the motor torque curve cross. Farm equipment manufacturers must make sure that this operating point is equal to or below the full-load torque of the motor.

Reducing the voltage to a permanent split-capacitor motor causes the torque-speed curve for the motor to shrink, Figure 16-35. The fan load curve and the motor torque-speed curve cross at a lower r/min. Therefore, the fan slows down. Further reducing the input voltage slows down the fan even more. The motor has less torque available as the voltage is reduced. Therefore,

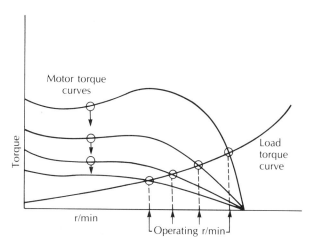

**Figure 16-35** The torque curve of the motor shrinks as the input voltage to the permanent split-capacitor motor is reduced.

this technique for varying the speed of an induction motor usually works only when powering fans.

The rotational speed of a 3-phase induction motor may be varied by changing the frequency to the motor. A special controller rectifies the input 60-Hz ac to dc. An electronic inverter changes the direct current back into alternating current, but at a different frequency. If the 60 Hz input frequency is changed to 40 Hz, the motor will turn more slowly. If the 60-Hz input is increased to 70 Hz, the motor will rotate faster. The voltage to the motor is also varied to prevent motor overheating due to too much current. Typically, speed variation may range from 20% to 110% of the motor nameplate full-load speed. For a motor rated at 1 725 r/min, this would be a speed variation of 345 r/min to 1 900 r/min. Much wider speed variation can be obtained, depending upon the application.

The electronic speed controller usually maintains a constant output torque from the motor. But, horsepower is proportional to torque times r/min, Equation 16.2. Therefore, the motor horsepower decreases as the r/min is reduced, and the motor horsepower increases as the r/min is increased. Variable-speed controllers for induction motors must be carefully matched to the type of load. These controllers are best suited to loads such as fans, blowers, and some pumps.

$$\text{Horsepower} = \frac{6.28 \times \text{torque} \times \text{r/min}}{33\,000} \quad \text{Eq. 16.2}$$

(torque is measured in pound-feet.)

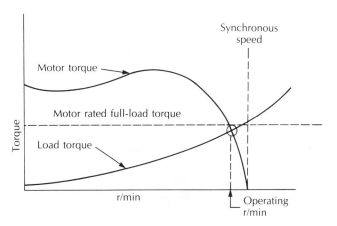

**Figure 16-34** The operating speed occurs where the motor torque-speed curve and the load torque curve cross.

Induction motors are available with multiple speeds. These motors change the number of magnetic stator poles. A 2-pole induction motor turns at about 3 450 r/min. A 4-pole motor turns at about 1 725 r/min. Equation 16.3 gives the approximate speed of an induction motor based upon the number of stator magnetic poles.

$$\text{Motor r/min} = \frac{2 \times 3\,450}{\text{Number of poles}} \quad \text{Eq. 16.3}$$

A 6-pole induction motor would turn at about 1 150 r/min.

$$\text{Motor r/min} = \frac{2 \times 3\,450}{6} = 1\,150 \text{ r/min}$$

A typical two-speed motor is capable of operating with four or six poles. Its two operating speeds are then 1 725 r/min and 1 150 r/min. If the motor could also be connected with eight poles, it could then be operated at 1 725 r/min, or 862 r/min.

## BEARINGS

The bearings hold the turning rotor or armature of the motor to the frame. Common bearings are the sleeve type and the ball type. Motors with sleeve bearings are designed to be operated with the shaft in the horizontal position. They are lubricated by a thin film of oil which is pulled into the space between the shaft and the bearings. Usually, there is an oil reservoir below the bearings, and a wick keeps oil next to the bearings. This oil reservoir must be mounted below the shaft. If the motor is mounted on a wall, the end bell of the motor should be rotated so that the oil reservoir is downward.

Motors with ball bearings are more expensive than those with sleeve bearings, but they can be mounted in any position. They are grease lubricated so there is no danger of lubricant spilling out of the bearing. Most ball bearings are prelubricated, and will run for years before it becomes necessary to open them and repack them with grease.

## BRAKING

Braking a machine quickly when it is turned off is sometimes desirable, particularly from a safety standpoint. Most farm machines experience high friction and, therefore, stop quite quickly when power is disconnected. Rotating machines with a great deal of weight have considerable turning inertia, and some type of braking is needed to bring them to a halt quickly.

One type of braking is simply a friction brake, similar to that on an automobile. A brake shoe contacts a steel drum or a disk contacts a plate. The friction brake can be added to the machine or motors with built-in brakes are available.

Dynamic braking is another method which is commonly used on ac motors. A special control and a source of dc power are required. Usually, these are supplied as a part of a dynamic control panel. When the ac power is disconnected, dc is applied to the field windings. This creates a strong magnetic field that does not rotate. As a result, the rotor comes to a halt quickly. The stronger the dc magnetic field, the faster the rotor and machine stop turning. Other methods of braking are also available.

## BELTS, CHAINS, AND COUPLINGS

Motors must be properly coupled to machines if efficient power transfer and long bearing life are to be achieved. Common types of couplings for farm equipment are V-belts, roller chains, and direct-drive couplers. V-belts are especially common because they are inexpensive and efficient, and require minimal maintenance. Proper belt tension and pulley alignment is important. Figure 16-36 shows how to check for proper belt tension. Push down hard on the belt. For every 12 in (305 mm) from the center of one shaft to the center of the other, the belt should depress about ¾ in (19 mm). For example, if the shafts are 18 in (457 mm) apart, the belt should depress about 1⅛ in (28.5 mm).

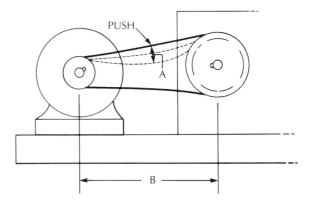

**Figure 16-36** The V-belt tension is correct when distance A is ¾ in (19 mm) for every 12 in (305 mm) of distance B.

Proper alignment is important to maximize power transfer and to minimize bearing, belt or chain wear. Figure 16-37 shows proper (and improper) alignment for belt and chain drives. Direct-drive couplings must be aligned properly to avoid vibration and wear on bearings. Figure 16-38 illustrates correct (and incorrect) direct-drive coupler alignment.

## Sizing a Pulley

Fans, pumps, and machinery are designed to operate at a certain speed, but this speed is often different from the r/min of the motor. R/min can be changed easily by changing the diameter of the pulley of a V-belt drive. The same relationship is true for gear and chain drives. Consider an example where there is a 4-in diameter pulley on a motor turning at 1 740 r/min, and a machine has an 8-in diameter pulley, Figure 16-39. The 8-in diameter pulley will make only ½ revolution when the 4-in diameter pulley makes one complete revolution. Therefore, the machine will have a shaft rotation only half as fast as the motor, or 870 r/min.

Equations 16.4, 16.5, and 16.6 are used to solve pulley-sizing problems.

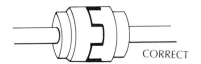

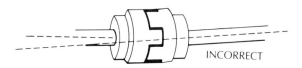

**Figure 16-38** Incorrect direct-drive coupler alignment can lead to early bearing failure.

$$\text{Motor pulley diameter} = \frac{\text{Machine r/min} \times \text{machine pulley diameter}}{\text{Motor r/min}} \quad \text{Eq. 16.4}$$

$$\text{Machine pulley diameter} = \frac{\text{Motor r/min} \times \text{motor pulley diameter}}{\text{Machine r/min}} \quad \text{Eq. 16.5}$$

$$\text{Machine r/min} = \frac{\text{Motor pulley diameter} \times \text{motor r/min}}{\text{Machine pulley diameter}} \quad \text{Eq. 16.6}$$

## Problem 16-1

An example will help to illustrate how a pulley is selected for a particular application. Consider a machine with a 12-in diameter pulley which is intended to rotate at 400 r/min. Select the proper size pulley for a motor with a full-load speed of 1 725 r/min.

## Solution

The problem is solved using Equation 16.4.

$$\text{Motor pulley diameter} = \frac{400 \text{ r/min} \times 12 \text{ in}}{1\ 725 \text{ r/min}} = 2.8 \text{ in}$$

Common pulley diameters are 2 in, 2½ in, and 3 in. If a 3-in diameter pulley is selected, the load will turn faster than 400 r/min.

$$\text{Machine r/min} = \frac{3 \text{ in} \times 1\ 725 \text{ r/min}}{12 \text{ in}} = 431 \text{ r/min}$$

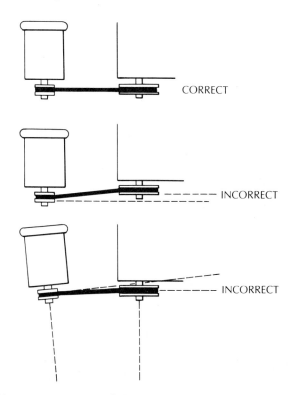

**Figure 16-37** Proper belt and chain alignment is important.

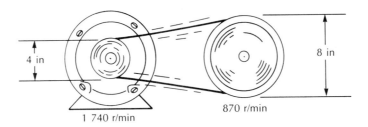

**Figure 16-39** R/min is changed by changing the diameter of the pulley on the motor and the machine.

If a 2½-in diameter pulley is selected, the machine will turn at 360 r/min. If the motor is oversized for the load, it will actually run faster than 1 725 r/min. Therefore, it would probably be wise to select a pulley which is a standard size smaller than 3 in in diameter or, in this case, 2½ in.

# REVIEW

Refer to the *National Electrical Code* when necessary to complete the following review material. Write your answers on a separate sheet of paper.

1. State the basic purpose of an electric motor.

2. Name the parts of the electric motor in this diagram.

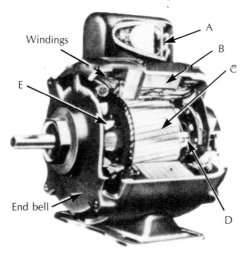

3. The motor in the diagram is
   a. 3 phase.
   b. single phase.
   c. 2 phase.
   d. split phase.

4. If the shaft of an electric motor turns at 1 725 r/min, and the motor operates from a 60-Hz ac supply, it is a(an)
   a. universal motor.
   b. permanent magnetic motor.
   c. synchronous motor.
   d. induction motor.

5. The motor with the highest starting torque from the list is a
   a. permanent split capacitor.
   b. shaded pole.
   c. two-value capacitor.
   d. split phase.

6. A 2-pole, 3-phase induction motor could have a shaft rotation of
   a. 3 450 r/min.
   b. 1 800 r/min.
   c. 1 725 r/min.
   d. 1 150 r/min.

7. An electric motor has a number 56 frame. The height of the shaft above the mounting surface is which of the following?
   a. 5.6 in
   b. 5.0 in
   c. 3.5 in
   d. 6.0 in

8. The type of capacitor used for a permanent split-capacitor motor is
   a. electrolytic.
   b. oil filled.
   c. variable.
   d. disc type.

9. A 4-pole synchronous motor operating from a 60-Hz ac supply will have a shaft rotation of
   a. 3 600 r/min.
   b. 3 450 r/min.
   c. 1 800 r/min.
   d. 1 725 r/min.

10. Will the 3-phase, 2-pole induction motor in the diagram have a shaft rotation that is clockwise or counterclockwise?

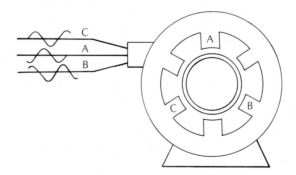

11. The shaft rotation of the motor in Problem No. 10 may be reversed by
    a. switching wires A and B.
    b. switching wires B and C.
    c. switching wires A and C.
    d. switching any two wires.

12. The induction motor in the accompanying diagram at the top of the next page is dual voltage, rates at 115/230 V. The motor is which of the following types?

a. Permanent split capacitor
b. 3-phase
c. Capacitor start, induction run
d. Split-phase start, induction run

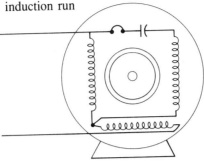

13. The motor in Problem No. 12 is connected for operation at
    a. 115 V.
    b. 230 V.
    c. 460 V.
    d. 277 V.

14. The starting current of a single-phase, ½-hp electric motor, at the instant the windings are first subjected to line voltage and the rotor is not turning yet, is approximately which of the following? (The motor draws 9.8 A full-load current at 115 V.)
    a. 9.8 A
    b. 24 A
    c. 58 A
    d. 82 A

15. The type of motor enclosure best suited for a farm feed-processing or feed-handling room is
    a. totally enclosed, fan cooled.
    b. open.
    c. dripproof.
    d. splashproof.

16. Examine the motor nameplate of Figure 16-8. If the motor is connected for 230-V operation, how many amperes will it draw fully loaded?

17. Refer to the diagram. Based on the torque-speed curve for the motor and fan,
    a. the motor is half loaded.
    b. the motor is overloaded.
    c. the fan will run at synchronous speed.
    d. there is insufficient accelerating torque to get the fan started.

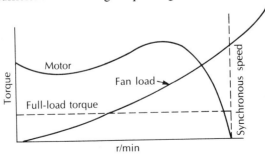

18. The motor in the following list not generally capable of having its shaft rotation variable is a
    a. direct-current motor.
    b. single-phase, capacitor start, induction run motor.
    c. permanent split-capacitor motor.
    d. universal motor.

19. The motor in the diagram can have the direction of rotation of the shaft reversed by
    a. changing to the reverse location the bracket holding the brushes.
    b. reversing lead wires T1 and T2.
    c. reversing lead wires T5 and T8.
    d. reversing the hot and neutral wire.

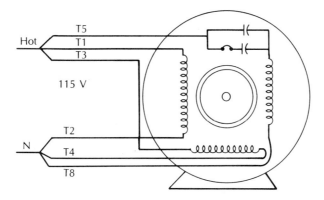

20. What is the type of single-phase motor illustrated in Problem No. 19?

21. The 3-phase motor in this illustration is
    a. wye connected.
    b. delta connected.
    c. improperly connected.

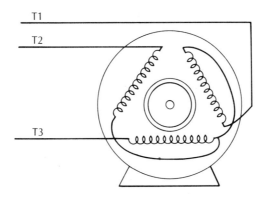

22. A 3-phase, 20-hp electric motor draws 54 A at full load if connected to a 230-V supply. If the motor is connected to a 460-V supply, it will
    a. develop 40 hp.
    b. draw 108 A.
    c. develop 10 hp.
    d. draw 27 A.

Unit 16   Electric Motors   339

23. If an induction motor is disconnected from the ac power supply, and the stator windings are energized with dc power, the motor will
    a. speed up.
    b. reverse and turn in the opposite direction.
    c. come to a stop quickly.
    d. coast to a halt.

24. If a motor is to be mounted with the shaft in the vertical position, the motor should have what type of bearings?

25. A fan is designed to operate at 1 160 r/min, and is driven by a 1 740-r/min motor through a V-belt drive. If the fan is equipped with a 6-in diameter pulley, the proper motor pulley diameter is which of the following?
    a. 2 in
    b. 3 in
    c. 4 in
    d. 8 in

# UNIT 17
# MOTOR CONTROL

## OBJECTIVES

After studying this unit, the student will be able to

- list several methods of controlling a motor.
- describe the operation of a magnetic motor starter.
- explain overcurrent protection of a motor and motor circuit.
- determine the size of the components of a motor circuit.
- determine the size of wire for a group of motors.
- wire a start-stop station or a single-contact control device.
- identify common control devices from their schematic symbols.
- wire a simple control circuit from a control ladder diagram.
- discuss protection of motor control circuits.
- name the NEMA enclosure types and give an example of their applications.

Electric motors present some special problems from the standpoint of control. An electric motor will try to provide the power required by a load, even if it results in self-destruction. Therefore, a motor must be protected from overloads. A motor draws up to six times as much current when the rotor is not turning as it does when operating at full speed. Control contacts and wires must be capable of carrying this high current during starting without causing damage or excessive voltage drop. If a motor should stall, the disconnect switch and control device must be capable of handling the high locked-rotor current. For example, a 100-hp, 460-V, 3-phase electric motor powering an irrigation pump draws 124 A when operating at full load. The locked-rotor current of the motor, however, is nearly 750 A, *NEC Table 430-151*. The motor starter and disconnect switch must be capable of interrupting 750 A in the event the motor stalls. *If the disconnect or controller is not rated for this high level of current, it could explode the instant it is opened.*

Motor circuits will be safe and motors will be protected provided the wiring procedures of *NEC Article 430* are applied. Proper sizing of motor overload protection will ensure many years of dependable service.

## TYPES OF MOTOR CONTROLLERS

A *controller* is simply a means of closing the circuit supplying power to an electrical motor, *NEC 430-81(a)*. Controlling may be accomplished manually or automatically. The simplest motor controller is an attachment plug and receptacle. This method is permitted only for portable motors rated at ⅓ hp and less, *NEC Section 430-81(c)*. Motors larger than ⅓ hp are permitted to be cord connected, but the starting and stopping of the motor must be accomplished by one of the means discussed next.

The branch-circuit protective device, such as a circuit breaker, may serve as the controller for stationary motors, not larger than ⅛ hp, that are normally allowed to operate continuously. An example is a clock motor, *NEC Section 430-81(b)*. These motors are designed with a high impedance, and they cannot be overloaded even if the rotor stalls.

A time-delay circuit breaker is permitted to serve as a motor controller, *NEC Section 430-83, Exception No. 2*. This is frequently the case with farm machinery. Circuit breakers are subject to failure after repeated on and off cycling and, therefore, they are not the best choice for a controller. It is also difficult to size a circuit breaker small enough to provide overload protection for the motor, and still not trip due to the high inrush starting current of the motor.

A fusible knife switch, Figure 17-1, can serve as a motor controller, provided the switch has a horsepower rating sufficient for the motor supplied, *NEC Section 430-90*. Time-delay fuses can usually be sized small enough to provide overload protection, and still not blow during starting. The knife switch must be rated in horsepower, *NEC Section 430-83*. The switch must also have a voltage rating sufficient for the circuit. A 250-V switch is required for motors operating at 120 V, 208 V, and 240 V. A 600-V switch is required for motors operating at 277 V or 480 V.

An ordinary snap switch rated in amperes or horsepower is permitted to be used as a controller for motors rated at not more than 2 hp, and for circuits operating at not more than 300 V, *NEC Section 430-83, Exception No. 1*.

The general-use snap switch must have an ampere rating twice that of the full-load current of the motor. A ⅓-hp motor, for example, draws 7.2 A full-load current at 115 V. A general-use snap switch with a 15-A rating would be required for this motor.

$$7.2 \text{ A} \times 2 = 14.4 \text{ A}$$

Switches rated for use only on alternating-current circuits are permitted to control motors with a full-load current of 80% of the current rating of the switch. Therefore, a 30-A ac switch is permitted to supply a motor which draws 24 A.

$$30 \text{ A} \times 0.8 = 24 \text{ A}$$

A 2-hp electric motor operating at 115 V draws 24 A.

Snap switches for motors are available which do provide overload protection, Figure 17-2. This type of switch, rather than an ordinary switch, is preferred for controlling motors up to 2 hp.

A manual motor starter rated in horsepower provides reliable control and overload protection for motors up to about 5 hp single phase, and 10 hp 3 phase, Figure 17-3. A thermally activated trip mechanism opens the motor circuit automatically if an overload occurs. The overload heaters are sized based upon the full-load current rating of the motor. Manual motor starters are available for both single- and 3-phase motors.

Magnetic motor starters use an electric solenoid coil to close the contacts and start the motor, Figure 17-4. These motor starters may be operated by a person actually activating a starting control, or they may be operated automatically. Only the magnetic motor starter can be controlled automatically. Magnetic motor starters can operate motors of up to several hundred horsepower.

**Figure 17-1  A fusible knife switch may be used as a motor controller.**

**Figure 17-2** A motor controller with a maximum rating of 2 hp (*Courtesy of Square D Company*)

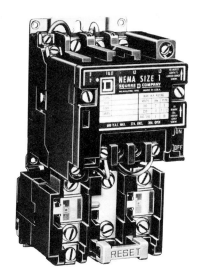

**Figure 17-4** Magnetic motor starter (*Courtesy of Square D Company*)

Thermally activated overload relays break the circuit to the coil of the starter if a motor overload occurs. The overload heaters are sized for the nameplate full-load current of the motor.

**Figure 17-3** Manual motor starter (*Courtesy of Square D Company*)

## MAGNETIC MOTOR STARTER CONSTRUCTION AND OPERATION

Understanding the operation and components of a magnetic motor starter is important when wiring a motor circuit, as well as when performing repairs when there is a malfunction. Only one manufacturer's motor starter is shown in the text, but all have essentially the same features.

The main contacts, Figure 17-5, are inside the main part of the starter. After many years of service, these contacts may become pitted and burned from the arcing as they open the circuit. Replacement kits for the main contacts are available.

Electrical power supplied to the coil of the solenoid closes the motor-starter contacts. The coil must be rated for the voltage of the control circuit electrical supply, Figure 17-6. Typical voltages are 24, 120, 208, 240, 277, and 480 volts. A coil is replaceable. If the coil becomes damaged, only the coil need be replaced. The steel solenoid core fits all coils made for the particular size and model of motor starter.

An auxiliary holding contact, Figure 17-7, is supplied with the motor starter. It is required when a start-stop station is used as the control device. The holding contact is replaceable should a short circuit occur in the control wiring, causing damage to the holding contact.

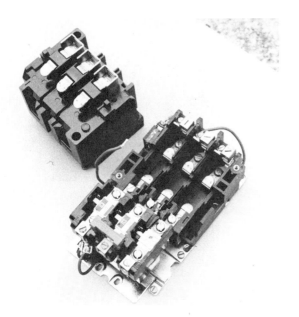

Figure 17-5 The main contacts in a motor starter can be replaced if they become pitted and burned.

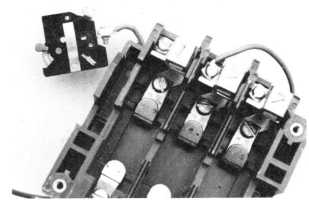

Figure 17-7 The holding contact or interlock may become damaged if a short circuit occurs in the control circuit. The holding contact can be replaced.

An overload relay section is added to the motor starter to sense motor current. This entire unit can be removed from the motor starter, Figure 17-8. A normally closed contact is usually sealed inside the overload section. Some models have one normally closed contact, while others have two or three. If this section becomes damaged, the entire section usually must be replaced.

Heating elements are added to this section sized for the nameplate full-load current of the motor. These overload heaters are replaceable.

## NEMA SIZES

The National Electrical Manufacturers Association (NEMA) has developed a standard numbering system for motor starter sizes. The same numbering system is used by all manufacturers. The sizes range from 00 to 8. The larger the number, the greater is the horsepower rating of

Figure 17-6 The magnetic coil of a motor starter must be rated for the particular control voltage. The coil can be replaced.

Figure 17-8 Overload thermal units or heaters must be sized for the full-load current of the motor.

the starter. The starters are rated according to the current they will be expected to carry continuously, and interrupt if the motor should stall. The motor current depends upon the supply voltage. Naturally then, a given size of motor starter could handle a higher horsepower motor at 460 V than at 230 V. NEMA sizes and horsepower ratings are given in Table 17-1.

Motor starters are available with two poles or three poles. The 2-pole motor starter is used for single-phase motors. Two-pole motor starters are usually available up to NEMA size 3. A 3-pole motor starter may be used for either a 3-phase motor or a single-phase motor. If a 3-pole starter is used for a single-phase motor, only two of the three poles are used. The third pole is simply ignored. Since 3-pole motor starters are more readily available than 2-pole starters, this is a common practice.

Single-phase and 3-phase horsepower ratings on a motor starter are not equivalent. NEMA sizes are equivalent, however, insofar as the current they will handle. Some 3-pole motor starters list both the single-phase and polyphase horsepower ratings, others do not. There is a simple rule that should be followed: *A 3-pole motor starter will handle a single-phase motor with a horsepower rating only half as large as a 3-phase motor.* For example, a NEMA size 2 3-pole motor starter will handle a 15-hp, 230-V, 3-phase motor. The same size motor starter is capable of handling only a 7½-hp, single-phase, 230-V motor. The reason is clear when the full-load current for each motor is compared from *NEC Tables 430-148* and *430-150*. This rule does not hold true exactly, however, for the NEMA small sizes 00 and 0, as shown in Table 17-1. For sizes 1 and larger, the horsepower essentially doubles as the size number is increased by 1.

Table 17-1 Motor horsepower and voltage ratings for NEMA size motor starters

| NEMA Size | Single phase | | Three phase | | |
|---|---|---|---|---|---|
| | 115 V | 230 V | 200 V | 230 V | 460 V |
| 00 | ⅓ | 1 | 1 ½ | 1 ½ | 2 |
| 0 | 1 | 2 | 3 | 3 | 5 |
| 1 | 2 | 3 | 7 ½ | 7 ½ | 10 |
| 1P | 3 | 5 | — | — | — |
| 2 | 3 | 7 ½ | 10 | 15 | 25 |
| 3 | 7 ½ | 15 | 25 | 30 | 50 |
| 4 | — | 25 | 40 | 50 | 100 |
| 5 | — | 50 | 75 | 100 | 200 |
| 6 | — | — | 150 | 200 | 400 |
| 7 | — | — | — | 300 | 600 |
| 8 | — | — | — | 450 | 900 |

## NEMA ENCLOSURE TYPES

The enclosure for housing the motor starter is selected for the type of environment in the area. NEMA has developed a standard numbering system for types of electrical equipment enclosures.

The enclosure type, except NEMA 1, is required to be marked on motor starter enclosures, *NEC Section 430-91*. This is a recent Code requirement; therefore, most motor starters presently in use are not so marked. *NEC Table 430-91* gives NEMA enclosure types for motor starters, listing the outdoor and indoor conditions for which they are suitable.

Common enclosures used for agricultural, commercial, and industrial locations are as follows.

**NEMA 1:** General-purpose enclosure used in any location that is dry and free from dust and flying flammable materials, Figure 17-9.

**NEMA 3:** Weather-resistant enclosure which is suitable for use outdoors. Not suitable for use in dusty locations.

**NEMA 4:** Watertight and dusttight enclosure suitable for outdoor locations as well as inside wet locations. Water can be sprayed directly on the enclosure without it leaking inside. Suitable for most agricultural locations provided corrosion is not a problem.

**NEMA 4X:** Watertight, dusttight, and corrosion-resistant enclosure suitable for outside and inside wet, dusty, and corrosive areas. Suitable for agricultural buildings. Available with stainless steel and nonmetallic enclosures, Figure 17-10.

Figure 17-9 NEMA 1, enclosure for dry and dust-free environments

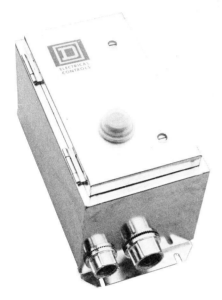

**Figure 17-10** NEMA 4X, corrosion-resistant enclosure for wet or dusty areas

**NEMA 7:** Explosionproof enclosure suitable for installation in Class I areas containing hazardous vapors. Must also be rated for the type of hazardous vapor, such as gasoline vapor, Group D, Figure 17-11

**NEMA 9:** Dust-Ignition-Proof enclosure suitable for installation in Class II hazardous areas, such as grain elevators. Must be rated for the type of dust, such as grain

**Figure 17-11** NEMA 7, explosionproof enclosure for areas where hazardous vapors, such as gasoline vapor, are present

**Figure 17-12** NEMA 9, dust-ignition-proof enclosure for areas where dust is suspended in the air and may cause a fire or explosion, such as a commercial grain elevator

dust, Group G, Figure 17-12. Some manufacturers build one enclosure rated as NEMA 7 and 9.

**NEMA 12:** Dusttight, driptight, and oil-retardant enclosure suitable for machine tools and areas where there are flammable flying particles, such as in cotton gins and textile processing and handling operations.

## SINGLE-MOTOR BRANCH CIRCUIT

A motor branch circuit consists of several different parts that must be sized properly. The motor is unique because it draws a very high starting current in comparison to the full-load running current. The motor also may stall or break down while in operation. Means must be provided to disconnect power, usually automatically, if a failure occurs. The components of most motor circuits are shown in a diagram at the beginning of *NEC Article 430*. Information supplied on the motor nameplate is necessary for sizing the circuit components.

- Branch-circuit disconnect
- Branch-circuit short-circuit and ground-fault protection
- Branch-circuit wires
- Motor controller
- Motor running overload protection
- Motor control circuit where a magnetic motor starter is used.

Sizing the motor branch circuit begins with the motor nameplate. Consider the motor nameplate of Figure 17-13. The motor is 3 phase, dual voltage. First, determine

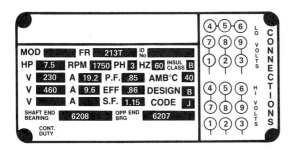

**Figure 17-13** Information required for wiring a motor branch circuit is contained on the motor nameplate.

the type of electrical supply available. Assume, in this case, that the electrical supply is 240 V, 3 phase. The motor full-load current therefore, will be 19.2 A.

The type of electrical equipment from which the motor circuit will originate must be determined. Common sources of electrical power for electric motors are:

1. Circuit-breaker panelboards
2. Fusible panelboards
3. Fusible disconnects tapped from a feeder, Figure 17-14
4. Fusible disconnects tapped from a bus duct

## MOTOR FULL – LOAD CURRENT

The motor full-load current that is used to determine the minimum size of motor circuit components is the current listed in *NEC Tables 430-148* and *430-150* (Tables 17-2 and 17-3 in this text), *NEC Section 430-6(a)*. However, this value should be checked against the motor nameplate ampere rating. Sometimes a farm motor may have an ampere rating higher than the value listed in the *NEC* Tables. In this situation, the higher of the values

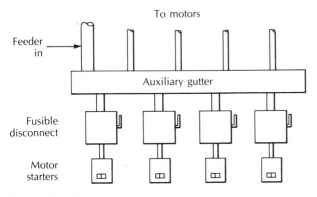

**Figure 17-14** A group of motors tapped from a single feeder, using fusible switches as disconnects

should be taken for sizing components, (See notes to Tables 17-2 and 17-3).

The nameplate current rating must be used for selecting the maximum size motor running overload protection. This is important because the overload heaters must be sized for a specific motor. If motors are changed, then the overload heaters may have to be changed.

The full-load current for a 3-phase, 208-V electric motor is not listed in *NEC Table 430-150*. The footnote at the bottom of the table tells how to obtain the correct value. Actually, a 208-V motor usually operates at about 200 V. The correct value is determined by multiplying the full-load current for a 230-V motor by 1.15. For example, a 200-V, 10-hp motor would draw 32 A.

**Table 17-2** (*NEC Table 430-148*) Full-load currents in amperes; single-phase alternating-current motors

The following values of full-load currents are for motors running at usual speeds and motors with normal torque characteristics. Motors built for especially low speeds or high torques may have higher full-load currents, and multi-speed motors will have full-load current varying with speed, in which case the name-plate current ratings shall be used.

To obtain full-load currents of 208- and 200-volt motors, increase corresponding 230-volt motor full-load currents by 10 and 15 percent, respectively.

The voltages listed are rated motor voltages. The currents listed shall be permitted for system voltage ranges of 110 to 120 and 220 to 240.

| HP | 115 V | 230 V |
|---|---|---|
| 1/6 | 4.4 | 2.2 |
| 1/4 | 5.8 | 2.9 |
| 1/3 | 7.2 | 3.6 |
| 1/2 | 9.8 | 4.9 |
| 3/4 | 13.8 | 6.9 |
| 1 | 16 | 8 |
| 1 1/2 | 20 | 10 |
| 2 | 24 | 12 |
| 3 | 34 | 17 |
| 5 | 56 | 28 |
| 7 1/2 | 80 | 40 |
| 10 | 100 | 50 |

Reprinted with permission from NFPA 70-1984, *National Electrical Code*®, Copyright © 1983, National Fire Protection Association, Quincy, Massachusetts 02269. This reprinted material is not the complete and official position of the NFPA on the referenced subject which is represented only by the standard in its entirety.

**Table 17-3** *(NEC Table 430-150)* **Full-load current\* 3-phase alternating-current motors**

| HP | Induction Type Squirrel-Cage and Wound-Rotor Amperes | | | | | Synchronous Type †Unity Power Factor Amperes | | | |
|---|---|---|---|---|---|---|---|---|---|
| | 115 V | 230 V | 460 V | 575 V | 2300 V | 230 V | 460 V | 575 V | 2300 V |
| ½ | 4 | 2 | 1 | .8 | | | | | |
| ¾ | 5.6 | 2.8 | 1.4 | 1.1 | | | | | |
| 1 | 7.2 | 3.6 | 1.8 | 1.4 | | | | | |
| 1½ | 10.4 | 5.2 | 2.6 | 2.1 | | | | | |
| 2 | 13.6 | 6.8 | 3.4 | 2.7 | | | | | |
| 3 | | 9.6 | 4.8 | 3.9 | | | | | |
| 5 | | 15.2 | 7.6 | 6.1 | | | | | |
| 7½ | | 22 | 11 | 9 | | | | | |
| 10 | | 28 | 14 | 11 | | | | | |
| 15 | | 42 | 21 | 17 | | | | | |
| 20 | | 54 | 27 | 22 | | | | | |
| 25 | | 68 | 34 | 27 | | 53 | 26 | 21 | |
| 30 | | 80 | 40 | 32 | | 63 | 32 | 26 | |
| 40 | | 104 | 52 | 41 | | 83 | 41 | 33 | |
| 50 | | 130 | 65 | 52 | | 104 | 52 | 42 | |
| 60 | | 154 | 77 | 62 | 16 | 123 | 61 | 49 | 12 |
| 75 | | 192 | 96 | 77 | 20 | 155 | 78 | 62 | 15 |
| 100 | | 248 | 124 | 99 | 26 | 202 | 101 | 81 | 20 |
| 125 | | 312 | 156 | 125 | 31 | 253 | 126 | 101 | 25 |
| 150 | | 360 | 180 | 144 | 37 | 302 | 151 | 121 | 30 |
| 200 | | 480 | 240 | 192 | 49 | 400 | 201 | 161 | 40 |

For full-load currents of 208- and 200-volt motors, increase the corresponding 230-volt motor full-load current by 10 and 15 percent, respectively.

\*These values of full-load current are for motors running at speeds usual for belted motors and motors with normal torque characteristics. Motors built for especially low speeds or high torques may require more running current, and multispeed motors will have full-load current varying with speed, in which case the nameplate current rating shall be used.

†For 90 and 80 percent power factor the above figures shall be multiplied by 1.1 and 1.25 respectively.

The voltages listed are rated motor voltages. The currents listed shall be permitted for system voltage ranges of 110 to 120, 220 to 240, 440 to 480, and 550 to 600 volts.

Reprinted with permission from NFPA 70-1984, *National Electrical Code®*, Copyright © 1983, National Fire Protection Association, Quincy, MA. This reprinted material is not the complete and official position of the NFPA on the referenced subject which is represented only by the standard in its entirety.

## DISCONNECT FOR A SINGLE MOTOR

A disconnect for a motor branch circuit must be capable of interrupting the locked-rotor current of the motor. This disconnecting means must disconnect both the motor and the controller from all ungrounded supply conductors, *NEC Section 430-103*. The disconnecting means must be located within sight from the motor controller, *NEC Section 430-102*. The definition of "in sight from" in *NEC Article 100* means that the controller is not more than 50 ft (15.24 m) from the disconnect, and that the controller is actually visible when standing at the disconnect location.

A fusible switch serving as the motor circuit disconnecting means must be rated in horsepower, *NEC Section 430-109*. A circuit breaker serving as a disconnecting means is not required to be rated in horsepower but, instead will be rated in amperes, *NEC Section 430-109*. This circuit breaker must have an ampere rating not less than 115% of the motor full-load current, *NEC Section 430-110(a)*.

There are exceptions to these rules for motor disconnects. Motors with a horsepower rating not greater than 1/8 hp may use the branch-circuit overcurrent device as the disconnect, *NEC Section 430-109, Exception No. 1*. If the motor is not larger than 2 hp, and operates at not more than 300 V, a general-use switch may serve as the disconnect, *NEC Section 430-109, Exception No. 2*. The switch must have an ampere rating at least twice that of the full-load current of the motor. An ac switch used on an ac circuit need only be rated at 1.25 times the full-load current of the motor.

Cord- and plug-connected motors may use an attachment plug and receptacle as the motor disconnect. This type of disconnect would be used for motors on portable or movable equipment used on the farm. The plug and receptacle should have a horsepower rating not smaller than the rating of the motor if it is intended to break power while operating, *NEC Section 430-109, Exception No. 5*. The current rating of the attachment plug and receptacle must be not less than 115% of the full-load current of the motor.

A fusible switch or a circuit breaker is permitted to serve as both the controller and disconnect, *NEC Section 430-111*. However, the requirements for motor controllers discussed previously must be met.

Assume that the motor of Figure 17-13 will have a fusible switch as the disconnect. The disconnect switch must have a rating marked on the switch of at least 7½ hp. However, if a circuit breaker serves as the disconnect, the rating will be in amperes. The minimum ampere rating of the circuit breaker is 1.15 times the full-load current, or 25 A. A 25-A or 30-A circuit breaker would be chosen. A circuit breaker rated at 30 A or more would most likely be chosen, because a 25-A breaker would probably trip due to the high starting current. The value of current from *NEC Table 430-150* is used, rather than the nameplate current of 19.2 A.

$$1.15 \times 22 \text{ A} = 25 \text{ A}$$

## BRANCH-CIRCUIT WIRES

The wires supplying a single motor shall have an ampere rating not less than 1.25 times the full-load current of the motor. For a multispeed motor, the highest normal full-load current shall be used, *NEC Section 430-22(a)*.

The minimum ampere rating of the circuit wires for the motor of Figure 17-13 is 28 A.

$$1.25 \times 22 \text{ A} = 28 \text{ A}$$

The minimum size copper branch-circuit wire is found in NEC Table 310-16 (included in Unit 9 of this text). If the wire is THWN copper, the minimum size is No. 10 AWG. The footnote at the bottom of NEC Table 310-16 which specifies the maximum rating overcurrent device for wire sizes No. 14, 12, and 10 AWG does not apply in the case of electric motor circuits, *NEC Section 210-19(a)*. Aluminum wire is seldom used for motor circuits. Most terminals in motor starters are rate only for copper wire.

## Problem 17-1

A single-phase, 5-hp, 230-V electric motor has a nameplate current rating of 26 A. Determine the minimum size copper THWN branch-circuit wire.

## Solution

Look up the full-load current of a 5-hp single-phase, 230-V motor in *NEC Table 430-148*. The current value is 28 A. The minimum ampere rating of the wire is 35 A.

$$1.25 \times 28 \text{ A} = 35 \text{ A}$$

## SHORT-CIRCUIT PROTECTION

A branch circuit supplying a single motor is only subjected to excessive current caused by motor overloads or by ground faults and short circuits. Sometimes, a single overcurrent protective device protects against all three conditions while, in many instances, overload protection and protection from ground faults and short circuits are provided separately. First, consider them separately, and size the short-circuit and ground-fault protection for the branch circuit. The overcurrent device is either a fuse or a circuit breaker placed at the beginning of the circuit, Figure 17-15.

The short-circuit and ground-fault protection must be capable of carrying the starting current of the motor, *NEC Section 430-52*. The maximum rating of a time-delay fuse or an inverse time circuit breaker is determined using *NEC Table 430-152* (Table 17-4 in this

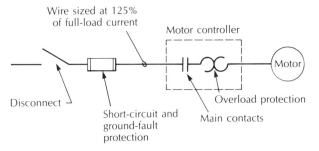

**Figure 17-15** Parts of a motor branch circuit

text). An *inverse time circuit breaker* is an ordinary circuit breaker which provides short-circuit protection, but it will carry an overload for a limited period of time. The code letter given on the nameplate of the motor is needed to determine the maximum size short-circuit and ground-fault protection. The motor of Figure 17-13 has a code letter J. This is an ac 3-phase squirrel-cage motor; therefore from *NEC Table 430-152*, the maximum size time-delay fuse is 175% of the motor full-load current, or 39 A. The motor full-load current must be taken from *NEC Table 430-150*. The maximum size inverse time circuit breaker would be 250% of the motor full-load current, or 55 A.

The motor short-circuit and ground-fault protective device should be sized as small as possible without causing nuisance tripping. Remember that the minimum size in the case of a circuit breaker is 1.15 times the full-load current of the motor, *NEC Section 430-58*. The motor of Figure 17-13 is shown in Figure 17-16 with all components sized within the limits of the *National Electrical Code*. A suggested procedure is to determine the overcurrent protection size, using the value determined in *NEC Table 430-152*. If the motor is not driving a hard-starting load, then choose the next smaller protective device than the value determined. If this is too small, then choose the next size larger. In no case is the fuse permitted to be sized larger than 225% of the full-load current, *NEC Section 430-52, Exception No. 2(b)*. For a circuit breaker, the maximum size is not permitted to exceed 400% for a motor drawing 100 A or less, and 300% for motors drawing more than 100 A, *NEC Section 430-52, Exception No. 2(c)*.

Fuse:

$$1.75 \times 22 \text{ A} = 38 \text{ A}$$

(Use a 35-A or 40-A fuse.)

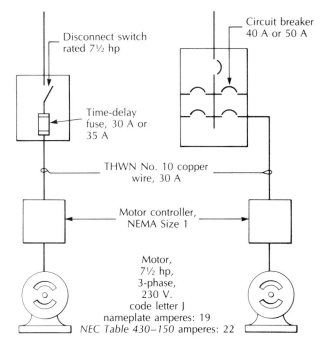

**Figure 17-16** Properly sized components of a motor circuit

Circuit breaker:

$$2.5 \times 22 \text{ A} = 55 \text{ A}$$

(Use a 50-A or 60-A breaker.)

## Problem 17-2

A single-phase, 5-hp 230-V electric motor with a nameplate full-load current of 26 A drives a normal-starting load and has a code letter G. Determine the proper size motor short-circuit protective time-delay fuse.

## Solution

Determine the full-load current from *NEC Table 430-148* and the fuse multiplier from *NEC Table 430-152*.

Full-load current: 28 A
Fuse multiplier: 175%
$1.75 \times 28 \text{ A} = 49 \text{ A}$

Use a 50-A fuse. The minimum wire size is No. 10 AWG copper THWN with a rating of 35 A.

**Table 17-4** *(NEC Table 430-152)* **Maximum rating or setting of motor branch-circuit short-circuit and ground-fault protective devices**

| Type of Motor | Percent of Full-Load Current | | | |
|---|---|---|---|---|
| | Nontime Delay Fuse | Dual Element (Time-Delay) Fuse | Instantaneous Trip Breaker | *Inverse Time Breaker |
| Single-phase, all types | | | | |
|     No code letter . . . . . . . | 300 | 175 | 700 | 250 |
| All ac single-phase and polyphase squirrel-cage and synchronous motors† with full-voltage, resistor or reactor starting: | | | | |
|     No code letter . . . . . . . | 300 | 175 | 700 | 250 |
|     Code letter F to V . . . . . . | 300 | 175 | 700 | 250 |
|     Code letter B to E . . . . . . | 250 | 175 | 700 | 200 |
|     Code letter A . . . . . . . | 150 | 150 | 700 | 150 |
| All ac squirrel-cage and synchronous motors† with autotransformer starting: | | | | |
|   Not more than 30 amps | | | | |
|     No code letter . . . . . . . | 250 | 175 | 700 | 200 |
|   More than 30 amps | | | | |
|     No code letter . . . . . . . | 200 | 175 | 700 | 200 |
|     Code letter F to V . . . . . . | 250 | 175 | 700 | 200 |
|     Code letter B to E . . . . . . | 200 | 175 | 700 | 200 |
|     Code letter A . . . . . . . | 150 | 150 | 700 | 150 |
| High-reactance squirrel-cage | | | | |
|   Not more than 30 amps | | | | |
|     No code letter . . . . . . . | 250 | 175 | 700 | 250 |
|   More than 30 amps | | | | |
|     No code letter . . . . . . . | 200 | 175 | 700 | 200 |
| Wound-rotor— | | | | |
|     No code letter . . . . . . . | 150 | 150 | 700 | 150 |
| Direct-current (constant voltage) | | | | |
|   No more than 50 hp | | | | |
|     No code letter . . . . . . . | 150 | 150 | 250 | 150 |
|   More than 50 hp | | | | |
|     No code letter . . . . . . . | 150 | 150 | 175 | 150 |

For explanation of Code Letter Marking, see Table 430-7(b).
For certain exceptions to the values specified, see Sections 430-52 through 430-54.

*The values given in the last column also cover the ratings of nonadjustable inverse time types of circuit breakers that may be modified as in Section 430-52.

†Synchronous motors of the low-torque, low-speed type (usually 450 rpm or lower), such as are used to drive reciprocating compressors, pumps, etc. that start unloaded, do not require a fuse rating or circuit-breaker setting in excess of 200 percent of full-load current.

Reprinted with permission from NFPA 70-1984, *National Electrical Code®*, Copyright © 1983, National Fire Protection Association, Quincy, MA. This reprinted material is not the complete and official position of the NFPA on the referenced subject which is represented only by the standard in its entirety.

## Problem 17-3

Determine the proper size inverse time circuit breaker to provide ground-fault and short-circuit protection.

## Solution

> Circuit breaker multiplier: 250%
> $2.5 \times 28 \text{ A} = 70 \text{ A}$

Use a 60-A breaker, if possible. If it is too small, then choose a 70-A breaker.

It is permissible to have an overcurrent device sized larger than the allowable ampere rating of the wire. For example, a 70-A circuit breaker can protect a No. 10 AWG copper THWN wire with a maximum rating of 35 A. It must be remembered that the purpose of this circuit breaker is to protect against ground faults and short circuits only. Overload protection not exceeding the ampere rating of the wire is provided usually in the form of thermal overload heaters in the motor starter. As long as properly sized overload protection is provided somewhere in the circuit, the short-circuit fuse or circuit-breaker protection is permitted to exceed the ampere rating of the motor branch-circuit wire.

## RUNNING OVERCURRENT PROTECTION

Electric motors are required to be protected against overload, *NEC Section 430-32*. Overload protection is usually provided as a device responsive to motor current or as a thermal protector integral with the motor. A device responsive to motor current could be a fuse, a circuit breaker, an overload heater, or a thermal reset switch in the motor housing. An automatically resetting thermal switch placed in the windings will sense winding temperature directly.

The service factor or temperature rise must be known from the motor nameplate when selecting the proper size motor overload protection. These are indicators of the amount of overload a motor can withstand. If a motor has a service factor of 1.15 or greater, the manufacturer has designed extra overload capacity into the motor. In this case, the overload protection may be sized as large as 125% of the *nameplate* full-load current. Internal heat is damaging to motor-winding insulation. A motor with a temperature rise of 40°C (104°F) or less has been designed to run relatively cool; therefore, it has greater overload capacity. The overload protection may be sized as large as 125% of the *nameplate* full-load current. A service factor of less than 1.15 or a temperature rise of more than 40°C (104°F) indicates little overload capacity. The overload protection under these circumstances is sized not larger than 115% of the *nameplate* full-load current.

Size the motor overload protective device at not more than these percents of nameplate full-load current:

- Service factor 1.15 or greater, 125%
- Temperature rise not greater than 40°C (104°F), 125%
- Service factor smaller than 1.15, 115%
- Temperature rise greater than 40°C (104°F), 115%

A time-delay fuse may serve as motor overload protection. Plug fuses or cartridge fuses are used for small motors. The fuse size is determined by selecting the proper multiplying factor, 1.15 or 1.25, based upon the service factor and temperature rise. Standard fuse sizes smaller than 30 A are listed in Unit 6.

## Problem 17-4

A ¾-hp, single-phase water pump motor is operated at 230 V and full-load current of 6.9 A. The motor has a service factor of 1.15 and a code letter K. Determine the maximum size time-delay fuses permitted to serve as running overload protection.

## Solution

The full-load current from *NEC Table 430-148* is 6.9 A. With a service factor of 1.15, the overload protective device may be 1.25 times the full-load current.

> $1.25 \times 6.9 \text{ A} = 8.6 \text{ A}$

Choose an 8-A fuse (refer to fuse sizes listed in Unit 6). If an 8-A fuse will not work, then the next larger size, 9 A, may be used provided it does not exceed 140% of the motor nameplate current, $1.4 \times 6.9 \text{ A} = 9.7 \text{ A}$, *NEC Section 430-34*.

Circuit breakers could be used as running overload protection, but they are not available in small-ampere sizes. If they are used for large motors, they will usually trip on starting if they are sized small enough to provide overload protection.

Magnetic and manual motor starters have an overload relay or trip mechanism which is activated by a heater sensitive to the motor current. Typical motor overload heaters are shown in Figure 17-17. The manufacturer of the motor starter provides a chart inside the motor starter listing the part number for thermal overload heater chart is used. A 3-phase motor nameplate full-load current for 230-V operation is 1.5 A. The proper overload heater to use is thermal unit No. B 2.40. The *NEC* allows this heater to be sized at 125% of the motor nameplate full-load current provided the service factor is 1.15 or larger, or the temperature rise is not greater than 40°C (104°F). The manufacturer has taken this into consideration when setting up the chart. If the service factor is less than 1.15 or if the temperature rise is greater than 40°C (104°F), then the heater will be 10% oversized. The motor is then vulnerable to burnout. If the motor service factor is less than 1.15, multiply the motor full-load current on the nameplate by 0.9 and use this new value to size the overload heater. For the example, multiplying 1.5 A by 0.9 gives 1.35 A. The overload heater corresponding to 1.35 A is a thermal unit No. B 2.10.

A thermal protector integral with the motor and installed by the manufacturer is permitted to serve as the overload protection, *NEC Section 430-32(a)(2)*. The manufacturer is required to size the thermal protector according to the multiplying factors in the *NEC*.

The number of overload protective devices required for a motor is specified in the *NEC*. If fuses are used, one fuse shall be placed in each ungrounded conductor supplying the motor, *NEC Section 430-36*. The number of thermal overload heaters is specified in *NEC Table 430-37*. The minimum is one for a single-phase motor, and three for a 3-phase motor.

The rules for sizing components for a single-motor branch circuit are summarized in Figure 17-19.

Figure 17-17 Overload thermal sensing units or heaters trip an overload relay if the motor current becomes excessive.

heaters. The heaters are sized according to the actual full-load current listed on the motor nameplate. Find the heater number from the manufacturer's list corresponding to the motor nameplate full-load current. A typical manufacturer's overload heat selection chart is shown in Figure 17-18.

An example will help to learn how the overload

Figure 17-18 Manufacturer's chart for selecting overload thermal sensing unit (*Courtesy of Square D Company*)

## Problem 17-5

A single-phase, 3-hp electric motor operates at 230 V, and draws 17 A full-load current. The service factor is 1.2 and the code letter is M. The motor disconnect is a fusible switch and the controller is a magnetic motor starter. Determine the following:

1. Minimum rating disconnect switch
2. Minimum NEMA size motor starter
3. Minimum size copper THW wire
4. Time-delay fuse size for short-circuit and ground-fault protection
5. Proper overload heater from the chart in Figure 17-18

## Solution

1. The disconnect switch must be rated in horsepower; therefore, the minimum is 3 hp.
2. Using Table 17-1, the minimum size motor starter for a single-phase, 3-hp motor is NEMA Size 1.

Unit 17  Motor Control  353

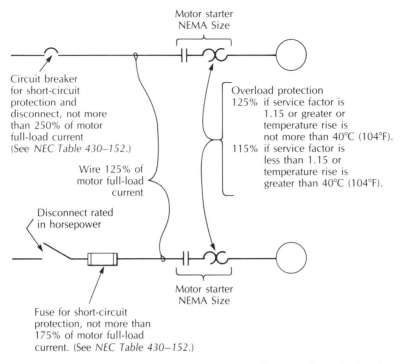

Figure 17-19 Selecting components for a single-motor branch circuit

3. Look up the motor full-load current in *NEC Table 430-148*. Multiply the current by 1.25.

$$1.25 \times 17 \text{ A} = 21 \text{ A}$$

From *NEC Table 310-16* (refer to Unit 9), the minimum copper THW wire size is No. 12 AWG.

4. The short-circuit time-delay fuse multiplier is found in *NEC Table 430-152* for a single-phase motor with the code letter M. Multiply the full-load current by 1.75.

$$1.75 \times 17 \text{ A} = 30 \text{ A}$$

Use a 30-A time-delay fuse.

5. The overload heater is based upon 125% of full-load current because the service factor is greater than 1.15. Therefore, look up the heater part number corresponding to a motor full-load current of 17 A. The proper thermal overload heater is thermal unit No. B 32.

## FEEDER SUPPLYING SEVERAL MOTORS

Groups of motors may be supplied by a single feeder. An example would be a feeder from the main service in a barn to the motor control center in a feed room. The feeder wires are required to have a minimum rating equal to the sum of the full-load current of all motors supplied, plus 25% of the full-load current of the largest motor, *NEC Section 430-24*. Consider the following example with single-phase, 230-V motors:

Silo unloader—5 hp, 28 A
Silo unloader—3 hp, 17 A
Conveyer—1 hp, 8 A
Bunk Feeder—2 hp, 12 A

The minimum size wire must have an ampere rating of 72 A.

$$28 + 17 + 8 + 12 + (0.25 \times 28) = 72 \text{ A}$$

Using copper THWN, the minimum size is No. 4 AWG.

Next, the disconnect must be selected for the motor feeder. The disconnect for a feeder serving a group of motors shall have a horsepower rating not less than the sum of the horsepower ratings of the motors served, *NEC Section 430-112*. For the example, a fusible disconnect switch must have a horsepower rating not less than 5 + 3 + 1 + 2 = 11 hp. The disconnect to be chosen would probably be rated at 15 hp.

If a circuit breaker serves as the disconnect, it must have a current rating not less than 115% of the sum of the full-load currents of all the motors served, *NEC Section 430-110(c)(2)*. For the example, this would be 75 A.

$$1.15 \times (28 + 17 + 8 + 12) = 1.15 \times 65 = 75 \text{ A}$$

The minimum size standard circuit breaker would be 80 A.

Typically, feeders serving groups of motors will be sized large enough to provide for future expansion. Therefore, the previous calculations serve only as a guide. To make sure that the feeder is adequate for present and future needs, the feeder in the example would not be sized for the minimum, but rather for at least 100 A. The wire size could possibly be either No. 3 AWG copper THW or No. 1 AWG THW aluminum. The overcurrent protection would then be sized according to the ampere rating of the feeder, *NEC Section 430-62(b)*.

The short-circuit protection for a feeder supplying a specific fixed motor load can be sized according to the rules of *NEC Section 430-62(a)*. This only applies to a fixed motor load that will not be changed in the future. Consider the previous example, assuming that circuit breakers serve as the disconnect and short-circuit protection for each motor. A fusible disconnect switch will serve as the short-circuit protection for the feeder. Figure 17-20 shows the sizes of wire and overcurrent protection for each circuit.

The maximum rating of short-circuit protection for the feeder is determined by taking the maximum size motor branch-circuit short-circuit device and adding to it the full-load current of all other motors served. The larg-

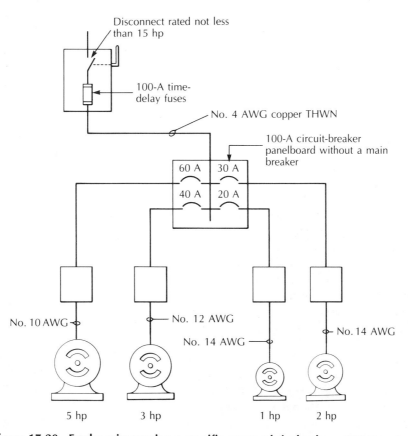

**Figure 17-20 Feeder wire serving a specific group of single-phase, 230-V motors**

Unit 17  Motor Control  355

est branch-circuit protection for the example is a 60-A circuit breaker for the 5-hp motor. Add to this 60 A the full-load currents for the 3-hp, 2-hp and 1-hp motors. The fuse size is 100 A maximum.

$$60 + 17 + 12 + 8 = 97 \text{ A}$$

It must be remembered that the feeder overcurrent device is permitted to exceed the ampere rating of the feeder only when the feeder is supplying a specific motor load. The panelboard is often used to serve other loads; therefore, the feeder then must be protected at its ampacity. For the example, a No. 3 AWG copper wire instead of a No. 4 should be chosen.

For group motor installations, dual-element, time-delay fuses for each motor may be sized at 125% of the motor full-load current rating. This provides motor branch-circuit protection, as well as running overload protection.

The main feeder dual-element, time-delay fuse may generally be sized at 1½ times the ampere rating of the largest motor of the group, plus the full-load current rating of the other motors of the group.

$$(1.5 \times 28) + 17 + 8 + 12 = 79 \text{ A}$$

(Use 80-A fuses.)
This fuse size would be a minimum.

## FEEDER TAPS

A common practice is to tap motor branch circuits directly from a motor feeder. This is permitted as long as the branch-circuit wire is terminated at an overcurrent device sized properly for the circuit, *NEC Section 430-28*. The branch-circuit wire size may be smaller than the feeder wire size. If the tap wire from the feeder to the overcurrent device is not more than 10 ft (3.05 m), and is enclosed in raceway, the branch-circuit wire may be as small as necessary to serve the motor load. However, if the tap is more than 10 ft (3.05 m) but not more than 25 ft (7.62 m) in length, the branch-circuit wire must have an ampere rating at least one-third the rating of the feeder conductors.

Consider the example of a No. 4 AWG copper THW feeder wire supplying the three motors of Figure 17-21. The tap wire from the feeder to the fusible disconnect for the ¾-hp and 10-hp, 230-V, single-phase motors is not required to be a minimum size because the tap is not more than 10 ft (3.05 m) long. However, the 1½-hp motor tap is more than 10 ft (3.05 m) long; therefore, it must have an ampere rating not less than one-third that of the feeder wire, or a minimum rating of 28 A. If the wires are copper THW, then the minimum size is No. 10 AWG. If the tap had not been more than 10 ft (3.05 m) long, then the 1½-hp motor circuit could have been wired with No. 14 AWG THW copper wire.

$$\frac{85 \text{ A}}{3} = 28\text{-A minimum for 1½-hp motor tap}$$

## MOTOR CONTROL CIRCUIT

A magnetic motor starter is operated with an electric solenoid coil. A diagram of the power flow to the motor and the control circuit is included with each magnetic motor starter.

A typical diagram is shown in Figure 17-22. The heavy lines show the power flow to the motor. The narrow lines belong to the control circuit which is prewired

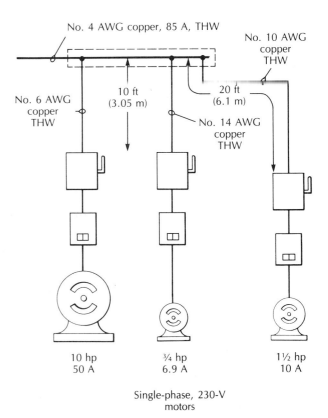

Figure 17-21 **Motor branch-circuit taps from a feeder. The motors are single phase.**

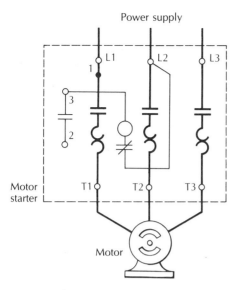

**Figure 17-22** Schematic diagram of a magnetic motor starter

by the manufacturer. The holding contact or interlock is the set of contacts between terminals 2 and 3. These contacts close at the same time the main contacts close. The control wire goes from terminal 3 to the coil which closes the contacts, then from the coil through a normally closed overload relay contact. Some motor starters have as many as three of these contacts in series. If the motor overloads, the thermal units will heat up and trip open this overload relay contact, breaking the control circuit and deenergizing the coil. This opens the motor circuit. From the normally closed overload relay contact, the control circuit wire goes to terminal L2. The terminals of the overload relay are shown in Figure 17-23.

**Figure 17-23** Overload relay terminals of a magnetic motor starter

Consider the situation where a simple switching device, such as a thermostat, is used to control the motor, Figure 17-24. When the thermostat closes, it must complete the circuit to energize the motor starter coil. Power is obtained from terminal 1, which is located next to terminal L1. The other side of the thermostat is connected to terminal 3. When the thermostat closes, electrical current flows from terminal 1 through the thermostat to the coil. From the coil, current flows through the overload relay contact to terminal L2. When the thermostat opens, the circuit is broken and the main contacts open. Only two wires are required when a simple switch device is used to control the motor starter; terminal 2 is not used.

A *start-stop station* (also called a *push-button station*) is another common device used to control a motor. The start and stop push buttons are *momentary contacts*. They immediately return to their original position after they have been pressed. Figure 17-25 shows a momentary contact push-button station.

A start-stop motor control circuit is shown in Figure 17-26. This requires a 3-wire control circuit. Current can flow from terminal 1 through the normally closed stop button. As soon as the start button is depressed, current flows to terminal 3 and then through the coil to complete the circuit. The main contacts and the holding contact between terminals 2 and 3 are now closed. When the start button is released the start contact opens, but current now flows from terminal 2 to terminal 3 through the

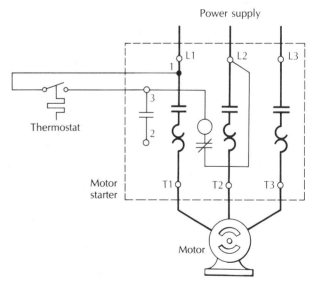

**Figure 17-24** Simple switch device, such as a thermostat, controlling a magnetic motor starter

Unit 17 Motor Control 357

**Figure 17-25** Start-stop station for a magnetic motor starter (*Courtesy of Square D Company*)

holding contact, keeping the coil energized and the contacts closed. Now the only way to stop the motor is to press the stop button or open the overload relay. Diagrams for wiring common motor control circuits are contained inside the motor starter. It is a good idea to put numbers on the control circuit wires to help keep them identified. Figure 17-27 shows the start-stop station wires connected to the motor starter. Compare Figure 17-27 with the diagram of Figure 17-26.

The diagram of Figure 17-26 is a schematic control wiring diagram for a motor control circuit. It is a little difficult to visualize the operation of the complete control circuit. Examine Figure 17-28 where two separate start-stop stations are wired independently to operate the motor. The schematic diagram can be confusing. A type of diagram which helps to visualize easily the components and operation of a control circuit is called a *ladder diagram*. The entire control circuit is drawn horizontally across the diagram between the two power wires L1 and L2.

Figure 17-29 is a ladder diagram for the start-stop control circuit of Figure 17-26. The flow of current can be traced easily from L1 to L2 through each component. Figure 17-30 is a ladder-type control diagram of the two start-stop stations of Figure 17-28. Stop buttons are wired in series so that any one of them can break the circuit. The start buttons must be wired in parallel so that any one of them can energize the coil. The holding contact is always wired in parallel with the start buttons.

## INTERLOCKING MOTORS

Two electric motors can be interlocked so that the second one cannot run unless the first one is running. This may be necessary to prevent plugging of a materials handling system. For example, if a conveyer stops, it may be desirable to shut down the silo unloader feeding

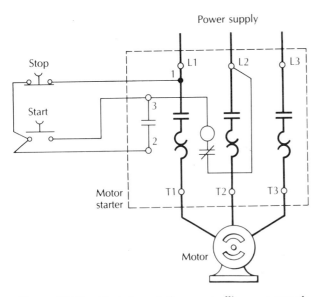

**Figure 17-26** Start-stop station controlling a magnetic motor starter

**Figure 17-27** Three wires are required to control a magnetic motor starter with a start-stop station.

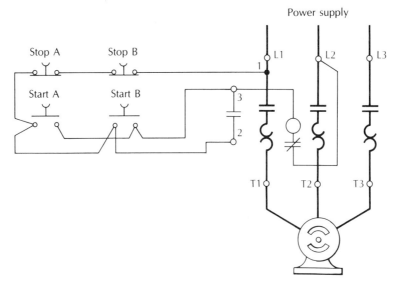

**Figure 17-28** Two start-stop stations controlling a magnetic motor starter.

the conveyer. Another common interlock technique is used to prevent two large motors from operating at the same time. This may be desirable to keep the electrical demand as low as possible in order to reduce the monthly electric bill. Extra interlock contacts can be added to a magnetic motor starter, Figure 17-31. Normally open and normally closed interlocking contacts are available. In Figure 17-32, motor B will operate only if motor A is operating. In Figure 17-33, motor D cannot operate if motor C is operating. In ladder diagrams, only the control circuits are shown. All parts labeled A are inside motor starter A, and similarly with parts B, C, and D. A confusing aspect of a ladder control diagram is that parts normally located physically together in the actual wiring may be located at any appropriate location on the ladder diagram.

## THERMAL PROTECTION IN MOTOR WINDINGS

A motor may have a thermal protector in the windings to sense motor overheating. This thermal protector may be required to be wired-in as a part of the control circuit wiring. In this case, the thermal protector is simply wired in series with the control circuit. If the thermal protector opens, the control circuit is broken and the motor stops. A ladder diagram of a thermal protector in a control circuit is shown in Figure 17-34.

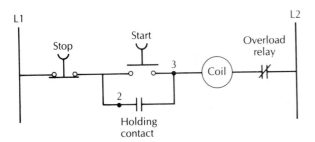

**Figure 17-29** Ladder diagram for a start-stop station controlling a magnetic motor starter

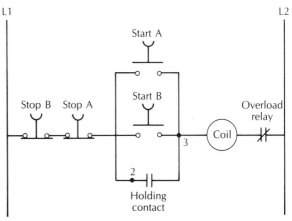

**Figure 17-30** Ladder diagram for two start-stop stations controlling a magnetic motor starter

Unit 17  Motor Control  359

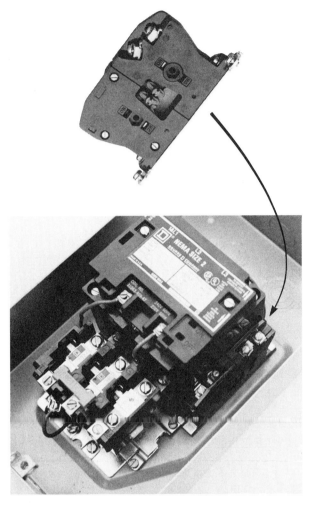

**Figure 17-31** Extra interlock contacts may be added to a magnetic motor starter to sequence or coordinate other motors.

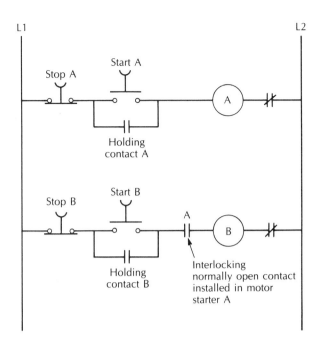

**Figure 17-32** Motor B is interlocked with motor A so that motor B cannot start unless motor A is running.

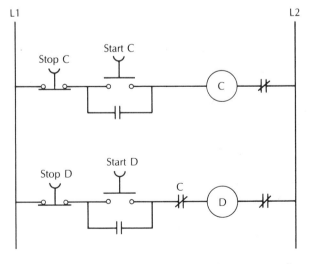

**Figure 17-33** Motor D is interlocked with motor C so that motor D cannot start if motor C is running.

## CONTROL DEVICES

Common control ladder diagrams and schematic wiring diagrams use some basic symbols and terminology. Switches and contacts are the basic control components, and they take several different forms. Switches and contacts may be operated mechanically by a timer or by some change in condition, such as temperature, pressure, flow, liquid level, or humidity.

Switches and contacts are described by the number of poles and the number of throws. The *number of poles* is the number of paths into a switch or contact. The *number of throws* is the number of paths leaving each pole.

It must also be known if the switch or contact is in the open or closed position when it is in the unactivated state. If the switch or contact is open when in the unactivated state, it is considered to be normally open (NO); when it is in the opposite position, it is said to be normally closed (NC). Common single-pole, single-throw

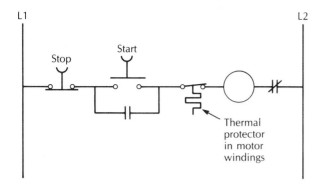

**Figure 17-34** A thermal protector in the motor windings may be wired in series with the control circuit.

(SPST) switch and contact symbols are shown in Figure 17-35.

A single-pole, double-throw (SPDT) switch or contact has one input wire and two output wires. One is normally open, and the other is normally closed. Many control devices use this type of contact so that one device can be used for many different purposes. Both output wires may be used, but often only one is used. Common single-pole, double-throw contacts are shown in Figure 17-36. A 3-way toggle switch, for example, is actually a single-pole, double-throw switch.

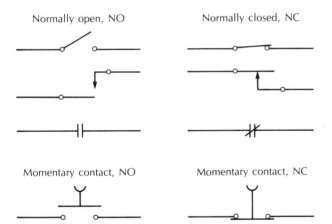

**Figure 17-35** Single-pole, single-throw (SPST) contact schematic symbols for control systems

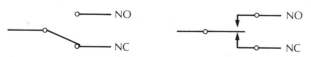

**Figure 17-36** Single-pole, double-throw (SPDT) contact schematic symbols for control systems

Another very common and versatile switch or contact arrangement is the double-pole, single-throw (DPST) contact. Some of the most common types are shown in Figure 17-37. The double-pole, double-throw (DPDT) contacts are common with control switches and timing relays, Figure 17-38.

The common types of switches and contacts are incorporated into various types of control devices. Only single-pole, single-throw contacts are shown in the following diagrams; however, the other types of contacts discussed previously are available. Control device schematic symbols that are frequently used are shown in Figure 17-39. A *limit switch* is a device which is activated when something presses against the lever arm or actuator. It may be of the two-position type or the momentary-contact type, Figure 17-40. Limit switches are frequently used to sense material flow in an automatic materials handling or processing system, such as a grain-drying system or an animal feeding system.

A current-sensing relay is a magnetic coil which closes a set of contacts when the current reaches a preset level, Figure 17-41. An application of a current-sensing relay is a feed grinder with an automatic conveyer sup-

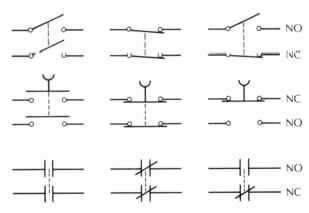

**Figure 17-37** Double-pole, single-throw (DPST) contact symbols for control systems

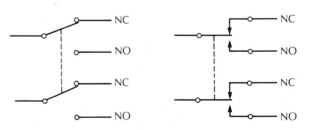

**Figure 17-38** Double-pole, double-throw (DPDT) contact symbols for control systems

Unit 17   Motor Control   361

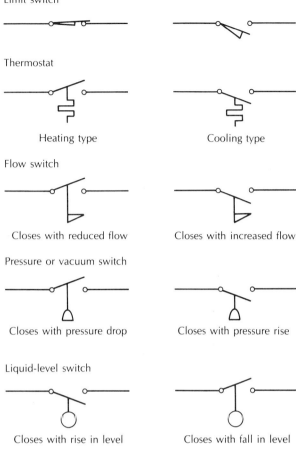

Figure 17-39 Control device schematic symbols

Figure 17-40 A limit switch is activated by making contact with solid material.

A typical timing relay may have an eight-pin socket, Figure 17-45. The timer voltage can be specified, but 120 V is common. This timing relay is a double-pole, double-throw type. Only the required contacts are used. Figure 17-46 shows a ladder diagram for two electric motors, one of which is operated with a timing relay.

Figure 17-41 A current-sensing relay will open or close a contact upon rise or fall of current flow to a motor or other equipment.

plying grain to the grinder. As long as the grinder is fully loaded, the conveyer is off. As soon as the grinder load-current drops, indicating the grinder is not full, the current-sensing relay closes the contacts and turns on the conveyer. A ladder diagram of the control system is shown in Figure 17-42.

Time-delay relays are used on many types of processes or machines where a time delay of a few seconds or minutes is desired. A typical example is a grain-dryer control. The time-delay relay consists of one or more contacts and a timing device. Once power is supplied to the timing relay, it begins timing. After the preset time period, the contact or switch is activated to the alternate position. The schematic diagram for a time-delay relay is shown in Figure 17-43. After the timing is complete, the switch moves in the direction of the arrow at the bottom of the schematic symbol. Some types of time-delay relays are shown in Figure 17-44.

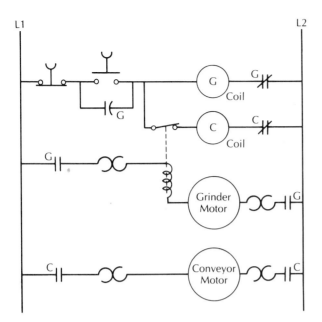

**Figure 17-42** Ladder diagram of a current-sensing relay that controls a conveyer supplying grain to a grinder

**Figure 17-44** Time-delay relays

The start-stop station activates motor A and the timer of the timing relay. After a 30-second delay, motor B operates.

## LOW-VOLTAGE CONTROL

A magnetic motor starter may be operated from a control transformer-supplied 24-V or 120-V system. This is of importance particularly for a large 460-V motor. A control transformer may step down the 460-V line voltage to 120 V. If the transformer is an integral part of the motor controller, it is permitted to be protected by fuses installed on the primary of the transformer, *NEC Section 430-72(c)*. Fuseholders are usually supplied as an integral part of the transformer. The wiring of a transformer-supplied control circuit is shown in Figure 17-47. The factory-installed jumper wire between terminals X2 and L2 must be removed. (The X2 marking on the motor controller can be seen in Figure 17-23.)

The control electrical supply may be completely external to the motor branch circuit. A typical example would be a programmable controller operating a group of motors. The programmable controller output would most

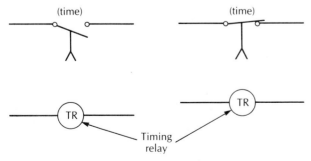

**Figure 17-43** Schematic symbol for a time-delay relay

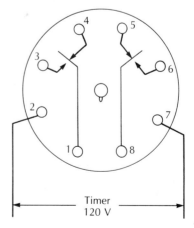

**Figure 17-45** Schematic wiring diagram of a time-delay relay with two single-pole, double-throw contacts

Unit 17   Motor Control   363

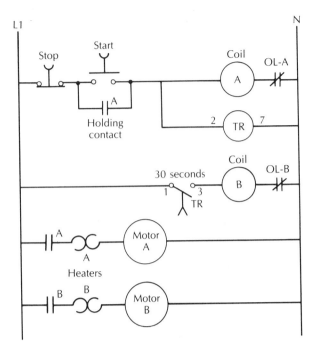

Figure 17-46 Time-delay relay will turn on motor B 30 seconds after motor A has been started. (Refer to Figure 17-45 to identify time-delay contacts.)

likely be 120 V to energize the motor starter coil. Figure 17-48 shows a motor starter operated from an external electrical supply.

## PROTECTING THE CONTROL CIRCUIT

The motor control circuit is considered a Class 1 remote-control circuit, and it falls within the scope of *NEC Article 725*, as well as *Part F* of *Article 430*. The motor control circuit wire may be tapped directly from the motor circuit wires, usually L1 and L2. Further, the control circuit wire carries only the current required by the coil in the motor starter; therefore, it is permitted to be a smaller wire size than is required for the motor circuit. The sizes of the motor control circuit wires are usually Nos. 14 or 12 AWG copper.

A start-stop station is often mounted in the cover of the motor controller. In this installation, the control circuit wires are completely within the motor controller. This wire need not be protected by a fuse, provided the motor circuit short-circuit and ground-fault protective device has a rating not more than the value in Column B of Table 17-5. Table 17-5 is derived from *NEC Section 430-72* and *NEC Table 430-72(b)*.

Consider the example of a 3-phase, 230-V, 15-hp motor which has a full-load current of 42 A, Figure 17-

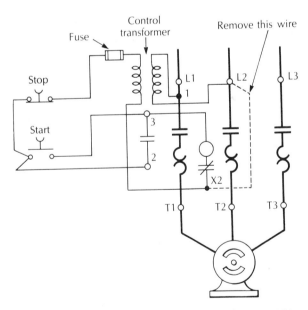

Figure 17-47 A control transformer steps down 480 V to 120 V for the control circuit.

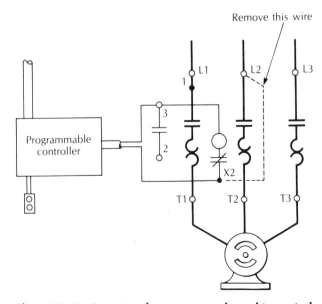

Figure 17-48 An external power source is used to control the motor.

364   Unit 17   Motor Control

Table 17-5 Maximum rating permitted for motor short-circuit and ground-fault protective device if supplemental protection is not provided for the control circuit wires (This table is derived from NEC Section 430-72 and NEC Table 430-72(b).

| Copper Control Wire Size, AWG | Maximum Ampere Rating of Motor Circuit Short-Circuit and Ground-Fault Protective Device | |
|---|---|---|
| | Wires Completely Within Motor Starter Enclosure Column B | Wires Extend Outside Motor Starter Enclosure Column C |
| 18 | 25 | 7 |
| 16 | 40 | 10 |
| 14 | 100 | 45 |
| 12 | 120 | 60 |
| 10 | 160 | 90 |

For other situations, see NEC Section 430-73, 725-12, or 725-35.

49. The motor circuit wire is No. 6 AWG THWN copper, and the motor circuit short-circuit protection is a 70-A time-delay fuse. The control circuit wire inside the controller enclosure is permitted to be No. 14 AWG, according to Column B of Table 17-5.

The motor circuit in the example has a second control station remote from the controller. The control wires are considered protected by the motor circuit short-circuit and ground-fault protective device provided it does not exceed the value of Column C of Table 17-5 for the size of control circuit wire in use. The minimum size wire to the remote station for the motor of Figure 17-49 is No. 10 AWG, unless separate overcurrent protection is provided for the control circuit. Fuse holders are available to add to the motor starter to protect the control circuit wires, Figure 17-50. Control circuit wire No. 18 AWG

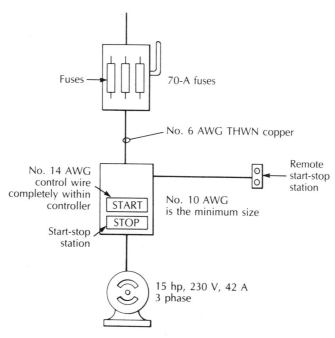

Figure 17-49 The control circuit wires may be smaller than the motor circuit wires and still be tapped directly from the main wires.

**Figure 17-50** The control circuit wires for large motors may require protection by a set of fuses in the motor starter.

and larger may be used if it is protected according to the wire ampacity as listed in *NEC Table 310-16*.

Consider an example where a remote start-stop station controls a 7½-hp, single-phase, 230-V electric motor with a full-load current of 40 A. The motor circuit wire is No. 8 AWG copper THW, and the short-circuit protection is an 80-A circuit breaker. If the control circuit wire is No. 14 AWG copper, then control circuit fuses are required, according to Column C of Table 17-5. Figure 17-51 shows the wiring of the control circuit with fuse protection added.

Motor control circuit conductors are required to be provided with suitable mechanical protection where they are subject to physical abuse, *NEC Section 430-73*. If an external source of power is used to control the motor, such as is shown in Figure 17-48, the disconnect for the motor circuit and control circuit must be adjacent, *NEC Section 430-74*. If they are not adjacent, a warning label must be placed at the motor controller indicating that two separate disconnects are required to shut off power in the controller. The location of the control circuit disconnect must be specified at the motor controllers so that maintenance personnel will know where to find it.

## REVERSING MOTOR STARTERS

Electric motors may be called upon to drive a machine which must be reversible. An example is a feed conveyer that operates in either direction to deliver feed to two batches of animals. A reversible motor starter is required to select the desired direction. For a 3-phase motor, only two lead wires to the motor must be reversed. However, for a single-phase motor, the starting winding leads must be reversed with respect to the running windings.

The least expensive means of reversing electric motors of small horsepower sizes is with a manually operated drum switch, Figure 17-52. The drum switch has a

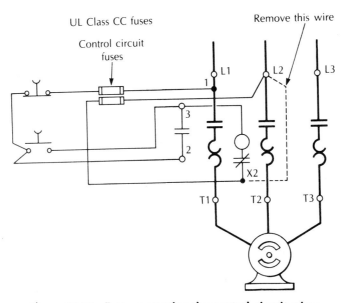

**Figure 17-51** Fuses protecting the control circuit wires

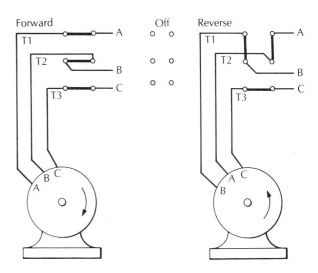

**Figure 17-53** A drum controller connected to reverse a 3-phase electric motor

**Figure 17-52** A manually operated drum controller can be used to reverse an electric motor.

forward, a reverse, and an off position. Figure 17-53 shows the wiring of one type of drum switch to reverse a 3-phase motor. Note the phase rotation at the motor for the forward and the reverse positions. A single-phase motor is more difficult to reverse.

Electric motors of large-horsepower sizes can only be reversed with a reversing magnetic motor starter. This type has two magnetic contactors, Figure 17-54. A three-function, forward-stop-reverse, push-button station is required. Figure 17-55 shows the wiring of a reversing magnetic starter for a 3-phase electric motor. The two magnetic contactors must be electrically or mechanically

**Figure 17-54** A reversing magnetic motor starter

Unit 17    Motor Control    367

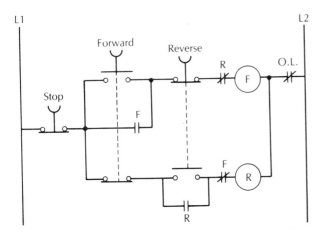

Figure 17-55 Ladder diagram of the control circuit for a reversing magnetic motor starter. The normally closed interlocking contacts prevent both contactors from being closed at the same time, causing a short circuit.

interlocked to prevent both from being closed at one time. Closing both at one time will cause a short circuit.

## COMBINATION MOTOR STARTER

The motor circuit disconnect short-circuit and ground-fault protection and the magnetic starter can be combined in one cabinet. These units are convenient when a fusible disconnect switch and magnetic motor starter will be installed at the same location. These units are called *combination motor starters*. A typical combination motor starter is shown in Figure 17-56.

Figure 17-56 A combination motor starter contains both the disconnect and the motor starter. (*Courtesy of Square D Company*)

## REVIEW

Refer to the *National Electrical Code* when necessary to complete the following review material. Write your answers on a separate sheet of paper.

1. A plug and receptacle is permitted to serve as a motor controller for portable motors rated at what horsepower and less?

2. The locked-rotor current of a 5-hp, 230-V, single-phase motor is how many amperes?

3. According to the motor starter nameplate in the illustration at the top of the next page, this motor starter may be used to control a 25-hp, 460-V motor. Explain which change must be made to make the motor starter suitable for controlling a 460-V motor.

4. The largest size single-phase, 230-V electric motor which is permitted to be controlled with the motor starter in the illustration at the top of the next page is how many horsepower?

5. In the upper left-hand corner of the motor starter nameplate in the illustration above are the numbers 3 and 2. These are the terminal numbers for which of the following?
   a. Main contacts
   b. Coil
   c. Holding contacts
   d. Overload relay

6. The minimum size motor starter required for a 100-hp, 460-V, 3-phase electric motor is what NEMA size?

7. A motor starter is to be installed within an area of an agricultural building subject to dusty, damp, and corrosive conditions. The type of enclosure suitable for these conditions is what NEMA enclosure?

8. The full-load current of a 3-hp, single-phase, 230-V electric motor is how many amperes?

9. The minimum size copper THW branch-circuit wire for a 10-hp, 230-V, 3-phase motor with a nameplate current of 24 A is No. ___?___ AWG.
   a. 12
   b. 10
   c. 8
   d. 6

10. The motor of Problem No. 9 has a service factor of 1.20, and code letter M. The maximum size time-delay fuse that can serve as running overcurrent protection is:
    a. 25 A
    b. 30 A
    c. 40 A
    d. 50 A

11. The motor circuit disconnect is required to be within sight from the controller, which means actually visible from the controller, and not more than how many feet (meters) from the controller?

12. A circuit breaker will serve as the disconnect, as well as the short-circuit and ground-fault protection for a motor circuit. The motor is rated 2 hp, single-phase, 230 V, 11 A with code letter H. The maximum size circuit breaker under normal starting conditions is:
    a. 15 A
    b. 20 A
    c. 30 A
    d. 40 A

13. The electric motor in Problem No. 12 has a service factor of 1.10. Using the overload thermal unit selection chart of Figure 17-18, the correct overload heater is thermal unit No.:
    a. B 17.5
    b. B 19.5
    c. B 22
    d. B 25

14. The maximum size time-delay fuse used for short-circuit and ground-fault protection for a 75-hp, 460-V, 3-phase motor, code letter M, under normal starting conditions is:
    a. 100 A
    b. 125 A
    c. 150 A
    d. 175 A

15. A 225-A, single-phase, feeder conductor No. 3/0 AWG THHN copper supplies several motors in a feed room of a beef farm. A 5-hp, 230-V motor is tapped directly from the feeder with a fusible disconnect switch 15 ft (4.57 m) from the point of the tap. The minimum size copper THHN tap conductor permitted is No. ___?___ AWG.
    a. 10
    b. 8
    c. 6
    d. 4

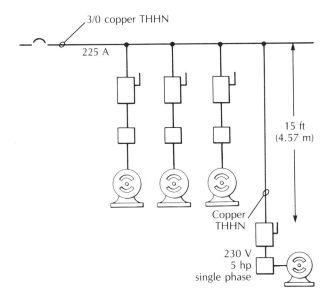

16. On a separate sheet of paper, redraw and complete the control wiring diagram to operate the magnetic motor starter with a start-stop station.

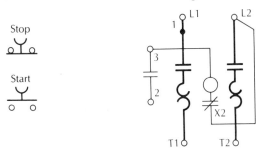

17. On a separate sheet of paper, redraw and complete the control wiring diagram to operate the magnetic motor starter with a pressure switch.

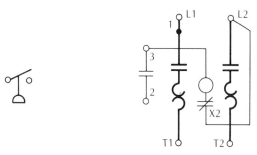

18. The schematic control diagram has a start-stop push-button station operating a magnetic motor starter with an emergency stop push button mounted on the machine. On a separate sheet of paper, draw a ladder control diagram for the circuit.

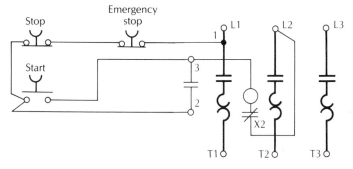

19. A magnetic motor starter coil is energized from a separate 115-V source. Show how to wire the control circuit. On a separate sheet of paper, redraw the motor controller and control circuit wires. A control circuit wire in the motor starter must be removed.

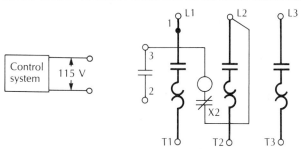

Unit 17   Motor Control   371

20. The start-stop push-button station turns on motor A and, after a time delay of 30 seconds, motor B is turned on by the time-delay relay. The time-delay is the type that instantly resets when power to the timer is interrupted. On a separate sheet of paper, draw a ladder diagram of the control system.

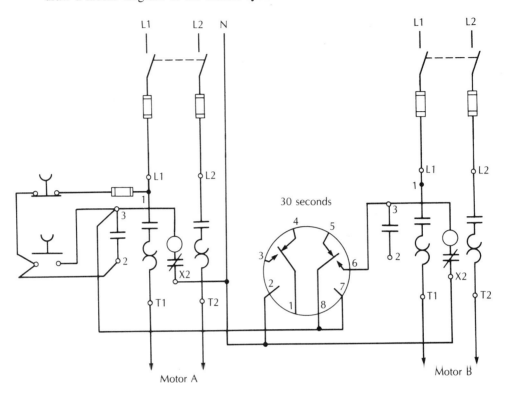

# UNIT 18
# CORD- AND PLUG-CONNECTED EQUIPMENT

## OBJECTIVES

After studying this unit, the student will be able to

- select an adequate type and size of flexible cord for supplying portable equipment.
- choose a cord plug and connector adequate for the current and voltage of the supplied equipment.
- select and install controllers for electric motors on portable equipment.
- install a flexible cord, plug, and connector in a manner so as to prevent strain on the cord and moisture from entering the wiring or connections.

Flexible cords are frequently used on farms to supply portable electrical equipment and permanently connected equipment which must be moved during normal operation. A top-unloading silo unloader is a typical example of permanently attached equipment which is moved during normal use, Figure 18-1. Cords supplying equipment in agricultural locations are required to be approved for hard usage, *NEC Section 547-3(b)*.

## CORD TYPES

Types of flexible electrical cords are listed in *NEC Table 400-4*. Basically, several types of cords are suitable for agricultural use. These types include junior hard-service cord Types SJ and SJT. These cords are permitted to be used for portable equipment in damp locations, and are classified by the *NEC* as hard-usage cords. They are available in sizes No. 18 to No. 10 AWG. They are available as 2-, 3-, or 4-conductor cord. If the cords are suitable for use in areas where they will be exposed to oil and grease, they will have the letter O in the type, such as SJO or SJTO. The conductors inside these cords are protected by a tough outer covering.

Judgment must be applied regarding the severity of the conditions of use for flexible cords. Most cords used for agricultural applications should be rated for the extra-hard usage designation. Hard-service cord fits this classification and carries the Types S or ST. These types are also suitable for use in damp locations. The oil-resistant types are SO and STO. Hard-service cord is available in sizes No. 18 to No. 2 AWG. Cords are also available with two or more conductors.

Portable electric heating equipment, such as infrared heat-lamp fixtures, may be called upon to operate continuously for days or weeks. Flexible cords also may be subjected to excessive heat from the heating unit as well as continuous current flow. These cords must be capable of withstanding elevated operating temperatures without deterioration of the conductor insulation or the outer jacket. These are classified as jacketed heater cords

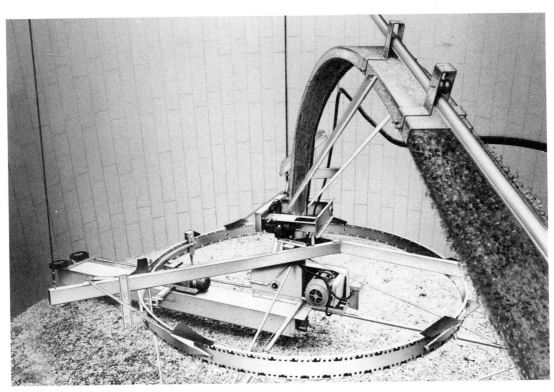

Figure 18-1 Many agricultural machines, such as this silo unloader, must be cord and plug connected. (*Courtesy of Jamesway Division, Butler Manufacturing Co.*)

Types HSJ and HSJO. Heater cords suitable for extra-hard usage are Types HS and HSO. These cords are available in sizes up to No. 10 AWG.

## FLEXIBLE CORD AMPERE RATINGS

The ampere ratings of flexible cords are listed in *NEC Table 400-5*. The table is summarized here as Table 18-1, and lists only those cords listed previously as considered suitable for agricultural applications. The ampere rating of some cords depends upon the number of current-carrying wires. Single-phase loads have two current-carrying wires; 3-phase loads have three current-carrying wires. Multiwire cords supplying several related loads must have the conductor ampere rating derated in accordance with the values in Table 18-2. If the neutral wire carries only the unbalanced load in a multiwire circuit, it is not considered as a current-carrying wire for the purpose of derating.

Portable cords smaller than No. 14 AWG are not recommended for agricultural applications; No. 12 AWG is recommended to minimize voltage drop. Sometimes, long cord lengths are required, and voltage drop must be considered.

## PLUGS AND RECEPTACLES

Electrical plugs, receptacles, and cord connectors are available in a variety of types and sizes. The National Electrical Manufacturers Association (NEMA) has developed standard configurations for plugs, connectors, and receptacles based upon voltage, current rating, and type of electrical supply. The most common plug and receptacle configurations for the locking and nonlocking types are shown in Unit 9 of this text. The proper size and plug configuration must be used for the desired voltage and current level. The *NEC* requires that plugs connected to circuits having different electrical supplies *not* be interchangeable, *NEC Section 210-7(f)*.

Receptacles are available for 120-V circuits that are rated at 20 A and will accept both 15-A and 20-A attachment plugs. A 30-A, 120-V attachment plug and receptacle are different from those for the 15-A and 20-A sizes, Figure 18-2. The 30-A type must be designed to handle the higher current. Motors on portable machinery must

374  Unit 18  Cord- and Plug-Connected Equipment

Table 18-1  Ampacity of hard-usage flexible cords

| Size AWG | S, ST, SO, STO, SJ, SJT, SJO, SJTO | | HSJ, HSJO, HS, HSO Heater Cords |
|---|---|---|---|
| | Current-carrying Wires | | |
| | 3 | 2 | |
| 18 | 7 | 10 | 10 |
| 16 | 10 | 13 | 15 |
| 14 | 15 | 18 | 20 |
| 12 | 20 | 25 | 30 |
| 10 | 25 | 30 | 35 |
| 8 | 35 | 40 | — |
| 6 | 45 | 55 | — |
| 4 | 60 | 70 | — |
| 2 | 80 | 95 | — |

Table 18-2  Derating factors for flexible cords with more than three current-carrying conductors

| Number of Conductors | Multiply Table 18-1 Amperes by: |
|---|---|
| 4 through 6 | 0.80 |
| 7 through 24 | 0.70 |
| 25 through 42 | 0.60 |
| 43 and above | 0.50 |

be cord- and plug-connected, and 240 V, 15 A is common. The design of this type of plug (250-V maximum) and receptacle must be different from the 120-V types (125-V maximum) to avoid the danger of accidental interconnections, Figure 18-3.

Locking-type receptacles and plugs are available to prevent vibration or similar movement from causing the plug to become disengaged from the receptacle. Figure 18-4 shows a typical locking-type attachment plug.

Attachment plugs generally are not made to be disengaged when carrying more than a few amperes. This is why an attachment plug is not permitted as a controller for motors larger than ⅓ hp, because the blades of the plug must be disengaged quickly to prevent arching. Special plugs are constructed which are rated to be disengaged under full-rated current.

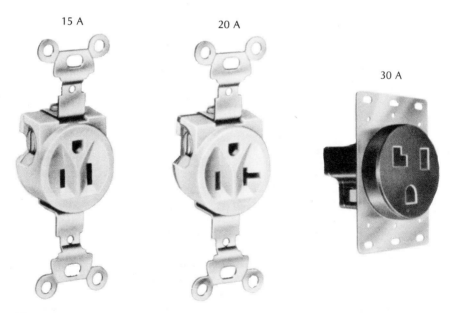

Figure 18-2  Nonlocking 125-V receptacle outlets rated at 15 A, 20 A, and 30 A

Unit 18  Cord- and Plug-Connected Equipment

**Figure 18-3** Receptacle outlet rated at 250 V, 15 A

Attachment plugs and cord connectors are not permitted to be of a type where live parts other than the blades can be exposed. They must be of the dead-front type, *NEC Section 410-56(e)*. The attachment plug of Figure 18-4 is of the dead-front type. The attachment plug must actually be disassembled to expose the wire connections.

## CORD-ATTACHED MOTORS

Portable equipment on farms must be cord attached to the power system. Except for motors rated at 1/3 hp or smaller, the plug and receptacle are not used as the controller. They are used as the means of disconnecting the motor from the power supply. A convenient method of protecting motor circuit wires as well as controlling a motor is by use of a properly sized fusible disconnect switch, Figure 18-5. The attachment plug and receptacle must have a minimum ampere rating of 125% of the full-load current of the motor. However, this plug generally is not capable of being disconnected when carrying current. The disconnect switch is used for that purpose.

## Problem 18-1

Consider a movable grain elevator at a grain storage and drying center powered with a 1-hp, 230-V, single-phase motor. Determine the size of the following.

1. Fusible disconnect and controller switch
2. Maximum size time-delay fuses for the switch
3. Required rating of plug and receptacle
4. Minimum size type STO extra hard-service cord with two hot wires and a grounding wire

## Solution

The electric motor on the grain elevator of Figure 18-5 is rated at 1 hp, 230 V, with a full-load current of

**Figure 18-4** Locking-type attachment plug

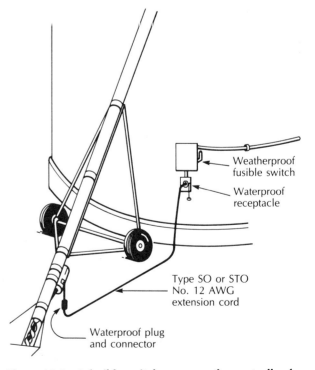

**Figure 18-5** A fusible switch serves as the controller for this portable grain elevator.

8 A. Assume, that the service factor is 1.15 and the motor nameplate current is the same as that found in *NEC Table 430-148*. Assume, also, that the motor does not have a code letter.

1. The disconnect must have a minimum single-phase, 1-hp rating.
2. Time-delay fuses can usually serve as short-circuit, ground-fault, and overload protection. Referring to Unit 17, size the fuses at 125% of the nameplate full-load current. Time-delay fuses rated at 10 A are required.

$$1.25 \times 8 \text{ A} = 10 \text{ A}$$

3. Some plugs and receptacles are rated in horsepower and, if so, the rating must be at least 1 hp. Usually they are rated in amperes and should be sized at not less than 1.25 times the motor full-load current. In this case, the minimum would be 10 A. Either the locking or nonlocking type may be chosen. Cords smaller than No. 14 AWG should not be used for such applications; No. 12 AWG is preferred. Therefore, the attachment plug and receptacle should have a rating of either 15 A or 20 A, 240 V, single phase.
4. From Unit 17, the motor circuit wire must have a rating of at least 125% of the motor full-load current. Therefore, the minimum wire rating is 10 A. The cord is Type STO, containing two hot wires and a grounding wire. The minimum size two-conductor cord suitable for a load of 10 A is No. 18 AWG. This type is much too small for an agricultural application. The recommended minimum size is No. 14 AWG; however, No. 12 AWG with a rating of 25 A is preferred.

## *OVERCURRENT PROTECTION FOR CORDS*

Flexible cords are usually selected for a particular application. However, many cords on farms are used for multiple applications with great variation in the loads they serve. For this reason, general-use extension cords should be rated at not less than 15 A, and preferably not less than 20 A. Cords attached to specific equipment, appliances, and portable lamps need only have a rating adequate for the loads they serve, based upon conductor sizing requirements in the cord. Consider, for example, an infrared heat-lamp fixture with three 250-W, 120-V lamps. The total load is 750 W, or 6.3 A. The heating is electric, and the load is continuous; therefore, the wire must be at least 125% of the full-load current, or 7.9 A. The minimum wire size for this infrared heat-lamp fixture is No. 18 AWG, Type HSJ.

A question which must be considered in the case of the infrared heat-lamp fixture is whether the No. 18 AWG cord plugged into a 15-A or 20-A circuit creates a hazard. The cord is adequate for the load; therefore, the only danger is from short circuits and ground faults. Actually, the cord is considered similar to a conductor tapped from a feeder wire. The tap wire may be of a smaller size than the feeder wire. The rules for plugging a cord into a circuit are found in *NEC Section 240-4*. Cord sizes No. 18 and No. 16 AWG are considered protected when plugged into a 15-A or 20-A circuit, provided they are approved as supply cords for portable lamps and equipment. Table 18-3 lists the highest circuit rating permitted for flexible cords supplying specific equipment and portable lamps.

Extension cords, on the other hand, may be used for unspecified loads. Flexible extension cords sizes No. 18 and No. 16 AWG frequently cause fires when they are misused for loads which draw more current than the cord can carry safely. The *NEC* requires supplemental overcurrent protection for listed extension cords No. 18 AWG, and for No. 16 AWG cords more than 25 ft (7.62 m) in length. No. 16 AWG extension cords in lengths of 25 ft (7.62 m) or less, and any length of sizes No. 14 and larger are not required to have supplemental overcurrent protection, *NEC Section 240-4, Exception No. 3*.

## *WIRE COLOR FOR CORDS*

Flexible cords are generally manufactured with specific colors of insulation on the individual wires similar

**Table 18-3 Maximum branch-circuit overcurrent rating when appliances, portable lamps, and equipment are to be cord attached to a circuit receptacle outlet without supplemental overcurrent protection for the cord**

| Flexible Cords Used for Appliances, Portable Lamps, and Equipment Copper Wire Size, AWG | Maximum Branch Circuit Overcurrent Protective Device Rating, Amperes |
|---|---|
| No. 18 and larger | 20-A circuit |
| No. 16 and larger | 30-A circuit |
| 20-A rated cord and larger | 40-A circuit |
| 20-A rated cord and larger | 50-A circuit |

to electrical cable. For a 2-wire nongrounding cord, the colors are white and black. A grounding cord has wires of the colors white, black, and green. A 3-wire cord has the colors white, black, red, and green. The green wire must be used as the grounding wire. A neutral wire *must* be white (or otherwise marked according to *NEC Sections 400-22,* and *200-6(c),* but a white wire is *permitted* to be hot. This is the case when 2-wire grounding cord is used for a 230-V motor. The black and white wires are hot, and the green wire is used for grounding.

Lamp cords and appliance cords do not always have individual insulation on each wire. The outer jacket serves as the conductor insulation. Alternative methods of identifying the neutral and grounding wires are devised. A white tracer braid may be used to identify the neutral wire. A white marker may be placed on the outer covering. The wire itself may be tinned to make it look white in color. These methods are described in *NEC Section 400-22.* The grounding wire must be identified with green insulation or green insulation with one or more yellow stripes. A green braid may be placed around the wire where the individual wires are not insulated, *NEC Section 400-23.*

## GROUNDING

Electrical equipment with exposed metal parts must be grounded *NEC Section 250-59.* Grounding is extremely important on farms where equipment is often used in damp or wet conditions. Good grounding usually provides humans and animals with adequate protection from electrical shock. A grounding wire and a grounding-type plug and cord receptacle attached to a grounded circuit will provide an adequate grounding path. The grounding terminal within the attachment plug and cord receptacle is identified with a green mark or with the letters G or Gr. The circuit supplying the flexible cord must also contain a grounding conductor if the equipment is to be adequately grounded. Circuits routinely supplying portable equipment must be rewired, if necessary, to provide an effective grounding path all the way back to the main service panel in the building.

## WET CONDITIONS

Cords used to supply portable agricultural equipment are often exposed to inclement weather or actually used in wet conditions. The cord attachment plug and connec-

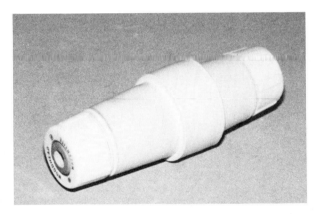

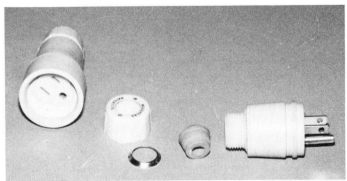

Figure 18-6 Waterproof attachment plug and connector for hard-service cord

tor must be watertight when used under these wet or damp conditions, *NEC Section 400-3*. Cord connectors are available that are completely watertight, Figure 18-6.

Receptacle outlets installed in outside damp or wet locations must be suitable for the conditions, and must be weatherproof when the attachment plug is not in place, *NEC Section 410-57(a)*. If wet conditions exist when the plug is in place, or if the plug and receptacle are exposed to the weather, a suitable protective cover is required over the connection. Figure 18-7 shows receptacle outlets, with circuit breakers as controllers, suitable for direct exposure to the weather.

## STRAIN RELIEF

Cords connected directly to outlet boxes or to attachment plugs and connectors shall be installed so as to prevent placing strain on terminal connections, *NEC Section 400-10*. Cord covering may be damaged in time by tension or flexing at connections. Strain-relief connectors are available to prevent tension on fittings or damage from flexing at connectors. Strain-relief connectors are often watertight. A typical strain-relief connector is shown in Figure 18-8.

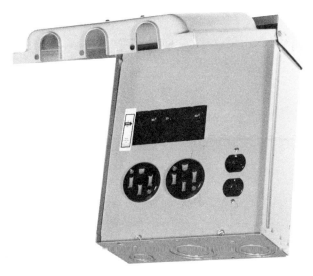

Figure 18-7 Receptacle outlets, with circuit breakers as controllers, suitable for direct exposure to the weather (*Courtesy of Midwest Electric Products, Inc.*)

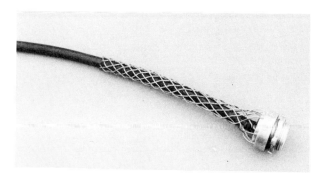

Figure 18-8 Strain-relief cord connector

## REVIEW

Refer to the *National Electrical Code* when necessary to complete the following review material. Write your answers on a separate sheet of paper.

1. Name a type of flexible cord classified for extra-hard usage and as oil resistant.

2. The minimum size Type HS flexible cord required for a 1 200-W, 120-V portable electric heater is No. ___?___ AWG.
   a. 18   c. 14
   b. 16   d. 12

3. Refer to Figure 9-49 (in Unit 9 of this text). In the following diagram, which attachment plug, A, B, C, or D, is suitable for an 18-A load at 120 V?

A      B      C      D

4. A livestock farmer installs portable heat-lamp fixtures in the lambing barn. Each fixture has No. 18 AWG Type HSJ cord. Is it permissible to plug these fixtures into a 20-A, 120-V circuit? Support your conclusion by stating the appropriate *NEC* reference.

5. The flexible cord used in an agricultural building where damp conditions exist must be what type of cord?

# UNIT 19
# READING WIRING DIAGRAMS AND BLUEPRINTS

## OBJECTIVES

After studying this unit, the student will be able to

- describe the basic components of a structural plan.
- identify electrical devices from a wiring plan or blueprint.
- determine from a wiring plan the number of wires and their location.
- recognize and identify a riser diagram.
- locate electrical equipment from a one-line diagram.

Electrical plans, blueprints, specifications, and wiring diagrams are required whenever directions must be communicated to other personnel. It is assumed that the installer is familiar with installation procedures. Therefore, an abbreviated system of symbols is used to communicate directions with a minimum of confusion. Standard symbols have been adopted and are used universally. Manufacturers use standard symbols when supplying wiring diagrams with equipment.

## ARCHITECTURAL SYMBOLS

Plans used in agriculture are usually in the form of simple line drawings of the structure layout. The actual construction materials are often not known until the builder has been selected. The owner and the builder then decide upon the actual construction materials. The owner will often select an equipment manufacturer to supply and install much of the equipment in the building. Once electrical equipment requirements are known, the electrical installer can finalize installation plans. When the owner serves as the general contractor, it is important for the electrical installer to maintain good communications with the equipment suppliers and the builder.

Agricultural installations may be handled by a general contractor, with architectural plans and specifications prepared in advance. In this case, the owner works with the architect and general contractor. Construction, plumbing, heating, electrical, and mechanical contractors work under the direction of the general contractor. Jobs of this type are usually let out for bids, with the work usually awarded to the low bidder. When the owner serves as the general contractor, and exact details of the job are not known in advance, the best the electrical contractor can do is provide an estimate of the cost of installation. With experience, the electrical contractor's estimate will be quite close to the actual cost of the installation.

Standard architectural symbols have been developed to help communicate construction details to the various contractors with a minimum of confusion. Figure 19-1 shows some common wall-construction symbols. It is necessary to know the type of wall, ceiling, and floor construction in order to plan the installation of the wiring. Plans with this type of detail are usually not available unless they have been prepared by an architect. However, most farm plans incorporate some of these details.

## ELECTRICAL SYMBOLS AND TYPES OF PLANS

Electrical construction plans or blueprints use standard electrical symbols to represent lighting fixtures, receptacle outlets, switches, and similar equipment and materials. The symbols with frequent application to agricultural plans are shown in Figure 19-2. The plan may contain a minimum amount of detail, showing only the type and location of electrical fixtures and outlets. The circuits and wiring may not be shown. Notes may be added to a plan to provide more detailed information on lighting fixtures and other outlets.

The wires may be drawn on a plan as shown in Figure 19-3. This type of plan indicates which outlets are on each circuit. The path the circuit follows is also shown. The actual installation is left up to the installer. If the wiring is cable in open-cavity walls and ceilings, the installer determines the most appropriate route from one outlet to the next. If the building is of masonry construction, the wiring will be in conduit. Conduit may then be placed in the floor from one outlet to the next.

Figure 19-4 shows the different methods of showing the wiring on a plan. The number and type of wires are indicated with cross marks on the wires. This is one of several wire coding systems. A long cross line is sometimes used, which indicates a neutral wire, while short lines indicate hot wires. An arrow indicates that a circuit wire returns to a panelboard. If there are several panelboards in a building, a letter or number is used to indicate which panelboard. The plans may specify if the wires are to be installed concealed in walls and ceilings, under the floor, or on a wall surface. Many plans do not make this distinction, and the choice is then up to the installer.

## SPECIFICATIONS

*Project specifications* spell out the details of the wiring installation. The specifications describe the wiring materials and methods to be used. The specifications explain details that are usually not contained on the plans. A completion schedule, billing and payment procedures, insurance requirements, and liabilities are usually contained in the specifications. Most farm wiring installations do not involve formal written specifications. They may be in the form of a mutual understanding. It is a good idea to have some specifications in writing. This information can be added to the plans in the form of notes, or on separate pages added to the plans.

Typical specifications for a farm building where damp corrosive conditions exist might be as follows.

1. All wiring shall be copper UF cable, or THWN in nonmetallic conduit.
2. All exposed surface wiring shall be in conduit.
3. Lighting fixtures shall be corrosion resistant, dusttight, and vaportight.
4. All exposed boxes shall be nonmetallic, dusttight, and vaportight with waterproof switch and receptacle covers.
5. Exposed conduit straps and fasteners shall be of corrosion-resistant material.
6. All concealed boxes shall be nonmetallic and dusttight with dusttight cable and conduit fittings.
7. Incandescent lighting fixtures shall be of nonmetallic construction with a glass waterproof and dusttight guard. The fixture shall have a minimum rating of 150 W.
8. Switches and receptacles shall be mounted 48 in (1.22 m) above the floor to the center of the outlet box.
9. All flexible conduit connections shall be of liquid-

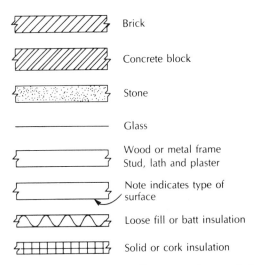

**Figure 19-1 Architectural wall-construction symbols**

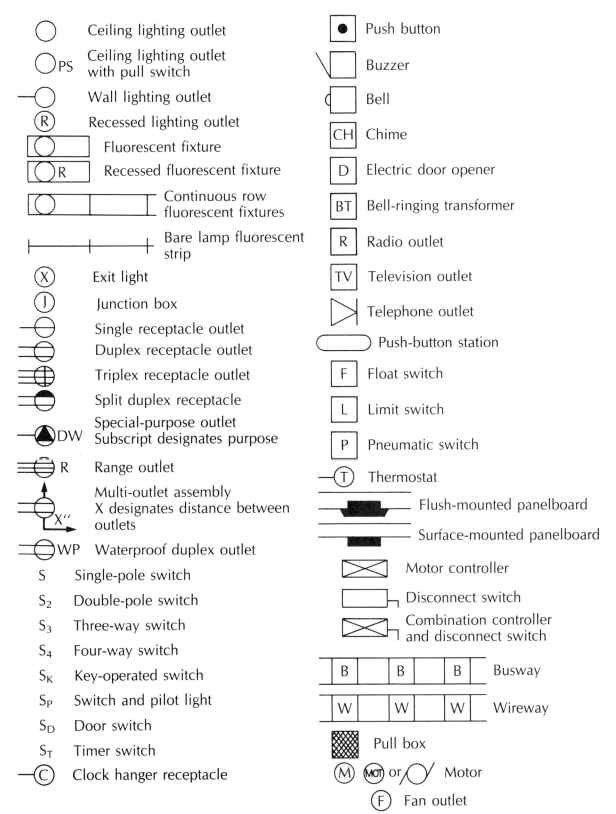

**Figure 19-2 Electrical wiring plan and blueprint symbols**

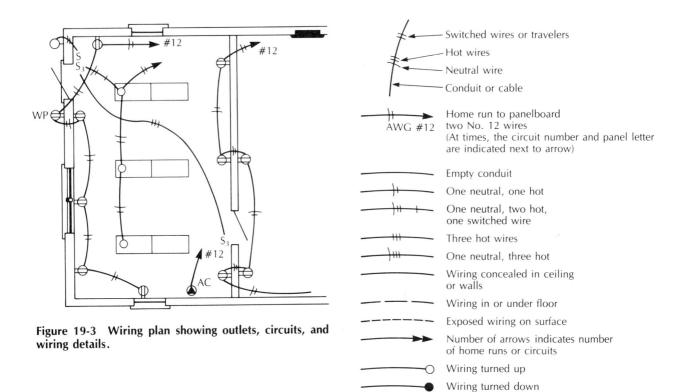

**Figure 19-3** Wiring plan showing outlets, circuits, and wiring details.

**Figure 19-4** Symbols used to represent circuit wires on a plan. On some plans, the number of wires is identified by cross lines, but no further attempt is made to identify the neutral.

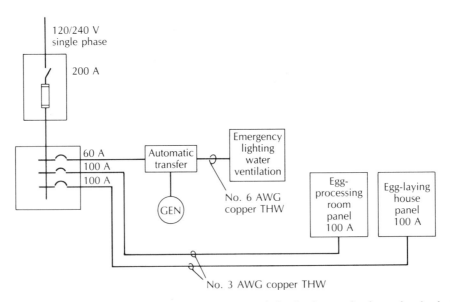

**Figure 19-5** Riser diagram for the service entrance and the feeders and subpanels of a building

384  Unit 19  Reading Wiring Diagrams and Blueprints

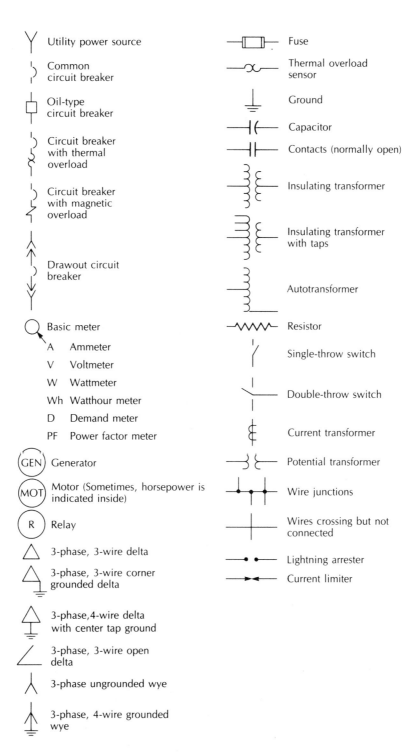

**Figure 19-6 Symbols for electrical system one-line diagrams**

tight flexible metal conduit with waterproof connectors.
10. Enclosures for panelboards and controllers shall be galvanized steel or another corrosion-resistant material.
11. Motors shall be of the totally enclosed fan-cooled type.
12. Motor starters shall have NEMA 4X enclosures.

## RISER DIAGRAMS

A *riser diagram* is a sketch of the main service equipment, and the main feeder wires and panelboards, Figure 19-5. The riser diagram shows at a glance the complete electrical distribution system in a building. The diagram is generally not drawn to scale, and the different panels and equipment may actually be in different rooms or areas of the building. It is necessary to examine the plans in order to determine the exact location of the equipment shown on the riser diagram.

The feeder wires from the service disconnect to the various subpanels in a building may not actually be shown on the plans. The plans may show only the locations of the equipment. It is the responsibility of the installer to run the feeder wires in the most appropriate manner. The wire size and type are shown on the riser diagram. Fuse and circuit-breaker sizes and types for the feeder and service equipment are also indicated on the riser diagram.

## ONE-LINE DIAGRAMS

*One-line diagrams* provide a means of representing a complete wiring system in a simple sketch. Only one line is drawn, and it can represent two, three, or four wires, depending upon the type of electrical system. One-line diagram symbols are shown in Figure 19-6; they are often used in riser diagrams.

One-line diagrams are useful for calculating and specifying sizes and types of materials to be used in an electrical system. A sketch can be made and notes added to specify materials. This sketch with notes can convey a clear idea to the installer. A typical one-line diagram of a 400-A, 3-phase, farm building service is shown in Figure 19-7.

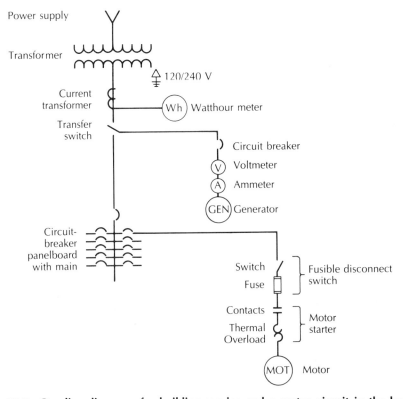

Figure 19-7 One-line diagram of a building service and a motor circuit in the building

# REVIEW

Refer to the *National Electrical Code* when necessary to complete the following review material. Write your answers on a separate sheet of paper.

1. The building in the plan illustrated with Problem No. 4 of this review is constructed of which of the following materials?
   a. Brick
   b. Concrete block
   c. Wood or metal frame
   d. Stone

2. How many circuits are shown on the plan of Figure 19-3?

3. How many lighting fixtures connected to the circuit of Figure 19-3 are wired with No. 14 AWG wire?

4. The wiring shown on the plan illustrated here is in conduit. Which run of wire is under the floor and does not contain a neutral wire?

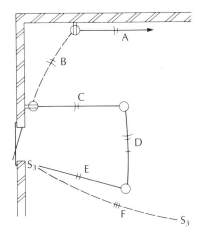

5. Which run in the plan illustrated above contains a neutral and a switched wire?

6. A service entrance consists of a 200-A, single-phase, fusible disconnect switch, followed by a 200-A circuit-breaker panelboard. Feeder wires extend from the circuit-breaker panelboard to a 100-A circuit-breaker panelboard in another area of the building. Draw a riser diagram for this electrical service, and identify the parts.

7. Draw a one-line diagram of a fusible disconnect switch, motor starter, and motor.

# UNIT 20
# GENERAL FARM WIRING

> **OBJECTIVES**
>
> After studying this unit, the student will be able to
>
> - list the basic electrical requirements of a farm shop.
> - wire a water pump circuit.
> - size water heater circuit wires.
> - wire a livestock waterer.
> - wire a fuel-dispensing pump.
> - list the electrical requirements of a general storage building.

## FARM SHOP

A good shop is important on every farm. Adequate electrical service must be available to operate the wide variety of electrical machinery. A minimum service considered adequate for a farm shop is 100 A. If the shop is electrically heated, then a 150-A service is preferred.

Adequate light is essential. Fluorescent lamps are most efficient, provided the shop is not cold in winter. Fluorescent lamps do not start well or operate efficiently when the bare lamps are exposed to cold air. Light must be provided over each work bench. Lamps should be located in the open areas of the shop to provide illumination when working on machinery. One lighting fixture should be provided for every 150 ft² (14 m²) of open area in the shop. Lighting should be on circuits separate from other outlets and equipment.

Receptacle outlets should be spaced not farther than every 4 ft (1.22 m) along work benches. All receptacle outlet circuits should be rated at 20 A. A receptacle outlet should be installed near each door, and at least every 15 ft (4.57 m) along the walls of the shop. Keep all receptacle outlets at least 18 in (457 mm) above the floor. Gasoline vapor is heavier than air and, if a spill occurs, it will hover at the floor until it escapes out the doors. *Fire hazard* from electrical sparks at a receptacle outlet is not a problem when the outlets are located at least 18 in (457 mm) above the floor. At least two 20-A receptacle circuits should be provided, Figure 20-1.

Electrical tools and equipment of ½ hp and larger should be provided with an individual circuit. Proper motor controllers must be provided. As a general rule, stationary equipment should be provided with an individual circuit. Typical stationary shop tools are:

Table saw                Air compressor
Radial arm saw           Hoist

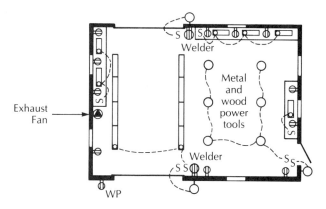

Figure 20-1 Suggested minimum wiring for a farm shop

| Jointer | Battery charger |
| Metal band saw | Lathe |
| Drill press | Arc welder |
| Grinder | |

The arc welder usually requires a 50-A, 240-V circuit. An outlet should be placed near the door so that it will be accessible to large machinery which cannot fit into the shop. A ventilation fan should be provided in the shop to vent vapors when welding in the winter. The shop is considered a dry area and, therefore, normal cable or conduit wiring is permitted, *NEC Section 547-1(c)*.

## WATER PUMP

The water pump may be located inside a building or in a separate well house, or a submersible pump may be located in the well. In the latter situation, a pressure tank is usually located in an adjacent building or in a separate well house, Figure 20-2. If the pump and pressure tank are located in a separate well house, an electric heater is usually provided to make sure the water does not freeze, Figure 20-3. A farm water pump should be provided with a 240-V electrical supply.

There is an advantage to locating the water system separate from the farm buildings. If the pump is remote, it is less likely to have power loss if a fire should occur. When the pump and pressure tank are located in a separate building, the power should come directly from the farm central distribution pole.

A receptacle outlet and a lighting outlet should be located near the pump to provide power for tools and light for servicing the pump.

All metal electrical equipment likely to become energized is required to be grounded, *NEC Sections 250-43*

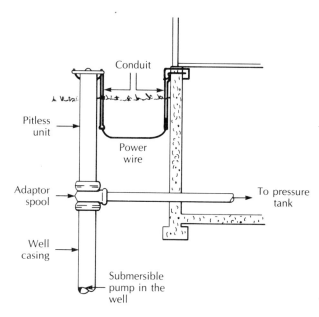

Figure 20-2 Wiring for a submersible pump in a well

and *430-142*. This includes a submersible pump in a well. If the pump is not grounded, then the earth acts as the grounding conductor, and this is not permitted by *NEC Section 250-91(c)*. The earth is a conductor acting only in parallel to the grounded neutral conductor and equipment grounding conductors. A high-resistance ground fault in the pump can leak current into the earth without opening an overcurrent device. This type of fault leakage has been found to be a source of stray voltage on some farms.

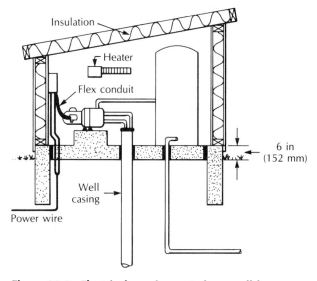

Figure 20-3 Electrical requirements for a well house

Unit 20 General Farm Wiring 389

# ELECTRIC WATER HEATERS

Modern water heaters have automatic temperature controls which maintain the water temperature between a range of 140°F and 180°F (60°C and 82°C). The *National Electrical Code* and Underwriters' Laboratories require that electric water heaters be equipped with a high-temperature limit device to disconnect power above a set temperature, *NEC Section 422-14*. UL states that the maximum cutoff temperature is 194°F (90°C). A pressure-relief valve in the hot-water output line is also required. Water heaters used for sanitizing milking equipment should be set at 170°F (77°C).

Electric water heaters for general use are available in sizes ranging from 20 gallons to 82 gallons. Typical sizes in gallons are 20, 30, 40, 42, 45, 50, 52, 66, and 82 gallons. Many water heaters contain one or two magnesium anodes (rods) which are always immersed in the water. These rods help to reduce corrosion.

The heating elements are generally rated at 236 V. However, they will operate properly in the range from 220 V to 240 V.

The *speed of recovery* is defined as the time required to bring the water temperature to satisfactory levels. Water heaters may contain one or two heating elements. The rating of each heating element in watts may be 750, 1 250, 1 500, 1 650, 2 000, 2 500, 3 000, or 3 500 watts.

Electric water heaters generally have two heating elements and two thermostats. A heating element consists of nickel-chrome (nichrome) alloy wire. The proper size and length of wire are used to conform to the voltage and wattage ratings of the element.

The nichrome wire is coiled and embedded (compacted) in magnesium oxide in a copper-clad, tubular housing. The magnesium oxide is heat resistant and is a good electrical insulator. Terminals are provided on one end of the heating element.

The top heating element is located about halfway up the tank (when the tank is in a vertical position). The bottom element is located close to the bottom of the tank. Many water heaters have controls which prevent both heating elements from operating at the same time.

## Water Heater Metering

The energy used by electric water heaters can be recorded in several ways. One method is to use a separate meter in addition to the regular kilowatthour meter. The off-peak method of heating the water in the storage tank uses a time switch to control the heating. This switch has an electric clock which operates a switch. The clock operates on a separate heater circuit between the off-peak hours of 7:00 PM and 11:00 AM, for example. As a result, the water is heated at this time only. For the remainder of the 24-hour day, depending upon requirements, either the separate circuit or the main circuit may be used. Large-capacity tanks should be used with this system.

A slight variation of the off-peak plan has most of the heating taking place at night. The rest of the heating occurs during the day. The rate charged to the customer for heating water is lower than the standard rate charged for the general use of electrical energy. Power suppliers often use a separate meter and radio control instead of a time clock; however, the basic operation is similar to the time-clock method. The power supplier can turn the water off on command at any time during the day. With this system, the customer's water heater is off only for a limited period of time.

## Water Heater Wire Size

The size of wire for a water heater is based upon the load demand of the water heater. This value must be determined for the water heater. Usually, only one heating unit will operate at any one time. Therefore, determining the wire size is based upon the wattage of the largest heating unit. The branch-circuit protection and wire size must be based upon 125% of the maximum water heater demand load, *NEC Section 422-14(b)*.

Assume that the water heater for a residence is to be connected for limited demand. This means that the 2 000-W element and the 3 000-W element (as stated in the specifications for that residence) cannot be energized at the same time. The maximum load demand is:

$$I = \frac{W}{V} = \frac{3000 \text{ W}}{240 \text{ V}} = 12.5 \text{ A}$$

A circuit breaker is permitted to be loaded to only 80% of its rating; therefore, a 20-A breaker is required. A 20-A circuit requires No. 12 AWG copper wire. A 2-pole circuit breaker is used for the water heater circuit. Some water heaters have a higher demand and require a 30-A circuit and No. 10 AWG wire.

Grounding of the water heater is extremely important. If a heater element develops a ground fault, the entire water-piping system may become energized. An effective grounding conductor will cause the circuit overcurrent device to blow or trip, thus preventing a shock hazard to humans and animals.

## LIVESTOCK WATERERS

Livestock waterers must be mounted to a firm concrete base. If the waterer is installed in an unheated barn, then it must be of the heating type. Power must be supplied to the heating element. A thermostat is provided in the waterer to regulate the temperature. This thermostat is usually adjustable. These waterers should be wired to a separate circuit which is turned completely off during the warm season. The waterers are usually wired from underground using No. 12 AWG copper Type UF cable. The cable is easily installed in the same trench with the water lines. The cable is run up through the same 4-in to 6-in trade size tile with the water pipe, Figure 20-4. Heat from the ground rises through this pipe in winter to help prevent the water pipe from freezing.

Grounding is extremely important at a stock waterer. A grounding electrode should also be provided at the waterer. If metal water piping is used, it can serve as the grounding electrode. If plastic water piping is used, then a ground rod should be driven.

The circuit for stock waterers is considered a continuous load and, therefore, must not be loaded beyond 80% of the circuit rating.

## FUEL-DISPENSING PUMPS

Gasoline vapors may become *explosive*, and special wiring for an underground tank is required in the area near the dispensing pump, the fill pipe, and the vent pipe. Overhead gravity storage tanks must be located a safe distance from electrical equipment and buildings. An overhead gravity storage gasoline tank should be located at least 100 ft (30.5 m) from a building. Furthermore, the ground level around the tank should be graded in such a way that leaking liquid fuel will not drain in the direction of buildings or electrical equipment.

Gasoline vapor is heavier than air and it will settle near the ground. Therefore, the area near the ground in the vicinity of the gasoline pump is the hazardous area. These areas are classified as Class I hazardous locations, *NEC Sections 500-4* and *514-2,* and all wiring within such areas must be rigid metal conduit or intermediate metal conduit (IMC), *NEC Section 501-4*. Electrical fittings shall be rated explosionproof, and all joints shall be made up tight with at least five threads fully engaged, *NEC Section 501-4(a)*. Liquid propane used as a vehicle fuel must be treated in a manner similar to gasoline.

These special wiring provisions apply to the hazardous classified area near the dispensing pump. The area within 3 ½ ft (1.1 m) in all directions of the dispensing pump and extending to the ground is considered the hazardous area. In addition, an area 18 in (457 mm) above the ground and extending out 20 ft (6.1 m) in all directions from the dispensing pump is considered part of the hazardous area, *NEC Table 514-2*. The area 18 in (457 mm) above ground and extending 10 ft (3.05 m) in all directions of an open-type fill pipe is part of the hazardous area. Similarly, the area 18 in (457 mm) above ground and extending 5 ft (1.52 m) in all directions of a vent pipe is considered a hazardous area. Only the wiring required for the pump should be within these areas. If it is necessary to install other wiring within these areas, it must be contained within rigid metal conduit or IMC, and all fittings and equipment must be rated for use in Class I hazardous locations and Group D vapors.

The wiring for the dispensing-pump motor must be single conductors in conduit extending to the source of supply. This conduit is usually run underground. A means must be provided to prevent vapors from moving through the conduit to an adjacent building. Therefore, a seal is required to prevent vapors from traveling through the conduit. The pump motor and controls provided as a part of the dispensing unit are installed within an explosionproof enclosure, and the wires are sealed at the factory. It is up to the wiring installer to install a seal in the conduit leaving the area of the pump. A seal is shown in Figure 20-5. The procedure for making an explosionproof seal is shown in Figure 20-6. A seal is also required at the point where the conduit emerges from the ground at the supply end. The conduit emerging from the ground must be rigid metal conduit or IMC.

Rigid metal conduit or IMC is not required to be run all the way back to the source of electrical supply, pro-

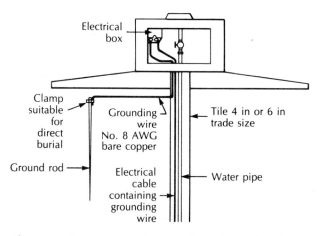

**Figure 20-4 Wiring and grounding of an electrically heated stock waterer**

Unit 20 General Farm Wiring 391

Figure 20-5 The gasoline-dispensing pump is wired with rigid metal conduit or IMC, and seals must be installed in the conduit.

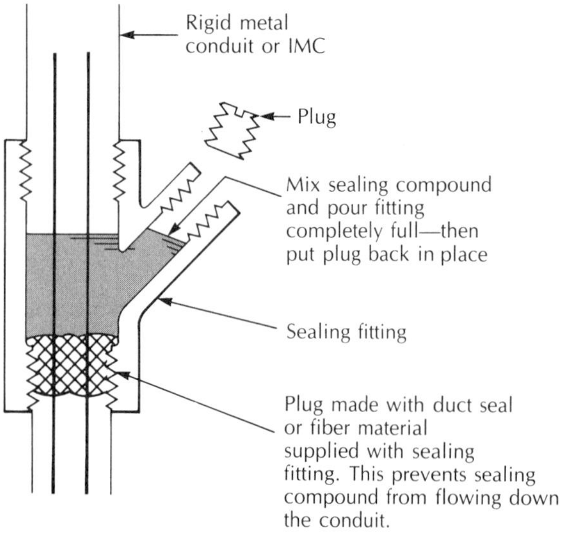

vided the conduit is at least 2 ft (610 mm) below the ground. Rigid nonmetallic conduit (PVC) may be run underground to minimize the cost of installation, *NEC Section 514-8*. A proper installation is shown in Figure 20-7.

Special bonding requirements apply to equipment in Class I hazardous locations such as a gasoline-dispensing pump. Double locknuts or a locknut and metal bushing are not considered acceptable bonding, *NEC Section 501-16(a)*. What this means is that a bonding jumper is required from the conduit entering the main service panel to the grounding terminal block, Figure 20-8. If a portion of the conduit run is rigid nonmetallic, then it will be necessary to run a grounding wire from the grounding terminal block to the dispensing pump enclosure.

A special requirement applies to the disconnect for a circuit supplying a gasoline-dispensing pump. All circuit wires must be disconnected simultaneously including the neutral if present, *NEC Section 514-5*. A 2-pole disconnect switch may be used, or special 2-pole circuit breakers are available which open the neutral and the ungrounded conductor, Figure 20-8.

Figure 20-6 Installation of a sealing fitting in a run of conduit

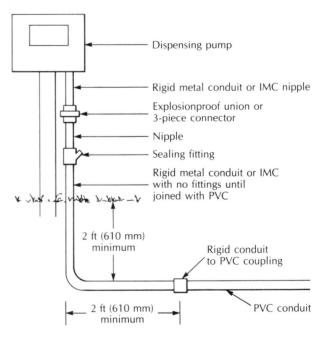

Figure 20-7 Rigid nonmetallic (PVC) conduit may be used to supply a gasoline-dispensing pump from underground. However, a grounding wire must be run if nonmetallic conduit is used.

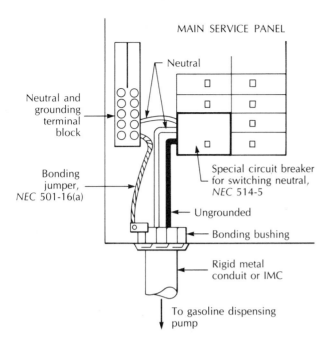

**Figure 20-8** Special bonding and disconnect requirements for a circuit supplying a gasoline-dispensing pump. The neutral wire is required to be opened with either a special circuit breaker or a separate 2-pole disconnect switch.

## STORAGE BUILDINGS

Storage buildings must be provided with adequate light for people working around machinery. These areas are usually cold and, therefore, incandescent lamps are preferred. The lights are on only for limited periods of time. Provide one 150-W lamp for every 250 ft$^2$ (23 m$^2$) of floor area. A 20-A receptacle circuit should be provided with an outlet every 40 ft (12.2 m). A welder outlet can be provided in the building if machinery repair is necessary.

The service-entrance requirements in a storage building depend upon the intended use of the building. A 100-A, 120/240-V, 3-wire service may be desirable. A 40-A, 50-A, or 60-A, 120/240-V, 3-wire service is usually adequate. If the conditions within the building are essentially dry and free from excessive dust, then normal wiring cable or conduit wiring can be used, *NEC Section 547-1(c)*. If animals are housed in an enclosed building, then wet conditions may exist and the wiring methods used should include Type UF cable or rigid nonmetallic (PVC) conduit, and sealed fixtures and outlets, *NEC Sections 547-1(a) and (b)*.

## RFVIFW

Refer to the *National Electrical Code* when necessary to complete the following review material. Write your answers on a separate sheet of paper.

1. List three basic electrical requirements for a farm shop.

2. Determine the electrical requirements for a well house with a ¾-hp, single-phase, 240-V water pump; a 1 000 W, 240-V resistance electrical heater; one 15-A general circuit for lights, and a receptacle outlet.

3. If the heating element in a stock waterer is rated at 800 W and 120 V, how many stock waterers can be connected to a 20-A circuit?

4. An electric water heater is rated at 4 500 W and 240 V. Determine the minimum size circuit wire for the water heater.

5. The circuit supplying a gasoline-dispensing pump must be wired using which of the following?
   a. NM cable
   b. UF cable
   c. Conduit
   d. Low-voltage wire

# UNIT 21
# WIRING FOR LIVESTOCK HOUSING

## OBJECTIVES

After studying this unit, the student will be able to

- discuss the wiring requirements for beef housing.
- discuss the wiring requirements for sheep housing.
- discuss the wiring requirements for horse housing and riding arenas.

## WIRING FOR BEEF HOUSING

A beef cattle feeding operation may vary from an open feedlot to environmentally controlled confinement housing, Figure 21-1. In an open feedlot, animals are in the open with little or no shelter. The primary electrical needs are for lighting, feeding, and watering. All-night lighting is important to help reduce animal injury and to increase feed intake. High rates of water and feed consumption result in high rates of weight gain.

Research has shown that beef animals gain in weight faster in winter when the area the animals are in is lighted at night to give them an artificially produced long-lighted day of about 18 hours. The feedlot or confinement barn should be lighted to a level of illumination of 10 footcandles (fc). A time clock can be used to control the lights. Efficient lighting, such as that produced by high-pressure sodium lamps, should be used to minimize energy cost.

An environmentally controlled confinement barn houses the maximum number of beef cattle in a minimum amount of space. The building is enclosed and insulated to conserve heat in winter, and to keep out heat in

Figure 21-1 Beef cattle are provided shelter and fed from an electrically powered feed bunk. (*Courtesy of Butler Manufacturing Co.*)

summer. Ventilation fans move air through the building to maintain a healthy environment. Some fans run continuously, while others are controlled by a thermostat. The animals are usually housed on a slotted floor with a waste-holding tank below. Waste falls through the slots in the floor and into the tank for later removal.

Beef cattle may be housed in a concrete feedlot with a partially open shelter. In arid areas, the shelter provides protection from the heat of the sun.

A lighting level of about 10 fc can be achieved by placing two 2-lamp, 40-W fluorescent fixtures about 12 ft (3.66 m) on center in the barn. The spacing can be nearly double if 400-W mercury-vapor lamps or 250-W high-pressure sodium lamps are used. Receptacle outlets should be located along the wall about 5 ft (1.52 m) above the floor. The wiring should be Type UF cable or rigid nonmetallic (PVC) conduit with watertight fixtures and outlets, if the building is sufficiently enclosed to cause high humidity in the barn at some time during the year, *NEC Section 547-1(a)*.

Calf operations must provide a dry, sheltered area for calving. The main requirement for growing healthy calves is dry, draft-free housing. Heat is not generally required except in extreme northern climates. Calves may be raised in ventilated confinement housing or semiopen housing, or in small, individual, covered pens called *hutches*. Adequate light is needed to care for the calves. A supply of warm water is required for preparing liquid milk replacer.

## WIRING FOR SHEEP HOUSING

Sheep housing ranges from confinement type to pasture grazing. Sheep are generally confined in a building in winter in northern climates and allowed to pasture or move around in an open lot in summer. Lighting outlets placed approximately 12 ft (3.66 m) on center will provide adequate lighting. Receptacle outlets should be located every 40 ft (12.2 m) along the wall.

Lambing takes place in the early spring, and heat is required. Lambing pens, 4 ft × 4 ft (1.2 m × 1.2 m), are usually located along one side of the housing area, Figure 21-2. A receptacle outlet for infrared heat lamps should be provided at each lambing stall, Figure 21-3. These outlets should be on a 20-A circuit. Six lambing pens can be served with one 20-A circuit.

Plenty of light must be provided in the sheep-shearing area. One electrical outlet is required for each shearing clipper. The clipper is suspended in air from a rope or cable to keep it ready for use, Figure 21-4.

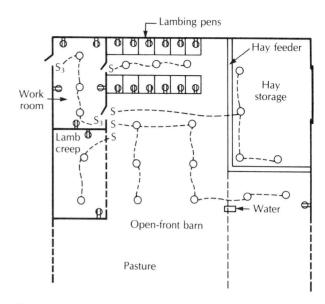

**Figure 21-2 A sheep barn, having one open side, with a pasture for grazing.**

## WIRING FOR HORSE HOUSING

Horses are maintained for pleasure riding and driving, purebred and equitation showing, jumping and hunting competition, working, and racing. Most horses are housed in individual pens or stalls, or in small groups, Figure 21-5. Larger breeders keep horses in open lots with shelters, and use electrically powered feeding systems.

A lighting outlet should be provided in each pen. Horses are often groomed and provided with care in the pen. An electrical outlet should be provided on the outside of every fourth pen in a row for animal- and

**Figure 21-3 A lambing pen with infrared lamp for heat**

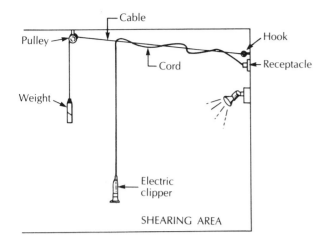

Figure 21-4  A sheep-shearing area must have plenty of light, and provisions for the shearing clipper.

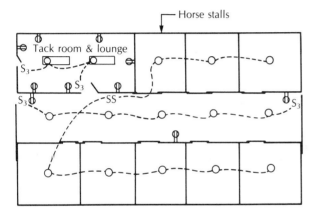

Figure 21-5  Horses may be housed in individual stalls.

building-maintenance power tools. Electrically heated waterers may be provided for each pen or every two pens.

An indoor riding arena must be provided with adequate illumination for pleasure riding and working the animals. The ceiling of such a building is usually 14 ft to 18 ft (4.27 m to 5.49 m) high. Incandescent, 200-W lamps placed 12 ft (3.66 m) on center will provide adequate illumination. Metal-halide or high-pressure sodium in low-bay fixtures provide a more efficient type of lighting. Fixtures rated at 400 W can be placed about 24 ft (7.32 m) on center, Figure 21-6.

A tack room must be provided for saddles, bridles, halters, harnesses, and similar equipment. The tack room is often heated, and must be kept dry. A damp tack room will often result in mold growth on the leather. A heated tack room can serve as a warm area for people working with horses in the winter. Fluorescent and incandescent lighting fixtures can be used in the tack room. An electric heater can be provided for use in winter.

Foaling often takes place in cold weather. A large stall is desirable, and heat may be needed in cold weather. An infrared quartz heater mounted high above the stall will throw heat down at the foal. This heater should be permanently mounted to the building.

The wiring in horse barns must be Type UF cable or rigid nonmetallic (PVC) conduit with watertight and dusttight outlets and fixtures if the area is damp and dusty, *NEC Section 547-1(a)*. This is often the case in the area where the horses are housed. Ceilings in horse pens must be high, and lighting fixtures must be provided with protective guards. All exposed wiring accessible to horses must be protected with rigid metal conduit. Horses tend to chew on parts of the stall if they are confined with minimal exercise. Wiring for waterers should be run underground to keep it unaccessible to the horses.

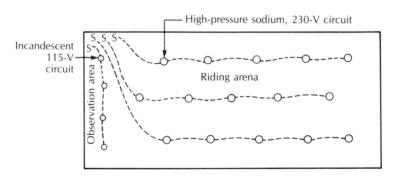

Figure 21-6  Lighting in a horse riding arena

# REVIEW

Refer to the *National Electrical Code* when necessary to complete the following review material. Write your answers on a separate sheet of paper.

1. Is heat essential for raising beef calves in northern climates?

2. Research has shown that a beef animal will gain weight faster in winter if lighted at night to what level of illumination, in footcandles?

3. Determine the maximum number of 120-V, 250-W infrared heat lamps that are permitted to be connected to a 15-A circuit in a lambing barn.

4. A horse riding arena is lighted with twelve high-pressure sodium fixtures rated at 240 V, and each drawing 1.9 A. Determine the minimum number of 20-A circuits required to supply these lighting fixtures.

5. A 3 200-W, 240-V electric quartz infrared heater is installed above a foaling stall. Is a 15-A circuit or a 20-A circuit required for this heater?

# UNIT 22
# DAIRY FARM WIRING

> **OBJECTIVES**
>
> After studying this unit, the student will be able to
>
> - define dairy housing terminology.
> - list the essential equipment used in milking barns.
> - locate the required lighting and receptacle outlets on a milkhouse plan.
> - state the electrical requirements for calf housing.
> - identify and solve some farm stray-voltage problems.

## TYPES OF DAIRY FACILITIES

Dairy milking facilities are of two basic types: stanchion barn, and milking parlor. In a *stanchion barn*, the cows are kept in individual stalls where they eat, drink, and rest. The operator takes the milking unit to the cow, and the milk is transported through a pipeline directly to the milk storage tank. Figure 22-1 shows a modification of the stanchion barn called a *tie stall* or *comfort stall*. The cow is held in place by a neck strap connected to a chain. The cow has considerable freedom to move about in a comfort stall. The electrically powered feed conveyer is mounted to the ceiling, and the waterers are located between stalls. The vacuum line to operate the milker and the pipeline for the milk are located above the stalls. The cows are usually allowed to go outside daily for exercise.

In the *milking parlor* system, the cows are brought to the milking area two or three times each day. The operator works in a pit so that it is not necessary to bend over while milking the cows, Figure 22-2. The vacuum line to

Figure 22-1 Dairy barn with tie stalls for milk cows. The vacuum line and the milk pipeline can be seen overhead. (*Courtesy of Butler Manufacturing Co.*)

operate the milkers is at the top. Milk goes to the pipeline below and drains to one end where it is pumped to

Figure 22-2  In a milking parlor, the operator stands in a pit and the cows are brought to the operator.

Figure 22-4  Dairy cows fed inside a covered free-stall barn (*Courtesy of Butler Manufacturing Co.*)

the milk tank. Heat is blown into the operator's pit through the openings near the floor.

The cows rest and feed with freedom to move about. The resting area is called a *free-stall barn*. The cows have freedom of choice to lie down in any stall. Feeding may be outside, as shown in Figure 22-3. The free stalls and feeding are commonly contained inside a single building, Figure 22-4. Air in the building is circulated with an inflatable air bag at the ceiling. The waste is removed with an electrically powered scraper pulled by a chain. Electrically heated waterers are installed at each end of the feeder.

Figure 22-3  Dairy cows fed outside in an electrically operated bunk feeder (*Courtesy of Butler Manufacturing Co.*)

## MILKING EQUIPMENT

A milking facility may vary from a minimum of mechanization to a completely electrically powered milking and feed handling system. The minimum equipment includes a vacuum pump to operate the milkers; a bulk milk tank and cooling compressor, Figure 22-5; and a water heater. If a pipeline is used to move the milk directly to the bulk tank, a second vacuum pump is usually required. A milk pump also is needed to lift the milk from the receiving jar to the bulk tank. An automatic washer is required if a pipeline milk handling system is used. This system is also used to wash the milking units automatically. An automatic washer is frequently used to wash the bulk milk tank.

The feed handling system is often electrically powered, but some dairy farmers prefer to deliver feed to the cows with a tractor and automatic unloading wagon. The waste in the stanchion barn is usually removed with a gutter cleaner.

## LIGHTING

Lighting fixtures must be arranged to provide about 50 fc of light in the area under the udder of the cow. This is necessary to properly inspect the cow and the condition of the milk before attaching the milker. One 150-W incandescent lighting outlet every 4 ft (1.22 m) should be provided behind the cows for a stanchion barn. Four-lamp, 40-W fluorescent fixtures should be installed every 8 ft (2.44 m) on center. Incandescent lamps are adequate in the feed alley when spaced every 12 ft

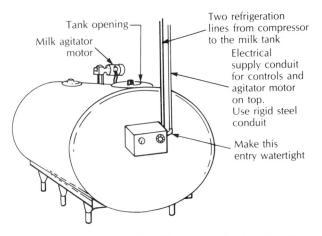

Figure 22-5 Milk in a bulk milk tank may be stored on the farm for up to two days.

(3.66 m). Receptacle outlets should be spaced throughout the stanchion barn both behind the cows as well as along the feed alley wall. Special circuits are needed when ventilation fans are to be installed in the barn.

The free-stall barn requires minimal light all night, but a higher level of light is required during times when cleaning and maintenance are taking place. These barns are usually cold, and fluorescent lamps are not suitable. Incandescent lamps are adequate, but have low efficiency. Several mercury-vapor lamps or high-pressure sodium lamps in low-bay fixtures will provide high levels of light at minimal operating cost. Receptacle outlets should be provided at each end of a free-stall barn with receptacles conveniently located throughout the barn. These receptacles should be located at least 5 ft (1.52 m) above the floor to prevent damage from animals. If the building is enclosed on all sides, moisture accumulation can become a problem. The wiring should be Type UF cable or rigid nonmetallic conduit. The outlets should be watertight and corrosion resistant, *NEC Section 547-1(a)*.

The milking parlor can be lighted with two continuous rows of 2-lamp fluorescent fixtures the entire length of the operator's pit. These fixtures must be located so as to shine light beneath the cow while preventing shadows cast by the operator, Figure 22-6. The minimum light required under the cow is 50 fc. The fluorescent fixtures in the milkhouse must be watertight and corrosion resistant. All wiring must be in either Type UF cable or rigid nonmetallic conduit. Water is sprayed in this area daily, and corrosion of fixtures and wiring materials can be a serious problem.

Lighting fixtures in the milkhouse must be located so that the milking equipment and the inside of the milk tank may be inspected. Overhead lights must be located to shine into the bulk tank. A 2-lamp, watertight, 40-W fluorescent fixture should be located over each wash vat, Figure 22-6. Lighting fixtures are not permitted to be located directly over openings into the bulk milk tank. Several receptacle outlets in watertight, corrosion-resistant boxes should be provided. A 20-A, 240-V electrical outlet with a knife-switch control or motor starter is required at the opening where the milk hauler flexible pipe enters the building. The pump on the milk tank truck is electrically powered, and the farmer must provide an outlet and controller. If automatic washing is used in the milking system, electrical power will have to be provided to the control unit. A schematic diagram of the wiring for a typical milk handling system is shown in Figure 22-7.

## HEATING AND COOLING

The bulk milk tank compressor, vacuum pumps, water heater, and electrical service are usually installed in a utility room. The heat from the milk can be used for heating the milking facility in winter. Sometimes, two water heaters are installed. A large-capacity water heater provides 170°F (77°C) water for washing milking equipment, while a small unit provides water at 120°F (49°C) for washing the cows' udders, Figure 22-8. A ventilation fan should be installed in the utility room, controlled by a cooling thermostat to turn on the fan if the temperature gets too high. A high temperature near the refrigeration condenser will decrease cooling efficiency. An office and a bathroom may also be provided in the milking facility.

## FEEDING EQUIPMENT

The primary feeds for dairy cows are alfalfa hay, corn silage, haylage, high-moisture kernel corn, and ground feed consisting mainly of small grain such as corn and oats, along with salt, minerals, and protein supplement. The type of feed used is variable from one farm to the next. The equipment consists primarily of conveyers and silo unloaders. The feed handling equipment is powered by electric motors ranging in size up to about 10 hp.

## CALF HOUSING

Calf housing ranges from a completely environmentally controlled building to individual calf hutches, Fig-

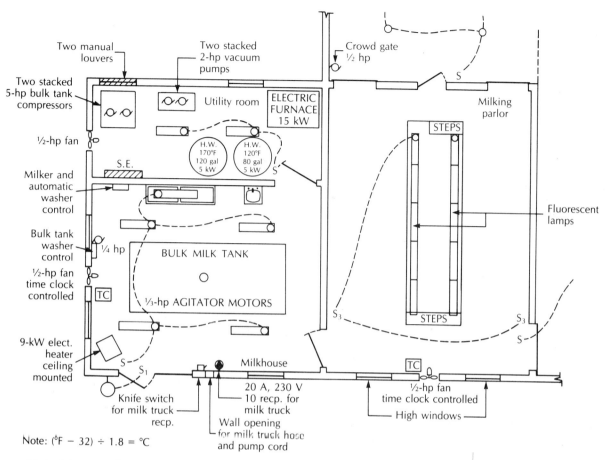

**Figure 22-6** Adequate placement of the lighting fixtures will provide for 50 fc at required locations in the milkhouse and milking parlor.

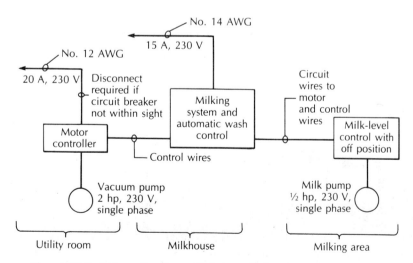

**Figure 22-7** Schematic diagram of the milk handling control system

Unit 22  Dairy Farm Wiring  401

**Figure 22-8** Two water heaters for high-temperature cleaning, and one at a lower temperature for washing cows

ure 22-9. The primary electrical requirements for calf hutches are adequate lighting and a receptacle outlet at each light pole. Light must be arranged to shine into the hutches so that calves can be inspected at night. Mercury-vapor lamps or high-pressure sodium fixtures should not be mounted too high on the pole, or light will not shine into the hutches.

**Figure 22-9** Calf hutches require adequate lighting and a receptacle outlet at each pole.

Environmentally controlled calf housing involves the addition of heat and ventilation. Adequate illumination must be provided. In calf stall areas, about one outlet (150-W incandescent) for every 60 ft$^2$ (5.58 m$^2$) of floor area is adequate. In calf pen areas, 10 fc are adequate, which can usually be obtained with one outlet per pen. It is best to plan for additional outlets over the service alley. Suggested lighting and receptacle outlets for a calf barn are shown in Figure 22-10.

In calf stall areas, at least one outlet should be placed on each wall. In calf pen areas, it may be most desirable to provide convenience outlets with each ceiling-mounted light fixture. An outlet should be within 25 ft (7.62 m) of point of use, whether on ceiling or on walls.

Calves do not give off enough body heat to maintain the temperature of a housing facility at the desired level (40°F to 50°F, or 4°C to 10°C) during the winter months, especially in northern climates. In a properly insulated calf building, the amount of heat required may generally be determined from the following: 3 W per square foot of glass, plus 1 ½ W per square foot of wall, plus 1 W per pound of calf. A forced-air, ceiling- or wall-mounted space heater works best. Although radiant heat space heaters can be used, caution must be exercised to not concentrate too much radiant energy directly onto the animals. Such space heaters should operate on individual circuits. For ventilation, provide 10 cubic feet per minute (cfm) for each 100 pounds of calf. Ten times this amount is needed in summer.

## FARM STRAY VOLTAGES

*Stray voltage* is a condition that may occur on a farm where a difference in voltage potential develops between two parts or areas. Several volts may exist between the concrete floor of a milking parlor and the metal stalls in the milking parlor. A cow standing on the floor suffers a shock when it touches the metal stall. A cow may experience an electrical shock when touching a stock waterer. Animals have a much lower resistance than humans and, therefore, they are more susceptible to shock than are humans.

Stray-voltage problems may be caused by the flow of alternating current on the power supplier's primary neutral wire. High current flowing on the primary neutral, or high resistance due to a bad connection, can cause voltage drop which is measured between the neutral wire and the earth on the farm. This is because the primary neutral and the farm secondary neutral are bonded together at the transformer.

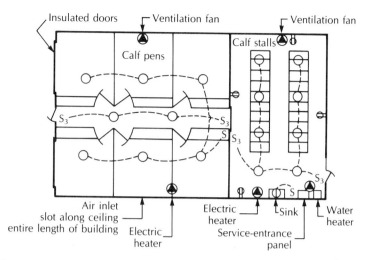

Figure 22-10 Suggested lighting and receptacle outlets for a calf barn

The earth acts as an electrical conductor which forms a parallel path to the neutral wires and grounding conductors. The source of ac at a farm is the power supplier's transformer. The neutral wire is grounded at the transformer. The flow of electrical current begins at the transformer, and it returns to the transformer. If a hot wire faults to earth, current flows through the earth to complete the circuit to the transformer. A voltage drop occurs at points where current enters and leaves the earth. This voltage drop is stray voltage, and it may be of sufficient magnitude to cause animals or even humans to suffer a shock. If equipment is properly grounded with an equipment grounding conductor, so little current flows into the earth that stray voltage is held to minor levels.

Neutral current can use the earth as a conductor parallel to a feeder neutral to a building, where the neutral is grounded at the supply and load ends. If abnormal resistance is present in this neutral due to a wire which is too small for the load and the length of the run, or corrosion at a connection, enough current may flow through the earth to cause an objectional level of stray voltage. It is important to balance the 120-V loads on farm feeder wires to minimize the load on the neutral. Electric motors should be operated at 240 V whenever possible to prevent high currents on the neutral during motor starting.

## Measuring Stray Voltage

Stray voltage is measured with a good-quality voltmeter which has a scale sensitive to fractions of a volt. The voltmeter must be of a type which will prevent dc voltage from being indicated on the ac voltage scale. The voltmeter can be connected between the neutral or grounded equipment at the barn and an isolated ground rod outside the barn, Figure 22-11.

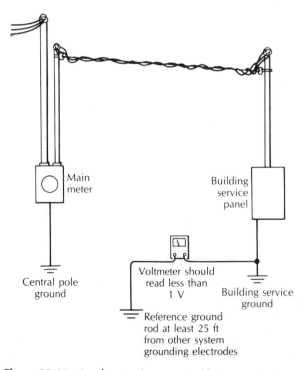

Figure 22-11 A voltmeter is connected between the barn neutral or grounded equipment and a reference ground rod to measure stray voltage.

If the farm main electrical disconnect is opened and the voltmeter reads a voltage, the source is coming from somewhere off the farm. The next step is to turn on as much 240-V load as possible while trying to eliminate or reduce the 120-V draw at any building on the farm. If the voltage reads higher every time another load is turned on, then the source is voltage drop somewhere on the primary neutral leading to the farm.

Next, go to every building on the farm and turn on a 120-V load. A portable hair dryer works well for this purpose because it applies about 10 A on the neutral when it is operating on high. If the voltmeter reading changes significantly either by increasing or decreasing, then the change is caused by voltage drop on the neutral carrying the load. A change in voltage reading of more than 0.5 V is considered too high, signaling a problem of resistance on that feeder neutral. Next, operate individual equipment to see if it is faulting to ground. Turning off the power to a circuit or an entire building will eliminate the source if a ground fault is the cause. A clue that there may be a ground fault is a steady level of stray voltage which is not affected by any of the previous tests. A ground fault at one building can cause a stray voltage at another building on the farm. If a submersible water pump in a well is not grounded, it may be the source of the problem when the pump is running.

## SOLVING STRAY-VOLTAGE PROBLEMS

Once the source or sources of stray voltage have been identified, the problem can be solved. It is possible for several sources to be present at the same time. If the major source is found to be off the farm, then the electric power supplier must be contacted to search for a possible cause on the supplier's distribution system.

It is not always practical for the power supplier to prevent some level of neutral-to-earth voltage at a farm. In such situations, a number of solutions can be considered. One solution is for the power supplier to break the bond between the primary neutral wire and the farm secondary neutral wire. This will prevent off-farm sources from getting onto the farm. However, this solution will do nothing to stop on-farm sources of stray voltage. Some power suppliers will install a solid-state, high-speed switch between the primary and secondary neutrals to ensure the safety that the bond is intended to provide. In some situations, the solution is to install a saturating reactor between the primary and secondary neutrals.

These and other solutions can be discussed with the power supplier if it is discovered that the problem originates on the primary side of the transformer.

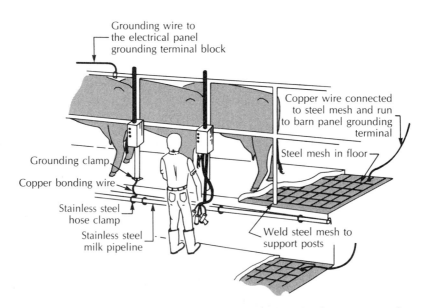

Figure 22-12 Grounding and bonding the equipment and the floor in a milking parlor to prevent the cows from experiencing an electrical shock from stray voltages

If the major stray-voltage source is found to originate on the farm side of the transformer, then the source can and should be eliminated. A ground fault can be found and the problem corrected. All equipment must be properly grounded with an equipment grounding conductor. Ground faults are not likely to cause stray voltage if the wiring system and equipment are properly grounded.

If a feeder neutral wire is found to be causing excessive voltage drop, then the source can be found and eliminated. A long feeder to a building may be carrying too much neutral current. A splice or termination in the neutral wire, especially if it is aluminum wire without an oxide inhibitor, may be the cause of abnormal resistance. The feeder to one building, for example, the dwelling, can cause a stray voltage at the barn. To prevent this situation, 4-wire feeders to all buildings on the farm can be installed from the main electrical distribution pole. When this is done, the neutral is not connected to the earth at the buildings, *NEC Section 250-24*.

An additional precaution is to place a grounded metal system in the floor. This method is recommended for all new construction of dairy milking facilities. The metal in the floor, as shown in Figure 22-12, is bonded to all the metal in the milking parlor and then connected to the grounding bus at the barn service panel. In addition to this measure, if the farm buildings are supplied with feeder wires where the grounding wire and the neutral wire are separated, stray voltage will be prevented from affecting cows while they are in the milking area. The bonding together of metal in the milking area is called an *equipotential plane*. This technique also should be considered when bonding around heated livestock waterers and livestock feeders.

## REVIEW

Refer to the *National Electrical Code* when necessary to complete the following review material. Write your answers on a separate sheet of paper.

1. Comfort stalls are found in a
    a. milking parlor.
    b. free-stall barn.
    c. stanchion milking barn.
    d. milkhouse.

2. The minimum electrically operated equipment essential for a dairy milking operation, other than a water heater, includes what two pieces of equipment?

3. What is the minimum recommended water temperature leaving the water heater for washing milk pipelines and equipment?

4. The area under the cow should be lighted to a level of illumination of how many footcandles (fc)?

5. Describe the type of lighting fixtures recommended for installation in a milking parlor.

6. A warm, environmentally controlled calf building with the stalls is 20 ft (6.1 m) wide and 30 ft (9.14 m) long. Determine the minimum recommended number of 150-W incandescent lamps to be installed.

7. Are fluorescent fixtures recommended for installation in an unheated free-stall barn? Explain.

8. Describe how the voltmeter should be connected to measure stray voltage in a barn.

9. Describe the common causes of on-farm stray-voltage problems.

10. Why does good equipment grounding help to prevent stray-voltage problems?

# UNIT 23
# FEEDING SYSTEM WIRING

> ## OBJECTIVES
>
> After studying this unit, the student will be able to
>
> - identify the types of feed storage silos and their electrical requirements.
> - identify the types of feed conveyers and their electrical requirements.
> - specify the type of wiring for feeding equipment.
> - specify the lighting and receptacle outlets for a feed handling center
> - describe interlock wiring of feed handling equipment for automatic operation.

## FEED STORAGES

Animal feeds must be stored and handled in such a way as to minimize the amount of labor required to handle the feed.

Dry hay is usually stored in bales and stacked in a barn or shelter. The hay storage area is usually dry, and it is only dusty when the barn or storage is being filled. The lights inside this area should be incandescent with glass covers to prevent the lamp from making contact with combustible materials. Receptacle outlets must be installed so that they will not be in direct contact with the hay. A receptacle outlet must be provided near the area where a portable elevator will be located for loading hay into the storage. A switch suitable for controlling a motor should be used to control this receptacle outlet.

Silos are used as often as possible to store animal feeds. *Silage,* which is chopped corn, and *haylage,* which is chopped hay, can be mechanically handled with a minimum of human labor. Silos are of two types; upright, and horizontal. The upright silo can be unloaded with an electrical unloader located either in the top or the bottom of the silo. A top-unloading silo is shown in Figure 18-1 (in Unit 18 of this text). A bottom-unloading silo is shown in Figure 23-1. Another type of silo uses a top unloader and a bottom conveyer, Figure 23-2, where both units are operated together. The top unloader dumps the silage down a hollow passage in the center of the silo, where it is removed by the conveyer at the bottom.

A horizontal (bunker or trench) silo is constructed at ground level, and a tractor is used for unloading. The electrical requirement in this system is adequate light to operate the tractor in the silo at night, Figure 23-3. The silage may be taken directly to animal feeders using a tractor and a self-unloading wagon, or it may be taken by a self-unloading wagon to an electrically powered feeder.

Small-grain and ground-type animal feeds are stored in upright metal storage bins with a built-in auger at the bottom for unloading. The motor powering this auger is

Figure 23-1  A bottom unloader is used for some silos.

Figure 23-2  This silo has a top unloader, but a conveyer in the bottom removes the silage. (*Courtesy of Butler Manufacturing Co.*)

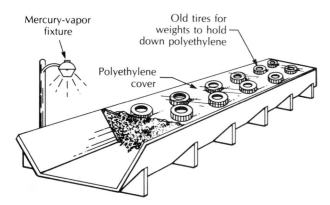

Figure 23-3  Adequate light is required for a horizontal (bunker or trench) silo.

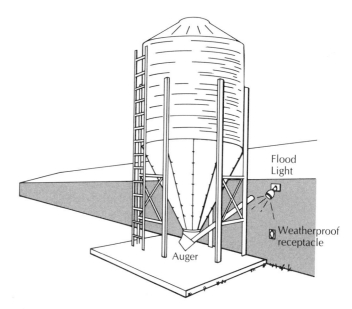

Figure 23-4  An electrically powered auger delivers feed into the building. Adequate light and a receptacle outlet should be provided.

usually located inside an adjacent building, Figure 23-4. A receptacle outlet must be provided near the bin for maintenance, and for the operation of a portable auger to load the bin. A light must be provided near the bottom of the bin for maintenance at night. A PAR-38 incandescent floodlight mounted on the side of the building will direct light at the bottom of the bin, where maintenance may be necessary.

It is important to be familiar with the types of storage and handling equipment in order to provide electrical service. Silo unloaders, either top- or bottom-unloading types, should be cord and plug connected to the motor starter. A bottom unloader is stationary during normal operation, but it must be movable for maintenance. The unloader is often removed from the silo during filling. A top unloader is suspended by cables from the top of the silo. A cord delivers power to the unloader. The unloader is moved downward a small amount each time it is used. An electric motor- or hand-operated winch at the outside bottom of the silo is used to lower the unloader.

## CONVEYERS

Feed conveyers include the following types: belt, flight and chain, screw or auger, pneumatic, and batch delivery.

## Screw or Auger

The *screw* or *auger conveyer* may be enclosed on the sides and open on the top or bottom, or it may be enclosed in a tube with openings. A major advantage of the auger conveyer is its ruggedness and low maintenance. However, power requirements are higher than they are for other types. In addition, the auger disturbs the material considerably as it is conveyed. This often causes light particles and heavy particles to separate.

## Flight and Chain

The *flight and chain conveyer* slides the material along a flat surface with a paddle or flight pulled by a chain. A major advantage of this type of conveyer is that large volumes of material can be moved quickly. A flight conveyer is particularly efficient at elevating materials up a gradual or steep incline, Figure 23-5. The feed-mixing hopper shown also weighs the feed. The power requirements are lower than they are for an auger.

Some flight conveyers are used to distribute feed in the bunk feeder in front of the animals. With this type, the conveyer has a reversible flight and chain which propels the conveyer back and forth along the feeder.

## Belt

*Belt conveyers* are used to distribute feed materials for long distances with fairly low power requirements,

Figure 23-6 A belt conveyer feeder is ideal for supplying long feeder lengths. (*Courtesy of Butler Manufacturing Co.*)

Figure 23-6. This type is also frequently used for short conveyers. A blade or brush is usually pulled along the moving belt by a cable to distribute the feed along the feeder.

## Pneumatic

Granular feed is sometimes moved pneumatically. In the *pneumatic conveyer,* a blower forces air through a tube and carries the feed with it. This method is used to deposit silage into an upright silo. A tractor is usually the prime mover for the silage blower, but electric motors are also used.

## Batch Process

The *batch process* uses a self-unloading truck or wagon to deliver the feed from the silo to the animal feeder. This type of system does not use electric power unless a vertical silo is part of the system. The batch process is used on large farms in combination with a horizontal bunker silo.

## Wiring for Conveyers

The wiring for conveyers and bunk feeders should be installed in conduit. The conduit can be fastened easily to the metal frame of the feeders. The connections to the motors should be flexible. Liquidtight flexible metal conduit provides a corrosion-resistant and waterproof

Figure 23-5 A flight conveyer is efficient at elevating materials up an incline. This feed-mixing hopper also weighs the feed. (*Courtesy of Butler Manufacturing Co.*)

wiring connection. The motors are exposed to the weather and subject to the entry of rodents and flying particles. Totally enclosed fan-cooled motors are best suited for these conditions. If the motor on the feeder or conveyer is not within sight from the controller, a disconnect switch is required to be installed adjacent to the motor, *NEC Section 430-86*.

## FEED PREPARATION

Several different types of feeds, such as silage, high-moisture corn, and protein supplement, may be combined and mixed as a batch before delivery to the animals. (See Figure 23-5.) Stationary mixers are usually electrically powered. The electrical connection to the mixer *must* be flexible. Liquidtight flexible metal conduit is preferred. The mixer is mounted on load cells which electronically weigh the material entering the mixer. The farm operator adds materials to the mixer until the desired amount of material results. If a rigid electrical connection were made to the mixer, the weighing system would not function properly. The wires under the mixer of Figure 23-5 connecting to the load cells are installed by the equipment manufacturer.

Roller mills are sometimes used to crush the kernels when high-moisture corn is added to a feed ration. A roller mill has lower power requirements than a hammer mill-type grinder. The high-moisture corn or dry corn is delivered to the roller mill, where it is crushed by passing it between rollers, Figure 23-7. Then the corn moves along a conveyer, which usually distributes it on top of the silage or puts it into a mixer.

Ground grain is often prepared on the farm. Grinding can be done in a batch process using a tractor-powered grinder and feed mixer. This process can be accomplished quickly, but high-power requirements are needed. Low-horsepower automatic grinders which blend several types of materials may be used. These grinders run automatically for extended periods of time. Typical farm units may range from 2 hp to 5 hp. Grain bins mounted above the mill often feed grain into the grinder by gravity.

## LIGHTING AND RECEPTACLES

The wiring in a feed preparation and distribution area must be made dusttight to prevent the entry of airborne materials. These areas are not generally considered hazardous, but a *fire* or *electrical shock hazard* can develop if feed materials are allowed to enter and accumulate in electrical enclosures. All covers must be tight and kept in place. Enclosure types NEMA 4, 4X, or 12 are preferred, but NEMA 1 enclosures are satisfactory except when dusty conditions exist. A NEMA 4, 4X, 9, or 12 enclosure should be used in areas where dry grain is ground into feed. Lighting fixtures, especially in these areas, must be in a dusttight fixture, unless they are sufficiently remote that dust accumulation on the fixture and lamps does not create a problem. Sufficient light must be placed in proper locations to illuminate areas where maintenance is required or where there are moving parts. Adequate light around machinery increases safety and reduces the chance of accidents. Receptacle outlets with dusttight covers should be located in all feed preparation areas to accommodate portable feed handling equipment and maintenance tools.

## SEQUENCING CONVEYERS

An automatic feed handling system consisting of several electrically powered components should be interlocked or sequenced to prevent plugging and damage. If a conveyer delivering silage from the silo to the bunk feeder fails, the silo unloader should be shut off. This is accomplished easily when magnetic motor starters are used. An interlock installed in the feed-conveyer starter must be closed before the silo unloader will start. If the conveyer stops, the interlock contact opens, and the silo unloader stops. The interlock wiring for a simple feed handling system is shown in Figure 23-8.

**Figure 23-7** A roller mill can be included in a feeding system to crush dry corn or high-moisture corn.

Unit 23 Feeding System Wiring 409

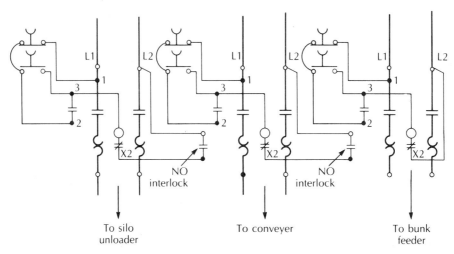

**Figure 23-8** These motors for a feeding system are interlocked. Thus, if the feed-bunk motor stops, the conveyer and silo-unloader motors will also stop.

## REVIEW

Refer to the *National Electrical Code* when necessary to complete the following review material. Write your answers on a separate sheet for paper.

1. Name three methods of unloading an upright silo.
2. If a very long conveyer is required, which type is most likely to be used?
3. What type of conveyer is used in the bunk feeder of Figure 21-1 (in Unit 21 of this text).
4. What type of enclosure is required for motors powering feed handling equipment?
5. If a feed preparation area is expected to be quite dusty, what motor-starter enclosure types are considered suitable?
6. Dry grain is ground into feed inside a building. Therefore, the lighting fixtures used in the area must be of what type?
7. What is the principal requirement when connecting the electrical supply wires to a motor on a feed mixer in a silo room? (See Figure 23-5.)
8. Which types of conveyers are most likely to move the greater amount of feed in the shortest amount of time?
9. Give an example of the use of a roller mill on a dairy farm or beef farm.

10. To complete the following diagram, show how to interlock the silo unloader so that it will shut down if the bunk-feeder motor shuts down. Show the complete wiring of the motor control system. Redraw the control system on a separate sheet of paper.

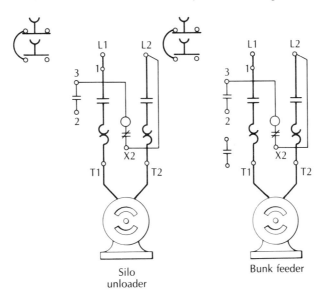

Silo unloader

Bunk feeder

# UNIT 24
# SWINE HOUSING WIRING

## OBJECTIVES

After studying this unit, the student will be able to

- describe the most important electrical requirements in swine housing.
- wire a swine feeding system.
- wire a ventilation fan.
- choose wiring materials for waste handling equipment.
- describe two methods of heating a farrowing building.

Pork production farms are variable in size and type of housing. It is important in a swine operation to hold costs to a minimum. The ideal market weight of a hog is about 225 lb (102 kg). It takes only 5½ months to raise a hog from birth to market weight in an efficient swine operation.

In a modern swine farm operation, the pregnant sows are kept in a separate gestation barn until they are ready to deliver. The sows are then moved to the farrowing and nursery barn. *Farrowing* is the process of giving birth to baby piglets. The sow is confined in a stall called a *crate*. The sow can stand up and lie down, but she cannot turn around. She is usually let out of the crate each day for exercise. The sow usually has from eight to twelve piglets in a litter. The sow must be confined in a crate so the chance of lying on the piglets is minimized. Heat is added to the crates to attract the piglets toward the warmth and away from the sow.

Once the piglets are weaned from the sow, they are moved from the nursery to group pens with other piglets of their approximate age. A separate building is used for the growing-finishing stage. The hogs are confined in a pen usually having a solid concrete floor on part of the pen and a slotted floor on one end, Figure 24-1. Waste is deposited on the slotted floor, where it drops through to either a holding tank below or a collection area. In some swine barns, the waste pit is flushed daily with water, with the waste water mixture going to a waste pond called a *lagoon*.

## TYPES OF HOUSING

One type of hog housing uses an open-front shelter and a pasture. This is called *open-front housing*. The hogs are segregated into small groups of different ages. One lighting outlet is required for each pen. incandescent lamps are adequate and, usually, 60-W bulbs provide

Figure 24-1 Hogs are usually confined in group pens with slotted floors.

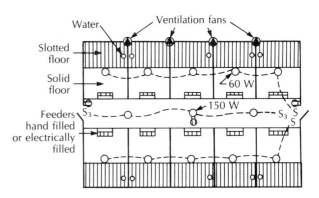

Figure 24-2 Suggested electrical outlets for confinement-type hog housing

sufficient light. Receptacle outlets should be installed about every 40 ft (12.2 m) along the feed alley. The outlets should be installed about 4 ft (1.22 m) above the floor to avoid hitting them with equipment which moves up and down the access alley.

A second type of housing is called *modified open-front housing*. In this type of housing, the hogs are confined in the building in group pens with a partially solid and partially slotted floor. Openings on one side of the building are controlled by the operator to provide adequate natural ventilation in summer and winter. The electrical lighting and receptacle requirements are the same as they are for the open-front type.

A third type of housing is called total confinement housing. In this type of housing, the hogs are maintained in controlled environment. Ventilation fans remove heat, moisture, and odors. This ventilation system is extremely essential. It must be reliable. Loss of ventilation on a hot summer day could result in the death of large numbers of hogs. Agricultural engineers at state universities, and others employed by ventilation equipment manufacturers, design the entire ventilation and control system and provide installation instructions. Wiring in these buildings should be in rigid nonmetallic conduit, using dusttight and watertight lighting fixtures, boxes, and receptacle covers. Type UF cable is also suitable for these areas, *NEC Section 547-1(a)*. The plan of Figure 24-2 shows a typical confinement-type hog finishing barn. One lighting outlet should be provided for each pen.

## FEEDING SYSTEM

Feeders for a pasture-type swine operation are often kept filled by delivery from a tractor and wagon or a truck. Hogs confined in a building are usually fed from an electrically filled feeder. However, sometimes manually filled feeders are used. Swine in a confinement barn are clean; therefore, the feed is often distributed on the floor. This provides more uniform access to the feed by all hogs in the pen. Feeders may be placed in each pen, however, supplied by an electrically operated filling system. A long, flexible auger can be used to move the feed from the outside storage bin to the feeders in the pens. A receptacle outlet and a floodlight should be provided on the outside of the building near the bin for maintenance purposes.

## VENTILATION SYSTEM

Most ventilation systems operate by exhausting air out of the building. An inlet is provided, usually on the opposite side of the buildings to allow fresh outside air to enter. The air inlet generally consists of a long, narrow slot opening along one side of the building. An agricultural engineer designs the flow rate of air required as a maximum in summer, and a minimum in winter. The fan controls include manual switches and thermostats which are used to select the proper ventilation rate between the maximum for summer and the minimum for winter. Some fans run continuously. As the inside temperature begins to rise, the thermostats successively turn on more fans.

The purpose of the ventilation system is to remove moisture, heat, and odors. The hogs, as well as all other animals and even people, produce heat and moisture. A person working hard will perspire. This moisture evaporates into the air and, if the humidity gets too high, an unhealthy environment results. The moisture condenses on everything. This is why *NEC Section 547-1(b)* is so concerned about confinement buildings. Moisture which condenses inside an electrical wiring system is extremely harmful.

In most climates, the hogs produce enough heat so that by the time the excess heat is removed the moisture problem has been eliminated. However, this is not true for young animals, such as baby piglets and calves. Therefore, heat is usually added to the buildings housing these young animals. Lights and machinery can also add significantly to the heat in a commercial building, but this is usually a minor factor in most farm buildings. Figure 24-3 shows the sources of heat and moisture in a farm building.

The operating requirements of an animal ventilation system are different from those for most commercial buildings. In an animal building, large volumes of air must be moved, but the difference in air pressure is small. Commercial building ventilation fans move air against a high pressure. For a commercial building, a centrifugal fan is required to achieve these high pressures. In a farm building, an axial or propeller fan is

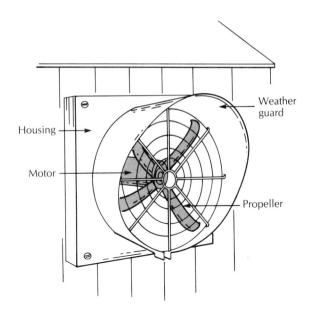

**Figure 24-4  An axial or propeller fan is used to move air through an animal building.**

used, Figure 24-4. These propeller fans are low in cost, and can move large volumes of air with a minimum use of electrical energy.

Most single-phase farm ventilation fans are controlled by a switch or a line-voltage thermostat. The thermostat acts as the controller. If the motor operates at 240 V, the thermostat usually opens only one of the hot wires to the fan. From Unit 17, recall that a disconnect is required within sight from the controller (thermostat) which will disconnect all hot conductors to the controller and motor. If the thermostat or controller is not within sight from the fan, a disconnect must be installed adjacent to the fan, *NEC Section 430-86*. The wiring in these buildings must be Type UF cable or rigid nonmetallic conduit with watertight and dusttight boxes and enclosures. The fan motors are usually located in the airstream leaving the building. This provides adequate cooling of the motor. The motor enclosure must be totally enclosed to keep out dust, but a totally enclosed, fan-cooled (TEFC) motor is not needed for this application.

No more than one fan should be placed on a single branch circuit. Every fan should be supplied by a separate branch circuit with individual overcurrent protection for each. If one fan circuit fails, the other fans will not be affected. A feeder may be run to a location where several fans are placed in a group. A small panelboard then houses the overcurrent protection for each fan.

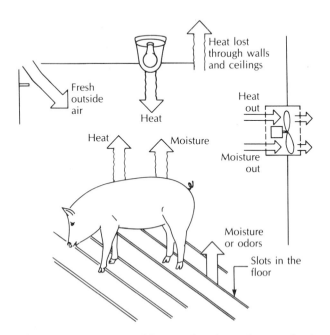

**Figure 24-3  Sources of heat and moisture in an animal building**

## WASTE HANDLING

An exhaust fan is frequently used to remove moisture and odors from the pit area when waste is stored in a tank beneath the hog house. Air is moved down through the slots in the floor and into the pit area. Corrosive gases may be a problem. Therefore, the fan wiring must be made tight and consist of corrosion-resistant materials. Type UF cable and rigid nonmetallic conduit are ideal, but the fittings must also be made of either nonmetallic or corrosion-resistant metal.

When waste is stored under the hog house, oxygen must be present in the liquid to prevent the development of excessive odors and dangerous quantities of hydrogen sulfide, a flammable poisonous gas. A motor-driven paddle wheel is frequently used to churn the liquid and thus expose it to oxygen in the air. The wiring in this area must be liquidtight and corrosion resistant.

> **CAUTION:** Personnel working in the area of this equipment should be aware of the possible presence of hydrogen sulfide.

## FARROWING

The farrowing building requires about one incandescent outlet for each 250 ft² (23 m²) of floor area. Receptacle outlets are needed above each farrowing crate for the possible connection of heat lamps for the baby piglets. Additional receptacle outlets are required for each 40 ft (12.2 m) of wall. Heating cable may be installed in the floor beneath the area where the piglets will rest. (Installation procedures for floor heating cable are covered in Unit 15.) A typical swine farrowing crate is shown in Figure 24-5. One 250-W, 120-V infrared heat lamp is required for each farrowing crate, Figure 24-6. Additional supplemental heat may be provided.

**Figure 24-5** Sow and piglets in a farrowing crate

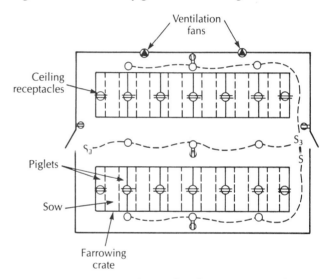

**Figure 24-6** Minimum electrical outlet requirements for a swine farrowing building. Additional heat may be provided.

## REVIEW

Refer to the *National Electrical Code* when necessary to complete the following review material. Write your answers on a separate sheet of paper.

1. A hog confinement building has the waste flushed daily to a lagoon. What is the most important electrical requirement in the building?

2. What type of ventilation system is used for a modified open-front hog barn?

3. What is a farrowing crate?

4. How many 250-W, 120-W infrared heat lamps are required for a building with 20 farrowing crates?

5. Describe the type of wiring required in the area of waste-handling equipment.

6. Ventilation fans used in swine housing, as well as most other animal housing, are of which type?

7. A mechanical feed-handling system for a hog barn is often done with what special type of auger?

8. A fan operates at 115 V and the motor draws 5.8 A. If the service factor is 1.15, determine the maximum size dual-element plug fuse which should be used for running overcurrent protection.

9. The replacement motor for a direct-drive axial fan should have an enclosure that is
   a. open.
   b. splashproof.
   c. totally enclosed, fan cooled.
   d. totally enclosed.

10. What length of time does it take to raise a hog from birth to market weight?

# UNIT 25
# POULTRY FARM WIRING

> ## OBJECTIVES
>
> After studying this unit, the student will be able to
>
> - specify the type of wiring materials used for a poultry building.
> - lay out the lighting outlets and specify the illumination level.
> - diagram the control system for the lighting in a confinement-type poultry egg-laying building.
> - install the wiring for equipment in the poultry building.

The major types of poultry farm operations are brooding, egg laying, broiler raising, and turkey raising. Broiler and turkey raising are most common in the southern United States. The birds are usually housed in semi-open shelters. Egg-laying operations usually use a completely enclosed building with no windows. A poultry operation presents problems of dust and fine material accumulation on motors and wiring. All wiring and materials must be dusttight and watertight. The materials should also be corrosion resistant.

## POULTRY EGG-LAYING BUILDING

Laying hens are usually confined in cages suspended from the ceiling, Figure 25-1. Each cage contains from two to four chickens. Feed and water are available to the hens at all times. The floor of the cage is on a slight incline to allow the eggs to roll to the outside of the cage. The feed may be distributed to the chickens by hand, or by an automatic feeder. The eggs may be gathered by hand, or a belt conveyer can be used to transport them to the sorting and packaging room.

Dust accumulation is a major problem, and all motors should be of the totally enclosed, fan-cooled type.

**Figure 25-1** Hens are confined in cages, and feed and eggs may be moved electrically.

Figure 25-2 shows the dust accumulation on the TEFC motors in a poultry building. Liquidtight flexible metal conduit should be used to make flexible connections to motors. PVC (rigid nonmetallic) conduit is best to use on surfaces which are unsuitable for attaching UF cable. Grounding of the equipment is extremely important, and a grounding wire must be run with the circuit wires when rigid nonmetallic conduit is used.

The proper amount of light in a poultry cage is essential in an egg-laying building. If there is too much light, the hens will pick at one another. Too much light is much worse than too little. The ideal level of illumination is 1 fc. The proper level of illumination can be obtained by placing 25-W incandescent lamps 12 ft (3.66 m) on center in the middle of the alley between the rows of cages, Figure 25-3. Covered lamps are preferred, but bare lamps of the 25-W size do not cause a fire hazard in the poultry house.

Egg laying can be controlled to some extent by the light and dark periods of the artificial day within the windowless poultry building. About 80% of the hens will lay an egg within 10 hours after the light has been turned on in the poultry building. A typical lighting schedule for a poultry egg-laying house is 18 hours of light and 6 hours of total darkness. If a poultry farmer plans to collect eggs beginning at 3 PM, the lights should be turned on 10 hours earlier, at 5 AM. This schedule ensures that most of the eggs will have been freshly laid at the time of gathering. Lighting schedules are also used to induce pullets to grow more quickly and uniformly so that they will begin laying eggs at the earliest possible age.

A time clock is used to control the lights in a poultry egg-laying house. A multipole timer can be used to control the lights directly, or a time clock can be used to control a lighting contractor. A typical lighting-control system is shown in Figure 25-4.

Ventilation is extremely important in a poultry egg-laying house, especially in the summer. A typical poultry building 50 ft (15.24 m) wide and 200 ft (61 m) long contains about 10 000 hens. If the ventilation fails in summer, the temperature in the building will rise rapidly. The oxygen in the building will become depleted, and hens will suffocate. Care must be taken to ensure reliable and adequate wiring for the ventilation fans.

The level of illumination in the egg cleaning and packaging area should be about 50 fc. Fluorescent lamps will provide good lighting in this area. The electrical requirements for equipment depend upon the specific equipment installed. Plenty of receptacle outlets should be provided for portable equipment that may be required.

A cold storage is necessary for the eggs once they have been cleaned, graded, and packaged. The requirements for wiring refrigeration equipment are contained in Unit 15.

Other types of poultry housing usually require a lighting outlet for every 400 ft$^2$ (37 m$^2$) of floor area. Receptacle outlets should be spaced conveniently throughout the building for maintenance purposes.

**Figure 25-2** Totally enclosed, fan-cooled motors are required inside a poultry building because of the dust accumulation.

## LIGHTING CIRCUIT PLANNING

The lighting circuits in a poultry building are usually at least 200 ft (61 m) long. Voltage drop on the circuit can become a problem. For this reason, the wire selected should be one size larger than required to satisfy the circuit requirements. Consider an example of a typical poultry egg-laying building. The building is 216 ft (66 m) long and 50 ft (15.24 m) wide. Four rows of cages are in the building. Determine the number of circuits required for lighting the poultry building.

Five aisles run through the building, and the lights will be located directly over each aisle. The lighting con-

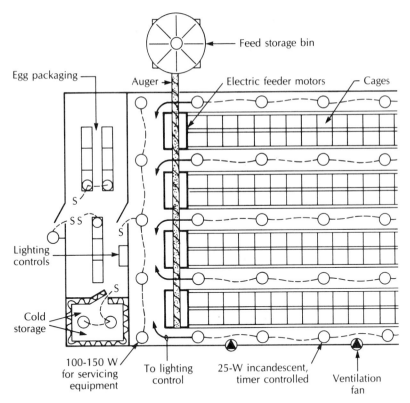

**Figure 25-3** Suggested electrical outlets for a poultry egg-laying building

sists of 25-W lamps spaced every 12 ft (3.66 m) on center. Therefore, there will be 18 lamps in each row.

$$\frac{216 \text{ ft}}{12 \text{ ft}} = 18 \text{ lamps} \qquad \begin{array}{c} \text{metric solution} \\ \frac{66 \text{ m}}{3.66 \text{ m}} = 18 \text{ lamps} \end{array}$$

Lamps of higher wattage may be used at some time in the future. Therefore, the circuit should be planned for extra capacity. If 25-W lamps are used, the maximum current flowing on the circuit will be 4 A.

$$\frac{18 \text{ lamps} \times 25 \text{ W}}{120 \text{ V}} = 3.8 \text{ A (round to 4 A)}$$

Even though this is a small load, the wire should be No. 12 AWG copper because of the length of the circuit, and the possibility of using lamps of higher wattage in the future.

## CHICK BROODING

Baby chicks require heat during the first few weeks after they hatch. Brooders are available in various types.

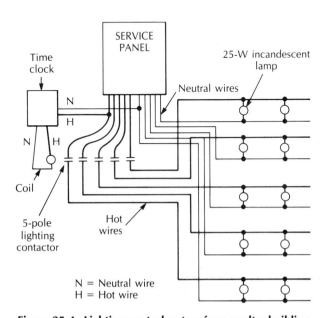

**Figure 25-4** Lighting-control system for a poultry building

Some are fueled with gas, and others are electrically powered. Heat lamps and quartz infrared heaters are used, but they allow heat to escape. Brooders often consist of a broad covering installed close to the floor with added heat inside the covered area. These are called *hover-type brooders*. A thermostat inside the brooder controls the temperature. The chicks may freely leave the hoover.

The chicks are usually separated into small groups of only a few hundred. If the group is too large, the chicks will huddle together, and some will suffocate. The hover-type brooder keeps the heat down at the floor level near the chicks, and spreads it out over a larger area than is possible with bare heat lamps. If the chicks are spread out, there is less chance that they will huddle together and suffocate. As the chicks grow, the hover is raised gradually and, eventually, removed. For brooders of the electric type, receptacle outlets should be installed at the ceiling above the location where the brooders will be placed.

## REVIEW

Refer to the *National Electrical Code* when necessary to complete the following review material. Write your answers on a separate sheet of paper.

1. In the cage of a poultry egg-laying building, which is more undesirable, too much light or too little light?

2. Electric hover-type chick brooders have a power rating of 800 W at 120 V. If there are 15 of these brooders in a building, how many 20-A circuits are required?

3. A poultry farmer plans to gather eggs at 10 AM. What would be the ideal time for the lights to be turned on in the laying building?

4. How many lighting fixtures should be installed in an egg-laying house where there are four aisles and the building is 132 ft (40.23 m) long?

5. What type of motor enclosure should be chosen for a broiler-chicken feeder inside a poultry building?

# UNIT 26
# GRAIN-DRYING AND STORAGE WIRING

---

**OBJECTIVES**

After studying this unit, the student will be able to

- identify the types of electrically powered equipment used for grain drying, handling, and storage.
- specify the proper wiring materials for grain-drying and storage installations.
- name three basic types of grain dryers.
- explain the operation of a basic grain-dryer electrical control system.
- explain the installation of a service entrance for a grain drying and storage center.

---

Inclement weather is a continual problem to farmers. The water content of grain must be kept below a certain level to prevent the grain from sprouting and growing in the storage bin. Grain which is not dry enough will spoil in storage due to mold and bacteria growth. The ideal moisture content varies somewhat from one type of grain to another, but 12% moisture content (wet basis) is a commonly acceptable level.

The farmer leaves the grain in the field as long as possible to allow for natural drying. This saves on farm energy use. But, if the farmer leaves the grain in the field too long, bad weather may set in before all of the grain is harvested. An early snow or heavy fall rains can prevent the harvesting of corn, soybeans, and navy beans. An unusually wet spring can delay planting, making in-the-field drying impossible before the fall weather becomes a problem. A farmer may even have to change the variety of corn to one with a shorter growing period if spring planting is delayed too long due to wet weather. Every year fewer farmers produce the abundant crops in the United States and Canada. As the number of farmers decreases the size of their acreages increases. Most farmers have so much grain to harvest that it is nearly impossible to achieve natural drying in the field. Some of the grain must be dried and stored on the farm or at a commercial elevator.

## GRAIN-DRYING AND STORAGE CENTERS

Small grain is usually stored on the farm in metal bins. The bin has an elevated perforated metal floor. An opening in the top of the roof is used for filling the bin. An electrically powered distributor is usually placed in the top of the bin to spread the grain evenly throughout the bin during filling. This distributor should be cord connected so that it can be moved from one bin to another.

Some of the grain may still spoil in storage even if it was dry initially when put into the bin. The moisture left in the grain may migrate throughout the grain as the seasons change. Therefore, a stirring auger is often used to mix the grain periodically while in storage, Figure 26-1.

Aeration is provided for metal long-term storage bins to prevent moisture migration and spoilage. *Aeration* is the forcing of cool dry air up through the grain. An aeration fan builds up air pressure in the space under the grain. The cool dry air passes through the perforated floor and moves up through the grain, removing any excess moisture which has accumulated. The grain can be aerated only when the outside humidity is low. If the humidity is high, moisture will be put into the grain rather than taken out. The aeration fan should be permanently connected to a manual disconnect switch, using liquidtight flexible metal conduit, Figure 26-2. The fan may be cord and plug connected, provided the cord is rated for extra-hard usage (Types SO or STO) and is protected from physical damage. A weatherproof receptacle is also necessary.

The grain bin is usually unloaded using electric augers. A large variety of augers is available, each using different techniques. Some grain bins have permanently installed unloading augers, while others have removable augers. The permanent type is usually wired direct, while the removable type must be cord and plug connected. A weatherproof receptacle and disconnect switch is provided for the connection and control of the bottom-unloading removable auger, Figure 26-3. When the bin is almost empty, a movable sweep auger may be placed in the bin to remove the grain crowned up along the

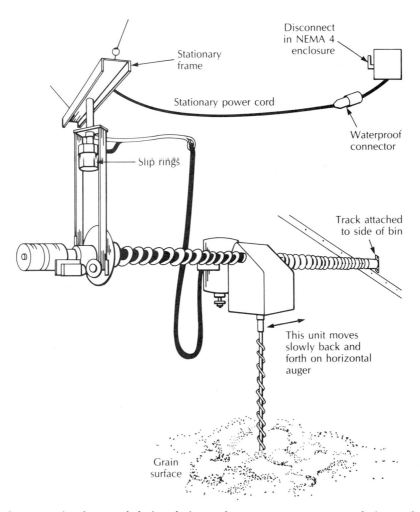

**Figure 26-1  A stirring auger is often used during drying and storage to prevent overdrying and moisture migration.**

422    Unit 26    *Grain-Drying and Storage Wiring*

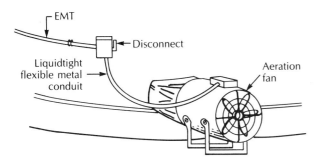

**Figure 26-2** A rainproof disconnect switch is used to control an aeration fan for a grain storage bin.

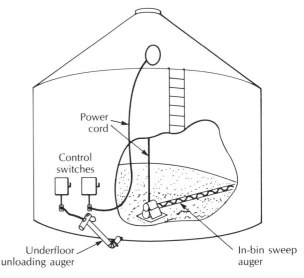

**Figure 26-4** Grain bin unloading augers

sides. This auger is cord and plug supplied, and it circulates around the bin, moving the grain to the center where it is removed by the bottom-unloading auger. This auger should be provided with a disconnect switch controller and a weatherproof receptacle on the outside of the bin, Figure 26-4. A permanently installed self-unloading system eliminates the need for installing a portable auger. Some farmers simply enter the bin and push the remaining grain to the center by hand.

Bucket elevators (sometimes called leg elevators) are used to elevate grain vertically (see Figure 26-10). The bucket elevator is an alternative method to the mobile grain elevator for loading grain into storage bins and dryers. A distributor at the top of the elevator is controlled by cables on the ground to permit the operator to direct the grain to any bin or dryer desired. The motor that powers the bucket elevator is at the top. The electrical installer must provide the wiring for the motor. It is not uncommon for a bucket elevator on a farm to extend more than 100 ft (30.5 m) into the air. Therefore, the conduit may be installed to the elevator before it is lifted into place to avoid having to work at a great height. Care must be taken, however, that the conduit is not damaged when the bucket elevator is lifted into place.

Other types of storages are also used. Many farmers store shelled corn in a large building with a concrete floor. The grain may be directly loaded into the storage from trucks or wagons, or it may be loaded using mobile grain elevators. The grain is removed using a tractor with a front-end bucket loader. A problem with this type of storage is the possibility of moisture migration up through the concrete floor into the grain. A layer of polyethylene should be place under the concrete to reduce moisture migration.

Ear corn is often stored on the farm in circular or rectangular cribs. Ear corn is bulky and hard to handle. However, in ear-corn storage, drying is accomplished naturally. Wind blows through the ear corn in storage and removes excess moisture. The only electrical requirements for ear-corn storage are adequate lighting for work after dark, and a weatherproof, switch-controlled receptacle to operate the flight and chain grain elevator. If the grain elevator is operated at 240 V, then an additional 120-V weatherproof receptacle should be provided for use with small tools.

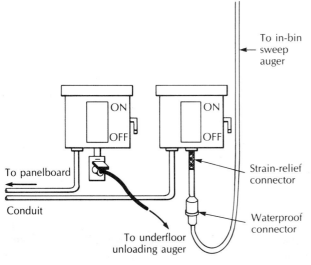

**Figure 26-3** Fusible switches to control the unloading augers

Unit 26 Grain-Drying and Storage Wiring

# WIRING GRAIN-HANDLING EQUIPMENT

Wiring for grain-drying and storage equipment installed inside a building may be wired with cable. Some dust and flying material is always present in the air; therefore, dusttight boxes and enclosures should be used. Type UF cable is preferred if cable is to be used.

Equipment is frequently moved from one area to another. Therefore, a cord and plug attachment is often used. The cord connector and attachment plug must be of a type that will prevent dust and moisture from entering the connection. The cord, Type SO or STO, should be rated for extra-hard usage. Strain-relief connectors should be used where cords enter permanently mounted boxes and motor enclosures.

The wiring on and between metal grain bins and permanently installed equipment must be in conduit. Rigid metal conduit, IMC, and EMT are the most ideal. EMT works best when running conduit around the side of a metal bin. EMT is easily formed to the contour of the bin. The conduit must not be attached tight against the bin if the bin is also to be used for drying. When this is the situation, the metal bin wall warms up, and the heat is moved into the conduit by conduction. Spacing the conduit a small distance from the bin will keep the conduit cool, Figure 26-5. Conduit can be run for short distances between bins, but adequate overhead clearances must be maintained. Wires can be run underground between bins, but they should be run in conduit, unless the area will be covered by a layer of concrete. If the rigid metal conduit or IMC is not protected by a concrete cover, it must be buried to a depth of at least 6 in (152 mm), *NEC Section 300-5*.

Conduit can be fastened to a metal bin with self-tapping metal screws. The self-drilling type may be used, or a properly sized starter hole is drilled before installing the screw. A sealing compound or washer that is not affected by heat should be used when the bin is also used as a dryer. This prevents air leaks in the bin. Heavy panels and equipment should be attached to the bin using bolts with lock washers, in addition to the sealing compound or sealing washers. Conduit may be supported to vertical elevators and similar equipment using beam clamps and similar connectors.

# GRAIN DRYERS

Three basic types of grain dryers are in-bin dryers, batch dryers, and continuous-flow dryers.

## In-bin Dryer

The *in-bin dryer* is common, and much the same as the batch dryer. A layer of damp grain, usually corn, is put into the bin and the dryer is started. As harvesting continues, grain is added until the bin is full. Drying continues until the grain at the top is dried to the desired moisture content. Then the heat is turned off and the grain is cooled by passing cool air through the grain. After the grain is cooled, it is removed from the dryer and placed in an adjacent storage bin. A farmer may have several in-bin dryers and several storage bins in a group. Grain may be stored in the bin dryer. The grain is continually mixed with a stirring auger during drying to prevent the grain at the bottom of the bin from becoming overheated, Figure 26-6.

Wiring and electrical motors and enclosures must be watertight when they are installed in a bin-type grain dryer. The humidity in the space above the grain is at 100%. This moisture must be prevented from entering enclosures. NEMA 4 and 4X enclosures should be installed in these areas. Cord connectors and plugs, as well as the connectors entering motors and enclosures, must be of the watertight type.

## Batch Dryer

A *batch dryer* is completely filled with grain and then started. Heated air moves from an internal chamber through the grain to the outside, Figure 26-7. Automatic controls are available to shut off the unit when drying is completed. Once the grain is completely dry, the dryer goes into a cooling cycle where cool air is forced through

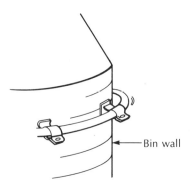

**Figure 26-5** Install conduit on the outside of an in-bin dryer so there will be space between the bin wall and the conduit.

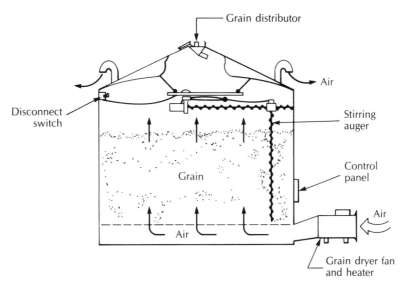

Figure 26-6 An in-bin grain dryer

the grain. The dryer is then emptied into a storage bin, and a new batch of grain is put into the dryer.

Sometimes cooling is done in the storage bin to reduce the amount of time it takes to dry a batch of grain. A batch dryer may have wheels, but this is only for transporting it to the farm. The dryer is permanently mounted once it is in place on the farm.

Batch dryers are available with 3-phase or single-phase motors. The dryers usually come completely wired and ready to operate. A nameplate usually gives the electrical service requirements. A large batch dryer may require a 200-A service. Some batch dryers contain receptacle outlets to operate up to two mobile augers. The service wiring to the dryer may be overhead or underground.

## Continuous-flow Dryer

The *continuous-flow dryer* is often similar in appearance to the batch dryer. However, there are also the bin type and the vertical type. The vertical type is generally used by commercial elevators. Undryed grain enters the top and moves continuously down through the dryer. By the time grain reaches the bottom, it has been dried to the proper moisture content and cooled. Automatic controls keep the hopper at the top filled with grain while an auger at the bottom moves the dried grain from the dryer to a storage bin.

A continuous-flow dryer has two or more air chambers. Heated air moves from the top chamber through the grain to the outside removing moisture. Cool air moves from the bottom chamber through the grain to the outside to cool the grain, Figure 26-8. A tall commercial dryer may have several chambers, all operating at a different temperature.

A small continuous-flow dryer comes completely wired and ready for operation except for the electrical supply. The dryer may be wired for single-phase or 3-phase service. The service wires may enter overhead or underground. Large commercial dryers are delivered in

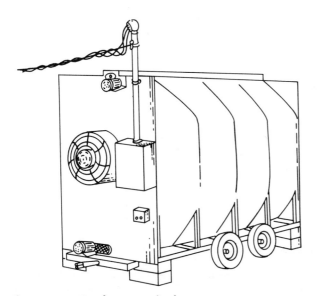

Figure 26-7 Batch-type grain dryer

Unit 26 Grain-Drying and Storage Wiring 425

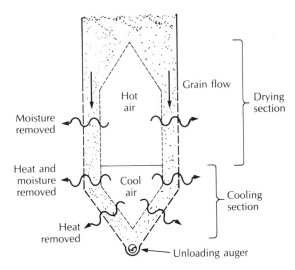

Figure 26-8 A continuous-flow grain dryer contains a heated-air drying section and a cool-air cooling section.

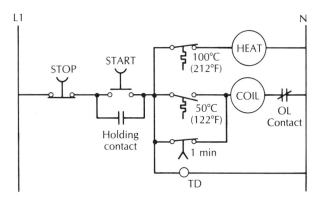

Figure 26-9 A simple grain-dryer control ladder diagram with a time-delay relay used to bypass the low-temperature limit thermostat during startup

sections and assembled on the site. The controls are usually prewired to the bottom section. A conduit must be run up the dryer to each fan, and control wires installed to each unit.

**CAUTION:** An electrical installer working high in the air on a continuous-flow dryer, grain bins or bucket elevator must always wear a safety harness to keep from falling

## CONTROLS

Most grain dryers are fired with liquid propane (LP) or natural gas. The fuel ignition system is activated when the fan is turned on, although the fan can be operated independently. A high-temperature limit thermostat is installed to prevent overheating of the grain. The high-limit thermostat usually turns off the heat. A low-limit thermostat shuts off the entire dryer if the temperature of the air drops too low. Usually this is an indication that the heat is off. When the dryer is first started, the low-temperature limit thermostat is open, thus preventing the unit from being started. A time-delay relay is required to bypass this thermostat for several minutes until the dryer has had time to heat up.

A simple grain-dryer control ladder diagram is shown in Figure 26-9. Automatic continuous-flow dryer control systems may be quite complex. A schematic wiring diagram and sometimes a control ladder diagram are supplied with each dryer. The farmer should have these diagrams available when troubleshooting is necessary.

## SERVICE EQUIPMENT

A grain-drying system should be provided with a separate service direct from the farm distribution pole. The overhead or underground wires must be adequately sized to minimize voltage drop. Extra capacity also should be planned at the time of installation if there is a chance that more equipment will be added. The grain-drying and storage center service entrance should be installed inside if a building is included as a part of the system. Sometimes, a small building is provided for the control system and other necessary equipment, Figure 26-10.

Figure 26-10 A small building can be provided to house the service equipment for the grain-drying and storage center.

If a building is not available, then the equipment is usually attached directly to the grain dryer or storage bin. Batch dryers and continuous-flow dryers usually have the main service equipment supplied as a part of the dryer.

Service equipment installed in the open must be raintight. A typical raintight service panel on an in-bin dryer is shown in Figure 26-11. A receptacle outlet for power tools and a lighting fixture should be installed on the grain bin near the control panel. Locate the lighting fixture near a ladder, if possible, to make lamp replacement easy.

The procedure for determining the size of service-entrance conductors for a grain-drying system is covered in Unit 4. Usually the service consists mostly of motor load; therefore, the service wires are treated as though they are a feeder to a group of motors. Some dryers use electric heat, and this load must be considered when determining the size of the service-entrance conductors.

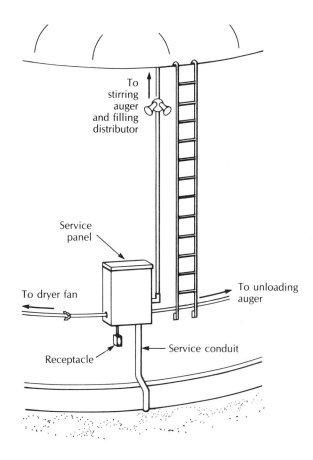

**Figure 26-11** Service equipment may be attached directly to the side of an in-bin grain dryer.

## REVIEW

Refer to the *National Electrical Code* when necessary to complete the following review material. Write your answers on a separate sheet of paper.

1. Describe the type of wiring which is suitable for electrical equipment attached to a metal grain bin or grain dryer.
2. Where is the electric motor usually located on a bucket or leg elevator?
3. How is electrical power supplied to a stirring auger in a grain bin?
4. Explain why it is not a good idea to mount electrical conduit in direct contact with a metal storage bin used for drying.
5. Why is sealing compound or a sealing washer desirable to use when attaching electrical equipment directly to a metal storage on drying bin?
6. Name the three basic types of grain dryers.
7. A continuous-flow dryer has at least two air chambers. Explain their purpose.
8. A disconnect switch used as a safety shutoff for the stirring auger is located inside the bin. What type of enclosure should it have?
9. What is the purpose of a low-temperature limit thermostat in a grain dryer?
10. A service panelboard mounted on the outside of a grain bin exposed to the weather must have what type of enclosure?

# UNIT 27
# IRRIGATION SYSTEM WIRING

## OBJECTIVES

After studying this unit, the student will be able to

- describe a center pivot irrigation system.
- describe the type of cable used on the irrigation machine.
- explain grounding requirements.
- determine the size of conductors to the irrigation machine.
- diagram and size all conductors for a pump and irrigation machine installation.

An irrigation system that is electrically driven must be wired in such a way that an electrical hazard is not created. Most systems are operated from a 480-V, 3-phase supply. The irrigation machine must be supplied with a grounded electrical system.

The electrical power supplier must be contacted well in advance of the installation to schedule the installation of the electrical supply. It is best to discuss the electrical supply with the power supplier before any firm plans are made. If new electrical lines must be constructed to supply the irrigation demand, part of the cost will be paid by the farmer. This cost may be too high, thus requiring the installation of a diesel-driven unit. A diesel-powered generator is then used to power the control system.

The wiring of an electrically driven irrigation machine is covered in *NEC Article 675*. The electrical wiring must be installed with proper materials and with extra care.

## CENTER PIVOT IRRIGATION

Electrically driven irrigation systems are frequently of the center pivot type. This type of system has an irrigation pipe which extends out radially from a central pivot point, Figure 27-1. Electric motors on drive units along the pipe propel the irrigation machine around the field. Alignment switches are used to turn on and off the individual drive motors. The motors must be totally enclosed, and all wiring must be watertight. The entire irrigation machine must be well grounded to prevent the creation of an electrical shock hazard.

A separate irrigation pump is used to supply the water needs as taken from a well, river, or lake. In some areas, a license is required before water can be taken from these sources for irrigation. The pump may be located next to the center pivot or it may be located at the edge of the field. It is best to locate the irrigation pump

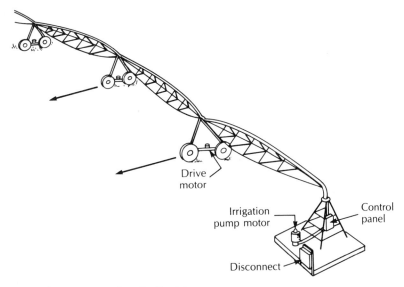

**Figure 27-1 Electrically driven center pivot irrigation system**

as close as possible to the source of electrical supply to minimize the length of run of wire. However, if the pump is at the side of the field, the water must be pumped through an underground main pipe to the center pivot. The power supplier may extend underground primary wires to a pad-mounted transformer near the center pivot, as shown in Figure 27-2.

The irrigation machine itself is provided with a control panel mounted on the center pivot stand. This unit is supplied with an underground conduit from the main control panel, Figure 27-3. The irrigation machine rotates while the main control panel is stationary; therefore, sealed collector rings, mounted just above the control panel, transfer power from the stationary part to the rotating part.

A special irrigation cable is required for the wiring on the irrigation machine, Figure 27-4. The cable is available from an irrigation equipment supplier. The construction of the cable is described in *NEC Section 675-4*. The cable shall be supported at intervals not exceeding 4 ft (1.22 m). The cable connectors used must be watertight and of the type that will make contact with the metal sheath around the cable.

## GROUNDING

The irrigation machine and all its metal equipment must be grounded to a grounding conductor and bonded together, *NEC Sections 675-12, 675-13, and 675-14*. This includes all electrical equipment on the machine, all electrical equipment associated with the machine, metallic junction boxes and enclosures, and control boxes and panels. A driven ground rod shall be provided at the center pivot, *NEC Section 675-15*.

The feeder wire from the main service to the center pivot irrigation machine must contain an equipment grounding conductor sized according to *NEC Table 250-95*. An equipment grounding wire is required even if the wires are contained within metal conduit, *NEC Section 675-12*.

As an example of grounding, assume that an irrigation machine has a demand load of 16 A. The wire supplying the machine would be only No. 12 AWG copper,

**Figure 27-2 Primary electrical power is provided to the pad-mounted transformer at the center pivot.** (*Courtesy of Tumac Industries Inc.*)

Figure 27-3 Control panel for a center pivot irrigation system (*Courtesy of Tumac Industries Inc.*)

according to *NEC Table 310-16*. But, No. 10 AWG will be used because of voltage drop along the irrigation machine. Furthermore, the circuit will be protected at 30 A. The minimum size grounding wire run with the feeder wires to the center pivot from the main service is sized according to *NEC Table 250-95*. The minimum size is No. 10 AWG.

## SIZING CONDUCTORS

For the center pivot-type machine, the motors on the irrigation machine do not run all at the same time. They are constantly turning on and shutting off. The conductor serving the motors is, therefore, permitted to be sized based upon the expected demand, *NEC Section 675-22*. The load is calculated as follows:

> 1.25 × full-load current of largest motor
> plus
> 0.6 × full-load current of all other motors

An example will help to illustrate how this procedure works. Assume that an irrigation machine has fourteen

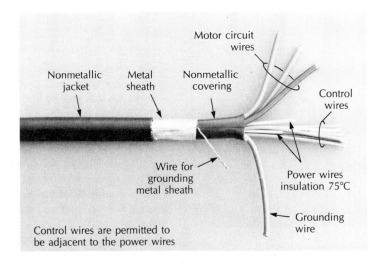

Figure 27-4 Special electrical cable is required for the wiring directly on the irrigation machine.

motors, each rated at 1 hp, and operates at 480 V. From *NEC Table 430-150*, these motors will each draw 1.8 A. Therefore, the total demand is calculated as follows:

```
1.25 × 1.8 A × 1 motor  =  2.3 A
0.6 × 1.8 A × 13 motors = 14.0 A
         Total demand   = 16.3 A
```

The wire must have a minimum ampacity of 16.3 A. This would call for No. 12 AWG copper, but No. 10 AWG would most likely be chosen because of voltage drop due to the long distances involved.

All of the irrigation machine motors are permitted to be on a single circuit provided that none of the motors draws more than 6 A and the circuit is protected at not more than 30 A. A controller is installed at each motor location with a tap conductor extending down to the motor. This tap may be not more than 25 ft (7.62 m) in length, and it must be at least No. 14 AWG wire of the type described in Figure 27-4.

## DISCONNECT AND CONTROLLER

A disconnect must be provided for the irrigation machine either at the center pivot or not more than 50 ft (15.24 m) from the center pivot, *NEC Section 675-8(b)*. The main disconnect, as well as the main controller, shall have a horsepower rating determined by the following procedure, *NEC Section 675-8(a)*.

```
2 × locked-rotor current of largest motor
               plus
0.8 × sum of nameplate ratings of all other
      motors
```

Consider the previous example of the fourteen motors, each rated at 1 hp and operated at 480 V. From *NEC Table 430-151*, the locked-rotor current of the motors is determined to be 10.8 A. Remember, the full-load current was 1.8 A.

```
2 × 10.8 A × 1 motor       = 21.6 A
0.8 × 1.8 A × 13 motors    = 18.7
Machine locked-rotor current = 40.3 A
```

In *NEC Table 430-151*, look up the equivalent horsepower in the 460-V column which corresponds to a locked-rotor current of 40.3 A. The conclusion is a 5-hp rating. The irrigation machine controller and the disconnect must have a rating of at least 5 hp. It is important to remember that a fusible disconnect switch must have a 600-V rating if it is to be used for a 460-V irrigation machine.

## LIGHTNING ARRESTERS

The metal irrigation machine is exposed in the open and is subject to lightning damage. The irrigation machine itself is required to be connected to a driven ground rod at the center pivot. This connection should be made so the grounding wire and the ground rod are not subject to physical damage. A No. 6 AWG copper grounding wire should be used to make this connection.

Lightning arresters should be installed to prevent or minimize damage to electrical wiring and components. A lightning arrester is recommended at every disconnect switch and at the center pivot control panel. This lightning arrester is connected to each power wire and to the grounding wire. A ground rod should be driven at every disconnect switch and connected to the grounding wire and the disconnect enclosure. A diagram of the grounding system is shown in Figure 27-5.

## IRRIGATION SERVICE

The total irrigation service consists of the pump motor, if electrically powered, and the irrigation machine. Consider a 75-hp, 460-V, 3-phase motor on an irrigation well. The requirements for sizing the main feeder and disconnect are found in *NEC Article 430*.

Size the feeder wire first. The wire must have an ampacity equal to the full-load current of all the motors, plus 25% of the full-load current of the largest motor. The irrigation machine motor current is the demand current determined earlier. That value is 16.3 A. The full-load current of the 75-hp, 460-V, 3-phase motor is 96 A.

```
Full-load current of all motors
                        = 16.3 + 96 = 112 A
              plus
0.25 × largest motor current
                        = 0.25 × 96 = 24 A
    Minimum feeder rating =         = 136 A
```

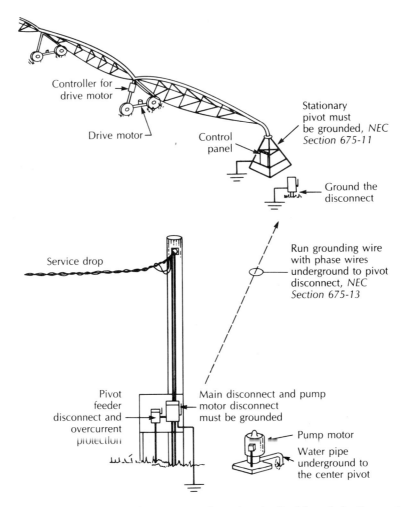

Figure 27-5 Grounding requirements for a center pivot electrically driven irrigation system and pump

This would call for No. 0 AWG THW copper wire. No. 2/0 or No. 3/0 AWG would likely be selected to prevent excessive voltage drop.

The disconnect must have a horsepower rating not less than the sum of the horsepower ratings of the motors supplied. For the irrigation machine the equivalent horsepower rating was determined to be 5 hp. Therefore, the minimum horsepower rating is 80 hp for the total load. Actually, the next standard size disconnect switch is rated at 100 hp. The feeder would most likely be protected by 200-A fuses. A diagram of the total electrical system is shown in Figure 27-6.

Irrigation systems other than the center pivot type are gaining in popularity. One such system is the lateral move system. The irrigation machine is usually electrically powered, and it moves laterally in a straight line across the field. This type of machine has a demand factor on the drive motors that is different from the center pivot machine. The rules for determining the continuous current rating of the machine and the locked-rotor current are found in *NEC Section 675-7*. The motor duty cycle is considered to be 100%, unless otherwise specified by the irrigation machine manufacturer.

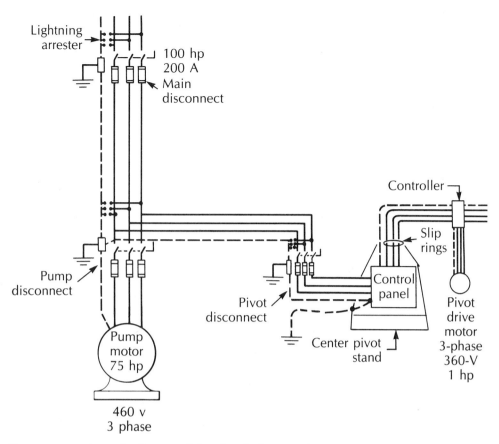

**Figure 27-6** Schematic diagram of the electrical system for a center pivot irrigation system

## REVIEW

Refer to the *National Electrical Code* when necessary to complete the following review material. Write your answers on a separate sheet of paper.

1. How is the center pivot irrigation system propelled around the field?

2. What is it about the flexible cable used on an irrigation machine that is different from the cable used in farm wiring?

3. If the wires from the center pivot disconnect switch to the center pivot control panel mounted on the pivot stand are run in rigid steel conduit, is a separate grounding wire required?

4. An irrigation machine is propelled by ten 3-phase, ¾-hp motors rated at 460 V. Determine the minimum current rating of the wire supplying the irrigation machine.

5. Determine the minimum equivalent horsepower rating of the irrigation machine described in Problem No. 4 of this review.

# INDEX

## A

AC, circuits, 30–32; *def.*, 15
Agricultural buildings, heating of, 310
Agricultural lamps, life, lumens, efficiency table, 232–233
Air conditioners for rooms, 314
Alternators. *See* Standby electrical power systems
Aluminum wire, 159, 161; *illus.*, 161
Aluminum wire, properties of, *table*, 19–20
Aluminum wire oxide inhibitor, *def.*, 159
Ammeter, clamp-on, *def.*, 17; *illus.*, 17
Ampacity, *table*, 375
Amperes, formulas for, 17
Animals, electrical shock, 2–3; hazards to equipment, 7; *illus.*, 3; wastes of, 6
Appliances, heating guides on, *table*, 256
Architectural symbols, 381–382; *illus.*, 382
AWG wire scale, *def.*, 18

## B

Balanced load, *def.*, 130
Balancing 120-V loads, 43; *illus.*, 43
Beef cattle, housing and wiring for, 394–395; *illus.*, 394
Bonding. *See* Grounding electrodes
Bonding service equipment, 81–82; *illus.*, 81, 82; main bonding jumper, *def.*, 81, 82, *illus.*, 82; multiple enclosures, *illus.*, 83
Branch circuits, color code on wire, 184–185
Branch circuits, single-motor, 346–347; *illus.*, 347
Branch circuits, wires in, 349
BTU, *def.*, 29

## C

Cable, installation of, 193–194; attic, *illus.*, 194; bored holes, *illus.*, 193; and boxes, *illus.*, 194; combination devices, *illus.*, 193; nonmetallic boxes, *illus.*, 194
Cable wiring, 161–162; *illus.*, 161, 162
Calf housing, 400, 402; *illus.*, 402, 403
Capacitance, AC circuits, 35–36; capacitor, *def.*, 35, *illus.*, 35; diagrams, 35–36
Cattle. *See* Beef cattle
Central distribution poles, 97–98; grounding and bonding, 97–98, *illus.*, 99; locating of, 100–102, *illus.*, 101; materials of conduits on, 97; overhead clearance, 98
Central point distribution, 94–98; advantages, 95; central distribution poles, with disconnects, 96–97, *illus.*, 97, without disconnects, 95; feeder taps, 95–96; lugs, 96, *illus.*, 96; standoff brackets, 95, *illus.*, 95; tap splices 95–96, *illus.*, 96
Chicks, Chickens. *See* Poultry
Circuit breakers, 114–115
Circuits, types of, 22; *illus.*, 22, 23
Clearances, overhead feeders, 150
Codes and standards, electrical, 7–9
Color, nature of, 226
Combination devices on single yoke, 192–193; *illus.*, 193
Combination motor starter, 368; *illus.*, 368
Conduction of heat, *def.*, 297
Conductors, dimensions of electrical, *table*, 171
Conduits, bending of, 197–198; *illus.*, 197, 198; *tables*, 198, 199
Conduits, number of wires in, 167, 174; *tables*, 168–170, 171–172, 173
Continuous current rating of fuses and circuit breakers, 116–117; *illus.*, 116, 117
Continuous load, *def.*, 116
Control circuit, protection for, 364–366; fuses, *illus.*, 366: large motors, *illus.*, 366; ratings, *table*, 365; 3-phase, 230-V, 15-hp motor, 364–365, *illus.*, 365
Control devices, schematic symbols for, 360–364; *illus.*, 361, 362, 363
Convenience, 3–4
Convection, *def.*, 297
Cooling, units of, 297
Copper wires, properties of, *table*, 19–20
Cords, electric, 162–163, *illus.*, 162, 163; overcurrent protection, 377, *table*, 377; strain relief, 379, *illus.*, 379; wet conditions, 378–379, *illus.*, 378, 379; wire colors, 377–378; types, 373–374, ampacity, *table*, 375, ampere ratings of flexible, 374, *table*, 375, silo unloader, 162, *illus.* 374
Corrosion, environments favoring, 6–7, 154
Corrosive materials, 154
Current, alternating, *def.*, 15, *illus.*, 15; amperes, *def.*, 17; ammeters, *def.*, 17; *illus.*, 17, 18; direct, *def.*, 15; flow of AC, sine wave, 15, *illus.*, 15; voltage graph, 15, *illus.*, 16

## D

Dairy barn, 57
Dairy farms, wiring for, 398–405; bunk feeder, *illus.*, 399; cooling, 400; feeding equipment, 400; free stall barn, *def.*, 399, *illus.*, 399; heating, 400, *illus.*, 400; lighting

Dairy farms, *cont.*
  in free stall, 400; lighting for milking, 399; location of lighting, 400, *illus.,* 401; milk, handling control system diagram, 399, *illus.,* 400, 401; milking parlor, *def.,* 398, *illus.,* 399, lighting for, 400; stanchion barn, *def.,* 398, *illus.,* 398
Day, photoperiodism, *def.,* 224
DC, *def.,* 15
Demand factors, table, 55
Derating factors, in wire, 155–159; illus., 157, 159; *tables,* 156, 157
Disconnects for single motor, 348–349
Distant loads, serving of, 136
Dual-voltage motors, *def.,* 326
Dust, 7, 177–178, 417–418
Dusttight wiring materials and installation, 417–418
Dwelling neutral, sizing of, 87–89

## E

Efficiency, *def.,* 37
Eggs. *See* Poultry
Electric boxes, 177–184; concealed wiring in hollow walls, 178, *illus.,* 179, 180; dust, 177–178, *illus.,* 177; FS, conduit outlet, 178; gangs, numbers of in boxes, 179; steel, 177–178, *illus.,* 178; *tables,* 181, 183; waterproof surface mount, 179, *illus.,* 180
Electric discharge lamps, *def.,* 233
Electric hazards from water and damp, 5–6
Electric plants, 258–259; grounding, 259; maintenance, 259; power failure alarms, 259–260
Electric shocks, 2–3, 219
Electric symbols, 382–386; circuit wires on plan, *illus.,* 384; electrical wiring plan and blueprint symbols, 383; plan types, *illus.,* 384; risers diagram, 384, 386; symbol chart, 385; wiring plan, *illus.,* 384
Electric systems, grounding of, 77–78
Electric systems, identification of, 51
Electric/mechanical conversion units, 26–27; *table,* 28
Electricity, heating with, baseboard heater, 298; ceiling cable, 299; disconnects, 305; electric furnace, 298; with fan, *illus.,* 299; floor cable, 299–300; high-temperature limit switch, 305–306; and infrared heaters, 298, circuit breakers, 303, heating contactor, 303, line-voltage thermostats, 303, *illus.,* 304, problems about, 303–304, quartz heaters, operation of, 304, *illus.,* 304; installation, 306, self-contained heater, 298; sizing of wires, 301–303, *illus.,* 302, supplementary overcurrent, 305; switch as disconnect, 305; thermostats, 300–301, *illus.,* 301
Electricity, nature of, atoms, structure of, *def.,* 13, 14, *illus.,* 14; battery terminals, 14, *illus.,* 14; free electrons, def., 13, 14, *illus.,* 14
Enclosures for motors, 331–332; illus., 331, 332
Equipment grounding, 216–217; *illus.,* 216, 217
Equipment grounding conductors, sizing of, 216–218
Eyes, protection of, 176; *illus.,* 176

## F

Farms, adequate wiring for, 1–4; ampere rating for, 154, *table,* 156; central distribution pole, *illus.,* 2; dwelling service entrance, sizing of, 85–89; fusible disconnect for outbuilding's power, *illus.,* 2; present loads, 2; reasons for spreadout of buildings, 5; shop, 388–389, *illus.,* 389; wiring, 4–5, *illus.,* 4, 6
Feed conveyors, 407–409; batch process, *def.,* 408; belt, 408, *illus.,* 408; flight and chain, *def.,* 408, *illus.,* 408; pneumatic, *def.,* 408; screw/auger, *def.,* 408; sequencing of, 409–410, *illus.,* 410; wiring for, 408–409
Feed preparation, 409; *illus.,* 409
Feed conductors, sizing, 140, 143; *table* 141–142
Feed system wiring, auger, 406–407, *illus.,* 407; dry hay, 406; haylage, *def.,* 406; lighting, 407–409; silage, *def.,* 406; silo, *def.,* 406, *illus.,* 407
Feeder disconnects, 144–145; inside, *illus.,* 145; outside, *illus.,* 145
Feeder grounding wire, sizing of, 145, 149
Feeder supplying several machines, 354–356; *illus.,* 355
Feeder taps, 147, 356; fuses in types of, 147, *illus.,* 148; *illus.,* 356
Feeder wire, 138–140
Feeder wire, grounded, 145–146; *illus.,* 146
Fixture loads, formulas, 242–243
Flexible metallic conduit, 166
Fluorescent fixtures, wiring of, 245–246; *illus.,* 245, 246
Fluorescent lamps, ballast, losses of, 238; bimetallic strip, 234; circuit, pre-heat of, 234; class P ballast, 237; cold, 235; color output, 236; DC circuit, 237; deluxe, red in, 236; diameters, 235; dimming circuits, 236; efficiency, 233; electron flow, 234; fixtures, 235; flashing circuit, 236; high-frequency circuit, 236; instant start, 234–235, circuit diagram, 235; lamp, life of, 235; light output vs. temperature, 235; power factor, 237–238; pre-heat types, 234, *illus.,* 234; rapid start circuit diagram, 235; sealed-glow switch, 234, *illus.,* 235; sound rating, 237; stroboscopic effect, 238; troubleshooting, 238; turning on and off, 235, 236
Flexible cords, 374
Flexible metal conduit, 166; *illus.,* 166
Four-wire delta systems, *def.,* 47–48, 49
Fuel-dispensing pumps, 391–393; *illus.,* 392, 393
Fuel-fired heaters, 307
Fuses, 107; *table,* 108–114
Fusible load centers, 76; *illus.,* 76
Fusible switch, *def.,* 58; *illus.,* 58

## G

Grain dryers, batch, *def.,* 424–425, *illus.,* 425; building for, *illus.,* 426; continuous-flow, *def.,* 425, 426, *illus.,* 426; in-bin, *def.,* 424, *illus.,* 424, 425; ladder diagram, time delay relay, 426; service equipment for, 426–427, *illus.,* 426, 427
Grain drying and storage centers, aeration, *def.,* 422; bins, 421–423, *illus.,* 423; bucket elevators, 423; ear corn, 423; shelled corn, 423; stirring by auger, reasons for, 422, *illus.,* 422; unloading augers, 422–423, *illus.,* 423; wiring for, 424
Grounding electrode, 78–79; AC table, 79
Grounding electrode conductor, *def.,* 78–79
Ground faults, 213–214; *illus.,* 213, 214
Ground fault circuit interrupters, 176, 219–221; *illus.,* 220

Grounding, 212–215; *illus.*, 214, 215
Grounding wires, bonding to grounded wire, 82–84, *illus.*, 84, 85; installing, 79–81, *illus.*, 80; sizing,

## H

Hand tools, electrical, 174–176; *illus.*, 175, 176
Hard hats, 176, *illus.*, 176
Heat, flow of, 30, 296; formulas, 29–30; pumps, 314, units, 29, 296, 297, conversion table, 297
Heaters, electric, 298–299
Heating cable, electric, in floors, 307–310, *illus.*, 309, 310
Hermetic compressor, *def.*, 311
Hertz, *def.*, 15
High-intensity discharge lamps, 238
High-temperature wire, 159, *table*, 160–161
Horses, wiring of housing for, 395–396, *illus.*, 396

## I

IMC, 164; and raceways, 197
Incandescent fixtures, wiring of, 243–244; *illus.*, 243
Incandescent lamps, 230–231; *illus.*, 231; *table*, 231
Inductive reactance, *def.*, 31
Infrared heaters, 298–299
Inrush currents, 120
Insulated conductors, 60–61
Insulators, open wiring on, 163
Interlocking motors, and sequencing, 358–359; *illus.*, 360
Interrupting rating, 117–119; and Code, 118–119; *illus.*, 118
IR drop, *def.*, 123
Irrigation systems, cable for, *illus.*, 430; center pivot, *def.*, 428, 429, diagram, 433, *illus.*, 429, 430, conductors, sizing of, 430–431; electric system, schematic diagram, 432; grounding, 429–430, *illus.*, 430; horsepower rating disconnects, 431; lightning arresters, 431, *illus.*, 432

## J

Junction boxes, sizing of. *See* Outlet and junction boxes

## K

$K_h$ factor, *def.*, 34; *illus.*, 34

## L

Lampholders, 192–193
Lamps, 229
Light, nature of, 225, types compared, *table*, 226; quality of, 228; units and measurement of, 226–228
Lighting, circuits, 242–243; controls, dimmer, 248, *illus.*, 248, lighting contactors, 247–248, diagram, 248, *illus.*, 247, photoelectric switches, 247, snap switches, 247, time clocks, 247, *illus.*, 247; design, 229–230; fixtures, 242
Lightning arresters, *def.*, 85, *illus.*, 87
Livestock barn, 56–57; *table*, 56
Liquidtight metallic conduit, 166–167; *illus.*, 166, 167
Load centers, common sizes, *def.*, 57, *table*, 60

Loads, turning on and off, 235, 236
Locked-rotor currents, 321–322
Low voltage control, 363–364, *illus.*, 364

## M

Magnetic motor starters, 342–344; *illus.*, 343, 344; *table*, 345
Main service at building, 5; *illus.*, 5
Main service at single building, 93–94
Main service, sizing of, 92–93; *table*, 94
Mechanical/electrical conversion units, 26–27; *table*, 28
Mercury-vapor lamps, 238–240
Metal-halide lamps, 241; *illus.*, 241
Metallic tubing, electrical (EMT), 199–201; bender size for raceway types, *table*, 200; bend radius, minimum, *illus.*, 201, *table*, 201; burr removal, 199, *illus.*, 200; couplings and connectors, *illus.*, 165, *illus.*, 200
Metal piping system, bonding of, 84–85
Metric units (SI), 10–11; and prefix table, 10
Motors, control of, 341–343; fusible knife switch, *illus.*, 342; magnetic starter, *illus.*, 343; manual starter, *illus.*, 343; types of controllers, 342–343, *illus.*, 343
Motors, control circuits for, 356–358; *illus.*, 357; overload relay terminals, *illus.*, 357; start-stop, *illus.*, 358, 359; thermostats, *illus.*, 376
Motors, cord-attached, 376–377; in grain elevator, *illus.*, 376
Motors, electric, alignment, 335; armatures, *def.*, 318, *illus.*, 319; bearings, 318, 331, 334; belts, tensions check for, 334, *illus.*, 334; braking, 334; capacitor-started, induction-run, 322, 324; code letters, 329; DC, *def.*, 319; dual voltage, *def.*, 326, efficiency, 330; fan, 318; frame, 317–318; frame numbers, 328; induction, *def.*, 319; locked-rotor current, 321–322; nameplates, 319, 326–327; permanent split-capacitor, 324; pulleys, sizing of, 335; repulsion-started, induction-run, 325; rotor, *def.*, 318, 320, *illus.*, 318; self-induction, 321; series-wound, 319; shaded-pole motors, 325; service factor, 328; single-phase, *def.*, 322, 323–325; sizes, agricultural, *illus.*, 330, *table*, 330; slip of motor, 321; soft-start, 332; squirrel cage, 320; stator, *def.*, 318; synchronous, *def.*, 319; terminal housing, 318; thermal protection, 330; 3-phase motors, 319–321; 2-value capacitor induction, 324–325; variable speed, 333–334; voltage, effects of changes in, *table*, 125; windings, thermal protection of, 359, *illus.*, 361
Motors, full-load current, 347; *illus.*, 347; *tables*, 347, 348

## N

NEC Tables, 220–40, 55; 250–94, 79; 250–95, 218; 310–16, 156; 310–17, 142; 310–18, 160; 310–19, 141; Note 8 to Tables 310–16, 310–17, 310–18, 310–19, 64; 346–12, 199; 347–8, 203; 370–6(a), 183; 370–6(b), 181; 430–148, 347; 430–150, 348; 430–151, 313; 430–152, 351; 3A, 3B, 3C in Chapter 9, 168–170; 4, Chapter 9, 173; 5, Chapter 9, 171–172; 8, Chapter 9, 19–20
NEMA enclosures for motors, 345–346; *illus.*, 345, 346
NEMA sizes for motors, 344–345; voltage ratings, *table*, 345
Neutral, sizing of, 61–63

Neutral wires, and conductor derating, 65
Neutral current, 43; *illus.,* 43
Neutral-to-earth voltage, 102, 402–403, 404; *See also* Stray voltage
Nonmetallic conduit, 163, 164–165; *illus.,* 166

## O

Ohm's Law, *def.,* 23; *illus.,* 24
One-circuit service, 76
One-line diagrams, 386; *illus.,* 386; symbol chart, 385
Open delta 3-phase system, 49–50; *illus.,* 49
Open neutral, 43–44
Oscilloscope, *def.,* 15
Outbuildings, grounding at, 147–149
Outlet box, free conductor
Outlet and junction boxes, cable clamp, *illus.,* 182; ceiling, *illus.,* 183; counting procedure, 181, *table,* 181; cubic inches per wire, 182–183, *table,* 183; fixture stud, *illus.,* 182; ganging of device boxes, 184, *illus.,* 184; grounding wires, 181–182, *illus.,* 182; masonry boxes, 184; recessed boxes, 184, *illus.,* 185; sizing of, 179–184; square boxes, 184, *illus.,* 184; volume per conductor, *table,* 181
Overcurrent conditions, types, 105–106; *illus.,* 105, 106
Overcurrent devices, standard ratings, 115–117; *illus.,* 117
Overcurrent protection, arcing, 104; devices, 197–115; feeder wires, 143, *illus.,* 145; heat, 104; location for feeders, *illus.,* 144; magnetic force, 104; overcurrent, 104
Overhead feeders, 150
Overhead spans, 150–151; maximum, *table,* 151
Overload, *def.,* 105

## P

Panelboards, 58, 114; *illus.,* 58, 59
Parallel circuits, 23–26; *illus.,* 23, 25
Parallel service conductors, sizing of, 63–65; *illus.,* 64; *table,* 64
Phase converters, autotransformer static, 263–264, *illus.,* 264; choice of, reasons for, 263–266; disconnection of, 269, *table,* 269; electrolytic capacitors, installation of, 270–271; grounding, 271; 480-V, 266; limits on, 262–263; motor controllers, 270; oil-filled capacitors, 265; overcurrent protection, 270; rotary phase, 265–266; starting torque, 262; static, 263–265; and 240-V safety switches, horsepower ratings, *illus.,* 269; use, 262; wiring for, 267–269
Pigs. *See* Swine
Pipes, cables for heating, 310
Plugs, 374–376; *illus.,* 375, 376
Polarity, 190–191
Poultry, wiring for housing of, 417–420; chicks, brooding of, 419–420; cleaning and packaging area for eggs, 418; cold storage, 418; dust, 417–418; eggs, collection of, *illus.,* 417; hens, 417–418; laying, control by light, 417–418; illumination levels, 418; lighting circuits, 418–419, *illus.,* 419; lighting control system, 418, *illus.,* 419; time clock, 418; totally-enclosed fan-cooled motors, for, *illus.,* 418; ventilation, 418

Power, and work, foot-pound, *def.,* 26; horsepower, *def.,* 26; joules, *def.,* 26; kWH, *def.,* 26–27, and meter, *illus.,* 27; watt, *def.,* 27; watt-hour, *def.,* 26
Power factor, *def.,* 32, 33–36; *table,* 33
Power meter, 34
Pull boxes, *def.,* 204; *illus.,* 204, 205
PVC conduit, 201–204; heating to bend, 203, *illus.,* 204; and hole saw, *illus.,* 202; *illus.,* 202; and metal boxes, *illus.,* 202; metal conduit, *illus.,* 203; nonmetallic boxes, 202, *illus.,* 201, 203; nonmetallic conduit, *illus.,* 201, *table,* 203; thermal expansion, 202, *illus.,* 203

## Q

Quartz incandescent lamp, 231, 233

## R

R value, of insulation, 296–297
Raceway wiring systems, 196–197, *illus.,* 196
Radiation, *def.,* 297
Raintight installations, 97; *illus.,* 98
Receptacle outlets, choices about, 189; circuit breakers, 192; configuration chart, 191; duplex, 190; grounding type, 189; multiwire, 191–192, *illus.,* 192; neutral wire, 190–192; split duplex, 191, *illus.,* 192. *See also* Plugs.
Refrigeration, 311–313; *illus.,* 312; *table,* 313
Reliability of electrical systems, 3
Resistance, *def.,* 19; 19–22; table, materials, 21
Resistive load in circuit, 31; *illus.,* 31
Response of fuse and circuit breaker, speed of, 119–121; vs. amount of current, 119; inrush current of starting motor, 120, *illus.,* 120; irreducible tripping time, *def.,* 119; mechanical limits, 119; and starting of motor, 120, *illus.,* 120; time, determining of for circuit breaker, 121, *illus.,* 121; time, determining of for fuse, 120, *illus.,* 120; time-current curves, 120, *illus.,* 120, 121
Reversing motor starters, 366–368; drum controller with connections, 367, *illus.,* 367; *illus.,* 367; ladder diagram, 368; large horsepower, 367–368; manual drum switch, 366–367, *illus.,* 367
Rigid metal conduit, 163, 197; *table,* 199
Roof, clearance of, 71, 73; *illus.,* 73
Running overcurrent protection, 352–354; *illus.,* 353, 354; *tables,* 353, 354

## S

Safety, animal and human, 2–3; *illus.,* 3
Safety switch, fusible, common sizes, 58–59; *table,* 60
SE wires, 67–71; *illus.,* 68; size table, 69
Series circuit, 22–25; *illus.,* 22–24
Service disconnects, sizing of, 59; *tables,* 60
Service-drop clearances, 73; *illus.,* 72, 73, 74
Service entrance, electric, demand load, 55, 59; farm building loads, 55–57, *illus.,* 56; purpose, 54–55; service equipment, choice of, 6, 57–58, hazardous environments, 5–6, load centers, 5, *def.,* 57, panelboards, 57–58, *illus.,* 6, safety switches, 58, *illus.,* 58

Service entrance, installation procedures, 65–67; SE cable, 66–67, *illus.*, 66–67, type R, *illus.*, 67, type U, *illus.*, 67
Service entrance, wires for, 67–70; *illus.*, 68, 69, 70, 71; *table*, 69
Service lateral, *def.*, 74
Service mast, 71–72; *illus.*, 71, 72; *table*, 71
Service wires, sizing of, 59–61; *illus.*, 62; *table*, 61
Settling particles, hazards of, 7
Sheep, wiring of housing for, 395; *illus.*, 395, 396
Short circuits, branch circuits, 349, *illus.*, 350; *def.*, 105; diagrams, 105; formulas, 105; inverse time circuit breaker, 349, *def.*, 350; maximum ratings, *table*, 351; overcurrent devices, 107–115; problems, 350–352; proper sizes of components, 350; sizing, 350
Sine wave, *def.*, 15
Single-phase electrical systems, 41–45; *illus.*, 42
Sodium lamps, 241–242; *illus.*, 241
Specifications, use, 382, 386
Standby electrical power systems, alternators, 252–253, 257–258, *illus.*, 253; alternator, sizing of, 255–257, *illus.*, 255, 256, single-phase electric motors, power requirements, *table*, 256, *table*, 255; alternator, wiring to, 258, *table*, 259; automatic starting of, 257, *illus.*, 257; critical circuits, origin of in, 260; engines and tractors, 253; grounding, 259, *table*, 259; indoor engine alternator, installation of, *illus.*, 257; installation, 257–258; transfer switch, 254–255, *illus.*, 254, 255; transfer switch, sizing of, 258
Storage buildings, 393
Stray voltage, 402–405, *def.*, 402–403; and AC on primary neutral, 402; and earth as conductor, 402–403; equipotential plane, *def.* 405; measuring, 403–404; in milking parlor, elimination of, 405, *illus.*, 405; neutral-to-earth voltage, 402–403, 404; sources of, 403, 405; voltmeter, *def.*, 403–404, *illus.*, 403
Subpanels, 144–145; panelboards with main lugs, *illus.*, 145
Swine, wiring of housing for, 412–415; fan propeller, 414, *illus.*, 414; farrowing, 415, *illus.*, 415; feeding systems, 413; group pens with slotted floor, 412, *illus.*, 413; heat sources, 414, *illus.*, 414; modified open-front housing, *def.*, 412–413; outlets, diagram, 413; piglets, 412; thermostats, 414; total confinement housing, *illus.*, 413; ventilation, 413–414; waste handling, 415
Switches, current-sensing relay, 361–362, diagram, 363, *illus.*, 362; DPDT, 361, diagram, 361; DPST, 361, diagram, 361; limit switches, *def.*, 361; normally closed, 360; normally open, 360; snap, 185–186; SPDT, 361, diagram, 361; SPST, 361, diagram, 361; switch loop, 187; time delay relays, 362; transfer, 252

## T

Tap conductors, 62–63; *illus.*, 62
Temperature scales, 296; *illus.*, 297
THHN wire, 154; *illus.*, 155
THW wire, 154; *illus.*, 155
Three-phase power, 44–46
Three-wire delta system, 47–48; *illus.*, 47, 48
Three-phase wye system, *def.*, 50–51; *illus.*, 50, 51

Three-piece connector, 97, *illus.*, 97
Three-wire circuit, 120 V in, 42
Three-wire single-phase service entrance conductors, 83–84; *illus.*, 84
Toggle (snap) switches, agricultural types, *illus.*, 186; *def.*, 185; double-pole, 186, 188, *illus.*, 190; 4-way, 188; lighting fixtures, with 4-wire switch, 188, *illus.*, 190; lighting fixtures, with 3-wire switches, 188, *illus.*, 188, 189; single-pole, diagram, 186, 3-way, 188
Transformers, autotransformers, 278; *illus.*, 278; buck and boost type, 283–284, *illus.*, 284; constant output voltage, 279; control, types of, 278; current, metering of, 99–100, *illus.*, 99, 100; dry-type, 277–278, *illus.*, 274; grounding, 290, *illus.*, 291; kVA rating, 277; insulating, 278, *illus.*, 278; load, kVA rating of, 284, *table*, 285; and magnetism around electric current, 274; mounting, 290; moving magnetic field, 274, *illus.*, 275; mutual induction, 274 oil-filled, 277; overcurrent protection, 286–287, *illus.*, 287; reasons for, 273–274; single-phase transformers, 281–283, *illus.*, 281, 282, 283; step-downs, 276; *illus.*, 276; step-ups, 276, *illus.*, 276; three-phase, 280, *illus.*, 280; turn ratios, 276; windings, connections of, 279–280, *illus.*, 279, 280; primary, *def.*, 276, secondary, *def.*, 276; windings, taps on, 281–282; wire size and protection, in transformers, 288–289; zig-zag/grounding, 278
Turkeys. *See* Poultry

## U

UF cable, 162; *illus.*, 162
Unbalanced loads, 62
Underground connections to meter, 75–76; *illus.*, 75
Underground feeders, 151m, 162
Underground services, 74–75; *illus.*, 74
Underground wires, current differences between
USE cable, 74–75; *illus.*, 74

## V

**Ventilation system**, *illus.*, 3
Volt-amperes, *def.*, 28
Voltage, 15–17, 44; *illus.*, 16
Voltage drop, examples, 123, *illus.*, 125; causes, 126–127, *illus.*, 126, 127; formulas, 123, 128; inductive reactance, 131; measuring, 126; permissible, 125, *illus.*, 125; in series feeders, 135; and skin effect, 131; *tables*, 132, 133, 134, 135
Voltage ratings of fuses and circuit breakers, 115–116; *illus.*, 115
Voltmeters, *def.*, 16; *illus.*, 16

## W

Waste, animal, 6
Water heaters, 76, 390, *illus.*, 76
Water pumps, 389–390; submersible in well, *illus.*, 389; well house, *illus.*, 389

439

Watertight wiring materials and installations, 205–206; incandescent light receptable, 206, *illus.*, 206; and waterproof UF cable connector, 206, *illus.*, 206
Wattmeters, 27
Watts, formula, *def.*, 27
Wires, ampere ratings of, 154–155; *table*, 156
Wires, color code of, 146–147
Wires, with conduit, 163–167; advantages, 163; conduit bodies, 163, *illus.*, 164; EMT couplings and connectors, 164, *illus.*, 165; flexible metal conduit fittings, 166; *illus.*, 166; FS boxes, 163, *illus.*, 165; installation in conduit, 199, *illus.*, 199; intermediate metallic conduit, 164; liquidtight flexible conduit, 166–167; *illus.*, 166, 167; metallic tubing, electrical (EMT), 164; PVC tubing, 164, *illus.*, 166; rigid metal, 163; rigid metallic conduit, 164; sizes, 163, 167; *table*, 164; types, 163–167
Wires, connections to devices, 195–196; *illus.*, 195, 196
Wires, dimensions, 18–19; *table*, 19–20
Wires, over long distances, 5
Wires, protection of, 161
Wires, sag of, 150
Wires, sizing to limit voltage drop, 127–131; air, carrying away of heat by, 130–131; conductors, circular mil areas, *table*, 129; high-temperature conductor insulation, 131; percentages, 129; and resistance values, copper and aluminum, *table*, 128; single-phase, with known voltage drop, 129; single-phase, with known wire size, 128; temperature, 128; three-phase, with known voltage drop, 128; three-phase with known wire size, 128; THW wire, 128–130; *illus.*, 129, 130
Wires, splicing, 176–177; *illus.*, 176, 177

## Z

Zig-zag grounding transformer, 47; *illus.*, 48

# ACKNOWLEDGMENTS

The authors and publisher wish to express their appreciation to the following organizations and individuals for generously supplying illustrations and information used in this book:

Amprobe Instrument Company, Lynbrook, NY

Baas Agrisystems, Inc., Falmouth, MI

Bussmann Mfg., McGraw-Edison Company, St. Louis, MO

Jamesway Division, Butler Mfg. Company, Fort Atkinson, WI

Michigan Committee on Rural Electrification, East Lansing, MI

Midwest Electric Products Company, Mankato, MN

Ronk Electrical Industries, Inc., Nokomis, IL

Square D Company, Milwaukee, WI

Tumac Industries, Inc., Colorado Springs, CO